Wireless Multimedia Communications and Networks

Wireless Multimedia Communications and Networks

Edited by **Elmer Dowse**

Willford Press

New York

Published by Willford Press,
118-35 Queens Blvd., Suite 400,
Forest Hills, NY 11375, USA
www.willfordpress.com

Wireless Multimedia Communications and Networks
Edited by Elmer Dowse

International Standard Book Number: 978-1-68285-197-5 (Hardback)

Printed in China.

Contents

Preface

Wireless multimedia networks have revolutionized the field of communications. It has brought an enormous shift in the various fields of engineering and management by providing diverse mediums and modes for interaction and communication. This book aims to compile some of the most significant concepts and topics on wireless communications and networks like antenna systems and design, multiple access schemes, ultra wide-band systems, vehicular networks, etc. This book includes contributions of experts and scientists from different parts of the globe which will provide innovative insights to all the students, researchers and professionals engaged in this field.

The information shared in this book is based on empirical researches made by veterans in this field of study. The elaborative information provided in this book will help the readers further their scope of knowledge leading to advancements in this field.

Finally, I would like to thank my fellow researchers who gave constructive feedback and my family members who supported me at every step of my research.

Editor

1

Analytical modeling of cognitive heterogeneous cellular networks over Nakagami-m fading

Fereidoun H Panahi* and Tomoaki Ohtsuki

Abstract

In this paper, we present a cognitive radio (CR)-based statistical framework for a two-tier heterogeneous cellular network (femto-macro network) to model the outage probability at any arbitrary secondary (femto) and primary (macro) user. A system model based on stochastic geometry (utilizing the spatial Poisson point process (PPP) theory) is applied to model the random locations and network topology of both secondary and primary users. A considerable performance improvement can be generally achieved by mitigating interference in result of applying the CR idea over the above model. Novel closed-form expressions are derived for the outage probability of any typical femto and macro user considering the Nakagami-m fading for each desired and interference links. We also study some important design factors which their important role in the determination of outage and interference cannot be ignored. We conduct simulations to validate our analytical results and evaluate the proposed schemes in terms of outage probability for different values of signal-to-interference-plus-noise-ratio (SINR) target.

Keywords: Stochastic geometry; Cognitive radio; Outage probability

1 Introduction

The best solution to the spectrum saturation and bandwidth availability problems in multi-tier cellular networks is to adopt technologies that make the most efficient use of existing spectrum through frequency reuse schemes [1,2]. In universal frequency reuse scheme, the existing spectrum can be aggressively and effectively reused by all of the coexisting network tiers. This will lead to higher spatial spectrum utilization and network usage capacity at the expense of an increased possibility of interference among network tiers and of a reduced quality of service (QoS). In multi-tier cellular networks, interference is increasingly becoming a major performance-limiting factor, and hence, interference modeling, coordination, and avoidance are the primary focus of interest for both the industry and academic communities. Applying the cognitive radio (CR) technology in multi-tier cellular networks to be aware of and adapt to communication environments, some of the above challenges can be tackled. In fact, CR is the key enabling technology for interference management and avoidance in multi-tier cellular networks [2,3].

* Correspondence: fereidoun@ohtsuki.ics.keio.ac.jp
Graduate School of Science and Technology, Keio University, 3-14-1 Hiyoshi, Kouhoku-ku, 223-8522 Yokohama, Japan

On the other hand, the aggregate interference environment is more complicated to model, and evaluating the performance of communication techniques in the presence of heterogeneous interference is challenging. For interference characterization, if the base stations (BSs) of the cellular network follow a regular grid (e.g., the traditional hexagonal grid model), then the signal-to-interference-plus-noise-ratio (SINR) characterization will be either intractable [3,4] or inaccurate due to unrealistic assumptions [5]. Moreover, as urban areas are built out, the BS infrastructure is becoming less like points on a hexagonal lattice and more random. Hence, the use of a hexagonal grid to model the BS locations is violated and is considered too idealized [6]. Furthermore, according to [4] and [6] for snapshots of a cellular network at different locations, the positions of the BSs with respect to each other follow random patterns due to the size and unpredictability of the BSs in these kind of networks. Therefore, the need for a powerful mathematical and statistical tool for modeling, analysis, and design of wireless networks with random topologies is quite obvious.

A new modeling approach called 'stochastic geometry' has been recently applied to the analysis of multi-tier cellular networks due to its ability to capture the topological randomness in the network and its aim at deriving accurate and tractable expressions for outage probability [3,6]. Stochastic geometry stems from applied probability and has a wide range of applications in the analysis and design of wireless networks in particular for modeling and analyzing systems with random channel access (e.g., ALOHA [7,8] and carrier sensing multiple access (CSMA) [9]), single- and multi-tier cellular networks [6], and networks with cognitive abilities [7,10]. This paper discusses this new theoretical model to provide a better understanding of the heterogeneous cellular networks of tomorrow and their challenges (interference modeling, coordination, and avoidance) that must be tackled in order for these networks to reach their potential. We focus on a two-tier femto-macro network where low-power and small-coverage local nodes (femto nodes) are distributed in the coverage of macro nodes. We provide an insight into the role of CR in interference mitigation in two-tier heterogeneous networks. We derive closed-form expressions for the outage probability of any typical femto and macro user in the network. We also study the effect of several important design factors which play vital roles in the determination of outage and interference.

The main contributions are therefore the following: (1) We analyze the Laplace transforms of all four types of aggregate interference between macro and CR femto networks (including the interference between macro nodes among themselves and femto nodes among themselves, the cross-interference from femto to macro network and vice versa) in perfect and imperfect spectrum sensing CR-based femto networks, considering simultaneously the Nakagami fading, the Poisson point process (PPP) model, and some important design factors which play vital roles in determination and mitigation of outage and interference. (2) This article provides an insight into the role of CR in interference mitigation in orthogonal frequency-division multiple-access (OFDMA) two-tier heterogeneous networks. (3) Tight closed-form expressions are derived for the outage probability of any typical femto and macro user considering the Nakagami fading (Rayleigh fading assumption is used in many studies to relax the difficulty of addressing a closed-form expression for the outage probability (see, e.g., [8] and [10])) for all the communication links (desired and interference links) with the possibility of using the CR ability for the femto network.

To the best of our knowledge, no closed-form expressions for the outage probability of the femto and macro users exist in the current literature without imposing the Rayleigh fading assumption. On the other hand, most of the available studies in this area are based on the existence of only one macro-BS (along with the macro users and the femto network) and the effect of considering multiple macro-BSs is ignored in the analysis of outage probability.

2 System model

2.1 Model description

We consider infinite spatially collocated macro-BS and femto-BS node heterogeneous networks (see Figure 1). It is assumed that the spatial distribution of the nodes is captured using two collocated and independent homogenous Poisson point processes (HPPPs) [6,7] i.e., Φ_M and Φ_F with intensities λ_M and λ_F, respectively. In other words, the locations of the macro-BS nodes constitute an HPPP Φ_M, where λ_M is the average number of the macro-BS nodes per unit area. Similar statement can be made for the HPPP formed by the femto-BS nodes Φ_F with intensity λ_F. According to superposition theorem [11], the overall node process over the network formed by both the macro-BS and femto-BS nodes also is an HPPP with intensity λ ($\lambda = \lambda_F + \lambda_M$). Furthermore, the macro and femto users are scattered over the plane according to some independent PPPs with different densities than λ_M and λ_F, respectively. However, our interference analysis is fundamentally concerned with the distribution of the transmitters (BSs).

Since femto-BSs are installed and maintained by the paying home users for better indoor performance, they are only accessible by their own mobile subscribers (femto users) (known as *closed-access* policy). On the other hand, macro-BSs can be accessed only by unauthorized users (macro users). In practice, macro network is deployed usually without awareness of the distributed femto network. To this end, wireless operators can consider giving priority to the macro users, and the femto network has to be self-optimized to mitigate its interference to the macro users. Motivated by this insight, the macro-BSs (along with the macro users) and the CR-enabled femto-BSs (along with the femto users) are analogous to primary and secondary systems in the CR model, respectively.

2.2 System structure

In OFDMA, the spectrum is orthogonally divided into time-frequency resource blocks (RBs), which increases flexibility in resource allocation, thereby allowing high spectral efficiency. As shown in Figure 2, we consider a spectrum of N RBs, out of which M ($M \leq N$) random RBs are idle or unoccupied by the macro users (primary system). With the CR capability, a femto-BS could actively acquire knowledge about its environment and access to the RBs without the aid of a macrocell in a decentralized fashion (clearly, no synchronization between the macro and femto network is needed any more) and automatically prevent disturbing the macro users [2].

1) As shown in Figure 2, each femto-BS's transmission strategy is divided into consecutive slots, each

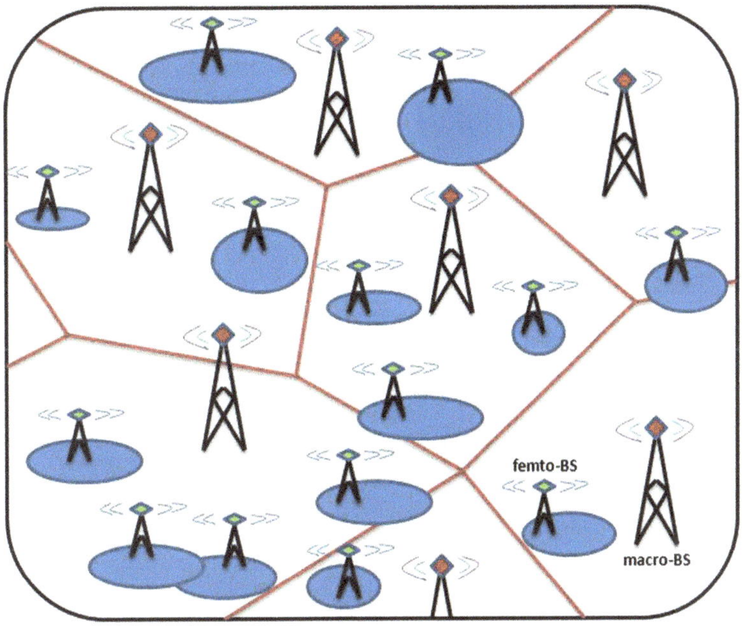

Figure 1 The heterogeneous model (femto and macro BSs).

having a duration of T. Each slot is divided into two consecutive stages, i.e., sensing and data transmission, with durations of T_S and T_D, respectively. Each femto-BS periodically senses the spectrum to identify which RBs are occupied by the macro network. Indeed, each femto-BS accomplishes sensing one RB in one unit slot T_{SRB} within T_S. Each femto-BS senses N_s RBs in sequence which are randomly selected from the N available RBs and detects its idle RB set. Clearly, the time required for sensing the N_s RBs is $T_s = T_{SRB}N_s$. Note that the femto-BSs cannot perform data transmission within the sensing time T_S. We assume that all femto-BSs are perfectly synchronized and have the same time as the sensing time. Methods for implementing a perfect synchronization among the femto-BSs are outside the scope of this paper; however, a set of possible candidates exist, including GPS synchronization, the wired backhaul (IEEE 1588), and leveraging synchronization signals broadcasted by the femto-BSs [12].

2) Each femto-BS senses the received interference power on each RB within the sensing duration.
   - If the received interference power on an RB exceeds a certain threshold, the RB is identified as being occupied by the macro network since all femto-BSs sense at the same time.
   - Otherwise, the RB is unoccupied by the macro network.
3) In the data transmission time (T_D), each femto-BS only allocates an unoccupied RB sensed in the sensing time to its user (by only utilizing these unoccupied RBs, *cross-tier* interference can be consequently avoided). Since the determination of each individual RB status as busy/idle is subject to (occasional) error, determined by the probability of (correct) detection of the presence of PUs' signals P_d

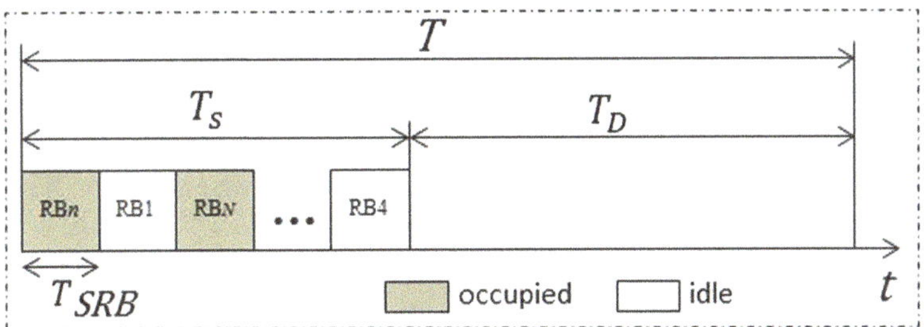

Figure 2 The CR femto-BS's transmission strategy in one time slot.

and probability of false alarm P_f (probability of falsely declaring an idle RB as busy), we study the effect of both the *ideal* detection, i.e., $P_d = 1$ and $P_f = 0$, and the cases involving imperfect sensing, i.e., $P_d \neq 1$ and $P_f \neq 0$ on the outage probabilities of femto and macro users.

As shown in Figure 3, the macro users' exclusion regions with radius D are used to guarantee that the femto-BSs will, on average, not generate an aggregate interference leading to the outage of macro (primary) users. As shown, for example, those femto-BSs located in the tagged macro user's exclusion region are not allowed to transmit data whether they pick the same RB as the tagged macro user or a different one. The tagged macro user is not disturbed by the femto-BSs transmitting on different RBs (from that of the tagged macro) even if they are inside its exclusion region. However, we deactivate them to prevent the tagged macro user from any harmful interference as a result of sudden appearance of them on the same RB in next time slots. We consider this law throughout the paper even in the case of the perfect sensing scenario. We also assume that the femto users are informed about the macro users' exclusion regions.

3 Stochastic geometry-based network configuration

3.1 Femto outage probability

We derive the probability of outage for a typical femto user (p_{OF}) defined as the probability that a randomly chosen femto user cannot achieve a target SINR θ considering a collocated spectrum sensing CR based femto network and macro-BSs as follows [6]:

$$p_{OF} = 1 - P[\text{SINR} > \theta] \tag{1}$$

In fact, the outage probability evaluates the cumulative distribution function (CDF) of SINR over the entire network. The experienced SINR by a typical femto user is calculated as

$$\text{SINR} = \frac{P_F h_F {r_F}^{-\alpha}}{\sigma^2 + I_{FB} + I_{MB}} \tag{2}$$

where P_F is the transmission power from the nearest femto-BS (tagged femto-BS) located in the random distance r_F from its tagged femto user (we assume that the tagged femto user under consideration is located at the origin), and α is the path-loss exponent. I_{FB} and I_{MB} are the aggregate interference power at the origin from the other femto-BSs and macro-BSs, respectively, and σ^2 is the noise power.

When the fading distribution is Nakagami-$m_{d,F}$ for the communication channel between the tagged femto user and its corresponding femto-BS, the probability density function (pdf) of the power fading coefficient $x = h_F$ is as follows

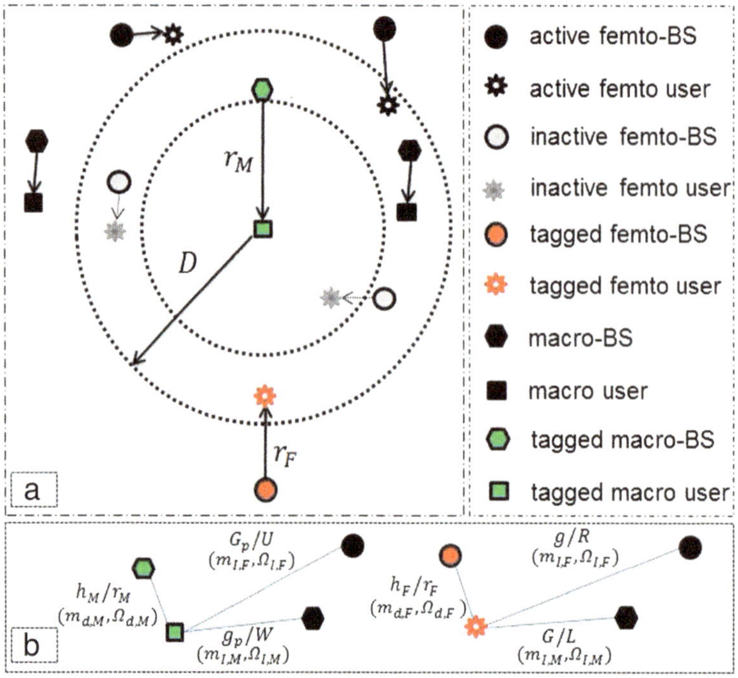

Figure 3 The macro users' exclusion regions. (a) The bipolar network model. **(b)** The link gains/distances, and Nakagami parameters.

$$p(x) = \frac{m_{d,F}{}^{m_{d,F}}}{\Omega_{d,F}{}^{m_{d,F}}\Gamma(m_{d,F})}(x)^{m_{d,F}-1}e^{-\frac{m_{d,F}x}{\Omega_{d,F}}}$$

where $m_{d,F}$ describes the severity of the fading for the desired link, and $\Omega_{d,F}$ is the average SNR. By changing $m_{d,F}$, we can get a variety of fading conditions. For example, if $m_{d,F} = 1$, then the desired link suffers from Rayleigh fading, and $m_{d,F} = \infty$ means that there is no fading. Note that we will represent the Nakagami fading parameters for each communication link, separately (see Figure 3b). Even though the fading power gain distributions are identical, we will still use different symbols to distinguish all the existing links for the sake of analytical convenience.

Here, we derive the complementary CDF (CCDF) of SINR as follows

$$\begin{aligned} P[\mathrm{SINR} > \theta] &= P\left[\frac{P_F h_F r_F{}^{-\alpha}}{\sigma^2 + I_{FB} + I_{MB}} > \theta\right] \\ &= P\left[h_F > (\sigma^2 + I_{FB} + I_{MB})\frac{\theta r_F{}^{\alpha}}{P_F}\right] \\ &= \mathrm{E}_I\left[P\left[h_F > (I)\frac{\theta r_F{}^{\alpha}}{P_F}\right]\right]. \end{aligned} \tag{3}$$

Considering the pdf of the power fading coefficient h_F and the incomplete gamma function (Note that $\Gamma(a,x) = \int_x^{\infty} t^{a-1}e^{-t}\mathrm{d}t$ indicates the incomplete gamma function), we have

$$P\left[h_F > (I)\frac{\theta r_F{}^{\alpha}}{P_F}\right] = \int_{\frac{\theta r_F{}^{\alpha}}{P_F}(I)}^{\infty} p(x)\mathrm{d}x$$

$$= \frac{-m_{d,F}{}^{m_{d,F}}(-1)^{m_{d,F}-1}}{(-m_{d,F})^{m_{d,F}}}\left[\frac{\Gamma\left[m_{d,F}, \frac{m_{d,F}}{\Omega_{d,F}}\left(\frac{\theta r_F{}^{\alpha}}{P_F}(I)\right)\right]}{\Gamma(m_{d,F})}\right]$$

where $I = \sigma^2 + I_{FB} + I_{MB}$.

We know that $\frac{\Gamma[m,my]}{\Gamma(m)} = e^{-my}\sum_{k=0}^{m-1}\frac{m^k}{k!}y^k$ (m is a positive integer). Therefore, $P\left[h_F > (I)\frac{\theta r_F{}^{\alpha}}{P_F}\right]$ is simplified as follows

$$\left[e^{-\frac{m_{d,F}}{\Omega_{d,F}}\left(\frac{\theta r_F{}^{\alpha}}{P_F}\right)I}\sum_{k=0}^{m_{d,F}-1}\frac{\left(\frac{m_{d,F}}{\Omega_{d,F}}\frac{\theta r_F{}^{\alpha}}{P_F}\right)^k}{k!}I^k\right] \tag{4}$$

From (3) and (4) we have

$$P[\mathrm{SINR} > \theta] = \int_0^{\infty}\left[e^{-\frac{m_{d,F}}{\Omega_{d,F}}\left(\frac{\theta r_F{}^{\alpha}}{P_F}\right)I}\sum_{k=0}^{m_{d,F}-1}\frac{\left(\frac{m_{d,F}\theta r_F{}^{\alpha}}{\Omega_{d,F}P_F}\right)^k}{k!}I^k\right]f_I(i)\mathrm{d}i. \tag{5}$$

Finally, the CCDF of SINR when the fading distribution is Nakagami-$m_{d,F}$ for the communication channel between the tagged femto user and its corresponding femto-BS is given by

$$P[\mathrm{SINR} > \theta] = \left[\sum_{k=0}^{m_{d,F}-1}\frac{(s)^k}{k!}(-1)^k\frac{d^k\mathcal{L}_I(s)}{ds^k}\right] \tag{6}$$

where $I = \sigma^2 + I_{FB} + I_{MB}$ and $s = \frac{m_{d,F}\theta r_F{}^{\alpha}}{\Omega_{d,F}P_F}$.

Proof: See Appendix 1.

Note that due to the assumption of independent PPPs for the femto and macro networks, the aggregate interference received from the femto-BSs is independent of the aggregate interference received from the macro-BSs [10]; therefore, we can write

$$\begin{aligned} \mathcal{L}_I(s) &= F_I(s) = \mathrm{E}[e^{-sI}] \\ &= \mathrm{E}\left[e^{-s(\sigma^2 + I_{FB} + I_{MB})}\right] \\ &= \mathrm{E}[e^{-sI_{FB}}]\mathrm{E}[e^{-sI_{MB}}]\mathrm{E}\left[e^{-s\sigma^2}\right] \\ &= \mathcal{L}_{I_{FB}}(s)\mathcal{L}_{I_{MB}}(s)e^{-s\sigma^2} \end{aligned} \tag{7}$$

Where $\mathcal{L}_{I_{FB}}(s)$ and $\mathcal{L}_{I_{MB}}(s)$ are the Laplace transform of random variables I_{FB} and I_{MB} evaluated at $s\left(s = \frac{m_{d,F}\theta r_F{}^{\alpha}}{\Omega_{d,F}P_F}\right)$ respectively. Therefore, (6) can be rewritten as follows

$$P[\mathrm{SINR} > \theta] = \left[\sum_{k=0}^{m_{d,F}-1}\frac{(s)^k}{k!}(-1)^k\frac{d^k\left(\mathcal{L}_{I_{FB}}(s)\mathcal{L}_{I_{MB}}(s)e^{-s\sigma^2}\right)}{ds^k}\right]. \tag{8}$$

3.1.1 Computation of $\mathcal{L}_{I_{FB}}(s)$ and $\mathcal{L}_{I_{MB}}(s)$ at the tagged femto user

We now calculate the closed-form expressions for $\mathcal{L}_{I_{FB}}(s)$ and $\mathcal{L}_{I_{MB}}(s)$ (at $s = \frac{m_{d,F}\theta r_F{}^{\alpha}}{\Omega_{d,F}P_F}$). The Laplace transform of the aggregate interference (from all the active femto-BSs except the tagged femto-BS denoted by fbs_0) is given as follows (Note: some of the femto-BSs located in the macro users' exclusion regions (with radius D) are deactivated, therefore, $\lambda'_F \leq \lambda_F$ (see Figure 3))

$$
\begin{aligned}
\mathcal{L}_{I_{FB}}(s) &= \mathrm{E}_{I_{FB}}[\exp(-sI_{FB})] \\
&= \mathrm{E}_{\Phi_F, g_i}\left[\exp\left(-s\sum_{i\in\Phi_F\setminus\{fbs_0\}} P_F g_i R_i^{-\alpha}\right)\right] \\
&= \mathrm{E}_{\Phi_F}\left[\prod_{i\in\Phi_F\setminus\{fbs_0\}} \mathrm{E}_{g_i}\left[\exp(-sP_F g_i R_i^{-\alpha})\right]\right]
\end{aligned}
\tag{9}
$$

where R_i is the distance of the ith interferer from the tagged femto receiver captured by the point process Φ_F. The interference channel gains g_i are assumed to be mutually independent and have identical pdfs (Nakagami- $m_{I,F}$). Each of the active interfering femto-BSs transmits with the same power P_F. Using the definition of the Generating functional [6] for the PPP, which states for some function $f(x)$ that $\mathrm{E}\left[\prod_{x\in\Phi} f(x)\right] = \exp(-\int_{\mathbb{R}^d} (1-f(x))\lambda \mathrm{d}x)$, (9) can be rewritten as

$$
\mathcal{L}_{I_{FB}}(s) = \exp\left\{-\mathrm{E}_g\left[\int_{r_F}^{\infty} (1-\exp(-sP_F g R^{-\alpha}))\lambda_I(R)\mathrm{d}R\right]\right\}
\tag{10}
$$

where we flipped the order of integration and expectation. Since the closest interfering femto-BS is at least at distance r_F from the tagged user, the integration limits are from r_F to ∞. In other words, interference is encountered from all the active femto-BSs located in the area $\mathbb{R}^d \backslash b(0, r_F)$ (where $b(x, y)$ is ball of radius y centered at point x). Therefore, not all the femto-BSs will contribute towards the aggregate interference, i.e., only those active femto-BSs which are outside the mentioned ball and at minimum satisfy all of the following conditions qualify as potential contributors.

Before explaining the following conditions ((a) and (b)), it is useful to translate the point process into polar coordinates. Therefore, according to [7] and [11] the intensity of the HPPP Φ_F is shown as

$$
\lambda_I(R) = \lambda_F' d R^{d-1} b_d
\tag{11}
$$

where R is the distance between an arbitrary femto-BS and the tagged femto receiver. b_d is the volume of a unit sphere in $\mathbb{R}^d$ ($b_d = \frac{\sqrt{\pi^d}}{\Gamma(1+d/2)}$, $\Gamma(x) = \int_0^{\infty} t^{x-1}e^{-t}\mathrm{d}t$ denotes the standard gamma function).

(a) Satisfying the aforementioned condition, any active femto-BS contributes towards the interference at the tagged femto receiver, if it picks the same RB as the tagged femto-BS to communicate with its user, (we show the probability of picking a same RB from a pool of all RBs as p_{RB} (the calculation of p_{RB} for this case is derived in Section 4, Scenario I, Case 1)).

(b) We assume that the CR femto-BSs employ a slotted ALOHA MAC (medium access control) protocol to schedule their transmission. Therefore, they only transmit with probability p_{tx} in the current time slot and defer the transmission with probability $1 - p_{tx}$.

Applying both conditions (a) and (b) which are indeed two independent thinning tools, the thinned point process of Φ_F is formed. In other words, Φ_F is subjected to two thinning processes each presenting an independent deletion/retention of each point of Φ_F with a fixed retention probability. We now formulize this operation, which will in fact result in reducing the number of interferers and hence lowering the outage probability at the tagged femto user. First, Φ_F is reconstructed by p_{RB}-thinning, where each point of Φ_F is retained with probability p_{RB} (to accommodate condition (a)), and then it is thinned again by applying the second independent thinning (to accommodate condition (b)). Thus, the intensity of the process (the number of interfering CR femto-BSs) becomes

$$
\lambda_I(R) = \lambda_F' d R^{d-1} b_d p_{RB} p_{tx}.
\tag{12}
$$

Now, (10) is rewritten as

$$
\exp\left\{-\mathrm{E}_g\left[\int_{r_F}^{\infty} (1-\exp(-sP_F g R^{-\alpha}))\lambda_F' b_d p_{RB} p_{tx} d R^{d-1}\mathrm{d}R\right]\right\}.
$$

By changing the variables $R^d \to x$ and then $x^{\frac{\alpha}{d}} \to y$, and following a few simple steps, the above expression is simplified as:

$$
\mathcal{L}_{I_{FB}}(s)_{|s=\frac{m_{d,F}\theta r_F{}^{\alpha}}{\Omega_{d,F}P_F}} = e^{r_F{}^d b_d p_{RB} p_{tx} \lambda_F' - \frac{d}{\alpha} b_d p_{RB} p_{tx} \lambda_F' \left(\frac{m_{d,F}}{\Omega_{d,F}}\theta\right)^{\frac{d}{\alpha}} r_F{}^d M(\theta,\alpha)}
\tag{13}
$$

where

$$
M(\theta,\alpha) = \mathrm{E}\left[(g)^{\frac{d}{\alpha}}\left(\Gamma\left(-\frac{d}{\alpha}, \frac{m_{d,F}}{\Omega_{d,F}}\theta g\right) - \Gamma\left(-\frac{d}{\alpha}\right)\right)\right].
\tag{14}
$$

Similar with the desired link, we consider the Nakagami power fading model for the femto interfering links as well (Nakagami fading links with equal parameters $m_{I,F}$ and $\Omega_{I,F}$). Then, following the derivations in Appendix 2, $M(\theta, \alpha)$ is expressed as

$$M(\theta,\alpha) = \left(\frac{-\Gamma\left(-\frac{d}{\alpha}\right) m_{I,F}{}^{m_{I,F}} \left(\frac{m_{d,F}}{\Omega_{d,F}}\theta\right)^{-\frac{d}{\alpha}}}{\Omega_{I,F}{}^{m_{I,F}}\Gamma(m_{I,F})} \right)$$

$$\times \left[\sum_{k=0}^{\infty} \frac{\left(\frac{m_{d,F}}{\Omega_{d,F}}\theta\right)^{k} \left(\frac{\Omega_{I,F}\Omega_{d,F}}{m_{I,F}\Omega_{d,F}+\Omega_{I,F}m_{d,F}\theta}\right)^{m_{I,F}+k}}{\Gamma\left(k+1-\frac{d}{\alpha}\right)} \Gamma(m_{I,F}+k) \right]. \tag{15}$$

Using the similar approach, the Laplace transform of the aggregate interference generated by the macro-BSs at the tagged femto user is obtained as follows

$$\begin{aligned} \mathcal{L}_{I_{MB}}(s) &= \mathrm{E}_{I_{MB}}[\exp(-sI_{MB})] \\ &= \mathrm{E}_{\Phi_M,G_i}\left[\exp\left(-s\sum_{i\in\Phi_M} P_M G_i L_i^{-\alpha}\right)\right] \\ &= \mathrm{E}_{\Phi_M}\left[\prod_{i\in\Phi_M} \mathrm{E}_{G_i}\left[\exp\left(-sP_M G_i L_i^{-\alpha}\right)\right]\right] \end{aligned} \tag{16}$$

where L_i is the distance of the ith interfering macro-BS from the tagged femto receiver captured by the point process Φ_M. The interference channel gains G_i are assumed to be mutually independent and have identical pdfs (Nakagami-$m_{I,M}$). Each of the active interfering macro-BSs transmits with the same power P_M. Again, using the definition of the Generating functional for the PPP, we can write

$$\mathcal{L}_{I_{MB}}(s) = \exp\left\{-\mathrm{E}_G\left[\int_0^{\infty}(1-\exp(-sP_M G L^{-\alpha}))\lambda_I(L)\mathrm{d}L\right]\right\}. \tag{17}$$

The interference is encountered from all the macro-BSs located in the area $\mathbb{R}^d \backslash b(0,\,0)$. It should be noted that not all the macro-BSs in $\mathbb{R}^d$ will contribute towards the aggregate interference, i.e., only those macro-BSs transmitting on the same RB as the tagged femto user qualify as potential interferers. The intensity of the HPPP Φ_M process can be therefore written as follows

$$\lambda_I(L) = \lambda_M^{'}\, dL^{d-1} b_d \tag{18}$$

where L is the distance between an arbitrary macro-BS and the tagged femto receiver and $\lambda_M^{'}$ is the intensity of those macro-BSs transmitting on the same RB as the tagged femto user at a time.

Now, (17) is rewritten as

$$\exp\left\{-\mathrm{E}_G\left[\int_0^{\infty}(1-\exp(-sP_M G L^{-\alpha}))\lambda_M^{'} L^{d-1} b_d d L^{d-1}\mathrm{d}L\right]\right\}.$$

By changing the variables and following a few simple steps, $\mathcal{L}_{I_{MB}}(s)$ is obtained as follows [11]

$$\mathcal{L}_{I_{MB}}(s)_{\left|s=\frac{m_{d,F}\theta r_F{}^{\alpha}}{\Omega_{d,F}\ P_F}\right.} = e^{-b_d\lambda_M^{'}\Gamma\left(1-\frac{d}{\alpha}\right)\left(\frac{P_M\frac{m_{d,F}}{\Omega_{d,F}}\theta}{P_F}\right)^{\frac{d}{\alpha}} r_F{}^{d}\mathrm{E}\left[(G)^{\frac{d}{\alpha}}\right]}. \tag{19}$$

Same as the desired link, we consider the Nakagami power fading model for the macro interfering links as well (Nakagami fading links with equal parameters $m_{I,M}$ and $\Omega_{I,M}$). Below, $\mathrm{E}\left[(G)^{\frac{d}{\alpha}}\right]$ is obtained using the definition of expectation and the standard gamma function

$$\mathrm{E}\left[(G)^{\frac{d}{\alpha}}\right] = \frac{1}{\Omega_{I,M}{}^{m_{I,M}}}\left[\left(\frac{\Omega_{I,M}}{m_{I,M}}\right)^{\frac{d}{\alpha}+m_{I,M}}\left(\frac{m_{I,M}{}^{m_{I,M}}}{\Gamma(m_{I,M})}\right)\right]\Gamma\left(\frac{d}{\alpha}+m_{I,M}\right). \tag{20}$$

3.1.2 Closed-form femto outage probability expression

Finally, from (1), (8), (13), and (19) and by replacing $s = \frac{m_{d,F}}{\Omega_{d,F}}\frac{\theta r_F{}^{\alpha}}{P_F}$, and $z = r_F{}^{\alpha}$, the closed-form expression for the outage probability of the tagged femto user is obtained as follows

$$p_{OF} = 1-\overbrace{\left[\sum_{k=0}^{m_{d,F}-1}\frac{(z)^k}{k!}(-1)^k\frac{d^k(\exp(l(z)))}{dz^k}\right]}^{P[\mathrm{SINR}>\theta]} \tag{21}$$

where

$\exp(l(z)) = e^{[\mathrm{A}]z^{\frac{d}{\alpha}}-\frac{m_{d,F}\theta\sigma^2}{\Omega_{d,F}P_F}z}$, and $z = r_F{}^{\alpha}$
and A is given by (22).

$$\mathrm{A} = \mathrm{A}_1 + \mathrm{A}_2 \tag{22}$$

where

$$\mathrm{A}_1 = b_d p_{RB} p_{tx}\lambda_F^{'} + \frac{d}{\alpha} b_d p_{RB} p_{tx}\lambda_F^{'}\left(\frac{\Gamma\left(-\frac{d}{\alpha}\right)m_{I,F}{}^{m_{I,F}}}{\Omega_{I,F}{}^{m_{I,F}}\Gamma(m_{I,F})}\right)$$

$$\times \left[\sum_{k=0}^{\infty}\frac{\left(\frac{m_{d,F}}{\Omega_{d,F}}\theta\right)^{k}\left(\frac{\Omega_{I,F}\Omega_{d,F}}{m_{I,F}\Omega_{d,F}+\Omega_{I,F}m_{d,F}\theta}\right)^{m_{I,F}+k}}{\Gamma\left(k+1-\frac{d}{\alpha}\right)}\Gamma(m_{I,F}+k)\right]$$

and

$$A_2 = -b_d\lambda'_M\Gamma\left(1-\frac{d}{\alpha}\right)\left(\frac{P_M\frac{m_{d,F}}{\Omega_{d,F}}\theta}{P_F}\right)^{\frac{d}{\alpha}} \times\frac{1}{\Omega_{I,M}{}^{m_{I,M}}}\left[\left(\frac{\Omega_{I,M}}{m_{I,M}}\right)^{\frac{d}{\alpha}+m_{I,M}}\left(\frac{m_{I,M}{}^{m_{I,M}}}{\Gamma(m_{I,M})}\right)\right]\Gamma\left(\frac{d}{\alpha}+m_{I,M}\right).$$

Note: $m_{d,F}$ is constrained to take integer values only, while $m_{I,F}$ and $m_{I,M}$ can take any value (bigger than 0.5). The restriction on $m_{d,F}$ is because of $m_{d,F}-1$, as an upper limit for the summation in (21). However, for non-integer $m_{d,F}$, by using the infinite series representation of incomplete gamma function, we also obtain an exact but infinite summation expression for the outage probability of the tagged femto user as in Appendix 3.

The probability of outage averaged over the plane is derived as

$$p_{\bar{o}F} = 1-\mathrm{E}_{r_F}\left[P[\mathrm{SINR} > \theta]\right]$$
$$p_{\bar{o}F} = 1-\int_0^\infty\left[P[\mathrm{SINR} > \theta]\right]f_{r_F}(r_F)\mathrm{d}r_F \tag{23}$$

where $f_{r_F}(r_F) = e^{-\lambda_F\pi r_F{}^2}2\pi\lambda_F r_F$ is the pdf of r_F [6].

Here, for example, we present the $p_{\bar{o}F}$ in closed-form expressions for three different values of $m_{d,F}$:

$$m_{d,F} = 1,\ \alpha = 4:\quad p_{\bar{o}F} = 1+\frac{\lambda_F\pi}{(\mathrm{A}-\lambda_F\pi)}$$

$$m_{d,F} = 2, \alpha = 4: p_{\bar{o}F} = 1+\frac{\lambda_F\pi\left[(\mathrm{A}-\lambda_F\pi)+\frac{\mathrm{A}}{2}\Gamma(2,0)\right]}{(\mathrm{A}-\lambda_F\pi)^2}$$

$$m_{d,F} = 3, \alpha = 4:$$

$$p_{\bar{o}F} = 1+\frac{\lambda_F\pi\left[(\mathrm{A}-\lambda_F\pi)^5+\frac{5\mathrm{A}}{8}\Gamma(2,0)(\mathrm{A}-\lambda_F\pi)^4+\frac{\mathrm{A}^2}{8}\Gamma(3,0)(\mathrm{A}-\lambda_F\pi)^3\right]}{(\mathrm{A}-\lambda_F\pi)^6}.$$

It should be noted that in all the above expressions $(\mathrm{A}-\lambda_F\pi) < 0$, and $\sigma^2 \to 0$ (operating in an interference-limited regime). For $m_{d,F} \geq 4$, obtaining $p_{\bar{o}F}$ is also quite easy due to the fact that $\exp(l(z))$ is a factor common to every term of the all derivatives of $\exp(l(z))$.

3.2 Macro outage probability

We derive the outage probability for a typical macro user (p_{OM}) defined as the probability that a randomly chosen macro user cannot achieve a target SINR γ, considering a collocated spectrum sensing CR-based femto network and macro-BSs as follows:

$$p_{OM} = 1-P[\mathrm{SINR} > \gamma]. \tag{24}$$

The experienced SINR by a typical macro user is calculated as

$$\mathrm{SINR} = \frac{P_M h_M r_M^{-\alpha}}{\sigma^2 + I_{MB} + I_{FB}} \tag{25}$$

where P_M is the transmission power from the nearest macro-BS located in the random distance r_M from its tagged macro user, and α is the path-loss exponent. I_{FB} and I_{MB} are the aggregate interference power to the tagged macro user (located at the origin) from the surrounding femto-BSs and macro-BSs, respectively, and σ^2 is the noise power.

When the fading distribution is Nakagami-$m_{d,M}$ for the communication channel between the tagged macro user and its corresponding macro-BS, the pdf of the power fading coefficient $y = h_M$ is as follows

$$p(y) = \frac{m_{d,M}{}^{m_{d,M}}}{\Omega_{d,M}{}^{m_{d,M}}\Gamma(m_{d,M})}(y)^{m_{d,M}-1}e^{-\frac{m_{d,M}y}{\Omega_{d,M}}}$$

where $m_{d,M}$ (a positive integer) is the fading parameter for the desired link, $\Omega_{d,M}$ is the average SNR. Next, the similar approach to obtain the outage probability of the tagged femto user is used for the outage probability calculation of the tagged macro user. Therefore, from (24) and (25) we have

$$P[\mathrm{SINR} > \gamma] = \left[\sum_{k=0}^{m_{d,M}-1}\frac{(s)^k}{k!}(-1)^k\frac{d^k\left(\mathcal{L}_{I_{FB}}(s)\mathcal{L}_{I_{MB}}(s)e^{-s\sigma^2}\right)}{ds^k}\right] \tag{26}$$

where $\mathcal{L}_{I_{FB}}(s)$ and $\mathcal{L}_{I_{MB}}(s)$ are the Laplace transform of random variables I_{FB} and I_{MB} evaluated at s $\left(s = \frac{m_{d,M}}{\Omega_{d,M}}\frac{\gamma r_M{}^\alpha}{P_M}\right)$, respectively. We now calculate the closed-form expressions for $\mathcal{L}_{I_{FB}}(s)$ and $\mathcal{L}_{I_{MB}}(s)$ (at $s = \frac{m_{d,M}}{\Omega_{d,M}}\frac{\gamma r_M{}^\alpha}{P_M}$).

3.2.1 Computation of $\mathcal{L}_{I_{MB}}(s)$ and $\mathcal{L}_{I_{FB}}(s)$ at the tagged macro user

Taking a similar approach to what we had before, $\mathcal{L}_{I_{MB}}(s)$ is obtained as follows

$$\mathcal{L}_{I_{MB}}(s) = \exp\left\{-\mathrm{E}_{g_p}\left[\int_{r_M}^\infty\left(1-\exp\left(-sP_M g_p W^{-\alpha}\right)\right)\lambda_I(W)\mathrm{d}W\right]\right\} \tag{27}$$

and

$$\lambda_I(W) = \lambda'_M\, dW^{d-1}b_d \tag{28}$$

in which W is the distance between an arbitrary macro-BS (captured by the point process Φ_M) and the tagged macro receiver, and λ'_M is the intensity of those macro-BSs transmitting on the same RB as the tagged macro user at a time. Similarly, the interference channel gains g_p (between the interfering macro-BSs and the tagged macro user) are assumed to be mutually independent and have identical pdfs (Nakagami-$m_{I,M}$). Each of the active interfering macro-BSs transmits with the same power P_M as for the tagged macro-BS. It should be noted that only those macro-BSs operating on the same RB as the tagged macro user and located in the area $\mathbb{R}^d \backslash b(0, r_M)$ make interference to the tagged macro user.

Same as before, (27) is simplified as

$$\mathcal{L}_{I_{MB}}(s)_{|s=\frac{m_{d,M}\gamma r_M{}^{\alpha}}{\Omega_{d,M}\ P_M}} = e^{r_M{}^d b_d \lambda'_M - \frac{d}{\alpha} b_d \lambda'_M \left(\frac{m_{d,M}}{\Omega_{d,M}}\gamma\right)^{\frac{d}{\alpha}} r_M{}^d V(\gamma,\alpha)} \tag{29}$$

in which $V(\gamma,\alpha) = \mathrm{E}\left[\left(g_p\right)^{\frac{d}{\alpha}}\left(\Gamma\left(-\frac{d}{\alpha}, \frac{m_{d,M}}{\Omega_{d,M}}\gamma g_p\right) - \Gamma\left(-\frac{d}{\alpha}\right)\right)\right]$.

The Nakagami power fading model is also considered for the interference links (Nakagami fading links with equal parameters $m_{I,M}$ and $\Omega_{I,M}$). The derivation of $V(\gamma, \alpha)$ is similar to the derivation of $M(\theta, \alpha)$ in Appendix 2. Therefore, $V(\gamma, \alpha)$ is expressed as follows

$$V(\gamma,\alpha) = \frac{-\Gamma\left(-\frac{d}{\alpha}\right) m_{I,M}{}^{m_{I,M}} \left(\frac{m_{d,M}}{\Omega_{d,M}}\gamma\right)^{-\frac{d}{\alpha}}}{\Omega_{I,M}{}^{m_{I,M}}\Gamma(m_{I,M})} \times \left[\sum_{k=0}^{\infty} \frac{\left(\frac{m_{d,M}}{\Omega_{d,M}}\gamma\right)^k \left(\frac{\Omega_{I,M}\Omega_{d,M}}{m_{I,M}\Omega_{d,M} + \Omega_{I,M} m_{d,M}\gamma}\right)^{m_{I,M}+k}}{\Gamma\left(k+1-\frac{d}{\alpha}\right)} \Gamma(m_{I,M}+k)\right] \tag{30}$$

Based on the same approach as described before, $\mathcal{L}_{I_{FB}}(s)$ is also given by

$$\mathcal{L}_{I_{FB}}(s) = \exp\left\{-\mathrm{E}_{G_p}\left[\int_{\mathrm{K}r_M}^{\infty} \left(1 - \exp\left(-sP_F G_p U^{-\alpha}\right)\right)\lambda_I(U)\mathrm{d}U\right]\right\}. \tag{31}$$

Since the closest interfering femto-BS is at least at distance $\mathrm{K}r_M$ from the tagged macro user, the integration limits are from $\mathrm{K}r_M$ to ∞. In other words, interference is encountered from all the femto-BSs located in the area $\mathbb{R}^d \backslash b(0, \mathrm{K}r_M)$ (see Figure 3). It should be noted that not all the femto-BSs outside this ball will contribute towards the aggregate interference, i.e., only those femto-BSs which are outside the mentioned ball and at minimum satisfy all of the following conditions are considered as potential contributors

(c) Satisfying the above condition, any arbitrary femto-BS contributes towards the interference at the tagged macro receiver, if it wrongly picks the same RB as the tagged macro-BS to communicate with its user (we show the probability of picking a same RB for data transmission from a pool of all RBs as p_{RB} (the calculation of p_{RB} for this case is seen in Section 4, Scenario II, Case 2)).

(d) Same as the condition (b) in subsection 3.1.1.

Applying these two independent thinning, the intensity of the process (the number of the interfering CR femto-BSs at the tagged macro user) becomes

$$\lambda_I(U) = \lambda'_F d U^{d-1} b_d p_{RB} p_{tx}. \tag{32}$$

Taking the similar approach, (31) is simplified as follows

$$\mathcal{L}_{I_{FB}}(s) = e^{\mathrm{K}^d r_M{}^d b_d p_{RB} p_{tx} \lambda'_F - \frac{d}{\alpha} b_d p_{RB} p_{tx} \lambda'_F \left(\frac{m_{d,M} P_F \gamma}{\Omega_{d,M} P_M}\right)^{\frac{d}{\alpha}} r_M{}^d T(\gamma,\alpha)} \tag{33}$$

in which $T(\gamma,\alpha) = \mathrm{E}\left[\left(G_p\right)^{\frac{d}{\alpha}}\left(\Gamma\left(-\frac{d}{\alpha}, \frac{\frac{m_{d,M}}{\Omega_{d,M}}\gamma P_F G_p}{\mathrm{K}^{\alpha} P_M}\right) - \Gamma\left(-\frac{d}{\alpha}\right)\right)\right]$.

The Nakagami power fading model is also applied to the interference links (Nakagami fading links with equal parameters $m_{I,F}$ and $\Omega_{I,F}$). The mathematical derivation of $T(\gamma, \alpha)$ is quite similar to the derivation of $M(\theta, \alpha)$ and $V(\gamma, \alpha)$ (see Appendix 2). Hence, $T(\gamma, \alpha)$ is expressed as follows

$$T(\gamma,\alpha) = \frac{-\Gamma\left(-\frac{d}{\alpha}\right) m_{I,F}{}^{m_{I,F}} \left(\frac{m_{d,M}}{\Omega_{d,M}\mathrm{K}^{\alpha}P_M}\gamma\right)^{-\frac{d}{\alpha}}}{\Omega_{I,F}{}^{m_{I,F}}\Gamma(m_{I,F})} \times \left[\sum_{k=0}^{\infty} \frac{\left(\frac{\frac{m_{d,M}}{\Omega_{d,M}}\gamma P_F}{\mathrm{K}^{\alpha}P_M}\right)^k \left(\frac{\mathrm{K}^{\alpha}P_M\Omega_{d,M}\Omega_{I,F}}{m_{I,F}\Omega_{d,M}\mathrm{K}^{\alpha}P_M + m_{d,M}\Omega_{I,F}P_F\gamma}\right)^{m_{I,F}+k}}{\Gamma\left(k+1-\frac{d}{\alpha}\right)} \Gamma(m_{I,F}+k)\right]. \tag{34}$$

3.2.2 Closed-form macro outage probability expression

Finally, from (24), (26), (29), and (33) and by replacing $s = \frac{m_{d,M}\gamma r_M{}^{\alpha}}{\Omega_{d,M}\ P_M}$, and $z = r_M{}^{\alpha}$, the closed-form expression for the outage probability of the tagged macro user is obtained as follows

$$p_{OM} = 1 - \left[\sum_{k=0}^{m_{d,M}-1} \frac{(z)^k}{k!}(-1)^k \frac{d^k(\exp(l(z)))}{dz^k}\right] \tag{35}$$

where

$\exp(l(z)) = e^{[\mathrm{B}]z^{\frac{d}{\alpha}} - \frac{m_{d,M}\gamma\sigma^2}{\Omega_{d,M}P_M}z}$ and $z = r_M{}^{\alpha}$
and B is given by (36).

$$\mathrm{B} = \mathrm{B}_1 + \mathrm{B}_2 \tag{36}$$

where

$$\mathrm{B}_1 = \mathrm{K}^d b_d p_{RB} p_{tx} \lambda_F' + \frac{d}{\alpha} b_d p_{RB} p_{tx} \lambda_F' \left(\frac{m_{d,M}}{\Omega_{d,M}}\frac{\gamma}{P_M}\right)^{\frac{d}{\alpha}} \times \left(\frac{\Gamma\left(-\frac{d}{\alpha}\right) {m_{I,F}}^{m_{I,F}} \left(\frac{m_{d,M}}{\Omega_{d,M}\mathrm{K}^\alpha P_M}\gamma\right)^{-\frac{d}{\alpha}}}{{\Omega_{I,F}}^{m_{I,F}}\Gamma(m_{I,F})}\right) \times \left[\sum_{k=0}^{\infty} \frac{\left(\frac{\frac{m_{d,M}}{\Omega_{d,M}}\gamma P_F}{\mathrm{K}^\alpha P_M}\right)^k \left(\frac{\mathrm{K}^\alpha P_M \Omega_{d,M}\Omega_{I,F}}{m_{I,F}\Omega_{d,M}\mathrm{K}^\alpha P_M + m_{d,M}\Omega_{I,F}P_F\gamma}\right)^{m_{I,F}+k}}{\Gamma\left(k+1-\frac{d}{\alpha}\right)} \Gamma(m_{I,F}+k)\right].$$

and

$$\mathrm{B}_2 = b_d \lambda_M' + \frac{d}{\alpha} b_d \lambda_M' \left(\frac{\Gamma\left(-\frac{d}{\alpha}\right){m_{I,M}}^{m_{I,M}}}{{\Omega_{I,M}}^{m_{I,M}}\Gamma(m_{I,M})}\right) \times \left[\sum_{k=0}^{\infty} \frac{\left(\frac{m_{d,M}}{\Omega_{d,M}}\gamma\right)^k \left(\frac{\Omega_{I,M}\Omega_{d,M}}{m_{I,M}\Omega_{d,M} + \Omega_{I,M}m_{d,M}\gamma}\right)^{m_{I,M}+k}}{\Gamma\left(k+1-\frac{d}{\alpha}\right)} \Gamma(m_{I,M}+k)\right].$$

The probability of outage averaged over the plane can also be derived as

$$p_{\bar{o}M} = 1 - \mathrm{E}_{r_M}[P[\mathrm{SINR} > \gamma]]$$
$$p_{\bar{o}M} = 1 - \int_0^\infty [P[\mathrm{SINR} > \gamma]] . f_{r_M}(r_M) \mathrm{d}r_M \tag{37}$$

where $f_{r_M}(r_M) = e^{-\lambda_M \pi {r_M}^2} 2\pi\lambda_M r_M$ is the pdf of r_M. Such as before, $p_{\bar{o}M}$ can be easily obtained for $m_{d,M} = 1, 2, 3, \ldots$.

3.3 Outage probability formulation under perfect and imperfect sensing

We now study the effect of both the *ideal* detection, i.e., $P_d = 1$ and $P_f = 0$, and the cases involving imperfect sensing, i.e., $P_d \neq 1$ and $P_f \neq 0$ on the outage probabilities of femto and macro users.

3.3.1 Scenario I

Ideal detection ($P_d = 1$ and $P_f = 0$):

3.3.1.1 (femto and macro outage probabilities) Each secondary node (femto-BS) has perfect knowledge of each primary (macro-BS) signaling. In other words, sensing at each femto-BS is done perfectly. Therefore, RBs occupied by the macro network are not chosen by any femo-BS for data transmission. In this case, the tagged femto user does not experience any interference from the macro-BSs. This assumption ($P_d = 1$ and $P_f = 0$) is reasonable in the scenarios where interference from other systems is assumed to be negligible.

Therefore, under this assumption, $\mathrm{E}[e^{-sI_{MB}}] = 1$ and $\mathcal{L}_{I_{MB}}(s) = 1$ (in (7)), consequently. Then, for the outage probability of the tagged femto user in (21), A is only equal to A_1 (see (22)). It should be noted that the calculation of p_{RB} for this case is presented in Section 4, Scenario I, Case 1.

Clearly, under this scenario and considering the above explanations, the tagged macro user does not experience any interference from the surrounding femto-BSs. Therefore, for the outage probability of the tagged macro user in (35), B is only equal to B_2 (see (36)).

3.3.2 Scenario II

Imperfect detection ($P_d \neq 1$ and $P_f \neq 0$):

3.3.2.1 (femto and macro outage probabilities) In this scenario, each secondary node (femto-BS) has imperfect knowledge of each primary (macro-BS) signaling. In other words, sensing at each femto-BS is done imperfectly and subject to (occasional) error. Therefore, RBs occupied by the macro network may be wrongly chosen by the femo-BSs, as well. Two cases can take place under the imperfect sensing scenario:

Case 1. The tagged femto-BS transmits data on an idle RB (in this case, the outage probability formulations (for both the tagged femto and macro users) are the same as in the perfect sensing scenario except for the calculation of p_{RB} (see Section 4, Scenario II, Case 1))

Case 2. The tagged femto-BS transmits on an occupied RB (in this case, the outage probability formulations for both the tagged femto and macro users are explained as follows and the calculation of p_{RB} is presented in Section 4, Scenario II, Case 2)

In Case 2, the tagged femto user can experience interference from both the active femto-BSs (which pick the same occupied RB as the tagged femto-BS) and macro-BSs. Then, for the outage probability of the tagged femto user in (21), A is equal to $\mathrm{A}_1 + \mathrm{A}_2$ (see (22)). Clearly, under this scenario and in Case 2, the tagged macro user can experience interference on its RB, from both the femto-BSs and macro-BSs. Indeed, for the outage probability of the tagged macro user in (35), B is equal to $\mathrm{B}_1 + \mathrm{B}_2$.

4 Resource block selection probability (p_{RB}) calculations under perfect and imperfect sensing

In this section, we discuss how the optimal values of the RB selection probability (p_{RB}) for a secondary transmitter (femto-BS) can be determined under each femto-BS's perfect and imperfect sensing scenarios.

4.1 Scenario I

Ideal detection for all the CR femto-BSs ($P_d = 1$ and $P_f = 0$) [13,14]:

Case 1. The tagged femto-BS assigns the ith idle RB to its femto user.

p_{RB}: The probability that the ith idle RB being selected for data transmission by any of the other active CR femto-BS [14].

$$p_{RB|M_s} = p_{idle}(M_s).\frac{\binom{M-1}{M_s-1}}{\binom{M}{M_s}}.\left(\frac{1}{M_s}\right) \tag{38}$$

where

$$p_{idle}(M_s) = \frac{\binom{M}{M_s}\binom{N-M}{N_s-M_s}}{\binom{N}{N_s}}.$$

The first term ($p_{idle}(M_s)$) indicates the probability of M_s idle RBs sensed by one femto-BS (during the sensing time, T_S) and the second term is the probability that the ith idle RB is inside the M_s idle RBs, and finally the third term indicates that the probability of selecting the ith idle RB (out of the M_s idle RBs) by that femto-BS is equal to $\frac{1}{M_s}$ (since each of the idle RBs within the M_s idle RBs has an equal probability of being chosen). It should be noted that each CR femto-BS will fail to access when $M_s = 0$ (the maximum value of M_s is equal to min{M, N_s}). Therefore, from (38),

$$\begin{aligned} p_{RB} &= p_{RB|(M_s\geq 1)} = \frac{1}{M}p_{idle}(M_s\geq 1) \\ &= \frac{1}{M}(1-p_{idle}(M_s = 0)). \end{aligned} \tag{39}$$

If the number of RBs sensed by one CR femto-BS (i.e., N_s) is more than or equal to $N - M + 1$, then at least we have one idle RB within the N_s detected RBs, i.e., $p_{idle}(M_s = 0) = 0$. On the other hand, if N_s is smaller than or equal to N-M, the CR femto-BS will fail to access when the RBs within the N_s sensed RBs are all occupied by the macro-BSs. Therefore, we can write

$$p_{idle}(M_s = 0) = \begin{cases} \dfrac{\binom{N\ \ M}{N_s}}{\binom{N}{N_s}}, & if\ N_s \leq N-M \\ 0, & if\ N_s \geq N-M+1. \end{cases} \tag{40}$$

From (39) and (40), the probability that a CR femto-BS selects the ith idle RB for data transmission is obtained as follows

$$p_{RB} = \begin{cases} \dfrac{1}{M}\left(1-\dfrac{\binom{N-M}{N_s}}{\binom{N}{N_s}}\right), & if\ N_s \leq N-M \\ \dfrac{1}{M}, & if\ N_s \geq N-M+1. \end{cases} \tag{41}$$

4.2 Scenario II

Imperfect detection for all the CR femto-BSs ($P_d \neq 1$ and $P_f \neq 0$) [13,14]:

Case 1. The tagged femto-BS assigns the ith idle RB to its femto user.

p_{RB}: The probability that the ith idle RB being selected for data transmission by any of the other active CR femto-BS [14].

We show the detection result indicator of the nth RBs by D_n ($n \in \{1, 2, ..., N\}$). If $D_n = 1$, the nth RB is detected as idle RB; otherwise, $D_n = 0$. The probability of one idle RB detected with no false alarm is $1 - P_f$ and the probability for an occupied RB detected as an idle RB is $1 - P_d$. In other words,

$$\Pr(D_n = 1) = \begin{cases} V_0 = 1-P_f, & \text{if } n\text{th RB is idle} \\ V_1 = 1-P_d, & \text{if } n\text{th RB is busy} \end{cases} \tag{42}$$

in which P_f is the false alarm probability and can be obtained as follows [14-16]

$$P_f(\tau) = \mathcal{Q}\left(\sqrt{2\eta+1}\mathcal{Q}^{-1}(P_d) + \sqrt{\tau f_s}\eta\right) \tag{43}$$

($\mathcal{Q}(x) = \frac{1}{\sqrt{2\pi}}\int_x^{\infty} \exp\left(\frac{-t^2}{2}\right)dt$ and P_d is the predefined detection probability. τ is the spectrum sensing time, f_s the sampling frequency, and η the received interference power on an RB to each femto-BS).

Indeed, the probability that the ith idle RB is detected with no false alarm by a CR femto-BS is $\Pr(D_i = 1) = V_0$.

To obtain the probability that the ith idle RB being selected for data transmission by one CR femto-BS (p_{RB}), first, we calculate the probability that the ith idle RB is sensed and included in the M_s idle RBs out of the N_s sensed RBs in the sensing period (T_S), and it is expressed as follows

$$\begin{aligned} &\Pr(\text{the } i\text{th idle RB is sensed} \mid M_s) \\ &= p_{idle}(M_s).\frac{\binom{M-1}{M_s-1}}{\binom{M}{M_s}} \end{aligned} \tag{44}$$

where

$$p_{idle}(M_s) = \frac{\binom{M}{M_s}\binom{N-M}{N_s-M_s}}{\binom{N}{N_s}}.$$

Conditioning on M_D (see Table 1) and M_s, the probability that the ith idle RB being detected as idle is obtained as follows

$$\begin{aligned}&\Pr(D_i = 1 \mid M_s, M_D)\\&= \Pr(D_i = 1).\Pr\left(\sum_{n\neq i,\ n\in\Phi} D_n = M_D-1 \mid M_s\right)\\&= V_0 \times \left[\sum_{m_{ID}=\max\{1,M_D-(N_s-M_s)\}}^{\min\{M_D,M_s\}} [\binom{M_s-1}{m_{ID}-1}(V_0)^{m_{ID}-1}(1-V_0)^{M_s-m_{ID}} .\binom{N_s-M_s}{m_{OD}}.(V_1)^{m_{OD}}(1-V_1)^{N_s-M_s-m_{OD}}]\right]\end{aligned} \tag{45}$$

in which Φ is the set of the detected RBs by one femto-BS (see Table 1 for the definitions of m_{ID} and m_{OD}). Replacing m_{OD} with $M_D - m_{ID}$, we have

$$\begin{aligned}&\Pr(D_i = 1 \mid M_s, M_D)\\&= \left[\sum_{m_{ID}=\max\{1,M_D-(N_s-M_s)\}}^{\min\{M_D,M_s\}} [\binom{M_s-1}{m_{ID}-1}(V_0)^{m_{ID}}(1-V_0)^{M_s-m_{ID}} .\binom{N_s-M_s}{M_D-m_{ID}}.(V_1)^{M_D-m_{ID}}(1-V_1)^{N_s-M_s-M_D+m_{ID}}]\right]\end{aligned} \tag{46}$$

Having the M_D detected idle RBs (including the ith idle RB), the probability of a CR femto-BS accessing the ith idle RB is equal to $\frac{1}{M_D}$. Thus, the probability that the ith idle RB is selected for data transmission by any CR femto-BS (under imperfect sensing scenario) is obtained as follows

$$\begin{aligned}p_{RB\mid M_s,M_D} &= \Pr(\text{the } i\text{th idle RB is sensed} \mid M_s)\\&\times \frac{1}{M_D} \times \Pr(D_i = 1 \mid M_s, M_D).\end{aligned} \tag{47}$$

Finally,

Table 1 Symbols used in Section 4

Symbol	Definition
N	Number of RBs
M	Number of idle RBs
N_s	Number of sensed RBs
M_s	Number of idle RBs within the N_s sensed RBs
M_D	Number of RBs detected as idle within the N_s sensed RBs
m_{ID}	Number of idle RBs (out of the M_D detected idle RBs) detected correctly. ($m_{ID} \in [\max\{1, M_s - (N_s - M_D)\}, \min\{M_s, M_D\}]$)
m_{OD}	Number of busy RBs (out of the M_D detected idle RBs) detected as idle. ($m_{OD} = M_D - m_{ID}$)

$$p_{RB} = \sum_{M_s=\max\{1,N_s-(N-M)\}}^{\min\{N_s,M\}} \sum_{M_D=1}^{N_s} p_{RB\mid M_s,M_D}. \tag{48}$$

Case 2. The tagged femto-BS assigns the ith busy RB (occupied by the macro network) to its femto user.

p_{RB}: The probability that the ith busy RB being selected for data transmission by any of the other active CR femto-BS.

To obtain the probability of the ith busy RB selected for data transmission by one femto-BS (p_{RB}), first, we calculate the probability that the ith busy RB is sensed and included in the $(N_s - M_s)$ busy RBs out of the N_s sensed RBs in the sensing period, and it is expressed as follows

$$\Pr(\text{ the } i\text{th busy RB is sensed} \mid (N_s-M_s)) = \frac{\binom{N-M-1}{N_s-M_s-1}\binom{M}{M_s}}{\binom{N}{N_s}}. \tag{49}$$

Conditioning on M_D and $(N_s - M_s)$, the probability that the ith busy RB being detected as idle is obtained as follows

$$\begin{aligned}&\Pr(D_i = 1 \mid (N_s-M_s), M_D)\\&= \Pr(D_i = 1).\Pr\left(\sum_{n\neq i,\ n\in\Phi} D_n = M_D-1 \mid (N_s-M_s)\right)\\&= V_1 \times \left[\sum_{m_{ID}=\max\{1,M_D-(N_s-M_s)\}}^{\min\{M_D,M_s\}} [\binom{M_s}{m_{ID}}(V_0)^{m_{ID}}(1-V_0)^{M_s-m_{ID}} .\binom{N_s-M_s-1}{m_{OD}-1}.(V_1)^{m_{OD}-1}(1-V_1)^{N_s-M_s-m_{OD}}]\right]\end{aligned} \tag{50}$$

in which Φ is the set of the detected RBs by a femto-BS. Again, replacing m_{OD} with $M_D - m_{ID}$, we have

$$\begin{aligned}&\Pr(D_i = 1 \mid (N_s-M_s), M_D)\\&= \left[\sum_{m_{ID}=\max\{1,M_D-(N_s-M_s)\}}^{\min\{M_D,M_s\}} [\binom{M_s}{m_{ID}}(V_0)^{m_{ID}}(1-V_0)^{M_s-m_{ID}} .\binom{N_s-M_s-1}{M_D-m_{ID}-1}.(V_1)^{M_D-m_{ID}}(1-V_1)^{N_s-M_s-M_D+m_{ID}}]\right]\end{aligned} \tag{51}$$

Having the M_D detected idle RBs (including the ith busy RB), the probability of a CR femto-BS accessing the ith busy RB is $\frac{1}{M_D}$. Thus, the probability that the ith busy RB is selected for data transmission by any CR femto-BS (under imperfect sensing scenario) is obtained as follows

$$\begin{aligned}p_{RB\mid(N_s-M_s),M_D} &= \Pr(\text{ the } i\text{th idle RB is sensed} \mid (N_s-M_s)) \times \frac{1}{M_D}\\&\times \Pr(D_i = 1 \mid (N_s-M_s, M_D)).\end{aligned} \tag{52}$$

Finally,

$$p_{RB} = \sum_{M_s=\max\{1,N_s-(N-M)\}}^{\min\{N_s,M\}} \sum_{M_D=1}^{N_s} p_{RB\,|\,(N_s-M_s),M_D}. \qquad (53)$$

5 Simulation results and discussions

Outage probability is considered as a QoS metric to meet a specific connection data rate requirement, or it can be utilized as a performance measure to evaluate the level of meeting the total demands of users in highly dense networks such as the heterogeneous cellular networks. By obtaining an exact closed-form expression for the system outage probability, we will be able to take a finer look at the system performance with the end goal of better understanding system design principles for the heterogeneous cellular networks.

Outage probability is affected by many parameters that have impact on the system performance. As mentioned before, cognition can be a potential solution for dynamic spectrum allocation which will adapt to the network geometry, solve the interference management issue, and reduce the outage probability. Therefore, the primary motivations for the femto-BSs to employ this cognition and hence, the p_{RB} and p_{tx} parameters in the cellular network are to provide beneficial coverage improvements for different types of users. In practical heterogeneous cellular networks such as LTE, these parameters are network-specific and not user-specific; thus, there needs to be some optimization techniques, which can be considered as future work, used to select parameters that can provide acceptable QoS for the majority of users and to meet the existing outage constraints.

On the other hand, outage probability is significantly affected by the performance of the spectrum sensing. The performance of the spectrum sensing and the determination of each individual RB as busy/idle is determined by two important parameters mentioned before, namely the probability of (correct) detection (P_d) and probability of false alarm (P_f). Considering the fact that p_{RB} is a direct function of P_f and P_d in the practical scenarios (see Section 4), a spectrum sensing with high P_f and low P_d results in a lower corresponding QoS and a higher observable outage probability. The ideal sensing procedure is the one with $P_f = 0$ and $P_d = 1$. In practice, however, P_d and P_f are related to each other through a receiver operating characteristic (ROC) curve [17], which is a fundamental attribute of each spectrum sensing system.

We will see later that the outage probability is also an increasing function of the SINR threshold as theoretically can be proved. The SINR threshold θ is generally under the control of the system designer and should be chosen reasonably. Clearly, setting a high value for θ maintains the signal quality but increases the outage probability and reduces the network utilization. Therefore, an upper bound of this threshold can be considered [18] in order to guarantee that outage probability is kept below a maximum value. On the other hand, lowering θ is desirable for better resource utilization. However, small θ allows more users to be admitted per RB which can make the system infeasible if the number of admitted users per RB exceeds a certain limit.

First, the accuracy of our analytical results for the downlink analysis in the proposed model is validated by simulations, as shown in Figures 4 and 5. The simulations which are built on Matlab platform are carried out to plot the curves of outage probability versus the SINR threshold for the tagged femto and macro user, as shown in Figures 4 and 5, respectively. The considered scenario is a two-tier network (exactly following the network model described in Sections 2 and 3) over an approximately 1×1 km square with the locations of different classes of BSs as realizations of independent PPPs of given densities, and the tagged users located at the center. To have an estimate of the outage probability at the tagged users, the simulation results are averaged over both the spatial PPP (500 different positions) and fading distribution (300 realizations), and are conducted using the parameters mentioned in the figures' captions. Analytical curves are compared with the simulations under both perfect and imperfect sensing. It is observed that the simulation results closely match our analytical model and the curves of analytical and simulation results match fairly well, which confirms our analysis. The plots exhibit slight discrepancies between analytical results and the corresponding simulation results which are mainly due

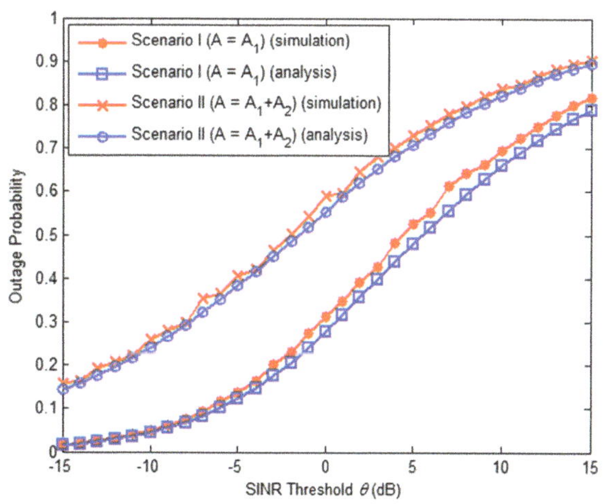

Figure 4 Simulation and analytical results for the outage probability of the tagged femto user under perfect and imperfect sensing. The system parameters are set as $P_F = 23$ dBm, $P_M = 43$ dBm, $\alpha = 4$, $m_{d,F} = m_{I,F} = m_{I,M} = 1$, $p_{RB} = 0.8$, $p_{tx} = 0.7$, K = 1, $\lambda'_M = 10$, $\lambda_F = 800$ and $\lambda'_F = 700$.

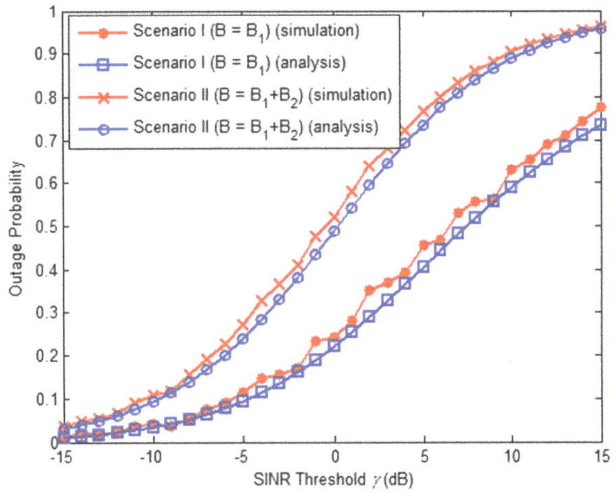

Figure 5 Simulation and analytical results for the outage probability of the tagged macro user under perfect and imperfect sensing. The system parameters are set as $P_F = 23$ dBm, $P_M = 43$ dBm, $\alpha = 4$, $m_{d,M} = m_{I,M} = m_{I,F} = 1$, $p_{RB} = 0.8$, $p_{tx} = 0.7$, K = 1, $\lambda'_F = 700$, $\lambda'_M = 10$ and $\lambda_M = 50$.

to the boundary effect and the independence assumption used in Section 3.

In Figure 6, the outage probability of the tagged femto user under perfect and imperfect spectrum sensing abilities for the CR femto-BSs is shown for different values of the target SINR θ on the horizontal axis. Our results show that the outage probability at the tagged femto receiver in the absence of a perfect spectrum sensing ability is considerably increased. Either in Scenario I, when all the femto-BSs employ perfect sensing to sense the RBs, or in Scenario II, Case 1, the tagged femto user does not experience any interference from the macro-BSs owing to the correct detection at the tagged femto-BS or choice of idle RBs for data transmission (RBs not occupied by the macro network). In this case, the interference seen by the tagged femto user is only the aggregate interference from the other femto-BSs which are transmitting on the same idle RB as the tagged femto. Clearly, the lowest outage probability is for the case of perfect sensing ($A = A_1$ in (22)). Now, let's consider the imperfect sensing scenario for the CR femto-BSs. Obviously, the tagged femto user is now subject to sensing error and therefore picking an occupied RB for its data transmission period. Under this condition, it may receive interference not only from the other femto-BSs which pick the same busy RB (due to the imperfect sensing), but also from those macro-BSs communicating with their own users on the same RB as the tagged femto. Therefore, the tagged femto user experiences an interference larger than before and consequently a significant increase in the outage probability. Moreover, the tagged femto user will face an outage with a higher probability whenever it picks an RB (occupied RB) already used by a larger number of macro users (a larger λ'_M) (see Figure 6).

Figure 7 depicts the outage probability of the tagged macro user for different values of the target SINR γ and different situations. Considering the results obtained in Figure 6, here, we also investigate the effect of employing the two different sensing scenarios for the CR femto-BSs on the outage probability of the tagged macro user. In the case of perfect sensing, the tagged macro user does not experience any interference from the femto-BSs because only those RBs sensed to be idle (RBs not occupied by the macro network) are always chosen for data transmission by the femto network. In this case,

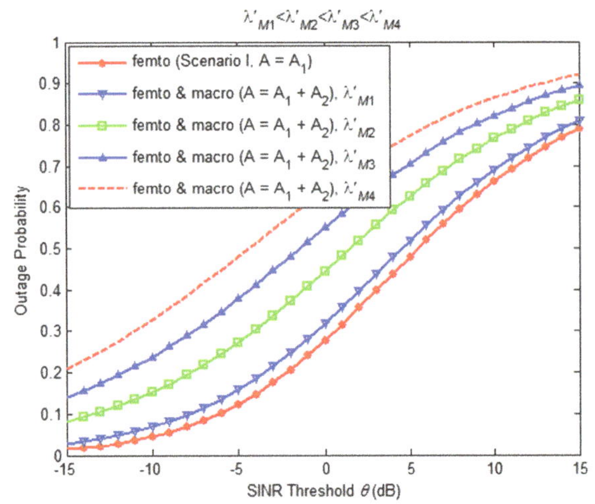

Figure 6 Outage probability of the tagged femto user under perfect and imperfect sensing with different values of λ'_M. The system parameters are set as $P_F = 23$ dBm, $P_M = 43$ dBm, $\alpha = 4$, $m_{d,F} = m_{I,F} = m_{I,M} = 1$, $p_{RB} = 0.8$, $p_{tx} = 0.7$, K = 1, $\lambda_F = 800$ and $\lambda'_F = 700$.

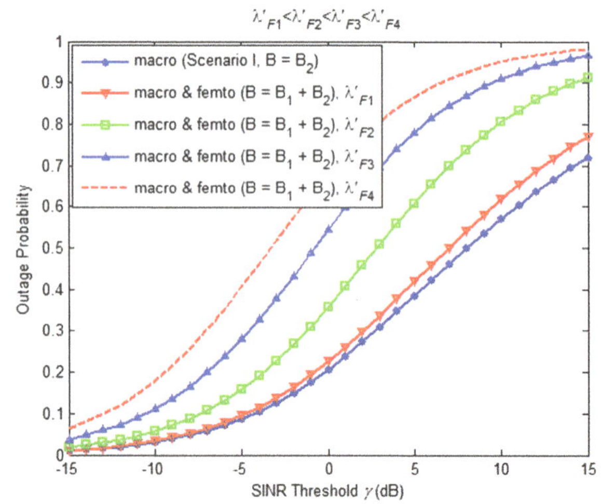

Figure 7 Outage probability of the tagged macro user under perfect and imperfect sensing with different values of λ'_F. The system parameters are set as $P_F = 23$ dBm, $P_M = 43$ dBm, $\alpha = 4$, $m_{d,M} = m_{I,M} = m_{I,F} = 1$, $p_{RB} = 0.8$, $p_{tx} = 0.7$, K = 1, $\lambda'_M = 10$ and $\lambda_M = 50$.

the interference observed at the tagged macro user is only the aggregate interference received from those macro-BSs transmitting on the same RB as the tagged macro. Clearly, the lowest outage probability is obtained for this case ($B = B_2$ in (36)). Now, the case of imperfect spectrum sensing of the CR femto-BS nodes is considered, when the femto-BSs are subject to sensing error and therefore the possibility of transmitting on the RB occupied by the tagged macro. Under this condition, the tagged macro user may receive interference not only from the other macro-BSs communicating with their own users over the same RB (due to the lack of RBs) as the tagged macro, but also from those femto-BSs which pick the same RB. Therefore, the tagged macro user experiences an interference larger than before and consequently a significant increase in the outage probability. Moreover, the tagged macro user will face an outage with a higher probability whenever its own RB is wrongly selected for data transmission by a larger number of femto-BSs (a larger $\lambda_F^{'}$).

Figure 8 illustrates the effect of K (in $D = Kr_M$) on the observed outage probability at the tagged macro user in the presence of both the macro and femto networks. Considering the previous explanations and Figure 3, let $\Phi_F^{'}$ include all the points (representing the femto-BSs) in Φ_F except the points inside the exclusion region D of the tagged macro user. Since $\Phi_F^{'} \subset \Phi_F$, the potential aggregate interference at the tagged macro user, caused by the active (considering slotted ALOHA) CR femto-BSs is less than that in the case with no D. Furthermore, as the exclusion region D becomes larger (when K = 10, for example), the probability of outage is significantly reduced. Indeed, the bigger the value of D, the closer the outage probability curve becomes to the black curve which represents the outage probability in the case when no overall interference from the femto network is observed at the tagged macro user due to the perfect sensing ability of the CR femto-BSs. However, the reduction in the outage probability can be less when the number of macro-BSs transmitting on the same RB as the tagged macro is larger (see Figure 8b).

For a commercial network, designers must find a way to achieve a lower probability of outage for a certain SINR as the minimum quality needed for a typical femto or macro user to experience an acceptable QoS. A common way to decrease the outage probability is to reduce the number of interfering BSs encountered at the users. This can be done through applying both the (a) and (b) conditions in the perfect sensing scenario (or both the (c) and (d) in the imperfect sensing scenario). As shown in Figure 9, under these constraints, the outage

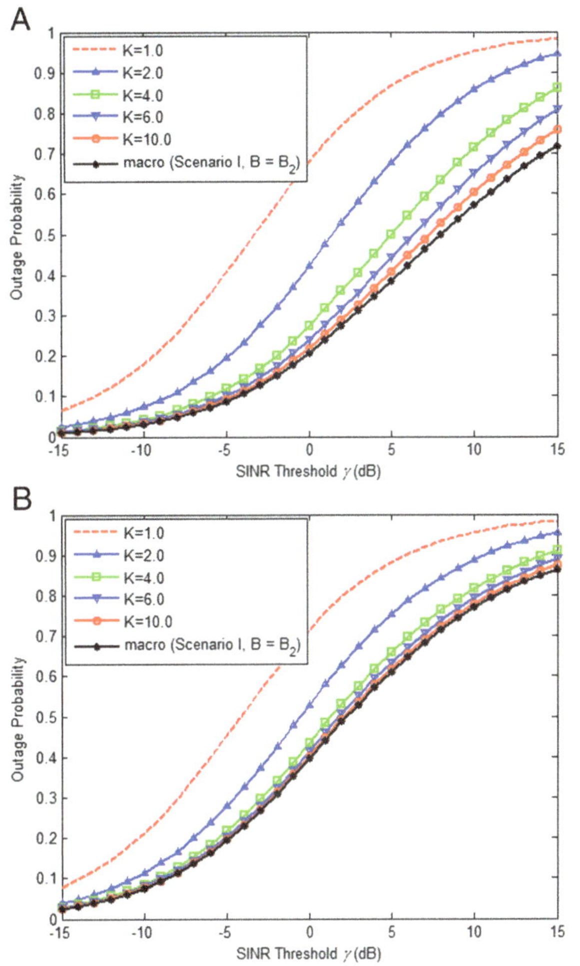

Figure 8 Effect of D ($D = Kr_M$, the exclusion region around the tagged macro user) on outage probability of the tagged macro user. The system parameters are set as $P_F = 23$ dBm, $P_M = 43$ dBm, $\alpha = 4$, $m_{d,M} = m_{I,M} = m_{I,F} = 1$, $p_{RB} = 0.8$, $p_{tx} = 0.7$, $\lambda_M = 50$, $\lambda_M^{'} = 10$ in **(A)** and $\lambda_M^{'} = 25$ in **(B)**.

probability is significantly reduced at the tagged femto user (note that the number of interfering macro-BSs at the tagged femto user is considered equal for all the curves in the figure). The goal is to see the effect of both p_{RB} and p_{tx} (these two parameters are employed at each CR femto-BS) on the outage probability of the tagged femto user. As can be seen, for the case when $p_{RB} = 1$ and $p_{tx} = 1$, outage occurs with higher probability. In other words, when all the existing femto-BSs (except those who are inside the macro users' exclusion regions) pick the same RB as the tagged femto ($p_{RB} = 1$) and when they all have data to transmit in the current time slot ($p_{tx} = 1$), the tagged femto user will experience the maximum value for the outage probability derived for different SINR targets. Clearly, a significant reduction in the outage probability is occurred for the smaller values of p_{RB} and p_{tx} (see Figure 9) (the smaller the values of

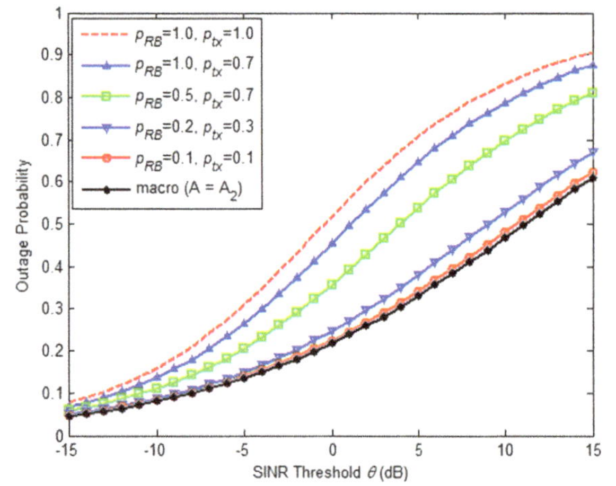

Figure 9 Effect of p_{RB} and p_{tx} (defined parameters for CR femto-BSs) on outage probability of the tagged femto user. The system parameters are set as $P_F = 23$ dBm, $P_M = 43$ dBm, $\alpha = 4$, $m_{d,F} = m_{I,F} = m_{I,M} = 1$, K = 1, $\lambda_F = 800$, $\lambda_F^{'} = 700$ and $\lambda_M^{'} = 10$.

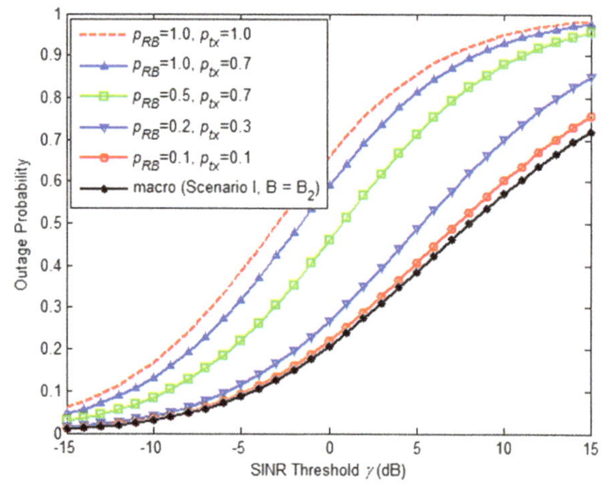

Figure 10 Effect of p_{RB} and p_{tx} (defined parameters for CR femto-BSs) on outage probability of the tagged macro user. The system parameters are set as $P_F = 23$ dBm, $P_M = 43$ dBm, $\alpha = 4$, $m_{d,F} = m_{I,F} = m_{I,M} = 1$, K = 1, $\lambda_F^{'} = 700$, $\lambda_M = 50$ and $\lambda_M^{'} = 10$.

p_{RB} and p_{tx}, the closer the outage probability curve becomes to the black curve which represents the outage probability in the case when the received interference at the tagged femto user is only the aggregate interference from the macro network). Indeed, this validates that many studies which do not consider these constraints over-estimate the interference encountered by a typical femto user.

It can be seen that even for high values of p_{tx}, the outage probability is relatively less than that in the case where there is no constraint on the femto-BS's transmission schedule ($p_{tx} = 1$). Also, it is obvious that in the presence of multiple RBs where each RB is picked with probability p_{RB}, the outage probability is further decreased. Hence, any practical heterogeneous network designed to satisfy both of the mentioned conditions can reap the benefit of opportunistic exploitation of spectrum, while possibly causing little or no harmful interference. The same story exists when the outage probability of the tagged macro user is investigated. Figure 10 shows the effect of p_{RB} and p_{tx} (parameters related to the CR femto-BSs) on the outage probability of the tagged macro user. For instance, when all the active femto-BSs select the same RB as the tagged macro ($p_{RB} = 1$) and when they all have data to transmit (on this busy RB) in the current time slot ($p_{tx} = 1$), the tagged macro user will experience the maximum value for the outage probability. It should be noted that the number of interfering macro-BSs at the tagged macro is considered equal for all curves in this figure.

Figure 11 plots the outage probability versus the SINR threshold θ for the tagged femto user. As $m_{d,F} = m_{I,F}$ increases, the outage probability of the tagged femto user decreases. This is because as the fading becomes less severe (the bigger the Nakagami-m parameter, the smaller fading) for both the desired and interfering links of the femto network, the received power at the tagged femto user from its corresponding femto-BS increases more compared to the total received power from all the interfering femto-BSs. It should be noted that $\beta_F = \frac{\Omega_{d,F}}{m_{d,F}} = \frac{\Omega_{I,F}}{m_{I,F}}$ and $\beta_M = \frac{\Omega_{d,M}}{m_{d,M}} = \frac{\Omega_{I,M}}{m_{I,M}}$ (the scale parameters) are fixed for all the curves. The interference links between the interfering macro-BSs and the tagged femto user are more likely to be severe in terms of fading compared to the femto network's desired and interfering links. In other words, the interfering signals received at the

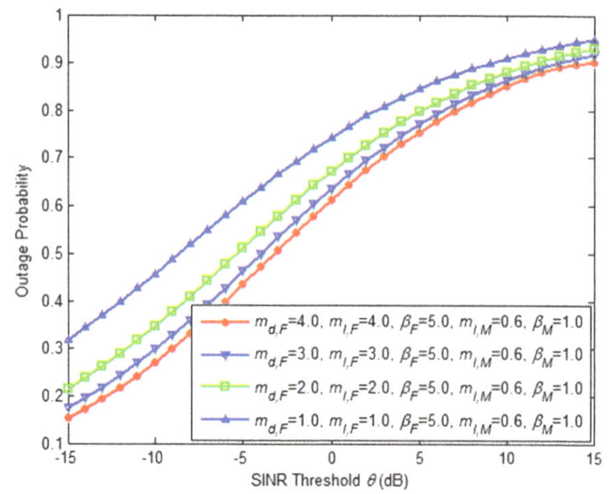

Figure 11 Impact of Nakagami parameter of m on outage probability of the tagged femto user. The system parameters are set as $P_F = 23$ dBm, $P_M = 43$ dBm, $\alpha = 4$, $p_{RB} = 0.8$, $p_{tx} = 0.7$, K = 1, $\lambda_F = 800$, $\lambda_F^{'} = 700$, $\lambda_M = 50$ and $\lambda_M^{'} = 10$.

tagged femto user originated by the macro-BSs suffer more serious fading than the received signals from the femto-BSs, in particular when we talk about indoor femto users. Similar arguments and statements can also be made about the outage probability of the tagged macro user observed in Figure 12.

In Figures 13 and 14, we analyze the performance of the authorized links (femto links) in terms of throughput (achievable with a simple ARQ scheme with error-free feedback) under both the perfect and imperfect sensing scenarios. Considering the slotted ALOHA scheme, we define the following terms [19,20],

Probabilistic link throughput (τ) of a femto user:

(i) In the half-duplex (HD) communication scenario: it is defined to be the success probability of a femto user (i.e., $1 - p_{OF}$) multiplied by the probability that the corresponding femto-BS actually transmits over a specific RB (i.e., $p_{RB}p_{tx}$), and the probability that the femto receiver actually receives over that RB (i.e., $1 - p_{RB}p_{tx}$)

(ii) In the full-duplex (FD) communication scenario: it is defined to be the success probability of a femto user (i.e., $1 - p_{OF}$) multiplied by the probability that the corresponding femto-BS actually transmits over a specific RB (i.e., $p_{RB}p_{tx}$).

Femto link throughput (T):

The femto link throughput is defined as the product of the probabilistic link throughput (τ) and the rate of transmission, i.e., $T = \tau \log(1 + \theta)$. Therefore, the femto link throughput for the half and full duplex cases is written as follows

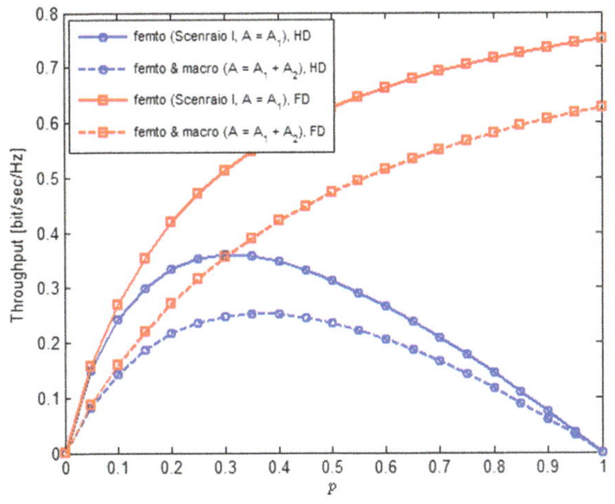

Figure 13 Femto link throughput of half and full duplex systems as a function of transmission probability over a specific RB (*p*). The system parameters are set as $P_F = 23$ dBm, $P_M = 43$ dBm, $\alpha = 4$, $m_{d,F} = m_{I,F} = m_{I,M} = 1$, K = 1, $\lambda'_M = 10$, $\lambda_F = 800$ and $\lambda'_F = 700$, and $\theta = 11$ dB.

$$T^{(half)} = p(1-p)(1-p_{OF})\log(1+\theta)$$
$$T^{(full)} = p(1-p_{OF})\log(1+\theta)$$

in which $p = p_{RB}p_{tx}$.

In Figure 13, the performance of half and full duplex systems are presented for the femto users. More specifically, the link throughput of any typical femto user (e.g. the link between the tagged femto user and its corresponding femto-BS) under perfect and imperfect spectrum sensing abilities for the CR femto-BSs is shown as a function of the transmission probability over a specific RB (i.e., $p = p_{RB}p_{tx}$). It can be seen that the

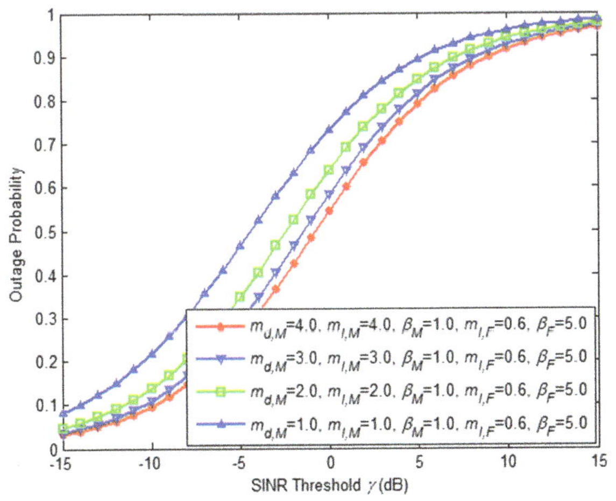

Figure 12 Impact of Nakagami parameter of *m* on outage probability of the tagged macro user. The system parameters are set as $P_F = 23$ dBm, $P_M = 43$ dBm, $\alpha = 4$, $p_{RB} = 0.8$, $p_{tx} = 0.7$, K = 1, $\lambda_F = 800$, $\lambda'_F = 700$, $\lambda_M = 50$ and $\lambda'_M = 10$.

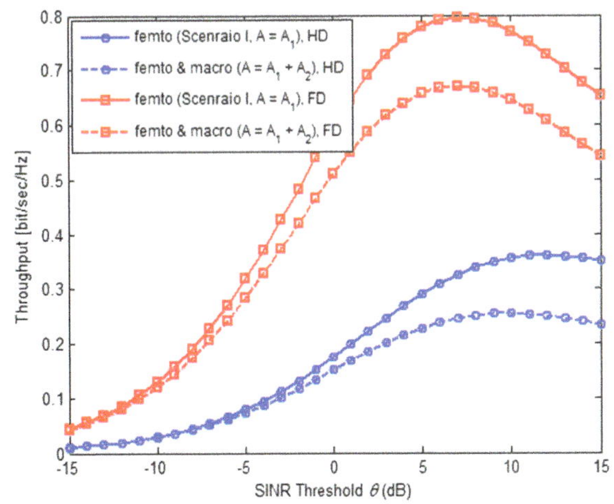

Figure 14 The femto link throughput of half and full duplex systems as a function of the target SINR. The system parameters are set as $P_F = 23$ dBm, $P_M = 43$ dBm, $\alpha = 4$, $m_{d,F} = m_{I,F} = m_{I,M} = 1$, K = 1, $\lambda'_M = 10$, $\lambda_F = 800$ and $\lambda'_F = 700$, and for each curve the obtained optimal p in Figure 13 is applied.

throughput achieved by the FD system is significantly higher, particularly when p is high. Regarding the performance of the HD system, for both the perfect and imperfect sensing cases, there is a unique optimal p which achieves the maximum throughput ($p = 0.3$ for the perfect and $p = 0.35$ for the imperfect sensing scenario). However, for high p, both throughput curves converge to zero due to over many transmissions and interferences on the RB. Obviously, for both the half and full duplex communications, a higher per-link throughput is achieved when the CR femto-BSs employ perfect sensing.

In Figure 14, the performance of half and full duplex systems are presented for femto users. More specifically, the link throughput of any typical femto user (e.g., the link between the tagged femto user and its corresponding femto-BS) under perfect and imperfect spectrum sensing abilities for the CR femto-BSs is shown as a function of the target SINR θ. It can be seen that the per-link throughput achieved by the FD system, for both the perfect and imperfect sensing scenarios, is significantly higher than the HD one. As it is seen, the link throughput curves are concave and there is an optimal point in each curve. With a high target SINR, we can transmit the user data with high spectral efficiency; however, the outage probability of this transmission is high, too. In contrast, with a low-target SINR, we can send many packets that include little information. In other words, a high reliable transmission can be experienced at low-target SINRs, while the minimum requirements for the transmission rate cannot be met.

6 Conclusion

In this paper, utilizing the spatial PPP theory, we presented a tractable model to derive the outage probability of a typical femto and macro user in a two-tier heterogeneous network which provides insight into system design guidelines. In other words, for the case of the node locations modeled by a PPP and Nakagami-m fading channels, we demonstrated the use of the CR-based framework to evaluate the outage probability at any arbitrary user. Exact closed-form expressions were obtained as a result. In addition, we observed that in the downlink analysis, the outage probability is a function of the network topology and several important system design parameters such as SINR target, exclusion regions, MAC (medium access control) mechanisms such as ALOHA (p_{tx}), and the RB selection constraint (p_{RB}) which is controlled by the spectrum sensing measurements.

7 Appendix 1

7.1 Proof of equation 6

Proof: From (5), we have

$$P[\mathrm{SINR} > \theta] = \int_0^\infty \left[e^{-\left(\frac{m_{d,F}\theta r_F{}^\alpha}{\Omega_{d,F}\,P_F}\right)I} \left[1 + \left(\frac{m_{d,F}}{\Omega_{d,F}}\frac{\theta r_F{}^\alpha}{P_F}\right) I + \frac{1}{2}\left(\frac{m_{d,F}}{\Omega_{d,F}}\frac{\theta r_F{}^\alpha}{P_F}\right)^2 I^2 + \ldots \right] \right] f_I(i)\mathrm{d}i$$

We then have

$$\left[\left[\left(\frac{m_{d,F}}{\Omega_{d,F}}\frac{\theta r_F{}^\alpha}{P_F}\right)^0 \int_0^\infty \left[e^{-sI}\left(I^0 f_I(i)\right)\right]\mathrm{d}i \right] + \left[\left(\frac{m_{d,F}}{\Omega_{d,F}}\frac{\theta r_F{}^\alpha}{P_F}\right)^1 \int_0^\infty \left[e^{-sI}\left(I^1 f_I(i)\right)\right]\mathrm{d}i \right] + \left[\frac{1}{2}\left(\frac{m_{d,F}}{\Omega_{d,F}}\frac{\theta r_F{}^\alpha}{P_F}\right)^2 \int_0^\infty \left[e^{-sI}\left(I^2 f_I(i)\right)\right]\mathrm{d}i \right] + \ldots \right]$$

where s in each integral is equal to $\frac{m_{d,F}}{\Omega_{d,F}}\frac{\theta r_F{}^\alpha}{P_F}$.

In each part, we have to calculate the Laplace transform of $(I^n f_I(i))$. We also know that $x^n f(x) \overset{\mathscr{L}}{\Leftrightarrow} (-1)^n \frac{d^n \mathscr{L}_f(s)}{ds^n}$. Therefore, we have

$$\left[\left[\left(\frac{m_{d,F}}{\Omega_{d,F}}\frac{\theta r_F{}^\alpha}{P_F}\right)^0 (-1)^0 \frac{d^0 \mathscr{L}_I(s)}{ds^0} \right] + \left[\left(\frac{m_{d,F}}{\Omega_{d,F}}\frac{\theta r_F{}^\alpha}{P_F}\right)^1 (-1)^1 \frac{d^1 \mathscr{L}_I(s)}{ds^1} \right] + \left[\frac{1}{2}\left(\frac{m_{d,F}}{\Omega_{d,F}}\frac{\theta r_F{}^\alpha}{P_F}\right)^2 (-1)^2 \frac{d^2 \mathscr{L}_I(s)}{ds^2} \right] + \ldots \right]$$

Hence, we have

$$P[\mathrm{SINR} > \theta] = \left[\sum_{k=0}^{m_{d,F}-1} \frac{(s)^k}{k!} (-1)^k \frac{d^k \mathscr{L}_I(s)}{ds^k} \right]$$

8 Appendix 2

8.1 Proof of equation 15

Proof: From (14), we have

$$M(\theta, \alpha) = \int_0^\infty \left[(g)^{\frac{d}{\alpha}} \left(\Gamma\left(-\frac{d}{\alpha}, \frac{m_{d,F}}{\Omega_{d,F}}\theta g\right) - \Gamma\left(-\frac{d}{\alpha}\right) \right) \right] \frac{m_{I,F}{}^{m_{I,F}}}{\Omega_{I,F}{}^{m_{I,F}}\Gamma(m_{I,F})} g^{m_{I,F}-1} e^{-\frac{m_{I,F}g}{\Omega_{I,F}}} \mathrm{d}g$$

We know that $\Gamma(a, x) + \gamma(a, x) = \Gamma(a)$, and $\gamma(a, x) = x^a \Gamma(a) e^{-x} \sum_{k=0}^{\infty} \frac{x^k}{\Gamma(a+k+1)}$, therefore, the above equation is simplified as follows

$$= \frac{-\Gamma\left(-\frac{d}{\alpha}\right) m_{I,F}{}^{m_{I,F}} \left(\frac{m_{d,F}}{\Omega_{d,F}}\theta\right)^{-\frac{d}{\alpha}}}{\Omega_{I,F}{}^{m_{I,F}}\Gamma(m_{I,F})} \int_0^{\infty} (g)^{m_{I,F}-1} e^{-g\left(\frac{m_{I,F}}{\Omega_{I,F}}+\frac{m_{d,F}\theta}{\Omega_{d,F}}\right)} \left[\sum_{k=0}^{\infty} \frac{\left(\frac{m_{d,F}}{\Omega_{d,F}}\theta\right)^k g^k}{\Gamma\left(-\frac{d}{\alpha}+k+1\right)}\right] \mathrm{d}g$$

$$= \frac{-\Gamma\left(-\frac{d}{\alpha}\right) m_{I,F}{}^{m_{I,F}} \left(\frac{m_{d,F}}{\Omega_{d,F}}\theta\right)^{-\frac{d}{\alpha}}}{\Omega_{I,F}{}^{m_{I,F}}\Gamma(m_{I,F})} \int_0^{\infty} (g)^{m_{I,F}-1} e^{-g\left(\frac{m_{I,F}}{\Omega_{I,F}}+\frac{m_{d,F}\theta}{\Omega_{d,F}}\right)} \left[\frac{\left(\frac{m_{d,F}}{\Omega_{d,F}}\theta\right)^0 g^0}{\Gamma\left(1-\frac{d}{\alpha}\right)} + \frac{\left(\frac{m_{d,F}}{\Omega_{d,F}}\theta\right)^1 g^1}{\Gamma\left(2-\frac{d}{\alpha}\right)} + \frac{\left(\frac{m_{d,F}}{\Omega_{d,F}}\theta\right)^2 g^2}{\Gamma\left(3-\frac{d}{\alpha}\right)} + \ldots\right] \mathrm{d}g$$

$$= \frac{-\Gamma\left(-\frac{d}{\alpha}\right) m_{I,F}{}^{m_{I,F}} \left(\frac{m_{d,F}}{\Omega_{d,F}}\theta\right)^{-\frac{d}{\alpha}}}{\Omega_{I,F}{}^{m_{I,F}}\Gamma(m_{I,F})} \left[\int_0^{\infty} \frac{(g)^{m_{I,F}-1} e^{-g\left(\frac{m_{I,F}}{\Omega_{I,F}}+\frac{m_{d,F}\theta}{\Omega_{d,F}}\right)}}{\Gamma\left(1-\frac{d}{\alpha}\right)} \mathrm{d}g + \int_0^{\infty} \frac{\left(\frac{m_{d,F}}{\Omega_{d,F}}\theta\right)^1 (g)^{(m_{I,F}+1)-1} e^{-g\left(\frac{m_{I,F}}{\Omega_{I,F}}+\frac{m_{d,F}\theta}{\Omega_{d,F}}\right)}}{\Gamma\left(2-\frac{d}{\alpha}\right)} \mathrm{d}g + \ldots\right]$$

$$= \frac{-\Gamma\left(-\frac{d}{\alpha}\right) m_{I,F}{}^{m_{I,F}} \left(\frac{m_{d,F}}{\Omega_{d,F}}\theta\right)^{-\frac{d}{\alpha}}}{\Omega_{I,F}{}^{m_{I,F}}\Gamma(m_{I,F})} \left[\frac{\left(\frac{m_{d,F}}{\Omega_{d,F}}\theta\right)^0 \left(\frac{\Omega_{I,F}\Omega_{d,F}}{m_{I,F}\Omega_{d,F}+\Omega_{I,F}m_{d,F}\theta}\right)^{m_{I,F}}}{\Gamma\left(1-\frac{d}{\alpha}\right)} \Gamma(m_{I,F}) + \frac{\left(\frac{m_{d,F}}{\Omega_{d,F}}\theta\right)^1 \left(\frac{\Omega_{I,F}\Omega_{d,F}}{m_{I,F}\Omega_{d,F}+\Omega_{I,F}m_{d,F}\theta}\right)^{m_{I,F}+1}}{\Gamma\left(2-\frac{d}{\alpha}\right)} \Gamma(m_{I,F}+1) + \ldots\right]$$

Hence, we have

$$M(\theta,\alpha) = \left(\frac{-\Gamma\left(-\frac{d}{\alpha}\right) m_{I,F}{}^{m_{I,F}} \left(\frac{m_{d,F}}{\Omega_{d,F}}\theta\right)^{-\frac{d}{\alpha}}}{\Omega_{I,F}{}^{m_{I,F}}\Gamma(m_{I,F})}\right) \times \left[\sum_{k=0}^{\infty} \frac{\left(\frac{m_{d,F}}{\Omega_{d,F}}\theta\right)^k \left(\frac{\Omega_{I,F}\Omega_{d,F}}{m_{I,F}\Omega_{d,F}+\Omega_{I,F}m_{d,F}\theta}\right)^{m_{I,F}+k}}{\Gamma\left(k+1-\frac{d}{\alpha}\right)} \Gamma(m_{I,F}+k)\right]$$

9 Appendix 3

9.1 Outage probability expression for non-integer $m_{d,F}$

Proof: From Section 3

For non-integer $m_{d,F}$, by using the infinite series representation of incomplete gamma function presented in Appendix 2 ($\frac{\Gamma(a,x)}{\Gamma(a)} = 1 - \frac{\gamma(a,x)}{\Gamma(a)}$, and $\gamma(a,x) = x^a\Gamma(a)e^{-x}\sum_{k=0}^{\infty}\frac{x^k}{\Gamma(a+k+1)}$), and following the exact procedure as that in Section 3 (from Eq. (1) to Eq. (21)) to obtain the closed-form outage probability expression for the tagged femto user, we obtain an exact but infinite summation expression for the outage probability. In other words, we use the infinite series representation of incomplete gamma not only for the interfering links calculations but also for the desired link to obtain an expression in which the Nakagami-m parameters of all the communication links ($m_{d,F}$, $m_{I,F}$ and $m_{I,M}$) can take any values (integer or non-integer) bigger than 0.5. Thus, considering the above explanations we have

$$p_{OF} = \sum_{k=0}^{\infty} \frac{(z)^{k+m_{d,F}}}{\Gamma(k+m_{d,F}+1)} (-1)^{k+m_{d,F}} \frac{d^{k+m_{d,F}}(\exp(l(z)))}{dz^{k+m_{d,F}}}$$

where

$\exp(l(z)) = e^{[\mathrm{A}]z^{\frac{d}{\alpha}} - \frac{m_{d,F}\theta\sigma^2}{\Omega_{d,F}P_F}z}$, and $z = r_F{}^{\alpha}$.

Operating in an interference-limited regime and using the Taylor series expansions of exponential functions, the above expression is simplified as follows

$$p_{OF} = \sum_{k=0}^{\infty} \frac{(z)^{k+m_{d,F}}}{\Gamma(k+m_{d,F}+1)} (-1)^{k+m_{d,F}} \frac{d^{k+m_{d,F}}\left(\sum_{n=0}^{\infty}\frac{\mathrm{A}^n z^{n\left(\frac{d}{\alpha}\right)}}{n!}\right)}{dz^{k+m_{d,F}}}$$

$$p_{OF} = \sum_{k=0}^{\infty} \frac{(z)^{k+m_{d,F}}}{\Gamma(k+m_{d,F}+1)} (-1)^{k+m_{d,F}} \frac{d^{k+m_{d,F}}\left(\frac{\mathrm{A}^0 z^0}{0!} + \frac{\mathrm{A}^1 z^{\left(\frac{d}{\alpha}\right)}}{1!} + \frac{\mathrm{A}^2 z^{2\left(\frac{d}{\alpha}\right)}}{2!} + \ldots\right)}{dz^{k+m_{d,F}}}$$

The fractional derivative, the arbitrary order derivative, is used here to simplify the above expression [21] (the notation $D^{\mu}(f(z)) = \frac{d^{\mu}(f(z))}{dz^{\mu}}$ denotes the μth order fractional derivative of the function $f(z)$ where $\mu > 0$). According to [21], $D^{\mu}(z^{\gamma})$ (where $\gamma > -1$) is equal to $\frac{\Gamma(\gamma+1)}{\Gamma(\gamma-\mu+1)}z^{\gamma-\mu}$, and we also have $D^{\mu}(cf(z)) = cD^{\mu}(f(z))$ (c is a constant value). Thus, we can write

$$p_{OF} = \sum_{k=0}^{\infty} \frac{(z)^{k+m_{d,F}}}{\Gamma(k+m_{d,F}+1)} (-1)^{k+m_{d,F}} \left[\frac{A^0\Gamma\left(0\left(\frac{d}{\alpha}\right)+1\right)z^{0\left(\frac{d}{\alpha}\right)-(k+m_{d,F})}}{0!\Gamma\left(0\left(\frac{d}{\alpha}\right)-(k+m_{d,F})+1\right)} + \frac{A^1\Gamma\left(1\left(\frac{d}{\alpha}\right)+1\right)z^{1\left(\frac{d}{\alpha}\right)-(k+m_{d,F})}}{1!\Gamma\left(1\left(\frac{d}{\alpha}\right)-(k+m_{d,F})+1\right)} + \frac{A^2\Gamma\left(2\left(\frac{d}{\alpha}\right)+1\right)z^{2\left(\frac{d}{\alpha}\right)-(k+m_{d,F})}}{2!\Gamma\left(2\left(\frac{d}{\alpha}\right)-(k+m_{d,F})+1\right)} + \ldots \right]$$

and finally,

$$p_{OF} = \sum_{k=0}^{\infty} \frac{(z)^{k+m_{d,F}}}{\Gamma(k+m_{d,F}+1)} (-1)^{k+m_{d,F}} \sum_{n=0}^{\infty} \frac{A^n}{n!} \frac{\Gamma\left(n\left(\frac{d}{\alpha}\right)+1\right)}{\Gamma\left(n\left(\frac{d}{\alpha}\right)-(k+m_{d,F})+1\right)} z^{n\left(\frac{d}{\alpha}\right)-(k+m_{d,F})}$$

where A is given by (22).

Similar expression can be obtained for the outage probability of the tagged macro user.

Competing interests

The authors declare that they have no competing interests.

References

1. J Andrews, H Claussen, M Dohler, S Rangan, M Reed, Femtocells: past, present, and future. IEEE J on Sel Areas in Comm **30**(3), 497–508 (2012)
2. S Cheng, S Lien, F Hu, K Chen, On exploiting cognitive radio to mitigate interference in macro/femto heterogeneous networks. IEEE Wireless Commun Mag **18**(3), 40–47 (2011)
3. H ElSawy, E Hossain, M Haenggi, Stochastic geometry for modeling, analysis, and design of multi-tier and cognitive cellular wireless networks: a survey. IEEE Commun Surveys & Tutorials **15**(3), 996–1019 (2013)
4. K Gilhousen, I Jacobs, R Padovani, AJ Viterbi, L Weaver, C Wheatley, On the capacity of a cellular CDMA system. IEEE Trans on Veh Technology **40**(2), 303–312 (1991)
5. J Xu, J Zhang, JG Andrews, On the accuracy of the Wyner model in cellular networks. IEEE Trans on Wireless Commun **10**(9), 3098–3109 (2011)
6. JG Andrews, F Baccelli, RK Ganti, A tractable approach to coverage and rate in cellular networks. IEEE Trans on Commun **59**(11), 3122–3134 (2011)
7. SAR Zaidi, DC McLernon, M Ghogho, *Outage probability analysis of cognitive radio networks under self-coexistence constraint*. (44th Annual Conf. on Inf. Sciences and Systems (CISS), 2010), pp.1. 6, 17–19
8. RK Ganti, M Haenggi, Interference and outage in clustered wireless ad hoc networks. IEEE Trans on Inf Theory **55**(9), 4067–4086 (2009)
9. A Hasan, JG Andrews, The guard zone in wireless ad hoc networks. IEEE Trans on Wireless Commun **4**(3), 897–906 (2007)
10. C-H Lee, M Haenggi, Interference and outage in Poisson cognitive networks. IEEE Trans on Wireless Commun **11**(4), 1392–1401 (2012)
11. M Haenggi, RK Gant, Interference in large wireless networks. Foundations and Trends in Networking **3**(2), 127–248 (2008). NOW Publisher
12. IEEE Instrumentation and Measurement Society, IEEE 1588 standard for a precision clock synchronization protocol for networked measurement and control systems. IEEE Std **1588**, (2008)
13. Q Chen, M Motani, WC Wong, YC Liang, Opportunistic spectrum access protocol for cognitive radio networks. in *Proc. IEEE International Conf. on Commun. (ICC)*. (2011), pp. 1–6
14. X Li, H Liu, S Roy, J Zhang, P Zhang, C Ghosh, Throughput analysis for a multi-user, multi-channel ALOHA cognitive radio system. IEEE Trans on Wireless Commun **11**(11), 3900–3909 (2012)
15. Y-C Liang, Y Zeng, E Peh, AT Hoang, Sensing-throughput tradeoff for cognitive radio networks. IEEE Trans on Wireless Commun **7**(4), 1326–1337 (2008)
16. L Luo, C Ghosh, S Roy, Joint optimization of spectrum sensing for cognitive radio networks. in *Proc. IEEE Global Telecommun. Conf. (GLOBECOM)*, (2010), pp. 1–5
17. Q Zhao, B Sadler, A survey of dynamic spectrum access: signal processing, networking, and regulatory policy. IEEE Sig Proc Magazine **24**(3), 79–89 (2007)
18. D Kim, On upper bounds of SIR-based call admission threshold in power-controlled DS-CDMA mobile systems. IEEE Commun Letters **6**(1), 13–15 (2002)
19. M Haenggi, Outage, local throughput, and capacity of random wireless networks. IEEE Trans on Wireless Commun **8**(8), 4350–4359 (2009)
20. J-M Dricot, G Ferrari, A Panahandeh, Fr Horlin , Ph. De Doncker, Probabilistic coexistence and throughput of cognitive dual-polarized networks. EURASIP Journal on Wireless Commun. and Networking, 2010:387625.
21. AA Kilbas, HM Srivastava, JJ Trujillo, *Theory and applications of fractional differential equations* (Elsevier B.V, Netherlands, 2006)

Experimental analysis of PAPR reduction technique using hybrid peak windowing in LTE system

Dongwan Kim[1,2] and Sunshin An[1*]

Abstract

Orthogonal frequency division multiplexing (OFDM) has the advantages of spectral efficiency in wideband communication systems such as 3GPP long-term evolution (LTE). However, OFDM suffers from a high peak-to-average power ratio (PAPR). Several schemes, such as peak windowing, clipping and filtering, have been proposed in order to reduce the PAPR in OFDM-based systems. However, these schemes reduce the peak signals regardless of the PAPR pattern; thus, while guaranteeing error vector magnitude (EVM) and adjacent channel leakage ratio (ACLR) performances they do not efficiently reduce the PAPR. In this paper, we propose a hybrid peak-windowing scheme that selects the optimum PAPR reduction scheme according to the PAPR pattern, and we evaluate its performance using a real LTE test bed. Our test results show that the proposed scheme outperforms the existing schemes in terms of EVM, ACLR, and power consumption.

Keywords: PAPR; CFR; Clipping and filtering; Peak windowing; Power consumption; LTE

1 Introduction

With the increasing demand for high data rate wireless communications, 3GPP long-term evolution (LTE) has become one of the dominant wireless access technologies. LTE is expected to result in significantly improved transmission rates and spectral efficiency compared to 3G systems such as wide code division multiple access (WCDMA). To enable better performance, the LTE standard adopts orthogonal frequency division multiplexing (OFDM), which is a transmission method employed for wideband communications because of its high spectral efficiency, robustness to inter-symbol interference (ISI), and simple receiver structure [1].

However, one of the major drawbacks of OFDM-based systems is the high peak-to-average-power ratio (PAPR) problem that is due to the summing of sub-carrier signals. The high PAPR signals over the dynamic range of a power amplifier (PA) and digital-to-analog converter (DAC) lead to non-linear distortions such as the spectral regrowth phenomena, in-band noise, and increased quantization errors. These results cause degradations in the adjacent channel leakage ratio (ACLR) and error vector magnitude (EVM) performance [2].

Several methods have been proposed to reduce PAPR in OFDM-based systems. Single-carrier frequency division multiple access (SC-FDMA) is a variant of OFDM-based systems in which the data symbols of each user are first modulated in the time domain and then discrete Fourier transform (DFT) spread across the data subcarriers to yield a signal that is commensurate with single-carrier modulation. Hence, in the absence of multiple superimposed subcarriers, the envelope fluctuations are less pronounced and the power efficiency is higher than for conventional OFDM-based systems. In addition, the most straightforward solution for alleviating the side effects of high PAPR is to guarantee sufficient headroom for the signal with the highest peak to be processed in the dynamic range. However, a large amount of headroom results in a significant reduction in power efficiency. Therefore, an attractive and cost-effective solution is to reduce the PAPR of the OFDM signal.

To reduce the high PAPR in OFDM, a number of approaches have been proposed. The existing solutions are broadly classified into two categories: receiver-dependent schemes and receiver-independent schemes. Receiver-dependent schemes transmit a PAPR reduction signal including additional information. At the receiver side, the

* Correspondence: sunshin@dsys.korea.ac.kr
[1]The School of Electrical Engineering, Korea University, Anam-dong, Seongbuk-Gu, Seoul 136-701, South Korea
Full list of author information is available at the end of the article

transmitted signal can be decoded using the additional information. Typically, the tone reservation (TR) [3], selective mapping (SLM) [4], and partial transmit sequence (PTS) [5] are receiver-dependent schemes. However, the inclusion of additional information in receiver-dependent schemes will cause a reduction in the data rate. Furthermore, to enable the use of additional information, the receiver structure should be modified to restore the PAPR reduction signals.

In contrast, receiver-independent schemes transmit a PAPR reduction signal without any additional information. The clipping and filtering (CAF) schemes [6,7] and the peak windowing (PW) schemes [8,9] are famous receiver-independent schemes. Because receiver-independent schemes operate only on the transmitter side, the schemes do not require any modification of the receiver structure and do not cause any reduction in data rate because of the absence of additional information. Thus, receiver-independent schemes are widely used in practice because of their convenience. However, in [6-9], it is seen that the original signals become distorted and there are degradations in the performance of certain parameters such as the ACLR and EVM because the receiver cannot know the distorted information. In [7], experimental results for the relationship between the average power, EVM, and ACLR for an OFDM-based system with CAF are presented. The CAF scheme is a simple and effective scheme that is used to reduce the PAPR; however, it still leads to performance degradation of the EVM and ACLR because of the peak regeneration phenomenon caused by its filtering operation. Another solution is the PW scheme, which can efficiently reduce high-PAPR signals compared to the CAF scheme. However, the PW scheme still cannot efficiently eliminate successive high-PAPR signals. In [8], weighting factors for each peak are obtained by solving matrix-based linear equations so that the overall window function reduces each peak to the clipping threshold. In [9], sequential asymmetric suppression peak windowing and weighted windowing were proposed. While both schemes can deal with successive peaks that may cause excessive attenuation or spectral regrowth in the PW scheme, [8,9] focus on eliminating successive peak signals, and therefore, they cannot be efficiently applied to other cases such as discontinuous peak signals.

In this paper, we propose a new PAPR reduction scheme called hybrid peak windowing (HPW). The proposed scheme minimizes the degradation of the ACLR and EVM performance by selecting the proper PAPR reduction scheme as the input signal PAPR pattern. By applying the proposed scheme, we can optimally eliminate high-PAPR signals while minimizing performance degradation. In addition, we evaluate the proposed scheme relative to the existing PAPR reduction schemes using a real LTE system test bed and prove that the proposed scheme exhibits better performance than the existing schemes.

The rest of this paper is organized as follows. In Section 2, we discuss the PAPR issue in LTE systems and the related LTE specifications. In Section 3, we introduce the existing PAPR reduction schemes and the proposed scheme. In Section 4, we present a performance analysis, and we conclude the paper in Section 5.

2 PAPR issue in LTE systems

2.1 Characteristics of OFDM signals

The complex baseband OFDM signal with N subcarriers is expressed as:

$$x(t) = \frac{1}{\sqrt{N}} \sum_{k=-\frac{N}{2}}^{\frac{N}{2}-1} X_k \cdot e^{j2\pi k\Delta ft},\ 0 \leq t \leq T \tag{1}$$

where X_k is the complex modulated data symbol on subcarrier k from a given binary phase-shift keying (BPSK) or M-ary quadrature amplitude modulation (M-QAM) constellation. The frequency spacing, Δf, which is chosen to maintain the orthogonality of subcarrier frequencies, is usually assumed to be $1/T$. Samples of $x(t)$ at $t = nT/NQ$, where Q is the oversampling factor, can be efficiently computed via the inverse discrete Fourier transform (IDFT) as $X = \left[X_0, \cdots, X_{\frac{N}{2}-1}, 0, X_{-1}, \cdots, X_{-\frac{N}{2}}\right]$. When N is large, the time signal, $x(t)$, exhibits a Gaussian distribution because the central limit theorem applies.

The OFDM signal is the sum of many independent signals modulated into sub-carriers of equal bandwidth, so it has a high peak-signal distribution compared to a single-carrier system such as WCDMA. The fluctuations can be characterized as a PAPR, which can be discretely approximated as:

$$\text{PAPR} = \frac{\underset{0 \leq n \leq (N-1)}{max} |x_n|^2}{E\{|x_n|^2\}} \tag{2}$$

Where x_n is the signal samples in the time domain. High-PAPR signals lead to a critical hardware implementation issue. For example, when large peak signals among x_n are processed out of the input dynamic range, it will result in non-linear distortions and decreased power efficiency of the PA. Therefore, high-PAPR values need to be reduced to improve power efficiency while minimizing non-linear distortion. However, there may be additional non-linear distortion because of intentional reductions of the PAPR.

In general, the PAPR is represented by the crest factor (CF), and we use both the PAPR and CF in this paper

(PAPR = CF^2). To measure the value of the CF, a complementary cumulative distribution function (CCDF) is used. The CCDF curve is characterized by the probability of exceeding a particular signal level with respect to the average power. In addition, we use a design criterion that is 0.001% of the CCDF value because that is a typical design constraint in practical communication hardware designs. In other words, we assume that signals whose probabilities are below 0.001% do not affect system performance because they rarely occur.

2.2 EVM and ACLR in the LTE standard

The performances of the EVM and ACLR are directly related to the PAPR reduction scheme. Before discussing PAPR reduction schemes, we first introduce the EVM and ACLR.

The EVM is defined as the ratio of the power of the error vector to the root mean square (RMS) power of the reference, and is represented as:

$$\mathrm{EVM}(\%) = \sqrt{\frac{P_{\mathrm{error}}}{P_{\mathrm{reference}}}}\cdot 100\,[\%] \tag{3}$$

where *P*error is the RMS power of the error vector and *P*reference is the power of the reference vector. In general, *P*reference is the original signal, $x(n)$. The EVM is closely related to the bit error rate (BER) and modulation error ratio, which is the ratio of the mean signal power to the mean error power. That is, a certain EVM level should be guaranteed to minimize the BER performance. In this paper, the EVM represents the degree of degradation of the modulation signal.

Out-of-band (OOB) radiation refers to unwanted radiation immediately outside the nominal channel resulting from the modulation process and non-linearity in the transmitter, excluding spurious radiation. The OOB radiation limit is specified in terms of a spectrum emission mask and the ACLR performance. The ACLR is defined as the ratio of the transmitted power to the power measured after a receiver filter in the adjacent channel, and is expressed as:

$$\mathrm{ACLR} = 10\log_{10}\left(\frac{\sum_{f_d\in B} S_y(f_d)}{\sum_{f_d\in B_{\mathrm{adj}}} S_y(f_d)}\right) \tag{4}$$

where B is the range of frequencies in the transmission channel, B_{adj} is the range of frequencies in the adjacent channel, and S_y is the power spectral density of the received signal. In addition, the ACLR should not exceed a certain threshold value because the OOB radiation can affect other channels as noise.

When the original signals are distorted intentionally to reduce peak signal components, the performance in terms of the EVM and ACLR will be degraded. Thus, the LTE standard defines the minimum requirements for the performance of the EVM and ACLR because they are critical metrics to guarantee system performance. The LTE standard defines technical specification (TS) 36.104 [10] and TS 36.141 [11] for the LTE system performance test. TS 36.104 establishes the minimum radio frequency (RF) characteristics and minimum performance requirements of the evolved universal terrestrial radio access (E-UTRA) base station (BS), and TS 36.141 establishes the RF test methods and conformance requirements for the E-UTRA BS. The minimum requirements given in TS 36.104 make no allowance for measurement uncertainty. In contrast, TS 36.141 defines test tolerances that are individually calculated for each test. However, TS 36.104 is used more broadly in LTE system performance evaluations because of its strict limitations, even though TS 36.141 is defined for RF test methods. Thus, in this paper, we use TS 36.104 for performance evaluation.

As described above, PAPR reduction schemes affect the EVM and ACLR performance. In TS 36.104 and TS 36.141, the related test specification for the EVM and ACLR are specified as the E-UTRA test model (E-TM) 1.1, E-TM 1.2, E-TM 2, E-TM 3.1, E-TM 3.2, and E-TM 3.3. E-TM 1.1 and E-TM 1.2 are related to the performance of the ACLR, and the measured value should be higher than the E-TM specification. Further, E-TM 2, E-TM 3.1, E-TM 3.2, and E-TM 3.3 are related to the performance of the EVM, and the measured value should be lower than the E-TM specification. The test purpose and specifications of the E-TMs are summarized in Table 1.

3 Proposed hybrid peak windowing scheme

The CAF scheme is the simplest scheme that is used for reducing peak signals. The CAF scheme is composed of a hard clipping procedure and a low-pass filtering (LPF) procedure. In hard clipping processing, the signals over a clipping threshold (Th) are eliminated, which is helpful for reducing the PAPR. However, this procedure generates sharp corners in the time domain. That is, the ACLR

Table 1 Purpose of test and specification of E-TM

E-TM	Test purpose	Specification (10 MHz) [10]
E-TM 1.1	ACLR	45 dBc
E-TM 1.2	ACLR	45 dBc
E-TM 2	64QAM (at min. power)	8%
E-TM 3.1	64QAM (at max. power)	8%
E-TM 3.2	QPSK	17.5%
	16QAM	12.5%
E-TM 3.3	QPSK	17.5%
	16QAM	12.5%

performance is degraded because the high-frequency noise components increase. To improve the ACLR performance, the CAF scheme passes the hard clipping output to an LPF. After processing with the LPF, high-frequency noise components are eliminated, and the CAF scheme realizes an improved ACLR performance. However, the CAF scheme generates a peak regeneration phenomenon, which is caused by the LPF operation after a hard clipping procedure. Thus, when we use the CAF scheme, we should set a lower clipping threshold than the target PAPR considering the peak regeneration phenomenon. Because of the lower clipping threshold relative to the target PAPR, the existing intentional signal distortion will be deteriorated and it will degrade the EVM performance. The CAF output, x_{caf}, is written as:

$$x_{\text{caf}}(n) = \{x(n)\cdot w_c(n)\} * h_{\text{LPF}}(n) \tag{5}$$

where $x(n)$ is the original input signal of the CAF scheme, $h_{\text{LPF}}(n)$ is the impulse response of the LPF used to eliminate OOB radiation, * is the convolution operator, and $w_c(n)$ is the hard clipping window function expressed as:

$$w_c(n) = \begin{cases} 1, \ |x(n)| < Th \\ \dfrac{Th}{|x(n)|}, \ |x(n)| \geq Th \end{cases} \tag{6}$$

The PW scheme multiplies peak signals with a specific window function in the presence of a peak signal. By multiplying a continuous window function, the PW scheme mitigates the performance degradation compared to the CAF scheme. In order to prevent OOB radiation, the PW scheme can control the window length L. However, increasing L can lead to degradation of the EVM performance because of the overlapping of the specific window function, which generates an abnormal peak cancellation window. The PW scheme is modeled as:

$$x_{\text{pw}}(n) = x(n)\cdot w_{\text{pw}}(n) \tag{7}$$

where $wpw\ (n)$ is a window function that can be described as:

$$w_{\text{pw}}(n) = 1 - \sum_{i=-\infty}^{\infty} h(i)\cdot w(n-n_i) \tag{8}$$

where $h(i)$ is the weight coefficient, $w(n)$ is a common symmetric window function, and n_i is the peak sample index whose value is defined as:

$$|x(n_i)| = \max_{n_{i-} \leq n \leq n_{i+}} |x(n)| \tag{9}$$

where n_i is the non-uniformly spaced sample index running over the specific set of samples, which exceeds the threshold Th. Further, n_{i-} represents a sample index on the rising edge of the signal, where it first exceeds the threshold Th, while n_{i+} represents a sample index, where the signal peak dips below the threshold Th. The window functions can be cosine, Hamming, Hanning, or Kaiser windows. The Kaiser window function is commonly used because it is simple to shape the spectrum by changing window length L and the shape factor β. Here, we considered Kaiser window with length of 31 taps and β is 5.

As described previously, to alleviate the peak regeneration phenomenon, the CAF scheme should set up a clipping threshold that is lower than the target PAPR level. Thus, the CAF scheme generates excessive degradation of the EVM performance. However, the CAF scheme eliminates peak signals with a fixed signal loss regardless of whether successive peak signals occur. In addition, the PW scheme eliminates peak signals smoothly by multiplying them by a specific window function. However, when the consecutive peak signals occur within half of the window length in the PW scheme, the window function can overlap. This leads to unintended distortions of the original signals and generates unintended distortion signals, even if the signal is under the clipping threshold. However, the PW scheme is usually preferred to the PAPR reduction scheme because it results in reduced performance degradation compared to the CAF scheme. In summary, using the same clipping threshold as the PW scheme, the CAF scheme cannot eliminate high-PAPR signals entirely because of the peak regeneration phenomenon. Furthermore, the PW scheme can deteriorate signal distortion in the case of successive peak signals.

To prevent unintended distortions, we propose a HPW scheme, which aims to determine the peak signal over the clipping threshold level and to suppress it while minimizing unintended distortion. The main goal of the HPW scheme is to select a proper PAPR reduction scheme as a peak signal pattern. When the single peak signal is detected in the half window length, the HPW scheme eliminates the peak signal using the PW scheme. However, when successive peak signals are detected within a half of the window length, the HPW scheme eliminates consecutive peak signals using the CAF scheme, and not the PW scheme. In addition, in the proposed scheme, an even-order filter structure is used to minimize peak regeneration that occurs during the filtering operation, and the delay between the PW scheme and the CAF scheme is designed to be the same.

Figure 1 depicts the block diagram for the proposed HPW scheme. The HPW scheme is composed of a peak detector, successive peak detector, PW block, and CAF block. In addition, the clipping threshold for the CAF block is set to be lower than the clipping threshold of the PW block because of the peak regeneration phenomenon. Consequently, the PW block generally has a better EVM performance than the CAF block, but this does not hold in the case of successive peak signals.

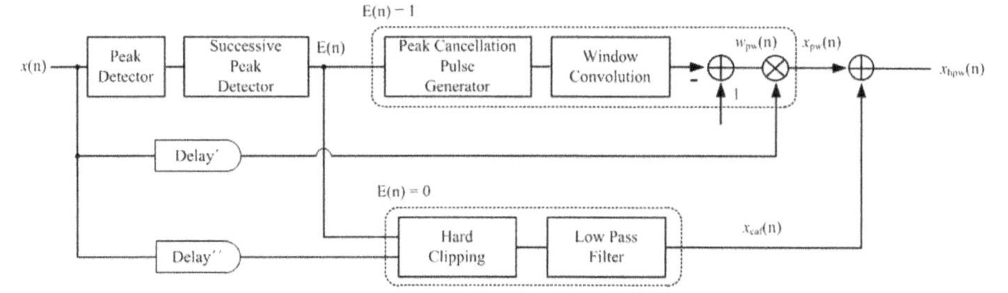

Figure 1 Block diagram of the hybrid peak windowing scheme.

When the original signals enter the HPW scheme, the peak detector first checks the peak signals above the clipping threshold. Then, the successive peak detector monitors whether or not the distance from the previous peak is less than half of the window length. If successive peak signals are detected within half of the window length, the successive peak detector sets the enabled signal to zero. Next, the HPW scheme processes the CAF scheme to prevent unintended excessive distortion. However, if successive peak signals are not detected within half of the window length, the enabled signal is set to one, and the HPW scheme processes the PW scheme, and not the CAF scheme. Thus, the correct PAPR reduction scheme is selected for the peak signal distribution. The output of the HPW scheme is modeled as:

$$x_{\mathrm{hpw}}(n) = \{1{-}E(n)\}\cdot x_{\mathrm{caf}}(n) + E(n)\cdot x_{\mathrm{pw}}(n) \qquad (10)$$

where $E(n)$ is the output of the successive peak detector.

4 Performance analysis

In this section, we compare the CCDF, ACLR, and EVM performance for the CAF scheme [7], PW scheme [8], and proposed HPW scheme. In the test bed for the performance evaluation, LTE test model streams that comply with the LTE standard specification are generated. The binary information bit streams are mapped to complex BPSK and M-QAM symbols. The output of the mapper is converted from serial to parallel and is processed using a 1024-point complex inverse fast Fourier transform (IFFT). The 1024-point complex-valued time-domain signals are followed by a guard interval (GI). After that, the outputs of the baseband test-model signals are processed by a digital transceiver board, and the output signals of the digital transceiver are processed by a DAC. The output signals of the DAC are up-converted by a local oscillator (LO), and those signals enter a laterally diffused metal-oxide semiconductor (LDMOS) PA. The final RF signals are decoded by performing signal analyzer measurements. In the performance evaluation, the system bandwidth is 10 MHz and the RF center frequency is 2.125 GHz. The PAPR reduction function is embedded in a digital transceiver board and implemented in a field-programmable gate array (FPGA). In addition, the input dynamic range of the PA is designed with a 6.5-dB PAPR level. Thus, the output PAPR of the PAPR reduction function is limited to 6.5 dB.

After PAPR reduction processing, the signal power level is adjusted. The peak signals have changing signal amplitudes due to intentional peak signal reduction, so the output of the PAPR reduction function is adjusted by a digital gain block (DGB). In addition, the input dynamic range of the DAC is maintained at a 6.5-dB PAPR signal level. The output power of the DAC is designed to be −10 dBm, and the output power of the PA is designed to be 46 dBm. Figure 2 depicts a block diagram of the LTE test bed employed for the performance evaluation.

First, we compared the FPGA resources of the HPW with those of the existing schemes in Table 2. In the implementation, we used Xilinx FPGA (XC7Z100; Xilinx Inc., San Jose, CA, USA), and the clock speed was 245.76 MHz.

As shown in Table 2, the CAF requires the smallest amount of FPGA resources, and more resources are required by PW and HPW when compared with those used by the CAF scheme. Unfortunately, because of its implementation complexity, the HPW scheme requires the most FPGA resources. As a result, the FPGA power consumption of HPW is the largest of the three schemes. However, compared with other schemes, the HPW scheme is characterized by the lowest PAPR while meeting the EVM and ACLR requirements; therefore, the PA power consumption of HPW will be the lowest. In the HPW scheme, the power consumed by the total system is reduced because of the power reduction of the PA despite the fact that the power consumed by the FPGA has increased. This will be discussed in detail at the end of this section.

The main goal of the PAPR reduction scheme is to increase the efficiency of the PA by limiting the peak signal level. Thus, the original CCDF without the PAPR reduction schemes is first measured to verify the original

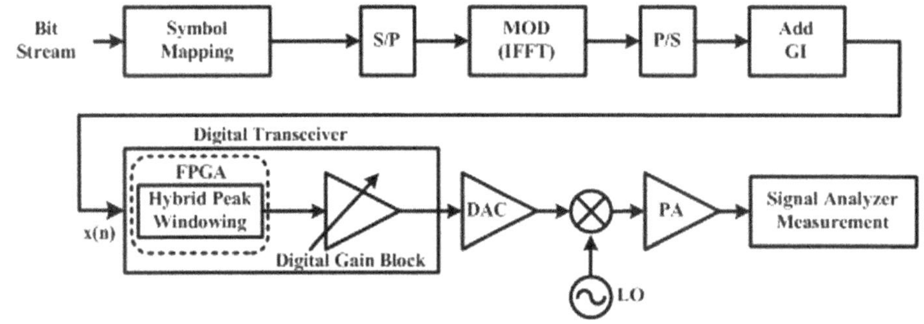

Figure 2 Test block diagram.

PAPR level. Figure 3 depicts the original CCDF without the PAPR reduction functions. As seen in Figure 3, the original CCDF value at 0.001% is 10.40 dB, and the original CCDF has a trend similar to a Gaussian distribution. The original CCDF value is measured at the DAC output to exclude the effects of the PA non-linear distortion. As seen in Figure 3, the original CCDF cannot satisfy the input dynamic range of the PA because the design constraint of the PA is a 6.5-dB PAPR. Consequently, it will degrade the ACLR and EVM performance because of the non-linear distortion of the PA and other active components. Thus, peak signals with PAPR values exceeding 6.5 dB should be eliminated to guarantee the input dynamic range of the PA. Figure 4 represents the degradation of the ACLR performance without the PAPR reduction scheme due to non-linear distortion. In Figure 4, we can verify that the value of the ACLR is 38.1 dBc. Thus, for this hardware design constraint, the ACLR performance cannot satisfy the LTE test specification without the PAPR reduction scheme because of the non-linear distortion of active components, especially the PA. In addition, Figure 5 represents the EVM performance of E-TM 3.1 without the PAPR reduction scheme. The value of the EVM is 2% and is caused by non-linear distortion of the DAC and PA. This error value contains other noise factors such as phase-locked loop (PLL) jitter noise and time-synchronous error. However, the EVM performance still satisfies the LTE test specification even though the non-linear in-band noise and inherent transceiver noise degrade the EVM performance.

As described above, the design constraint of the PA is set to a 6.5-dB input dynamic range. Thus, peak signals above the 6.5-dB PAPR level are eliminated by the HPW scheme, and the desired CCDF is obtained, as shown in Figure 6. The value of the CCDF at 0.001% is 6.5 dB, so the input dynamic range of the PA satisfies our design metric with the HPW scheme. After applying the HPW scheme, the ACLR and EVM performance that satisfied the LTE test specification are shown in Figures 7 and 8, respectively. As seen in Figure 7, an ACLR performance value of 63.6 dBc is obtained, which satisfies E-TM 1.1. The improved result is because of the reduced PAPR level achieved by the HPW scheme, which implies that the non-linear distortion components are eliminated by the HPW scheme. In addition, an EVM performance value of 5.8% for E-TM 3.1 is obtained, as shown in Figure 8. That is, both the inherent system noise and intentional distortion by the HPW scheme degrade the EVM performance. However, the EVM performance for the HPW scheme still satisfies the E-TM 3.1 specification. In particular, the E-TM 3.1 test specification applies to most high-degree modulation schemes such as 64 QAM at maximum power, and this specification is regarded as the strictest and one of the most important specifications of all the EVM test specifications.

Table 2 FPGA resources comparison between HPW and existing schemes

	Flip flop	Look up table		Memory	DSP48
		Logic	Shift register	(36 kbit)	(25 × 18)
CAF	2,012	328	54	8	21
PW	10,512	7,341	74	72	46
HPW	10,968	7,512	74	72	46

The proposed HPW scheme is compared with the CAF [7] and PW [8] schemes. Table 3 summarizes the performance comparison between the proposed HPW scheme and the existing schemes. As seen in Table 3, the HPW scheme outperforms the CAF and PW schemes in all E-TMs in terms of the EVM and ACLR performance. As described above, both the CAF and PW schemes eliminate the PAPR component regardless of the PAPR signal pattern, which may lead to excessive intentional distortion.

In particular, we verify the differences in EVM performance between the HPW scheme and the existing schemes. We also check the difference in the performance according to E-TM 3.2 and E-TM 3.3, even though they have the same modulation scheme. Because the PAPR pattern in each E-TM is different, the side effects of a PAPR reduction scheme such as EVM degradation will be different for each E-TM. We then verified that the HPW scheme minimizes the side effects of intentional signal distortion by selecting

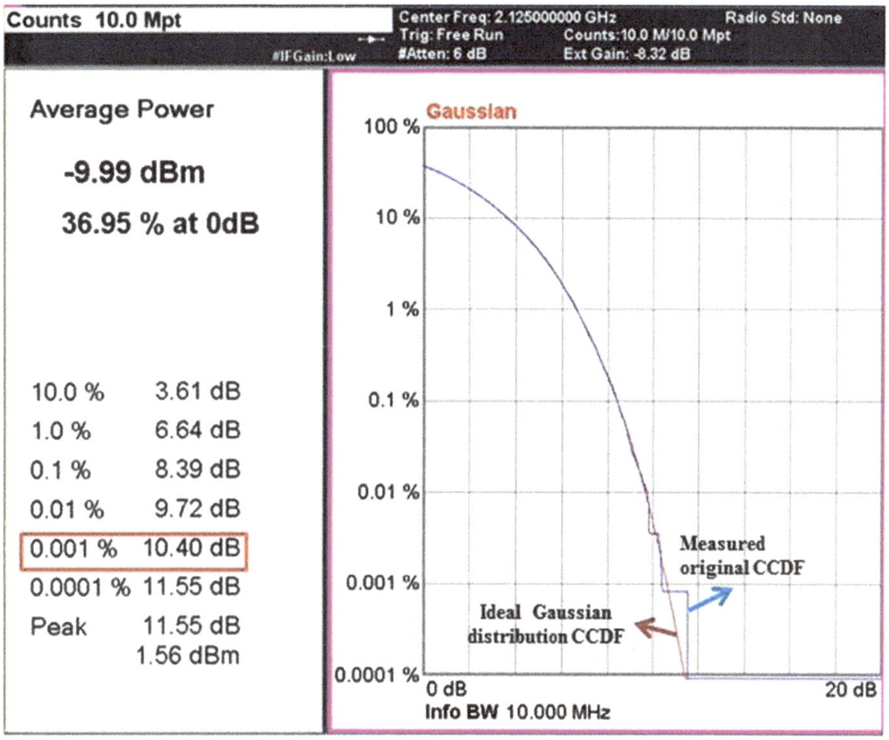

Figure 3 CCDF of the original LTE test model signal.

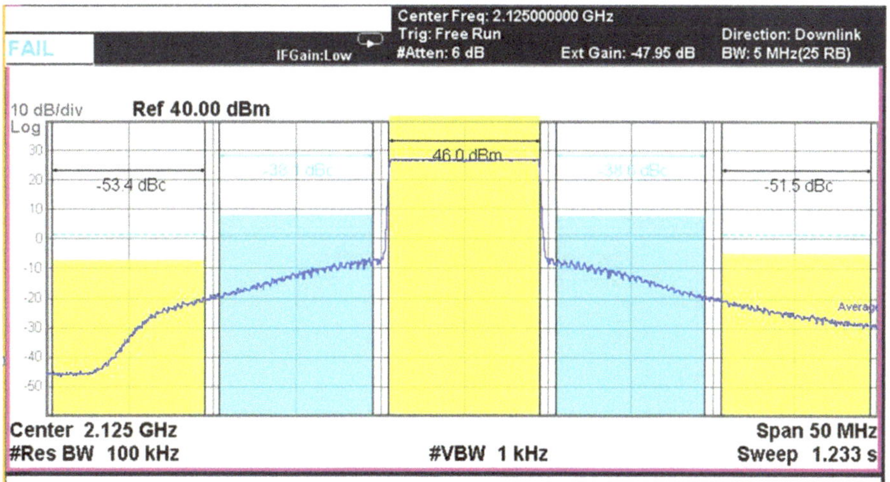

Offset Freq	Integ BW	Lower dBc	Lower dBm	Upper dBc	Upper dBm	Filter
10.00 MHz	9.015 MHz	-38.09	7.913	-38.56	7.446	OFF
20.00 MHz	9.015 MHz	-53.40	-7.398	-51.52	-5.515	OFF

Figure 4 ACLR performance without the PAPR reduction scheme.

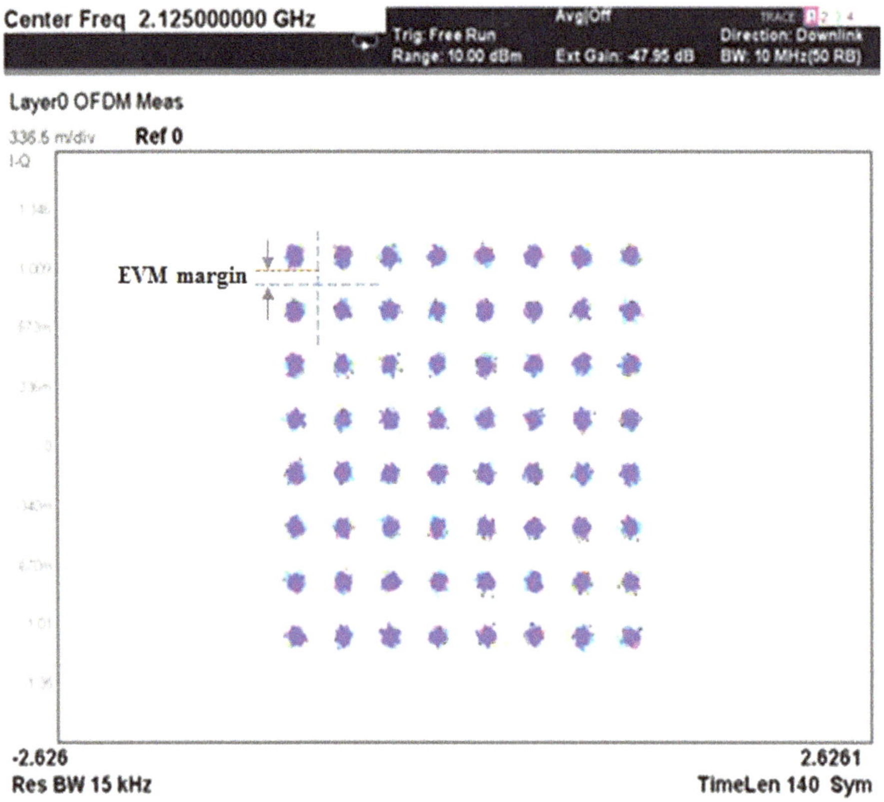

Figure 5 EVM performance without the PAPR reduction scheme.

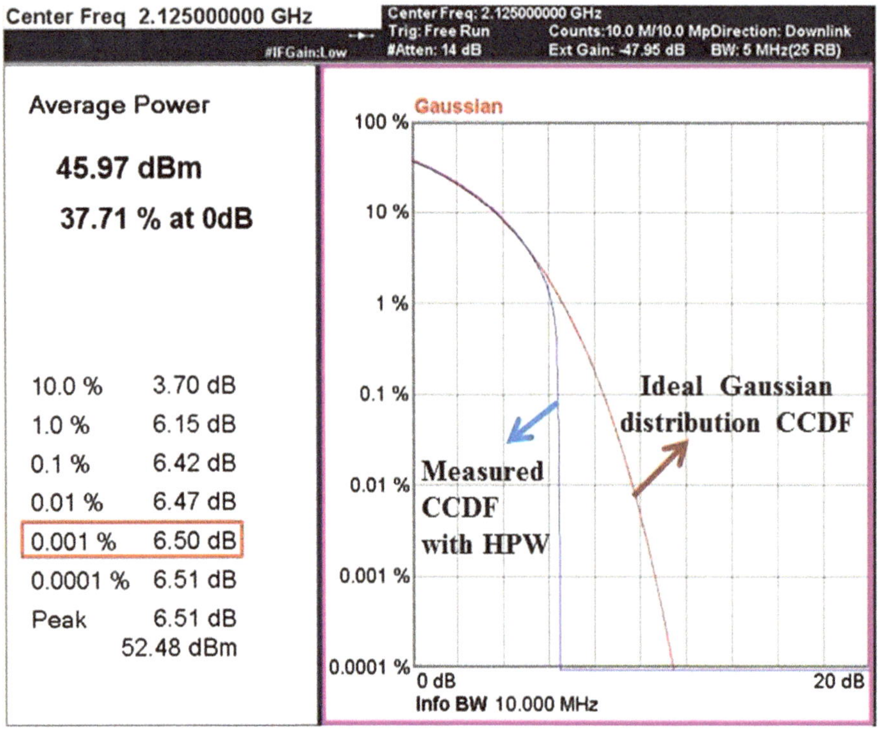

Figure 6 CCDF performance with the HPW scheme.

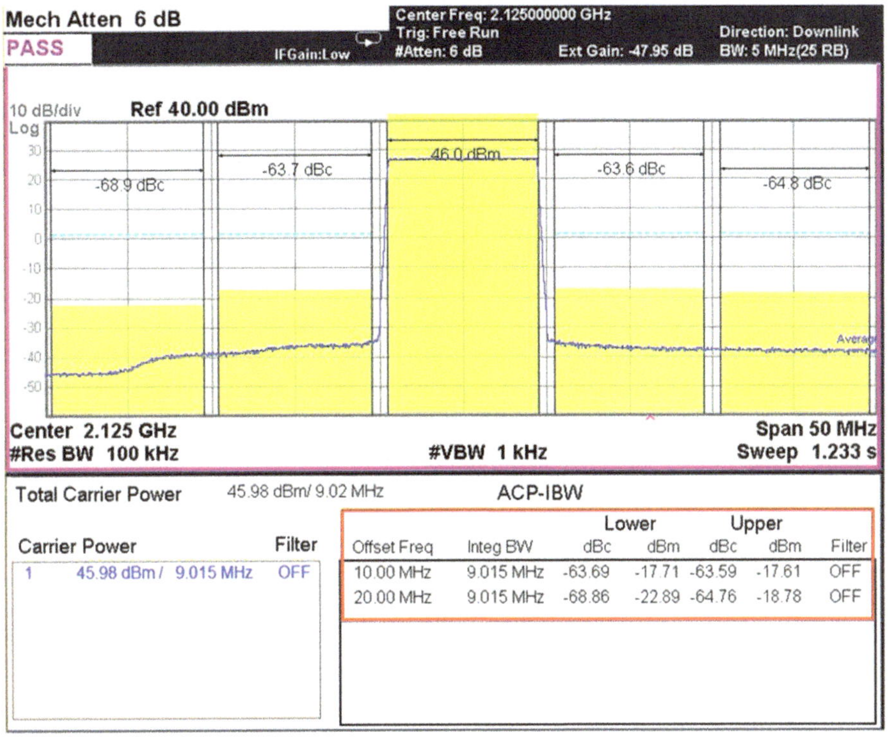

Figure 7 ACLR performance with the HPW scheme.

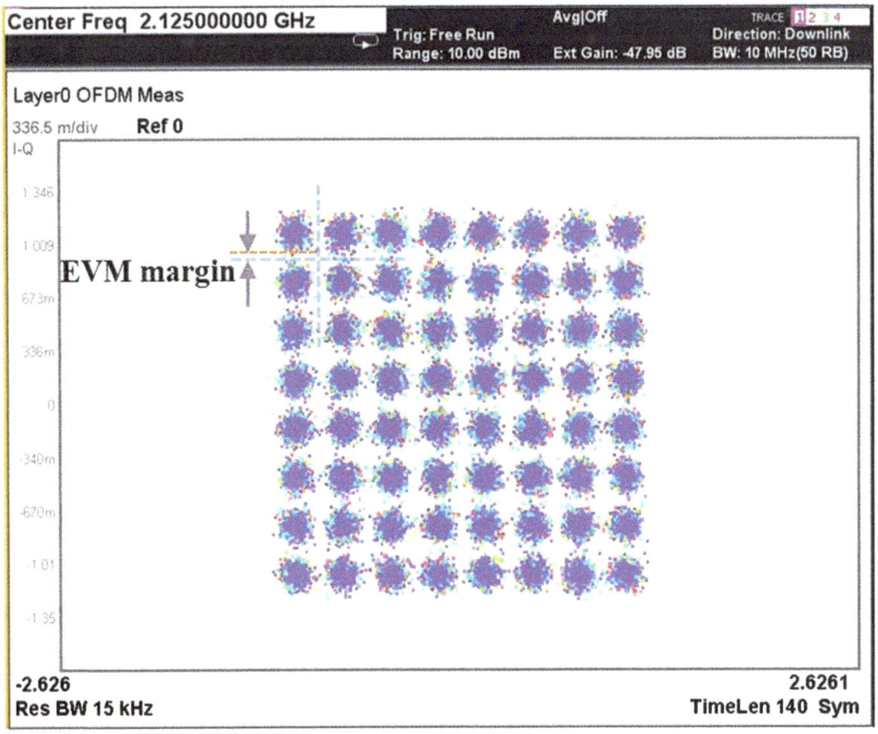

Figure 8 EVM performance with the HPW scheme.

Table 3 Performance comparison for E-TM at 6.5-dB PAPR

E-TM	HPW	CAF [7]	PW [8]
E-TM 1.1	63.6 dBc	61.9 dBc	62.3 dBc
E-TM 1.2	63.5 dBc	62.1 dBc	62.2 dBc
E-TM 2	0.8%	1.7%	1.4%
E-TM 3.1	5.8%	7.3%	6.8%
E-TM 3.2	4.2%	4.7%	4.8%
	7.8%	8.5%	8.4%
E-TM 3.3	10.7%	11.4%	11.2%
	4.2%	4.7%	4.6%

the proper PAPR reduction scheme for the peak signal distribution for all of the E-TMs.

As shown in Tables 1 and 3, the ACLR performance still has a large margin to the specification requirements compared to that of EVM. Therefore, because the critical side effect is EVM, which resulted in degraded performances with the PAPR reduction scheme, we measured the amount by which the EVM degraded with the PAPR reduction level, while still meeting the ACLR requirements. When verifying the minimum PAPR level while satisfying the critical EVM specification, we can reduce the PAPR level to 5.82 dB using the HPW scheme. However, when we use the CAF scheme and PW scheme at a 5.82 dB PAPR level, we obtain EVM performance values of 10.24% and 9.81%, which do not satisfy E-TM 3.1.

Figure 9 depicts the results of the EVM performance for HPW, PW, and CAF schemes using the PAPR value while satisfying the ACLR specification. As seen in Figure 9, the HPW scheme always outperforms PW and CAF schemes for the same PAPR value. We also conducted a test to verify the minimum PAPR level while satisfying EVM standards. As seen in Figure 9, the HPW, PW, and CAF schemes can reduce the PAPR level to 5.82, 6.25, and 6.38

Table 4 Power consumption comparison

	HPW	PW	CAF
Digital transceiver	31.12	31	30.81
(FPGA)	(15.02)	(14.9)	(14.71)
PA	132.72	137.31	138.59
Total power	147.74	152.21	153.30

dB, respectively, while satisfying the LTE standards. This means that when we use HPW, we can reduce PAPR to as low as 5.82 dB, enabling us to design a more energy-efficient PA. The relation between the energy efficiency of the PA and the input PAPR is given in Equation 11, which is derived from reference [12]:

$$\eta = G \exp(-g\mathrm{PAPR}_i) \tag{11}$$

where the efficiency, η, is in [%] and the PAPR_i is in [dB]. G and g are 90.7% and 0.1202, respectively, and indicate the ideal peak efficiency and slope of the PA's efficiency, respectively. Therefore, the HPW scheme can reduce the power consumption of a PA by much lower than other schemes by controlling the drain voltage of the PA based on the possible reduction level of the PAPR. Consequently, the HPW scheme enhances the PA efficiency by 2.27% and 2.93% when compared to the PW and CAF schemes, respectively. The PA accounts for a large proportion of the power consumption in the overall communication system [13], and reducing the power consumption of a PA is one of the most efficient schemes in the implementation of low-power communication systems. As described in Table 2, even though HPW increases the power consumption of the FPGA because of its implementation complexity, HPW can further reduce the power consumption of a PA depending on its low PAPR characteristics. This leads to

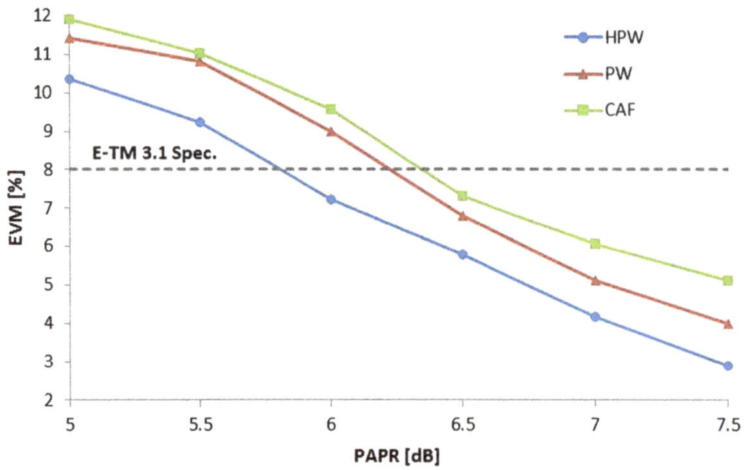

Figure 9 EVM performances vs. PAPR.

an overall power reduction in the communication system. The experimental results for the power consumption are summarized in Table 4, which shows that the HPW scheme is responsible for the largest power consumption in digital transceivers because of increasing FPGA power. However, the HPW scheme has the lowest power consumption in PAs, and its total power is reduced by 5.87 and 4.59 W when compared to the CAF and PW schemes, respectively.

5 Conclusions

In this paper, we proposed a new HPW scheme for OFDM systems. With the proposed HPW scheme, we realized improved performance by selecting a suitable PAPR reduction scheme for the signal amplitude distribution. Even though the performance in terms of the CCDF, EVM, and ACLR are related to many other hardware component characteristics as design metrics, we have demonstrated that the proposed HPW scheme outperforms the CAF and PW schemes. This was achieved using a real LTE system test bed. We also demonstrated that the power consumption could be reduced by the proposed HPW scheme. Because the HPW scheme has the smallest PAPR level compared to others, while still meeting the LTE standard requirements, there was a power reduction of the PA depending on the minimum PAPR. Even though the power consumption of the FPGA resources had increased, the power efficiency of the overall communication system increases because the power reduction of PA is much greater.

Competing interests

The authors declare that they have no competing interests.

Acknowledgements

This work was supported partially by the National Research Foundation (NRF) of Korea grant funded by the Korea government (MEST) (no. NRF2012K1A3A1-A09026959).

Author details

[1]The School of Electrical Engineering, Korea University, Anam-dong, Seongbuk-Gu, Seoul 136-701, South Korea. [2]Samsung Electronics Co., Ltd., Suwon-si, Gyeonggi-do 443-742, South Korea.

References

1. 3GPP TR 36.913, Requirements for further advancements for Evolved Universal Terrestrial Radio Access (E-UTRA), v.8.0.1, March 2009.
2. E Costa, M Midrio, S Pupolin, Impact of amplifier nonlinearities on OFDM transmission system performance. IEEE Commun Lett **3**, 37–9 (1999)
3. J Tellado, *Peak to average power reduction for multicarrier modulation, Ph.D. dissertation, Stanford Univ*, 2000
4. RW Bäuml, RFH Fisher, JB Huber, Reducing the peak-to-average power ratio of multicarrier modulation by selected mapping. Elect Lett **32**(22), 2056–7 (1996)
5. SH Müller, JB Huber, OFDM with reduced peak-to-average power ratio by optimum combination of partial transmit sequences. Elect Lett **33**(5), 368–9 (1997)
6. X Li, LJ Cimini Jr, Effects of clipping and filtering on the performance of OFDM. IEEE Comm Lett **2**, 131–3 (1998)
7. T Lee, H Ochiai, Experimental analysis of clipping and filtering effects on OFDM systems, in *Proc. of the 2010 IEEE Int. Conf. on Commun. (ICC 2010)* (South Africa, Cape Town, 2010)
8. S Cha, M Park, S Lee, K-J Bang, D Hong, A new PAPR reduction technique for OFDM systems using advanced peak windowing method. IEEE Trans Consum Electron **54**(2), 405–10 (2008)
9. G Chen, R Ansaru, Y Yao, Improved peak windowing for PAPR reduction in OFDM, in *IEEE 69th VTC Spring*, 2009, pp. 1–5
10. 3GPP specification 36.104, http://www.3gpp.org/DynaReport/36104.htm
11. 3GPP specification 36.141, http://www.3gpp.org/DynaReport/36141.htm
12. D Wulich, Definition of efficient PAPR in OFDM. IEEE Commun Lett **9**, 832–4 (2005)
13. Z Hasan, H Boostanimehr, VK Bhargava, Green cellular networks: a survey, some research issues and challenges. IEEE Commun Surv Tuts **13**(4), 524–40 (2011). Fourth Quarter

3

An efficient MIMO scheme with signal space diversity for future mobile communications

Zhanji Wu[*†] and Xiang Gao[†]

Abstract

An efficient wireless transmission scheme with the signal space diversity (SSD) is proposed to improve the performance of multiple-input multiple-output (MIMO) systems in fading channels. By introducing the rotated modulation and space-time component interleaver, the proposed scheme jointly optimizes channel coding, modulation, and MIMO and can improve the link reliability and energy efficiency. An optimum spatial component interleaver is proposed to maximize the MIMO achievable rate. Based on the average mutual information (AMI)-maximization criterion, the optimal rotation angles of real-valued signal and complex-valued QAM signal are investigated for the MIMO scheme. For the iterative demapping and decoding (ID) scheme, a simple genetic algorithm (GA) to search binary convolution code (BCC) is also put forward to match the rotated modulation. Simulation results show that the optimized BCC-coded MIMO scheme with SSD-ID outperforms the turbo-coded MIMO scheme with bit-interleaved coded modulation (BICM)-ID by 1.4 dB signal-to-noise ratio (SNR) gain, while the new scheme has much lower complexity. So, the proposed scheme is simple, efficient, and promising for future wireless communication systems.

Keywords: Quadrature amplitude modulation (QAM); Signal space diversity (SSD); Multiple-input multiple-output (MIMO); Iterative demodulation and decoding (ID); Extrinsic information transfer chart (EXIT)

1 Introduction

Wireless communications have made a great progress in the recent few years. By introducing more advanced technology, 5G will provide higher spectral efficiency, more spectrum resources, and more reliability to meet the growing demand for mobile traffic [1].

Bit-interleaved coded modulation (BICM) is a bandwidth-efficient coded modulation scheme which increases the time diversity in fading channels [2,3]. For its iterative version, BICM with iterative demapping and decoding (BICM-ID), the extrinsic information is transferred between the channel decoder and the soft-in-soft-out demapper, which is like the serial turbo decoder. Multiple-input multiple-output (MIMO) scheme is the extension of the coding theory on the space domain, so it is also named space-time coding (STC) [4]. Foschini proposed a layered space-time (LST) architecture to process multidimensional signals in the space domain [5]. The BICM-LST is a conventional spectral-efficient spatial multiplexing technology to deal with MIMO fading channels, and the BICM-threaded layered space-time (TLST) with a cyclic-shift spiral spatial interleaver is regarded as the most efficient method, because the cyclic-shift spatial interleaver introduces effective space diversity for the codeword on each layer [6]. In general, the BICM-LST can be viewed as the serial concatenation of the channel coding, modulation, and spatial layered multiplexing. Because BICM-LST exhibits a robust diversity performance on fading channels, it is widely deployed by wireless communication standards.

As for the bandwidth-efficient quadrature amplitude modulation (QAM), uncoded rotated multidimensional modulation schemes over independent Rayleigh fading channels were studied in [7] for the single-input single-output (SISO) scheme. Different from the other well-known diversity (time, frequency, code, space), it has an intrinsic modulation diversity, which is named signal space diversity (SSD). Through the combination of constellation rotation and component interleaver, the schemes can achieve very high modulation diversity, and the error performance over fading channels can approach that on the additive white Gaussian noise (AWGN)

*Correspondence: wuzhanji@bupt.edu.cn
†Equal contributors
School of Telecommunication Engineering, Beijing University of Posts and Telecommunications, Xitucheng Road 10, Beijing, China

channels. SSD schemes for SISO system have been extensively researched. In [8], SSD is introduced to the BICM by the means of modifications to the QAM constellation mapper and demapper so as to improve the BICM performance of QAM constellations for broadcasting applications. In [9], a LDPC-coded SSD scheme for multi-level modulation was presented. N.F. Kiyani and J.H. Weber studied the rotated-MPSK SISO BICM-ID system [10,12], which focused on two-dimensional multiphase shift keying (MPSK) scheme. In [11], the performance analysis of BICM-ID with SSD in fading channels is presented. In [13], the extension of BICM-SSD schemes with a non-binary code was proposed. We also proposed coded orthogonal frequency division multiplexing (OFDM) systems with SSD in [14,15]. In [16], the schemes combining SSD with SISO-coded BICM and BICM-ID systems were investigated. It provided a new criterion for determining the optimal rotation angle by maximizing the average mutual information (AMI). For the optimization of BICM-ID system, it proved that SSD can mitigate the different-slope problem of the demapper's extrinsic information transfer (EXIT) curve under different channels. However, finding well-matched channel codes for given labeling in BICM-ID system with SSD is still a big challenge.

The combination of signal rotation and space-time coding in MIMO system can effectively improve the diversity gain. In order to achieve full diversity, the quasi-orthogonal space time block codes (QOSTBC) with rotating the constellations of half of the complex symbols has been widely discussed in [17-20]. Some specific optimal rotation angles and corresponding optimization criterions for QAM and phase shift keying (PSK) constellations are provided. A rotation-based method that aims at maximizing the minimum distance in the space-time constellation is proposed in [17]. The proposed scheme shows good improvement of the codes compared to their non-rotated counterparts. In [18], the authors considered the design of rotated QOSTBC for the MISO system. The code designs are based mainly on the rank and the determinant criteria, and the optimal rotation angle $\pi/6$ can provide full diversity and the optimal coding gain. In [19], the authors proposed to design the signal constellations properly to ensure that the resulting quasi-orthogonal STBCs can guarantee to achieve the full diversity. The optimal rotation angles are determined by maximizing the diversity product. A novel method to exactly derive the coding gain of QSTBC as a function of the rotation angle and the minimum Euclidean distance of two-dimensional constellations is proposed in [20]. A coded MIMO scheme for block-fading channels was proposed in [21], which consists of a channel code and a space-time code. The space-time code is designed based on SSD technique, which allows full spatial multiplexing MIMO transmission and achieves full space diversity. In [22], the uncoded SSD scheme was extended to V-BLAST MIMO systems in order to achieve the maximum diversity gain without additional power or bandwidth consumption. An improved turbo-coded SSD scheme was proposed for MIMO-OFDM BICM system in [23], and the linear minimum mean square error (LMMSE) equalization is utilized for the non-ID MIMO detection. In general, the research of SSD technique in coded MIMO systems is still on the original stage. There are still many open problems. For instance, the optimal rotation angles in current research mainly depend on the maximum product distance introduced in [7]. Unfortunately, this criterion is only valid for the SISO system in high signal-to-noise ratio (SNR) region. As for the coded MIMO scheme, when powerful forward error-correction codes (FECs) are considered, actual SNR can be quite low. Hence, the angle values applied to uncoded SISO system do not lead to the best error performance for the coded modulation MIMO scheme. What is more, current research works mainly focus on local optimizations. For example, most proposed MIMO systems with SSD are only an extension of SISO-SSD system, and all are based on the conventional non-precoding transmitter. The channel coding, QAM modulation, and STC are independent with each other, which is just a straightforward serial concatenation. Hence, the performance of the BICM-LST is still rather far away from the MIMO fading channel capacity. For example, a near-capacity BICM-LST scheme was proposed in [24], which allows the iterative processing of the list sphere detection (LSD) and turbo decoding, but simulation results indicate that the gaps to the MIMO capacity are still more than 2 dB. As each individual optimization becomes mature so far, from the philosophy, it is high time to optimize these key technical elements jointly so as to improve the overall performance.

An improved coded MIMO system based on SSD is proposed for the jointly optimization of constellation rotation angle, spatial component interleaver, and the matching of channel coding and labeling, which is named joint coding and modulation diversity (JCMD), where the terminology 'coding' refers to both the channel coding and space-time coding. Firstly, in order to maximize the MIMO achievable rate, an optimum spatial component interleaver is proposed. Secondly, based on the AMI-maximization criterion, the optimal rotation angles of real-valued signal and complex-valued QAM signal are investigated for MIMO schemes, which are different from the SISO scheme. Thirdly, for the JCMD-ID scheme, a simple genetic algorithm (GA) to search binary convolution code (BCC) is put forward to match the rotated QAM modulation. Simulation results show that the optimized BCC-coded JCMD-ID MIMO scheme outperforms the turbo-coded BICM-ID MIMO scheme in [24] by 1.4 dB

SNR gain, while the new scheme has much lower complexity.

Throughout this paper, we use bold letters to represent vectors or matrices. $(\cdot)^T$ and $(\cdot)^H$ represent transposition and conjugate transposition, respectively. SNR $= E_s/N_0$, where E_s denotes the average symbol energy per receive antenna and $N_0 = 2\sigma^2$ denotes the variance of the complex Gaussian noise.

The paper is organized as follows. An improved JCMD MIMO scheme is proposed in Section 2. Theoretical analysis about the achievable rate of JCMD-MIMO for rotated real-valued signals is given in Section 3. Based on the AMI analysis, the optimal rotation angles for JCMD and JCMD-ID MIMO systems are presented in Section 4. Section 5 introduces an outer convolutional code search method for the optimization of JCMD-ID system with optimal rotation angle. Simulation results are presented in Section 6 on fast fading channels. Concluding remarks are offered in Section 7.

2 System model

Based on the BICM-TLST scheme, a system model of N_L-layer MIMO-JCMD-ID is shown in Figure 1. Perfect channel state information (CSI) is assumed to be known at both the transmitter and the receiver. In Figure 1, the iterative feedback processing is depicted in dashed lines. Without loss of generality, a rank-L $N_R \times N_T$ MIMO system with L nonzero eigenvalues is assumed, where N_R and N_T are the number of receive and transmit antennas, respectively, and $N_L \leq L \leq \min\{N_R, N_T\}$. In the transmitter, K information bits $\mathbf{B} = (b_1, b_2, \ldots, b_K)^T$ are encoded and interleaved to yield the coded bit sequence $\mathbf{C} = (c_1, c_2, \ldots, c_N)^T$. Afterwards, m-tuple coded bits are mapped to a complex symbol $x_k = x_k(I) + \mathrm{j} \cdot x_k(Q)$, which is chosen from a 2^m-ary rotated QAM constellation set $\chi = \{\hat{x}_1, \hat{x}_2, \ldots, \hat{x}_{2^m}\}$ according to some optimal angle. Each symbol x_k has one Q-component $x_k(Q)$ and one I-component $x_k(I)$. The rotated mapped symbol sequence is first mapped onto N_L layers in a round robin manner. Afterwards, the conventional cyclic-shift spatial interleaver of TLST is used for different symbols on N_L layers to exploit both the space and time diversity as the following, because the cyclic-shift spatial interleaver allows the codewords to be distributed on all layers:

$$w_k^l = x_k^i, \quad l = (i + k - 2) \bmod N_L + 1, \tag{1}$$

where w_k^l denotes the kth ($k \in \mathbf{N}^+$) symbol at the lth ($l \in [1, N_L]$) layer after the spatial interleaver, and x_k^i denotes the kth symbol at the ith ($i \in [1, N_L]$) layer before the interleaver. Then, a spatial Q-component interleaver is applied for the Q components of N_L symbols at the same instant as the following:

$$z_k^n(Q) = w_k^l(Q), \quad n = N_L - l + 1, \tag{2}$$

where z_k^n denotes the kth symbol at the nth ($n \in [1, N_L]$) layer after the spatial Q interleaver. Thus, I components keep the same layer order as before, and just Q components change the layer order. The spatial Q-component interleaver is used to make the fading of I component and that of the Q component as uncorrelated as possible in the space domain. Thus, the modulation diversity of the proposed scheme is further extended to the spatial dimension. Actually, the Q-component spatial interleaver can be different from the reverse interleaver in Equation 2. For example, it can be a cyclic-shift interleaver as the following,

$$z_k^n(Q) = w_k^l(Q), \quad n = (l \bmod N_L) + 1. \tag{3}$$

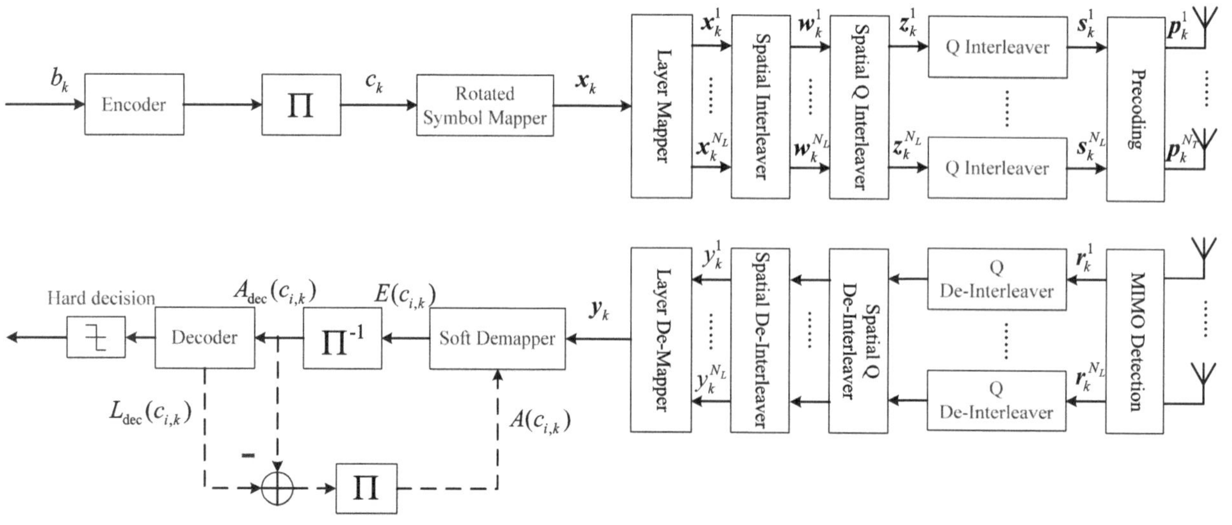

Figure 1 System model for MIMO-JCMD-ID with perfect CSI. The iterative feedback processing is depicted in dashed lines.

If perfect CSI is known, we prove that the reverse interleaver is better than other interleavers through the later theoretical analysis and computer simulations. If CSI is unknown at the transmitter, the cyclic-shift interleaver in Equation 3 can be used.

In order to make the fading of I component and that of the Q component as uncorrelated as possible in the time domain, after the spatial Q interleaving, Q components of the mapped symbols in each layer are interleaved through a time-domain pseudo S-random interleaver to reconstruct a new symbol vector $\mathbf{s}_k = \left[s_k^1 \cdots s_k^{N_L}\right]^T$, where s_k^l denotes the kth symbol at the lth layer after the component interleaving. Afterwards, the symbols are mapped onto N_T transmit antennas via the spatial precoding and then transmitted.

The ideal precoding matrix can be obtained by singular value decomposition (SVD), which can divide MIMO channel into parallel independent SISO channels. According to the SVD criterion, the $N_R \times N_T$ MIMO channel matrix $\mathbf{H}_k$ can be decomposed as

$$\mathbf{H}_k = \mathbf{U}_k \mathbf{D}_k {\mathbf{V}_k}^H, \tag{4}$$

where the $N_R \times N_R$ matrix $\mathbf{U}_k$ and the $N_T \times N_T$ matrix $\mathbf{V}_k$ are unitary matrices. $\mathbf{D}_k$ is a $N_R \times N_T$ non-negative diagonal matrix with N_L nonzero descending-order singular values, $\sqrt{\rho_1} \geq \sqrt{\rho_2} \geq \ldots \geq \sqrt{\rho_{N_L}} > 0$, where ρ_i is the ith largest eigenvalue of $\mathbf{H}_k \cdot {\mathbf{H}_k}^H$. Thus, the SVD-based linear precoding is performed as the following:

$$\mathbf{p}_k = \mathbf{V}_k \cdot \mathbf{s}_k. \tag{5}$$

In the receiver, the corresponding detection matrix is $\mathbf{U}_k^H$. The precoding and detection process can be expressed as linear transformations as shown in Equation 6.

$$\begin{aligned} \mathbf{r}_k &= \mathbf{U}_k^H \cdot (\mathbf{H}_k \cdot \mathbf{p}_k + \mathbf{n}_k) \\ &= \mathbf{U}_k^H \cdot \mathbf{H}_k \cdot \mathbf{V}_k \cdot \mathbf{s}_k + \mathbf{U}_k^H \cdot \mathbf{n}_k \\ &= \mathbf{D}_k \cdot \mathbf{s}_k + {\mathbf{n}'}_k, \end{aligned} \tag{6}$$

where $\mathbf{r}_k = \left[r_k^1 \cdots r_k^{N_R}\right]^T$ denotes the received symbol vector, $\mathbf{n}_k$ and ${\mathbf{n}'}_k$ are column vectors of N_R complex Gaussian random variables with mean zero and variance $\sigma^2 = \frac{N_0}{2}$. Thus, due to SVD, the MIMO channel can be viewed as N_L parallel fading channels, and for lth layer, the kth received symbol that corresponds to s_k^l in the transmitter can be expressed as

$$r_k^l = \sqrt{\rho_l} \cdot s_k^l + {n'}_k^l. \tag{7}$$

After the corresponding Q-component de-interleaving in time domain for each layer and spatial Q-component de-interleaving, the kth received symbol on lth layer is reconstructed as y_k^l that corresponds to x_k^l in the transmitter. For y_k^l, the fading coefficients of I-component $\lambda_k(I)$ and Q-component $\lambda_k(Q)$ are different, which can be expressed as

$$\begin{aligned} y_k^l(I) &= \lambda_k^l(I) x_k^l(I) + n_k^l(I) \\ y_k^l(Q) &= \lambda_k^l(Q) x_k^l(Q) + n_k^l(Q). \end{aligned} \tag{8}$$

Assuming that the Q interleaver is long enough, each coordinate of the symbol after the Q de-interleaving can be regarded as suffering from independent fading coefficients. The equivalent coded modulation (CM) channel for JCMD system before sending to the soft demapper can be modeled as

$$Y_k^l(\eta) = \Lambda_k^l(\eta) X_k^l(\eta) + N_k^l(\eta), \eta \in \{I, Q\}, l \in [1, N_L], \tag{9}$$

where $N_k^l(I)$ and $N_k^l(Q)$ are identically independently distributed (i.i.d.) Gaussian noise random variables with zero mean and variance of $\sigma^2 = \frac{N_0}{2}$. For MIMO fading channels, $\Lambda_k^l(I)$ and $\Lambda_k^l(Q)$ are singular values of corresponding sub-channels. This is in net contrast with respect to SISO scheme where the fading coefficients are Rayleigh distributed. That means the modulation diversity of proposed scheme is further extended to the spatial dimension. By denoting $\mathbf{X} = \left[\mathbf{X}_1, \ldots, \mathbf{X}_{N_L}\right]^T$, $\mathbf{Y} = \left[\mathbf{Y}_1, \ldots, \mathbf{Y}_{N_L}\right]^T$, $\mathbf{N} = \left[\mathbf{N}_1, \ldots, \mathbf{N}_{N_L}\right]^T$, and $\Lambda = \mathrm{diag}\left(\Lambda_1, \ldots, \Lambda_{N_L}\right)$ representing a $(2N_L \times 2N_L)$ diagonal matrix, the channel model in Equation 9 can be written in the matrix form as $\mathbf{Y} = \Lambda\mathbf{X} + \mathbf{N}$, where $\mathbf{X}_l = \left[X_k^l(I), X_k^l(Q)\right]$, $\mathbf{Y}_l = \left[Y_k^l(I), Y_k^l(Q)\right]$, $\mathbf{N}_l = \left[N_k^l(I), N_k^l(Q)\right]$, and $\Lambda_l = \mathrm{diag}\left(\Lambda_k^l(I), \Lambda_k^l(Q)\right)$.

After that, symbols on multiple layers are reassembled as symbol streams $\mathbf{y} = \left[y_1, y_2 \cdots\right]^T$ through the layer demapping. A serial concatenation of a soft-in-soft-out rotated symbol demapper and a channel decoder are employed to approach the maximum likelihood (ML) receiver performance. For JCMD-ID, the iterative demapping and decoding scheme is an application of the turbo decoder principle. The soft demapper calculates the extrinsic value $E(c_{i,k})$ of bit $c_{i,k}$ which corresponds to the ith ($i = 1, 2, \ldots, m$) bit of the received symbol y_k as follows:

$$E(c_{i,k}) = L(c_{i,k}) - A(c_{i,k}), \tag{10}$$

where $L(c_{i,k}) \triangleq \ln \frac{P(c_{i,k}=0|y_k)}{P(c_{i,k}=1|y_k)}$ is the log-likelihood ratio (LLR) and $A(c_{i,k}) \triangleq \ln \frac{P(c_{i,k}=0)}{P(c_{i,k}=1)}$ is *a priori* L-value. Based on Bayes' theorem, we can write

$$E(c_{i,k}) = \ln \frac{\sum\limits_{\substack{\hat{x}\in\chi,\\ \hat{c}_i=0}} P(y_k|x_k=\hat{x}) \exp\left\{\sum\limits_{\substack{j=1,\\ j\neq i}}^{m}\left[(-1)^{\hat{c}_j}\frac{A(c_{j,k})}{2}\right]\right\}}{\sum\limits_{\substack{\hat{x}\in\chi,\\ \hat{c}=1}} P(y_k|x_k=\hat{x}) \exp\left\{\sum\limits_{\substack{j=1,\\ j\neq i}}^{m}\left[(-1)^{\hat{c}_j}\frac{A(c_{j,k})}{2}\right]\right\}}, \tag{11}$$

where $\hat{c}_i$ is the ith bit corresponding to symbol $\hat{x}$. For the fading channel, the conditional probability is given by

$$P(y_k|x_k=\hat{x}) = \frac{1}{2\pi\sigma^2}\exp\left(-\frac{\left(y_k^I-\lambda_k^I\hat{x}^I\right)^2+\left(y_k^Q-\lambda_k^Q\hat{x}^Q\right)^2}{2\sigma^2}\right). \tag{12}$$

To the complexity, by applying Max-Log-MAP algorithm, Equation 11 can be simplified as

$$E(c_{i,k}) = \max_{\substack{\hat{x}\in\chi,\\ \hat{c}_{i,k}=0}}\left\{\Omega_k(\hat{x}) - \sum_{\substack{j=1,\\ j\neq i}}^{m}\left[(-1)^{\hat{c}_{j,k}}\frac{A(c_{j,k})}{2}\right]\right\} - \max_{\substack{\hat{x}\in\chi,\\ \hat{c}_{i,k}=1}}\left\{\Omega_k(\hat{x}) - \sum_{\substack{j=1,\\ j\neq i}}^{m}\left[(-1)^{\hat{c}_{j,k}}\frac{A(c_{j,k})}{2}\right]\right\}, \tag{13}$$

where $\Omega_k(\hat{x}) = -\frac{\left(y_k^I-\lambda_k^I\hat{x}^I\right)^2+\left(y_k^Q-\lambda_k^Q\hat{x}^Q\right)^2}{2\sigma^2}$

For the JCMD system without the iterative demapping and decoding, $A(c_{i,k}) = 0$. Finally, the decoder can utilize the extrinsic values to decode information bits.

In the transmitter, compared with the conventional BICM, the JCMD scheme introduces extra constellation rotation and Q-component interleavers. Constellation rotation does not increase the complexity, because the rotated symbol mapping can be implemented through look-up table operations as the same as the conventional modulation without rotation. Q-component interleavers also can be implemented by the low-complexity index-based look-up table operations.

In the receiver, the soft rotated demapping operation of JCMD system is the same as that of BICM system, which is shown in Equation 11. Q-component de-interleavers also can be implemented by the simple reverse index-based look-up table operations.

3 Theoretical analysis of the achievable rate for rotated real-valued signals

Firstly, since real-valued signals are elementary, we analyze the real-valued transmit signals in MIMO systems.

Lemma 1. For any constant rank-2 MIMO fading channel with real-valued transmit signals, the constant achievable rate of JCMD-MIMO is not less than that of BICM-MIMO with the conventional uniform power allocation. If and only if the constant eigenvalues are identical, both of them are equal, otherwise the former is greater than the latter.

Proof. A simple rank-2 MIMO case with two eigenvalues ρ_1 and ρ_2 is illustrated in Figure 2. Due to the well-known SVD, BICM-MIMO can be viewed as two parallel fading channels with two eigenvalues ρ_1 and ρ_2 for spatial layer 1 and layer 2, respectively. Two fading amplitude coefficients of layer 1 and layer 2 are the corresponding singular values $\sqrt{\rho_1}$ and $\sqrt{\rho_2}$, respectively, as shown in the left half of Figure 2. Thus, given the same transmit power $\frac{P}{2}$ on each layer for one real-valued symbol, the received symbol power of layer 1 and that of layer 2 are $\frac{\rho_1 P}{2}$ and $\frac{\rho_2 P}{2}$, respectively, where P is the total transmit power for two layers. So, according to Shannon's theory, the achievable rate of BICM-MIMO is shown as the following:

$$\begin{aligned} C_1 &= \frac{W}{2}\cdot\log_2\left[\left(1+\frac{\rho_1 P}{2\sigma^2}\right)\left(1+\frac{\rho_2 P}{2\sigma^2}\right)\right] \\ &= \frac{W}{2}\cdot\log_2\left(1+\frac{\rho_1+\rho_2}{2\sigma^2}P+\frac{\rho_1\rho_2}{4\sigma^4}P^2\right), \end{aligned} \tag{14}$$

where W is the channel bandwidth, σ^2 is the variance of AWGN. As we know, in order to achieve the capacity in the Equation 14, the transmit signals should be Gaussian distributed. Note that the rotation does not change the achievable rate for the BICM-MIMO scheme without the I/Q-component interleaver.

For the JCMD-MIMO scheme, we consider the $\frac{\pi}{4}$-rotated real-valued transmit signal, which is rotated by $\frac{\pi}{4}$ compared with the conventional real-valued signal. Due to the orthogonal modulation, the transmit powers of I component and that of Q component on each layer are both $\frac{P}{4}$. For JCMD-MIMO, after the spatial Q-component deinterleaver at the receiver, the fading amplitude coefficient of I component is different from that of Q component in one symbol, that is to say, one is $\sqrt{\rho_1}$, and another is $\sqrt{\rho_2}$. Thus, the received power of I component is also different from that of Q component, that is to say, one is $\frac{\rho_1 P}{4}$, and another is $\frac{\rho_2 P}{4}$. So, the total received symbol power is $\frac{(\rho_1+\rho_2)P}{4}$ both for layer 1 and layer 2. Thus, JCMD-MIMO can be viewed as two parallel fading channels with identical fading amplitude coefficient $\sqrt{\frac{\rho_1+\rho_2}{2}}$ for both layer 1 and layer 2, as shown in the right half of Figure 2. In the receiver, after the phase compensation, the received signal also can be viewed as the real-valued signal with the power

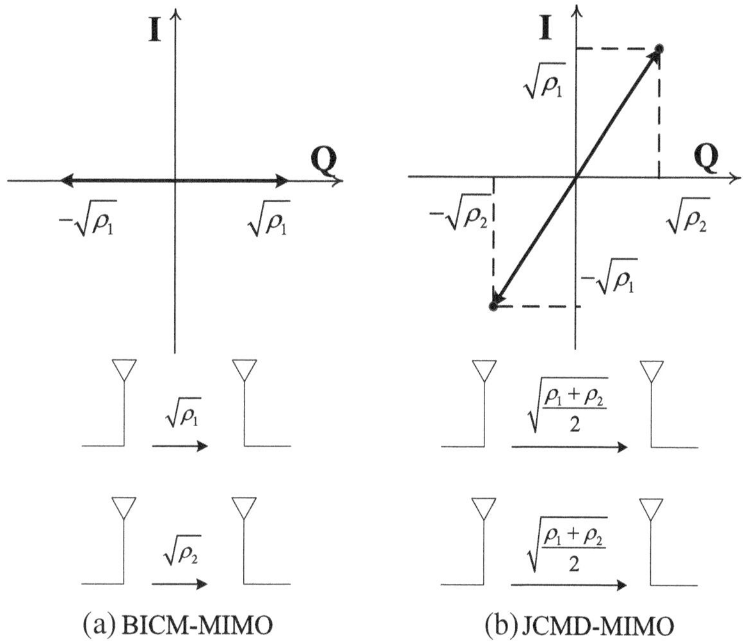

Figure 2 Achievable rates of conventional BICM-MIMO and JCMD-MIMO system. (**a**) BICM-MIMO can be viewed as two parallel fading channels with two eigenvalues ρ_1 and ρ_2. (**b**) JCMD-MIMO can be viewed as two parallel fading channels with identical fading amplitude coefficient $\sqrt{\frac{\rho_1+\rho_2}{2}}$ for both layers.

$\frac{(\rho_1+\rho_2)P}{4}$. Therefore, the achievable rate of JCMD-MIMO is shown as the following:

$$C_2 = \frac{W}{2}\cdot\log_2\left[1+\frac{(\rho_1+\rho_2)}{4\sigma^2}P\right]^2 = \frac{W}{2}\cdot\log_2\left[1+\frac{\rho_1+\rho_2}{2\sigma^2}P+\frac{(\rho_1+\rho_2)^2}{16\sigma^4}P^2\right]. \quad (15)$$

So, comparing Equation 14 with 15, it is easy to come to the conclusion: $C_2 \geq C_1$

If and only if $\rho_2 = \rho_1$, $C_2 = C_1$. □

Furthermore, we will prove that $\frac{\pi}{4}$ is the optimum rotation angle for real-valued transmit signals in the JCMD-MIMO scheme. Let us consider a general θ-rotated real-valued transmit signal, which is rotated by θ compared with the conventional real-valued signal. Thus, the transmit power of I component and that of Q component on layer 1 are $\frac{P}{2}\cos^2\theta$ and $\frac{P}{2}\sin^2\theta$, respectively, and the total transmit power on layer 1 is also $\frac{P}{2}$. Therefore, the received power of I component and that of Q component are $\frac{\rho_2 P}{2}\cos^2\theta$ and $\frac{\rho_1 P}{2}\sin^2\theta$, respectively. So, the total received symbol power on layer 1 is $\frac{(\rho_1\sin^2\theta+\rho_2\cos^2\theta)}{2}P$. Likewise, the total received symbol power on layer 2 is $\frac{(\rho_2\sin^2\theta+\rho_1\cos^2\theta)}{2}P$. Thus, we can get the following achievable rate:

$$C(\theta)=\frac{W}{2}\cdot\log_2\left\{\left[1+\frac{(\rho_1\sin^2\theta+\rho_2\cos^2\theta)}{2\sigma^2}P\right]\cdot\left[1+\frac{(\rho_1\cos^2\theta+\rho_2\sin^2\theta)}{2\sigma^2}P\right]\right\} = \frac{W}{2}\cdot\log_2\left[1+\frac{\rho_1+\rho_2}{2\sigma^2}P+\left(\rho_1\rho_2+\frac{\sin^2(2\theta)(\rho_1-\rho_2)^2}{4}\right)\frac{P^2}{4\sigma^4}\right]. \quad (16)$$

Obviously, when $\theta = 0$, the achievable rate is minimum as Equation 14; and when $\theta = \frac{\pi}{4}$, the achievable rate is maximum as Equation 15.

According to Lemma 1, we can come to the following theorem:

Theorem 1. For any actual rank-2 MIMO fading channel with real-valued transmit signals, the ergodic achievable rate of JCMD-MIMO is greater than that of BICM-MIMO with the conventional uniform power allocation.

Proof. For any actual rank-2 MIMO fading channel, the ergodic achievable rate is the mathematical average expectation of the above constant achievable rate over all possible channel eigenvalue realization. Generally speaking, the equal eigenvalue realization is only a small probability event. The case that the channel eigenvalues are different must be existed. According to Lemma 1, the ergodic achievable rate of JCMD-MIMO is greater than that of BICM-MIMO with the conventional uniform power allocation. □

Generally speaking, assuming a rank-L MIMO with a descending-order eigenvalue vector $\bar{\rho} = \{\rho_1, \rho_2, \ldots, \rho_L\}$, Q-component interleaver only changes the order of Q components on L layers to another eigenvalue vector $\bar{\zeta} = \{\zeta_1, \zeta_2, \ldots, \zeta_L\}$, where $\bar{\zeta}$ is just another arrangement order of $\bar{\rho}$ corresponding to the output order of Q-component spatial interleaver. Hence, due to the orthogonal modulation, JCMD-MIMO can be viewed as L parallel fading channels with an eigenvalue vector $\bar{\upsilon} = \frac{\bar{\rho}+\bar{\zeta}}{2}$. So, the achievable rate of rank-L JCMD-MIMO is shown as the following:

$$C_L(\bar{\zeta}) = \frac{W}{2} \cdot \sum_{i=1}^{L} \log_2\left[1 + \frac{(\rho_i + \xi_i)}{2\sigma^2 L}P\right] = \frac{W}{2} \cdot \log_2\left[\prod_{i=1}^{L}\left(1 + \frac{(\rho_i + \xi_i)}{2\sigma^2 L}P\right)\right], \quad (17)$$

where P is the total transmit power for L layers. Hence, the optimum problem of the achievable rate is to find the optimum $\bar{\zeta}$ so as to maximize C_L. We can reach the following theorem:

Theorem 2. To maximize the achievable rate of rank-L JCMD-MIMO with a descending-order eigenvalue vector $\bar{\rho} = \{\rho_1, \rho_2, \ldots, \rho_L\}$, the optimum Q-component interleaver vector $\overline{\omega} = \{\rho_L, \rho_{L-1}, \ldots, \rho_1\}$, that is to say, $\overline{\omega}$ should be in ascending order, which is just in reverse order of $\bar{\rho}$.

Proof. Assuming someone claims to find a non-increasing-order vector $\bar{\rho}'$ other than $\overline{\omega}$ to have maximum achievable rate $C_L(\bar{\rho}')$, we must can find a pair of $\{\rho_i', \rho_{i+1}'\}$ from $\bar{\rho}'$ to satisfy $\rho_i' > \rho_{i+1}'$. And then, we can construct a new vector $\eta = \{\rho_1', \rho_2', \ldots, \rho_{i-1}', \rho_{i+1}', \rho_i', \rho_{i+2}', \ldots, \rho_L'\}$, which only changes the order of $\{\rho_i', \rho_{i+1}'\}$ from $\bar{\rho}'$. So, we can compute the difference of $\frac{C_L(\eta) - C_L(\bar{\rho}')}{W/2}$,

$$\begin{aligned}\frac{C_L(\eta) - C_L(\bar{\rho}')}{W/2} &= \sum_{k=1}^{L}\log_2\left[1+\frac{(\rho_k+\eta_k)}{2\sigma^2 L}P\right] - \sum_{k=1}^{L}\log_2\left[1+\frac{(\rho_k+\rho_k')}{2\sigma^2 L}P\right] \\ &= \log_2\left[1 + \frac{(\rho_i + \eta_i)}{2\sigma^2 L}P\right] + \log_2\left[1 + \frac{(\rho_{i+1} + \eta_{i+1})}{2\sigma^2 L}P\right] \\ &\quad - \log_2\left[1 + \frac{(\rho_i + \rho_i')}{2\sigma^2 L}P\right] - \log_2\left[1 + \frac{(\rho_{i+1} + \rho_{i+1}')}{2\sigma^2 L}P\right] \\ &= \log_2\left[1 + \frac{(\rho_i + \rho_{i+1}')}{2\sigma^2 L}P\right] + \log_2\left[1 + \frac{(\rho_{i+1} + \rho_i')}{2\sigma^2 L}P\right] \\ &\quad - \log_2\left[1 + \frac{(\rho_i + \rho_i')}{2\sigma^2 L}P\right] - \log_2\left[1 + \frac{(\rho_{i+1} + \rho_{i+1}')}{2\sigma^2 L}P\right].\end{aligned} \quad (18)$$

Let a constant $c = \frac{P}{2\sigma^2 L}$, and then,

$$\frac{C_L(\eta) - C_L(\bar{\rho}')}{W/2} = \log_2 \frac{M_1}{M_2}, \quad (19)$$

where

$$\begin{aligned}M_1 &= 1 + c\left(\rho_i + \rho_i' + \rho_{i+1} + \rho_{i+1}'\right) \\ &\quad + c^2\left(\rho_i + \rho_{i+1}'\right)\left(\rho_{i+1} + \rho_i'\right) > 0,\end{aligned} \quad (20)$$

$$\begin{aligned}M_2 &= 1 + c\left(\rho_i + \rho_i' + \rho_{i+1} + \rho_{i+1}'\right) \\ &\quad + c^2\left(\rho_i + \rho_i'\right)\left(\rho_{i+1} + \rho_{i+1}'\right) > 0.\end{aligned} \quad (21)$$

So, the difference

$$M_1 - M_2 = c^2(\rho_i - \rho_{i+1})(\rho_i' - \rho_{i+1}'). \quad (22)$$

Because $\rho_i > \rho_{i+1}$ and $\rho_i' > \rho_{i+1}'$.

So, $M_1 - M_2 > 0$; Thus, $\frac{M_1}{M_2} > 1$; and then, $C_L(\eta) - C_L(\bar{\rho}') > 0$.

That is to say, we find a counter-example η to have more achievable rate than $C_L(\bar{\rho}')$, which contradicts the assumption of maximum achievable rate $C_L(\bar{\rho}')$. Therefore, $\overline{\omega}$ should be in ascending order, which is just in reverse order of $\bar{\rho}$. □

So, in Section 2, the reverse Q-component spatial interleaver is applied as Equation 2 so as to have the maximum achievable rate. In fact, if L is an even number, the rank-L JCMD-MIMO with the reverse Q-component spatial interleaver can be viewed as $\frac{L}{2}$ parallel pairs of rank-2 JCMD-MIMO with eigenvalues $\{\rho_i, \rho_{L+1-i}\}$, where $i \in \left[1, \frac{L}{2}\right]$. Likewise, if L is an odd number, it can be viewed as the parallel combination of one SISO fading channel with eigenvalue $\left\{\rho_{\frac{L+1}{2}}\right\}$ and $\frac{L-1}{2}$ pairs of rank-2 JCMD-MIMO with eigenvalues $\{\rho_i, \rho_{L+1-i}\}$, where $i \in \left[1, \frac{L-1}{2}\right]$. Thus, the largest eigenvalue layer couples with the smallest eigenvalue layer, the second largest eigenvalue layer couples with the second smallest eigenvalue layer, and so on. Actually, BICM-MIMO is also one special case of JCMD-MIMO when $\bar{\zeta} = \bar{\rho}$.

Theorem 3. As for the achievable rate of a rank-L JCMD-MIMO with a descending-order eigenvalue vector $\bar{\rho} = \{\rho_1, \rho_2, \ldots, \rho_L\}$, the upper bound is $C_L(\overline{\omega})$, where $\overline{\omega}$ is in reverse order of $\bar{\rho}$, and the lower bound is the BICM-MIMO achievable rate $C_L(\bar{\rho})$.

Proof. According to Theorem 2, the maximum achievable rate is $C_L(\bar{\omega})$, we can get $C \leq C_L(\bar{\omega})$. In addition, $C_L(\bar{\rho})$ is the minimum achievable rate, which can be proved by the similar math skill as follows.

Assuming someone claims to find a non-descending-order vector $\bar{\rho}'$ other than $\overline{\rho}$ to have minimum achievable rate $C_L(\bar{\rho}')$, we must can find a pair of $\{\rho_i', \rho_{i+1}'\}$ from $\bar{\rho}'$ to satisfy $\rho_i' < \rho_{i+1}'$. And then, we can construct a new vector $\boldsymbol{\eta} = \left\{\rho_1', \rho_2', \ldots, \rho_{i-1}', \rho_{i+1}', \rho_i', \rho_{i+2}', \ldots, \rho_L'\right\}$, which only changes the order of $\left\{\rho_i', \rho_{i+1}'\right\}$ from $\bar{\boldsymbol{\rho}}'$. So, we can compute the difference of $\frac{C_L(\boldsymbol{\eta})-C_L(\bar{\boldsymbol{\rho}}')}{W/2}$,

$$\begin{aligned}\frac{C_L(\boldsymbol{\eta})-C_L(\bar{\boldsymbol{\rho}}')}{W/2} &= \sum_{k=1}^{L}\log_2\left[1+\frac{(\rho_k+\eta_k)}{2\sigma^2L}P\right]-\sum_{k=1}^{L}\log_2\left[1+\frac{(\rho_k+\rho_k')}{2\sigma^2L}P\right]\\ &= \log_2\left[1+\frac{(\rho_i+\eta_i)}{2\sigma^2L}P\right]+\log_2\left[1+\frac{(\rho_{i+1}+\eta_{i+1})}{2\sigma^2L}P\right]\\ &\quad -\log_2\left[1+\frac{(\rho_i+\rho_i')}{2\sigma^2L}P\right]-\log_2\left[1+\frac{(\rho_{i+1}+\rho_{i+1}')}{2\sigma^2L}P\right]\\ &= \log_2\left[1+\frac{(\rho_i+\rho_{i+1}')}{2\sigma^2L}P\right]+\log_2\left[1+\frac{(\rho_{i+1}+\rho_i')}{2\sigma^2L}P\right]\\ &\quad -\log_2\left[1+\frac{(\rho_i+\rho_i')}{2\sigma^2L}P\right]-\log_2\left[1+\frac{(\rho_{i+1}+\rho_{i+1}')}{2\sigma^2L}P\right].\end{aligned} \tag{23}$$

Let a constant $c = \frac{P}{2\sigma^2L}$, and then,

$$\frac{C_L(\boldsymbol{\eta}) - C_L(\bar{\boldsymbol{\rho}}')}{W/2} = \log_2\frac{M_1}{M_2}, \tag{24}$$

where

$$\begin{aligned}M_1 &= 1+c\left(\rho_i+\rho_i'+\rho_{i+1}+\rho_{i+1}'\right)\\ &\quad + c^2\left(\rho_i+\rho_{i+1}'\right)\left(\rho_{i+1}+\rho_i'\right) > 0,\end{aligned} \tag{25}$$

$$\begin{aligned}M_2 &= 1+c\left(\rho_i+\rho_i'+\rho_{i+1}+\rho_{i+1}'\right)\\ &\quad + c^2\left(\rho_i+\rho_i'\right)\left(\rho_{i+1}+\rho_{i+1}'\right) > 0.\end{aligned} \tag{26}$$

So, the difference

$$M_1 - M_2 = c^2(\rho_i-\rho_{i+1})\left(\rho_i'-\rho_{i+1}'\right). \tag{27}$$

Because $\rho_i > \rho_{i+1}$ and $\rho_i' < \rho_{i+1}'$.

So, $M_1 - M_2 < 0$; Thus, $\frac{M_1}{M_2} < 1$; and then, $C_L(\boldsymbol{\eta}) - C_L(\bar{\boldsymbol{\rho}}') < 0$.

That is to say, we find a counter-example $\boldsymbol{\eta}$ to have small achievable rate than $C_L(\bar{\boldsymbol{\rho}}')$, which contradicts the assumption of minimum achievable rate $C_L(\bar{\boldsymbol{\rho}}')$. Therefore, $C_L(\bar{\boldsymbol{\rho}})$ is the minimum achievable rate. □

4 AMI analysis for complex-valued QAM signals

For practical communication systems, the transmit signal $\mathbf{X}$ usually belongs to a finite alphabet (constellation signal set), such as the complex-valued QAM signal. AMI also varies with the rotation angle of QAM signals in the MIMO system. Assuming equiprobable QAM constellation inputs and the independent Rayleigh fading channel, the AMI of the coded modulation (CM) MIMO system before the demapper is called CM-AMI, which is irrelevant to the labeling and defined as [2]

$$\begin{aligned}I_{\mathrm{CM}} &= I\left(\mathbf{X};\mathbf{Y}|\Lambda\right)\\ &= \sum_{l=1}^{N_L}\left\{m - E_{\mathbf{x}_l,\mathbf{y}_l,\lambda_l}\left[\log_2\frac{\sum_{\hat{\mathbf{x}}\in\chi}P\left(\mathbf{y}_l|\hat{\mathbf{x}},\lambda_l\right)}{P(\mathbf{y}_l|\mathbf{x}_l,\lambda_l)}\right]\right\}.\end{aligned} \tag{28}$$

The AMI after the demapper is called BICM-AMI, which is relevant to the labeling and defined as [2]

$$\begin{aligned}I_{\mathrm{BICM}} &= \sum_{k=1}^{m} I\left(C_k;\mathbf{Y}|\boldsymbol{\Lambda}\right)\\ &= \sum_{l=1}^{N_L}\left\{m-\sum_{k=1}^{m}E_{c_k,\mathbf{y}_l,\lambda_l}\left[\log_2\frac{\sum_{\hat{\mathbf{x}}\in\chi}p(\mathbf{y}_l|\hat{\mathbf{x}},\lambda_l)}{\sum_{\hat{\mathbf{x}}\in\chi,\hat{c}_k=c_k}p(\mathbf{y}_l|\hat{\mathbf{x}},\lambda_l)}\right]\right\}.\end{aligned} \tag{29}$$

AMI analysis is an effective method to calculate its achievable rate. From Equation 28, because $I\left(\mathbf{X};\mathbf{Y}|\Lambda\right)$ is independent from labeling, it implies that the AMI of CM system is not related to the labeling. However, for the BICM system, the BICM-AMI strongly depends on the labeling. Extensive literature has proven that Gray labeling is optimal for the non-ID BICM system in the AWGN channel [25]. Monte Carlo simulation techniques are useful tools to get the expectation of a complex function by ergodicity of the random variables [26]. The expectation operations in Equations 28 and 29 can be evaluated by using the Monte Carlo simulation techniques. Based on Equation 9, we can get the received symbol before the soft demapper by generating the random coded bits c_k (corresponding to the modulated symbol x_k^l on each layer), the random fading coefficient λ_k^l (SVD of i.i.d. Rayleigh-distributed random channel matrix $\mathbf{H}_k$), and the i.i.d. Gaussian noise random variable n_k^l. Thus, by using the Monte Carlo simulation techniques, we can estimate the expectation values of Equations 28 and 29 by ergodicity of (c_k, $\mathbf{H}_k$, n_k^l), where $P\left(\mathbf{y}_l|\mathbf{x}_l,\lambda_l\right)$ can be calculated as Equation 12. As for complex-valued QAM signals, the optimum rotation angle usually is different from the value of the real-valued signals, and we can get it by maximizing AMI.

Figure 3 shows how 2 × 2 MIMO CM-AMI varies with the rotation angle θ for real-valued binary phase shift keying (BPSK) symbols. Since $I\left(\mathbf{X};\mathbf{Y}|\boldsymbol{\Lambda}\right) = I\left(C_k;\mathbf{Y}|\boldsymbol{\Lambda}\right)$, the I_{CM} and I_{BICM} are the same for BPSK. Both for low and high SNR, e.g., at SNR = −5 dB and SNR= −15 dB, the optimal rotation angles are always $\theta = 45°$, which coincides with our analysis in Section 3.

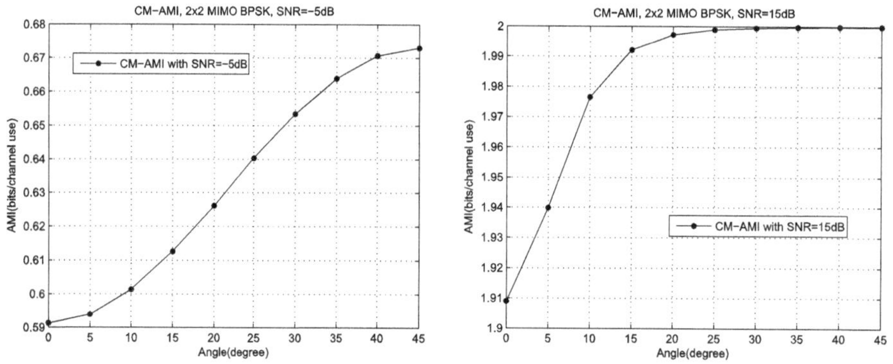

(a) SNR=-5dB (b) SNR=15dB

Figure 3 2 × 2 MIMO CM-AMI I_{CM} vs. rotation angle θ for BPSK over Rayleigh fading channels. (**a**) SNR = −5 dB. (**b**) SNR = 15 dB.

Figures 4 and 5 show how 4 × 4 MIMO CM-AMI and BICM-AMI vary with the rotation angle θ for QPSK and 16QAM, respectively. The AMIs of proposed systems with two kinds of spatial Q-component interleavers (reverse and cyclic-shift) are plotted in the same figures. Examples of low SNR and high SNR are presented. In Figures 4 and 5, for both spatial Q interleaving algorithms, the optimal rotation angles just have slight difference in high SNR. For CM-AMI, the optimal angle of QPSK at SNR = −3 dB is 45°, while the optimal angle at SNR = 11 dB is about 30° and 29° for the cyclic-shift interleaver and the reverse interleaver, respectively. For BICM-AMI, the optimal angle of QPSK at SNR = −3 dB is 0°, while the optimal angle at SNR = 11 dB is about 26° and 27° for the cyclic-shift interleaver and the reverse interleaver, respectively.

Based on the AMI maximization criterion, the optimal angle corresponds to the maximum AMI value. Figure 6 shows the CM-AMI and BICM-AMI curves that correspond to the optimal angles in Figures 4 and 5 for 4 × 4 MIMO systems. In Figure 6, CM-AMI and BICM-AMI for the reverse interleaver described in Equation 2 are always not less than that of the cyclic-shift interleaver described in Equation 3 both for QPSK and 16QAM. These observations concide well with Theorem 2.

Based on maximizing AMI, we can also obtain the relationship between optimal angle and SNR. The optimal angles for CM and BICM systems vary with SNR for QPSK and 16QAM, which is plotted in Figure 7. The optimal rotation angle of CM is always bigger than that of BICM. When SNR increases, the optimal angle of CM becomes smaller, but the optimal angle of BICM becomes bigger.

For a given modulation order m, according to the AMI value I corresponding to optimal rotation angle in Figure 6, the optimal code rate for a given SNR can be calculated as $R = \frac{I}{m \cdot N_L}$. Therefore, we can get the relationship between the optimal code rate R and SNR, which is shown in Figure 8.

Thus, for a given operating SNR, we can obtain the optimum rotation angle in Figure 7 and the optimal

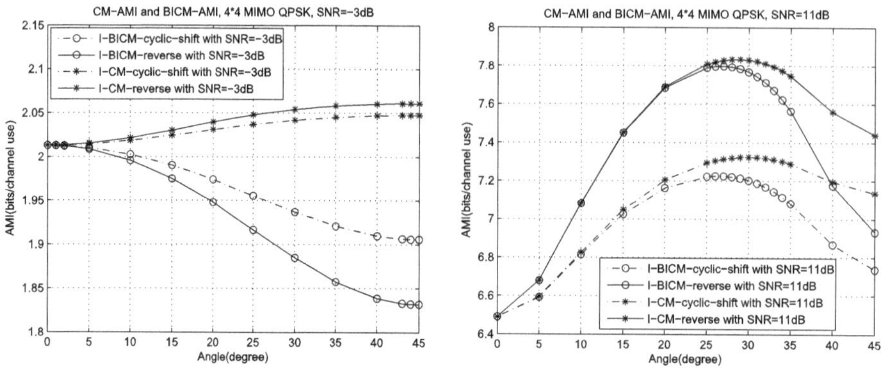

(a) SNR=-3dB (b) SNR=11dB

Figure 4 4 × 4 MIMO CM-AMI I_{CM} and BICM-AMI I_{BICM} vs. rotation angle θ for QPSK over Rayleigh fading channels. (**a**) SNR = −3 dB. (**b**) SNR = 11 dB.

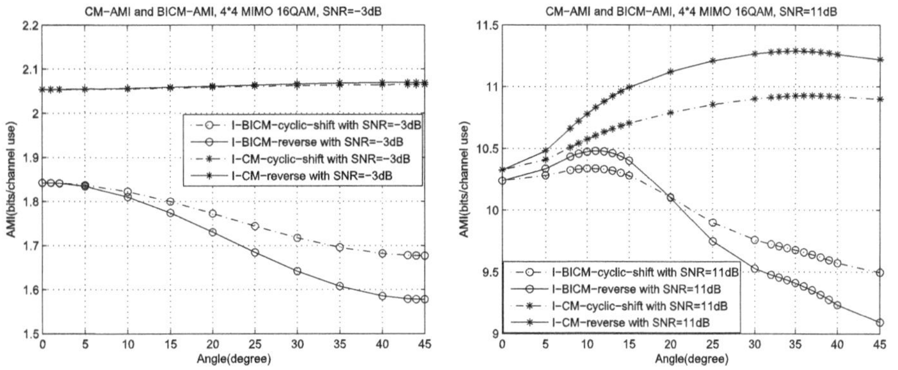

(a) SNR=-3dB (b) SNR=11dB

Figure 5 4 × 4 MIMO CM-AMI I_{CM} and BICM-AMI I_{BICM} vs. rotation angle θ for 16QAM over Rayleigh fading channels. (**a**) SNR = −3 dB. (**b**) SNR = 11 dB.

code rate in Figure 8. Furthermore, we can get the relationship between code rate R and optimal rotation angle, which is shown in Figure 9 for QPSK and 16QAM. It provides a good reference to select an optimal rotation angle for a given code rate. For instance, for code rate = 0.5, the optimal angles for 4 × 4 MIMO-BICM QPSK and 16QAM are 18° and 0°, respectively. According to Figures 7 and 9, at low SNR or low code rate, non-rotation is the best for BICM, which implies that the channel coding dominates the BICM performance, while 45° rotation is the best for CM, which indicates that the rotation symmetry affects the CM performance. However, at high SNR or high code rate, the signal space diversity dominates the performance both for BICM and CM, so BICM-AMI can approach CM-AMI, and the optimum angle of BICM with the optimum reverse spatial interleaver is also close to that of CM with the same interleaver. Note that the rotation angle is just related to the code rate and modulation. So, in practice, the rotation symbol mapper can be implemented through a look-up table that is calculated and stored in advance for a given modulation order and code rate, which is the same as the conventional symbol mapper. Hence, it does not introduce additional processing complexity and delay.

5 Code design for JCMD-ID system

For the non-ID BICM system, Gray mapping has been proved to be optimal. Based on the optimal rotation angle

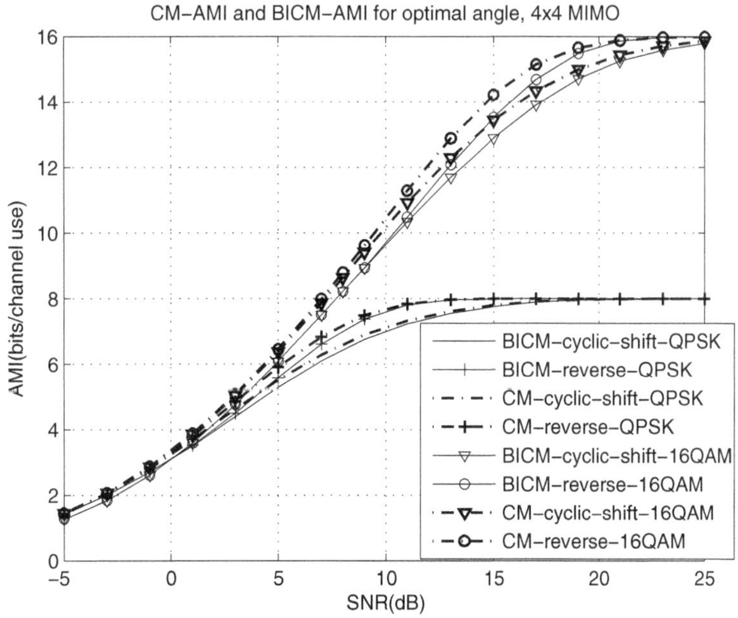

Figure 6 CM-AMI and BICM-AMI corresponding to the optimal angles for 4 × 4 MIMO.

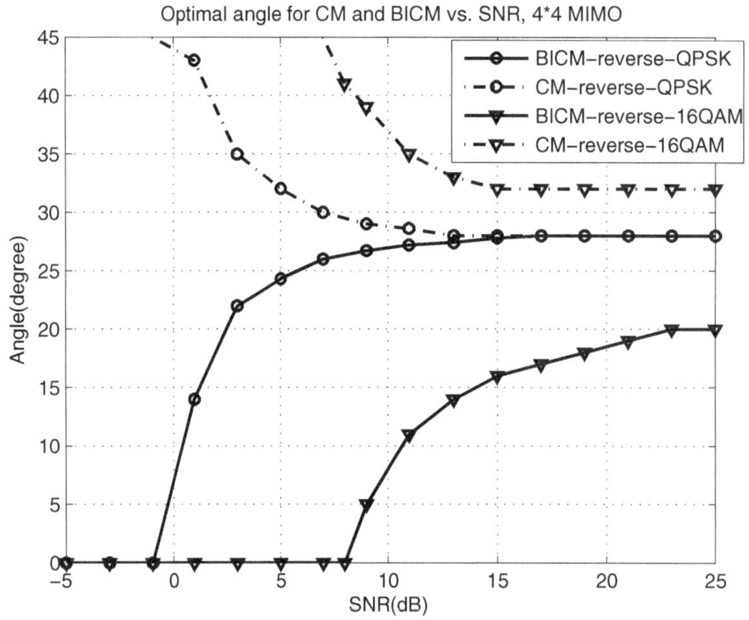

Figure 7 Optimal angle θ vs. SNR for QPSK and 16QAM over 4 × 4 Rayleigh fading channels.

obtained by maximizing BICM-AMI and Gray mapping, the non-ID JCMD system only needs to consider the optimization of channel codes to achieve excellent performance. For the JCMD-ID system, besides the optimal rotation angle, it is also very crucial to choose a pair of a well-matched labeling and an outer channel code by some joint optimization.

The convergence behavior of the iterative demodulation and decoding can be analyzed by the EXIT chart method, which can describe the flow of extrinsic information between the demodulator and the decoder [27-29]. Several advantages of EXIT charts are summarized in [29]. The inputs to the demapper are the noise-corrupted channel observations and the *a priori* knowledge $A(c_k)$ on

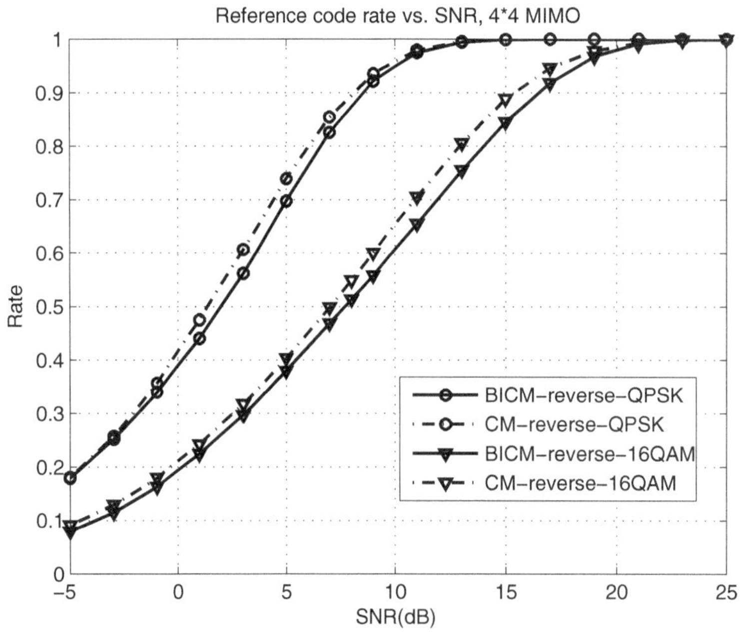

Figure 8 Code rate vs. SNR for QPSK and 16QAM over 4 × 4 Rayleigh fading channels.

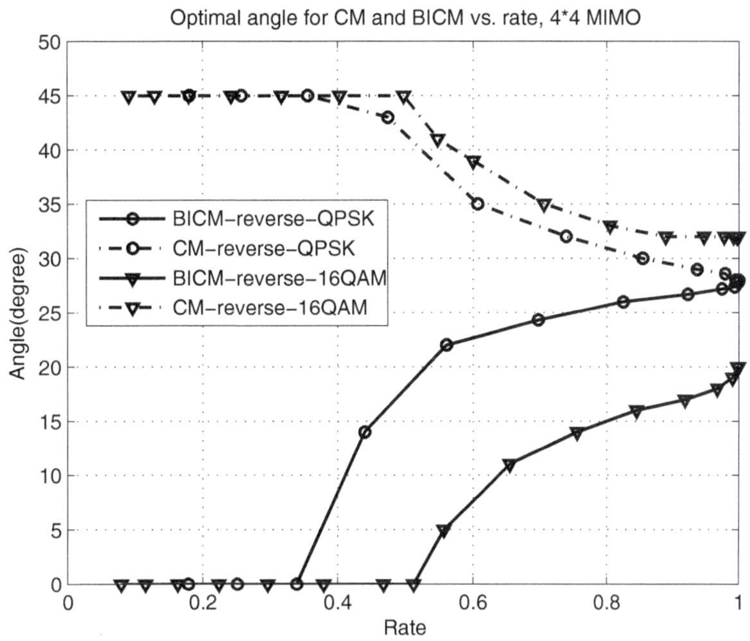

Figure 9 Optimal angle θ vs. code rate $\boldsymbol{R}$ for QPSK and 16QAM over 4 × 4 Rayleigh fading channels.

the unmapped bits. The demapper outputs channel and extrinsic information $E(c_k)$. According to Equations 12 and 19 in [27], the mutual information $I_{A1} = I\left(c_k; A\left(c_k\right)\right)$ $(0 \leq I_{A1} \leq 1)$ between transmitted unmapped bits and the L-values $A(c_k)$ can be written as

$$I_{A1} = \frac{1}{2} \sum_{x=-1,1} \int_{-\infty}^{\infty} p_{A(c_k)}\left(\xi \,|c_k = x\right) \cdot \log_2 \times \frac{2p_{A(c_k)}\left(\xi \,|c_k = x\right)}{p_{A(c_k)}\left(\xi \,|c_k = -1\right) + p_{A(c_k)}\left(\xi \,|c_k = 1\right)} d\xi. \tag{30}$$

The mutual information $I_{E1} = I\left(c_k; E\left(c_k\right)\right)$ $(0 \leq I_{E1} \leq 1)$ can be written as

$$I_{E1} = \frac{1}{2} \sum_{x=-1,1} \int_{-\infty}^{\infty} p_{E(c_k)}\left(\xi \,|c_k = x\right) \cdot \log_2 \times \frac{2p_{E(c_k)}\left(\xi \,|c_k = x\right)}{p_{E(c_k)}\left(\xi \,|c_k = -1\right) + p_{E(c_k)}\left(\xi \,|c_k = 1\right)} d\xi. \tag{31}$$

In [27], it turns out that the *a priori* input $A(c_k)$ is almost Gaussian distributed. Additionally, large interleavers keep the a priori L-values $A(c_k)$ fairly uncorrelated over many iterations. Hence, it seems appropriate to model the *a priori* input $A(c_k)$ by applying an independent Gaussian random variable with variance σ_{A1}^2 and mean zero (see Equations 1 to 10 in [27]). Thus, the conditional probability density function can be written as

$$p_{A(c_k)}\left(\xi \,|c_k = x\right) = \frac{1}{\sqrt{2\pi}\sigma_{A1}} \exp\left[-\frac{\left(\xi - \frac{\sigma_{A1}^2}{2} x\right)^2}{2\sigma_{A1}^2}\right] \tag{32}$$

The mutual information $I_{E1} = I\left(c_k; E\left(c_k\right)\right)$ can be viewed as a function of I_{A1}, SNR, and the rotation angle θ, i.e,

$$I_{E_1} = T_1(I_{A_1}, SNR, \theta). \tag{33}$$

The function T_1 can not be expressed in a closed form. For a given value of the input mutual information $I_{A1} = I\left(c_k; A\left(c_k\right)\right)$ $(0 \leq I_{A1} \leq 1)$ to the demodulator, to compute $T_1(I_{A_1}, SNR, \theta)$, the distributions $p_{E(c_k)}\left(\xi \,|c_k = x\right)$ are most conveniently determined by the Monte Carlo simulation (histogram measurements) proposed in [27,28]. The convenient method has been verified to allow an accurate prediction of the SNR-decoding threshold with low complexity [27].

On the other hand, the extrinsic transfer characteristics of the decoder can describe the input/output relationship between the input I_{A2} and the output I_{E2}, which is independent of SNR value. It can be computed by assuming the *a priori* input to be Gaussian distributed and applying the same method as used for the demapper characteristic T_1. Therefore, the transfer characteristic of the decoder is denoted by

$$I_{E_2} = T_2(I_{A_2}). \tag{34}$$

Qiuliang et al. [16] verified that the SSD technique can affect the demapper's EXIT curve for the SISO BICM-ID SSD system. For the MIMO system, we analyze the effect of the SSD technique to different demapper labelings. Different 16QAM labelings (natural, Gray, and reference [16]) are shown in Figure 10. For the 1/2-rate coded 16QAM 4×4 MIMO system, the corresponding 16QAM EXIT curves with 45° rotation and non-rotation are depicted in Figure 11. The doping technique is used for error-floor removal [30] and the doping rate $P = 100$. For Natural and Gray labelings, the demappers' EXIT curves with rotation have a larger slope than that without rotation. For the referenced labeling, the demapper's EXIT curve with rotation is always above the curve without rotation.

Demapper-matched code design is very crucial for the JCMD-ID MIMO system with rotation. In order to approach the capacity, based on EXIT chart, we propose an optimization method of outer channel codes to match with a given demapper for the JCMD-ID system. We choose the BCC as the channel code. For a given SNR, if two EXIT curves of the demapper and the decoder do not intersect, the iterative decoding can converge, otherwise it cannot converge. Thus, the SNR which makes the two EXIT curves tangent is the SNR convergence threshold, which is also called pinch-off SNR. The objective of JCMD-ID optimization is to find the outer channel code that has the lowest SNR convergence threshold to match with the demapper.

We define the optimization function as follows.

$$\alpha(\mathbf{G}(N_{\text{Reg}}), SNR) = \min_{I_n \in [0,1], n=1,2,\ldots,\text{Num}} \left[T_1(I_n) - T_2^{-1}(I_n) \right], \tag{35}$$

$$\overline{\text{SNR}} = \arg\min_{\text{SNR}} \left[\alpha \left(\mathbf{G} \left(N_{\text{Reg}} \right), \text{SNR} \right) > 0 \right], \tag{36}$$

where $\mathbf{G}(N_{\text{Reg}})$ denotes the generator polynomial of BCC with N_{Reg} registers. Num is the number of selected statistical samples. $\overline{\text{SNR}}$ is the pinch-off SNR. For $\frac{1}{N_{\text{out}}}$ rate BCC, the objective is to find the optimum $\mathbf{G}_{\text{Opt.}}(N_{\text{Reg}})$ with the lowest pinch-off SNR from $\left(2^{N_{\text{Reg}}+1} - 1\right)^{N_{\text{out}}}$ generator polynomial candidates. However, the exhaustive searching of a channel code to well match the demapper's EXIT curve is trivial, especially for N_{Reg} is large.

GA is an efficient optimization algorithm, which is stochastic search techniques based on the mechanism of natural selection and natural genetics [31]. In GA, a genetic representation is required for the individuals in a population. Generator polynomials of BCC $\mathbf{G}(N_{\text{Reg}}) = \left[\mathbf{g}_1, \mathbf{g}_2, \ldots, \mathbf{g}_{N_{\text{out}}}\right]_2$ inherently provides a $(N_{\text{Reg}}+1) \times N_{\text{out}}$-length binary string $\mathbf{S}_g = < \mathbf{g}_1, \mathbf{g}_2, \ldots, \mathbf{g}_{N_{\text{out}}} >$. $\mathbf{g}_i$ is the $(N_{\text{Reg}} + 1)$-length binary generator polynomial corresponding to ith $(1 \le i \le N_{\text{out}})$ output. Based on the genetic algorithm, an optimization method is proposed as follows.

Step 1: Initial population. Set the current iterative number (number of generations) $N_{\text{pop}} = 0$, the number of candidate generator polynomials (population size) N_g, the maximum iterative number $N_{\max}$, crossover probability P_c, and mutation probability P_m. N_g binary strings $\mathbf{S}_g^i = < \mathbf{g}_1^i, \mathbf{g}_2^i, \ldots, \mathbf{g}_{N_{\text{out}}}^i >$ that correspond to the candidate polynomials are randomly initialized, which are denoted by a set $\mathbf{C}^{N_{\text{pop}}}$, where $\mathbf{g}_1^i, \mathbf{g}_2^i, \ldots, \mathbf{g}_{N_{\text{out}}}^i \in \left[1, 2^{N_{\text{Reg}}+1}\right), 1 \le i \le N_g$.
Step 2: Selection. Reduce the SNR by small steps Δ_{SNR} (e.g., 0.1 dB) and compute the pinch-off SNRs $\overline{\text{SNR}_i}$ of all the candidate generator polynomials $\mathbf{G}_i(N_{\text{Reg}}) = \left[\mathbf{g}_1^i, \mathbf{g}_2^i, \ldots, \mathbf{g}_{N_{\text{out}}}^i\right]_2$ in population, where $< \mathbf{g}_1^i, \mathbf{g}_2^i, \ldots, \mathbf{g}_{N_{\text{out}}}^i > \in \mathbf{C}^{N_{\text{pop}}}, 1 \le i \le N_g$. We use the pinch-off SNR to measure the fitness. The fitness function is associated with the maximum pinch-off SNR $\overline{\text{SNR}}_{\max} = \max_{1 \le i \le N_g} \left\{\overline{\text{SNR}_i}\right\}$, shown as

$$f\left(\mathbf{S}_g^i\right) = \overline{\text{SNR}}_{\max}(\text{dB}) + 0.1 - \overline{\text{SNR}_i}(\text{dB}). \tag{37}$$

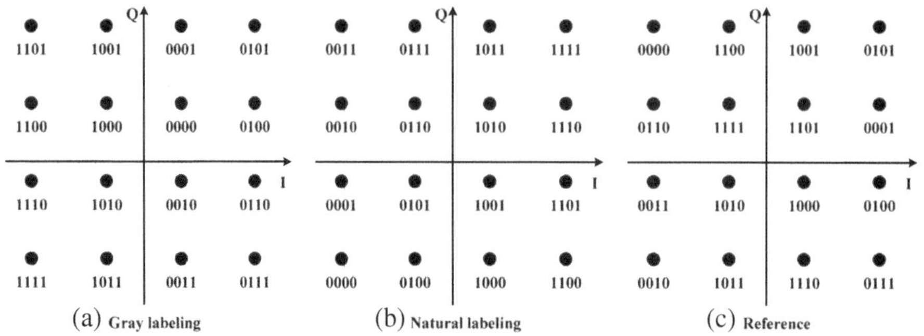

Figure 10 16QAM labelings. (**a**) Gray labeling. (**b**) Natural labeling. (**c**) The reference labeling in [16].

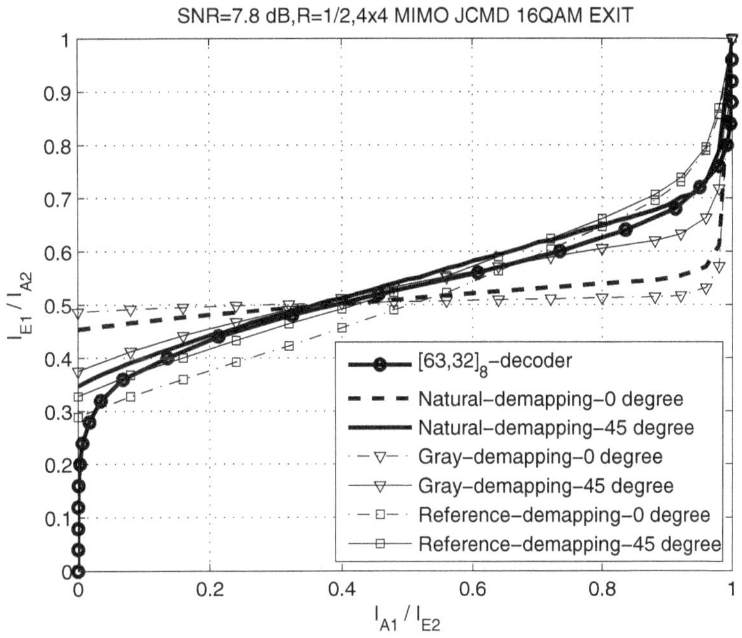

Figure 11 EXIT chart analysis for 16QAM at SNR = 7.8 dB.

N_g individuals are selected to breed a new generation with probability proportional to the fitness value.

The probability that $\mathbf{S}_g^i$ is selected is $P(i) = \frac{f\left(\mathbf{S}_g^i\right)}{\sum_{k=1}^{N_g} f\left(\mathbf{S}_g^k\right)}$.

Based on roulette wheel selection (RWS) algorithm, the tth $(1 \leq t \leq N_g)$ individual selection follows the steps below.

A. Generate a uniform random number $\chi(t)$, $\chi(t) \in [0, 1]$.

B. If $\sum_{i=0}^{k-1} P(i) \leq \chi(t) < \sum_{i=1}^{k} P(i)(1 \leq k \leq N_g)$, $P(0) = 0$, $\mathbf{S}_g^k$ is selected.

Step 3: Crossover. For two adjacent selected individuals, a random number κ_c from the range $[0, 1]$ is generated. Only when $\kappa_c < P_c$, the crossover operator is carried out. The crossover point is selected randomly. All bits beyond that point in either string are swapped between the two parent individuals, and then two children individuals are obtained.

Step 4: Mutation. For each individuals, a random number κ_m from the range $[0, 1]$ is generated. Only when $\kappa_m < P_m$, the mutation operator is carried out through one bit flip at random mutation position.

Step 5: Judgment. $N_{\text{pop}} = N_{\text{pop}} + 1$. After step $2 \sim 4$, a new population $\mathbf{C}^{N_{\text{pop}}+1}$ is formed. If $N_{\text{pop}} < N_{\max}$, then go to step 2, otherwise stop. The generator polynomial with the lowest pinch-off SNR in $\mathbf{C}^{N_{\text{pop}}+1}$ is chosen as the optimum one.

Using the method above, a rate-half BCC code with $N_{\text{Reg}} = 5$ is optimized for the natural labeling 16QAM JCMD-ID system. Based on the AMI analysis in Section 4, the optimal rotation angle for 0.5 rate 16QAM JCMD-ID system is 45°. The optimal generator polynomial is $[63,32]_8$. As shown in Figure 11, the natural labeling demapper's EXIT curve with 45° rotation keeps a narrow open tunnel with that of BCC decoder, while its non-rotation curve intersects with the decoder, which shows the effect of rotation to the ID system. The other two labeling demappers' EXIT curves with and without rotation always intersects with the decoder, so the Gray and reference labelings are not suitable for the BCC decoder. Therefore, the natural labeling with 45° rotation matches well with the $[63,32]_8$ BCC code and has the best performance.

6 Simulation result

6.1 Results of non-ID JCMD system on fast fading channels

For the non-ID JCMD MIMO system, the optimal Gray labeling and powerful DVB-T2 LDPC coding are used to achieve excellent performance [32]. The optimal rotation angle is obtained by maximizing BICM-AMI in Figure 9. The size of coded block N is 64,800 bits. For the LDPC decoder, the log-belief propagation (BP) algorithm with 30 maximum iterations is utilized. In order to ensure the fairness of the comparison, the SVD precoding is implemented for both the conventional BICM MIMO system and proposed JCMD MIMO system.

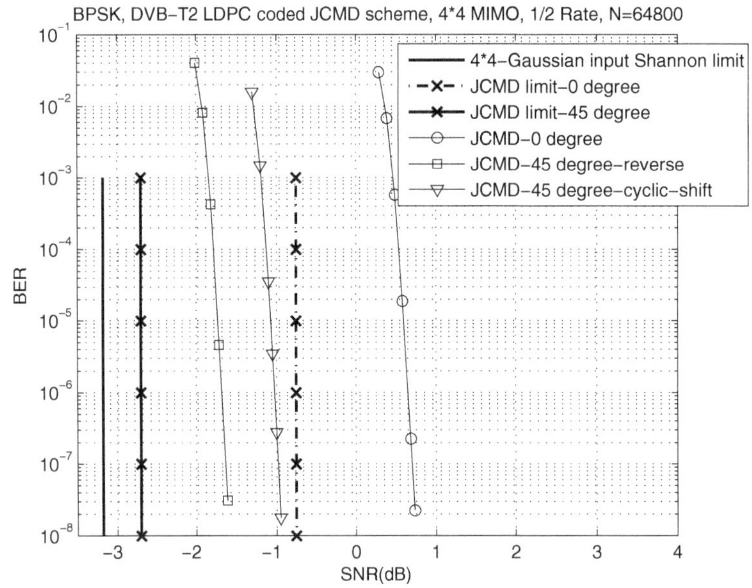

Figure 12 BER performance of rate-1/2 DVB-T2 LDPC coded, BPSK 4 × 4 JCMD MIMO systems on fast fading channels.

Figure 12 shows the bit error rate (BER) performance comparisons of rate-1/2 LDPC-coded BPSK MIMO systems on the independent fast Rayleigh fading channel. The channel fading coefficients on each symbol are independent identical distributed Rayleigh random variables with the variance 1. BPSK is also a simple real-valued signal, so the optimal rotation angle is chosen as 45°. For 4 × 4 MIMO systems, JCMD with 45° rotation obtains significant SNR gains as compared with JCMD without rotation. For the target BER $= 10^{-5}$, it can achieve 2.2 dB SNR gain. Note that the spatial interleaver, spatial Q-component interleaver, and time Q-component interleaver are all

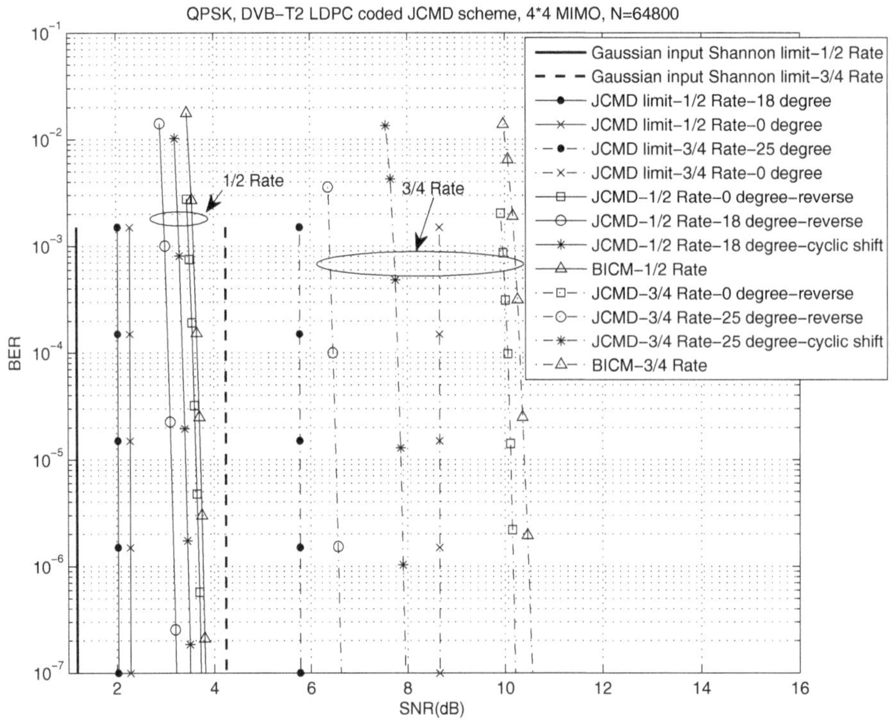

Figure 13 BER performance of rate-1/2 and 3/4 DVB-T2 LDPC coded, QPSK 4 × 4 JCMD MIMO systems on fast fading channels.

Table 1 SNR gains and gaps to the capacity for JCMD 4 × 4 MIMO systems

Parameters	Gains compared with BICM system (dB)	Gaps to the JCMD/Gaussian input Shannon limit (dB)
BPSK, 1/2-rate	2.2	1.0/1.5
QPSK, 1/2-rate	0.6	1.1/1.9
QPSK, 3/4-rate	3.9	0.7/2.3

implemented in both JCMD scheme with and without rotation. That means the performance gain of JCMD does not exactly come from the interleavers compared with BICM scheme. We can still obtain significant gain through the optimal rotation and the matching of channel coding and labeling. Furthermore, the 4 × 4 JCMD MIMO system with the reverse spatial Q interleaver can obtain about 0.6 dB SNR gain compared with that with the cyclic-shift spatial Q interleaver. The results coincide well with the above analysis in Section 3.

For the QPSK 4×4 MIMO system, the optimal angles for 1/2 and 3/4 rate are 18° and 25°, respectively. As shown in Figure 13, JCMD systems with the optimal rotation angle can obtain 0.6 and 3.9 dB SNR gains compared with the conventional BICM systems employing the ideal SVD precoding for low rate 1/2 and high rate 3/4, respectively. Given the optimal rotation angle, JCMD-MIMO with the reverse interleaver obtains significant 0.3 and 1.3 dB SNR gains at BER $= 10^{-5}$ compared to that with the cyclic-shift interleaver for the code rate 1/2 and 3/4, respectively. Meanwhile, the proposed 1/2 rate JCMD system with the reverse interleaver is only about 1.1 dB SNR gap to the JCMD limit for 4 × 4 MIMO. For the 3/4 rate JCMD MIMO, the gap to the JCMD limit is reduced to 0.7 dB for 4 × 4 MIMO. SNR gains and gaps are summarized in Table 1. From the results, SNR gain becomes bigger for higher code rate.

6.2 Results of JCMD-ID system on fast fading channels

For the JCMD-ID 16QAM MIMO system, the optimal rotation angle is obtained by maximizing CM-AMI in Figure 9, and it is 45° at 1/2 rate for 4 × 4 MIMO systems. In order to confirm our optimization method for JCMD-ID MIMO system, simulations are carried out with proposed $[63,32]_8$ BCC-coded JCMD-ID scheme on fast fading channels for 4 × 4 MIMO systems. The size of coded block $N = 64,800$ bits and the maximal global iterative number is 30. The powerful 1/2 rate DVB-T2 LDPC-coded Gray-labeling BICM and BICM-ID schemes with the same block size are also simulated as the reference, and the same ideal SVD precoding method is used for the conventional BICM and BICM-ID schemes and the proposed JCMD-ID scheme.

The BER performance comparisons are shown in Figure 14.

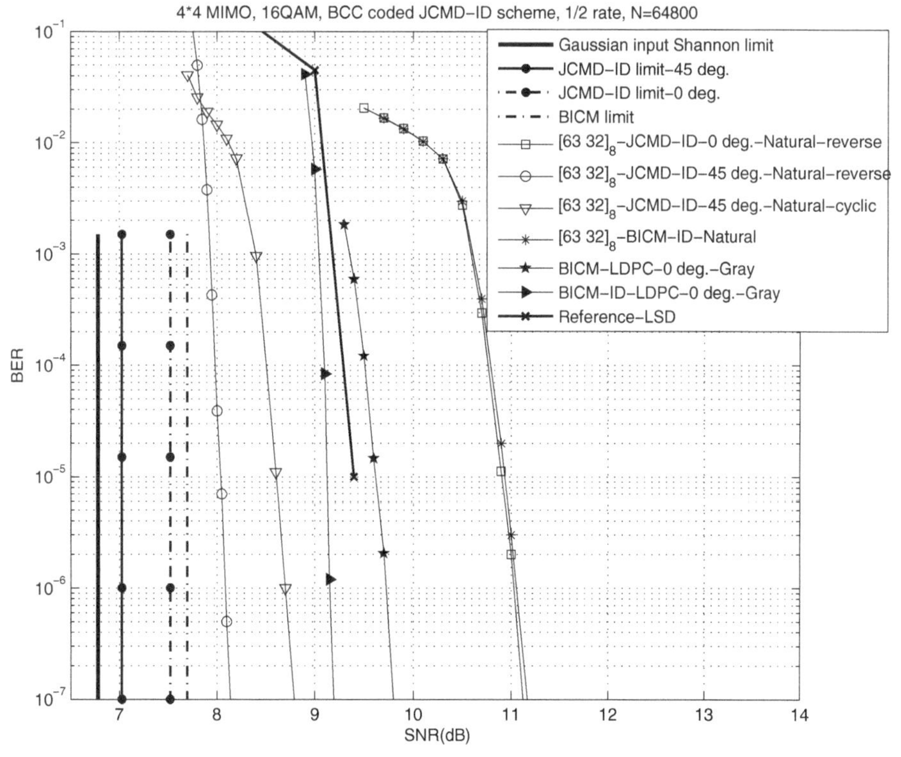

Figure 14 BER performance of rate-1/2, 16QAM 4 × 4 JCMD-ID MIMO systems on fast fading channels.

Table 2 Average decoding complexity for each information bit

Operation	Optimized BCC	Turbo used in [24]	DVB-T2 LDPC
Additions	331	816	997.5
Max ops.	158	288	0
Look-ups	0	0	735

For the 4 × 4 MIMO system with natural labeling, the optimized BCC $[63,32]_8$ coded JCMD-ID scheme with 45° rotation and reverse Q interleaver exhibits excellent performance, which is only 1.3 and 1.1 dB away from the Gaussian-input capacity and JCMD-ID limit, respectively. It can obtain about 2.9 dB SNR gain compared with the BCC-coded BICM-ID scheme and JCMD-ID without rotation, which coincides with the above EXIT analysis. In addition, the scheme with the reverse interleaver obtains 0.4 dB SNR gain at BER $= 10^{-5}$ compared with that with cyclic-shift Q interleaver, which also proves the above analysis. Furthermore, the optimized JCMD-ID scheme also outperforms the DVB-T2 LDPC-coded BICM-ID scheme and the turbo-coded BICM-ID scheme in [24] by 0.9 and 1.4 dB gains, respectively.

For the JCMD-ID and conventional BICM-ID schemes, the main complexity lies in the channel decoding. The Max-Log-MAP algorithm is a simplified algorithm for the decoding of turbo code and BCC. For each decoding iteration, it requires $10 \times 2^{N_{\text{Reg}}} + 11$ additions and $5 \times 2^{N_{\text{Reg}}} - 2$ maximum operations [33]. The BCC decoding only needs one iteration (one Max-Log-MAP operation), while turbo decoding needs eight iterations (16 Max-Log-MAP operations). For the Log-BP algorithm of LDPC decoding, each iteration requires $Md_c\,(d_c - 1) + Nd_v\,(d_v - 1) + Nd_c$ additions and Md_c^2 look-up table operations, where M is the number of the parity bits, and d_c and d_v are the average degree distributions of check nodes and variable nodes, respectively [34]. LDPC decoding requires 30 iterations. Thus, we can obtain the average decoding complexity comparison for each information bit, as shown in Table 2. Assuming the equal complexity of the three operations (addition, max, look-up), the decoding complexity of BCC is just is 44.3% and 28.2% of turbo and LDPC decoding, respectively. Therefore, the optimized BCC has much lower complexity compared with the DVB-T2 LDPC and turbo codes. This indicates that the optimized scheme with BCC coding obtains much better performance with much lower complexity.

7 Conclusions

A high-spectral-efficient JCMD scheme over MIMO fading channels is proposed. By jointly optimizing the component interleaver, the rotation modulation, and the BCC code, this scheme exhibits excellent performance. An optimum spatial component interleaver is proposed to maximize the achievable rate. For real-valued signals, we prove that the achievable rate of JCMD MIMO is greater than that of the conventional BICM MIMO scheme and $\frac{\pi}{4}$ is the optimal rotation angle. For the rotated QAM, the optimal rotation angles are investigated for the MIMO system according to the maximizing AMI criterion. For the JCMD-ID MIMO system, a simple GA-based search algorithm of BCC generator polynomials is also proposed to match the rotation modulation. Simulation results prove that this new scheme can significantly outperform the conventional turbo-coded BICM-ID scheme over MIMO fading channels by 1.4 dB SNR gain, while it has much lower complexity. In a word, the proposed JCMD scheme is simple, efficient, and robust for the future wireless communication systems.

Competing interests
The authors declare that they have no competing interests.

Acknowledgements
This work is sponsored by the National Natural Science Fund (61171101), National Great Science Specific Project (2009ZX03003-011-03) of People's Republic of China and 2014 Doctorial Innovation Fund of BUPT (CX201426) and the Fundamental Research Funds for the Central Universities.

References

1. C Felita, M Suryanegara, in *Proceedings of International Conference on QiR (Quality in Research): 25-28 June 2013; Yogyakarta*. 5 g key technologies: identifying innovation opportunity (IEEE, Piscataway, 2013), pp. 235–238
2. G Caire, G Taricco, E Biglieri, Bit-interleaved coded modulation. IEEE Trans. Inf. Theory. **44**(3), 927–946 (1998)
3. X Li, JA Ritcey, Bit-interleaved coded modulation with iterative decoding. IEEE Commun. Lett. **1**(6), 169–171 (1997)
4. B Vucetic, J Yuan, *Space-Time Coding*. (John Wiley & Sons, Inc., England, 2003)
5. G Foschini, Layered space-time architecture for wireless communication in a fading environment when using multi-element antennas. Bell Labs Tech. J. **1**(2), 41–59 (1996)
6. HE Gamal, AR Hammons, The layered space-time architecture: a new perspective. IEEE Trans. Inf. Theory. **1**, 2321–2334 (2001)
7. J Boutros, E Viterbo, Signal space diversity: a power and bandwidth efficient diversity technique for the Rayleigh fading channel. IEEE Trans. Inf. Theory. **44**(4), 1453–1467 (1998)
8. CA Nour, C Douillard, in *Proceedings of 5th International Symposium on Turbo Codes and Related Topics: 1-5 Sept. 2008; Lausanne*. Improving BICM performance of QAM constellations for broadcasting applications (IEEE, Piscataway, 2008), pp. 55–60
9. NF Kiyani, UH Rizvi, JH Weber, GJM Janssen, in *Proceedings of IEEE Wireless Communications and Networking Conference: 11-15 March 2007; Kowloon*. Optimized rotations for LDPC-coded MPSK constellations with signal space diversity (IEEE, Piscataway, 2007), pp. 677–681
10. NF Kiyani, JH Weber, in *Proceedings of IEEE Symposium on Communications and Vehicular Technology in the Benelux: 15-15 Nov. 2007; Delft*. OFDM with BICM-ID and rotated MPSK constellations and signal space diversity (IEEE, Piscataway, NJ, USA, 2007), pp. 1–4
11. NF Kiyani, JH Weber, Exit chart analysis of iterative demodulation and decoding of MPSK constellations with signal space diversity. J. Commun. **3**(3), 43–50 (2008)
12. NH Tran, HH Nguyen, T Le-Ngoc, Performance of BICM-ID with signal space diversity. IEEE Trans. Wireless Commun. **6**(5), 1732–1742 (2007)
13. M Zhenzhou, S Zhiping, Z Chong, Z Zhongpei, in *Proceedings of International Conference on Communications, Circuits and Systems: 25-27*

May 2008; Fujian. Design of signal space diversity based on non-binary LDPC code (IEEE, Piscataway, 2008), pp. 31–34
14. W Zhanji, W Wenbo, Improved coding-rotated-modulation orthogonal frequency division multiplexing system. IET Commun. **6**(3), 272–280 (2012)
15. W Zhanji, W Wenbo, in *Proceedings of 2010 Global Mobile Congress: 18-19 Oct. 2010; Shanghai*. A novel joint-coding-modulation-diversity OFDM system (IEEE, Piscataway, 2010), pp. 1–6
16. X Qiuliang, S Jian, P Kewu, Y Fang, W Zhaocheng, Coded modulation with signal space diversity. IEEE Trans. Wireless Commun. **10**(2), 660–668 (2011)
17. N Sharma, CB Papadias, Improved quasi-orthogonal codes through constellation rotation. IEEE Trans. Wireless Commun. **51**(3), 332–335 (2003)
18. W An Zhong, Z Jian Kang, in *Proceedings of 11th Canadian Workshop on Information Theory (CWIT): 13-15 May 2009; Ottawa*. Novel rotation angle for quasi-orthogonal space-time block codes (IEEE, Piscataway, 2009), pp. 213–216
19. W Su, X-G Xia, Signal constellations for quasi-orthogonal space time block codes with full diversity. IEEE Trans. Inf. Theory. **50**(10), 2331–2347 (2004)
20. D Dung Ngoc, C Tellambura, in *Proceedings of IEEE Global Telecommunications Conference (GLOBECOM): Dec. 2005; St. Louis*. Optimal rotations for quasi-orthogonal STBC with two-dimensional constellations (IEEE, Piscataway, 2005), pp. 2317–2321
21. L Yueqian, M Salehi, in *Proceedings of 46th Annual Conference on Information Sciences and Systems (CISS): 21-23 March 2012; Princeton*. Coded MIMO systems with modulation diversity for block-fading channels (IEEE, Piscataway, 2012), pp. 1–5
22. H Lee, A Paulraj, MIMO systems based on modulation diversity. IEEE Trans. Wireless Commun. **58**(12), 3045–3049 (2010)
23. H Soon Up, C Jinyong, J Sungho, R Ho Jin, S Jongsoo, in *Proceedings of IEEE International Symposium on Broadband Multimedia Systems and Broadcasting: 13-15 May 2009; Bilbao*. Performance evaluation of MIMO-OFDM with signal space diversity over frequency selective channels (IEEE, Piscataway, NJ, USA, 2009), pp. 1–5
24. BM Hochwald, S ten Brink, Achieving near-capacity on a multiple-antenna channel. IEEE Trans. Commun. **51**(3), 389–399 (2003)
25. SY Goff, Signal constellation for bit-interleaved coded modulation. IEEE Trans. Inf. Theory. **49**(1), 307–313 (2003)
26. FM Gardner, JD Baker, *Simulation Techniques: Models of Communication Signal and Processes*. (Wiley, New York, 1995)
27. S ten Brink, Convergence behavior of iteratively decoded parallel concatenated codes. IEEE Transa. Commun. **49**(10), 1727–1737 (2001)
28. S ten Brink, Designing iterative decoding schemes with the extrinsic information transfer chart. AEU Int. J. Electron. Commun. **54**(6), 389–398 (2000)
29. A Ashikhmin, G Kramer, S ten Brink, Extrinsic information transfer functions: model and erasure channel properties. IEEE Trans. Inf. Theory. **50**(11), 2657–2673 (2004)
30. S Pfletschinger, F Sanzi, Error floor removal for bit-interleaved coded modulation with iterative detection. IEEE Trans. Wireless Commun. **5**(11), 3174–3181 (2006)
31. P Guo, W Xuezhi, H Yingshi, in *Proceedings of 3rd International Conference on Biomedical Engineering and Informatics: 16-18 Oct. 2010; Yantai*. The enhanced genetic algorithms for the optimization design (IEEE, Piscataway, 2010), pp. 2990–2994
32. Frame structure channel coding and modulation for a second generation digital terrestrial television broadcasting system (DVB-T2). ETSI Std. DVB Document A122 [S] (June 2008)
33. P Robertson, E Villebrun, P Hoeher, in *Proceedings of IEEE International Conference on Communications: 18-22 Jun 1995; Seattle*. A comparison of optimal and sub-optimal map decoding algorithms operating in the log domain, (1995), pp. 1009–1013
34. C Jinghu, A Dholakia, E Eleftheriou, MPC Fossorier, Reduced-complexity decoding of LDPC codes. IEEE Trans. Commun. **53**(8), 1288–1299 (2005)

4

Dynamic switching off algorithms for pico base stations in heterogeneous cellular networks

Jie Wu[1], Shi Jin[1*], Lei Jiang[2] and Gang Wang[2]

Abstract

The densification of pico base stations (PBSs) in heterogenous cellular networks (Hetnets) causes redundant energy consumption of the whole network during low traffic periods. In order to address this problem, we take advantage of the traffic load fluctuation over time and area to adapt the necessary resource to the actual traffic demand by introducing sleep mode to the PBSs. To achieve this target, we propose two centralized algorithms (i.e., the heuristic algorithm and the progressive algorithm) in this paper to dynamically switch off the unnecessary PBSs, both of which can track the traffic variation well. We design a utility function of some key factors considered particularly for PBSs in Hetnets. These algorithms rely on the utility function to switch off the redundant PBSs, enabling the working mode of all PBSs be reconfigured periodically. The progressive algorithm is proposed to overcome the inefficiency of the heuristic algorithm. The simulations demonstrate that the execution time of the progressive algorithm is at most one third of that of the heuristic algorithm, which enables the network to respond to the traffic variation more promptly. Besides, the progressive algorithm can switch off more PBSs than the heuristic algorithm while slightly affects the network blocking probability, which indicates that the progressive algorithm has a better potential to save energy. Moreover, simulations also reveal that some key parameters all have nonnegligible influence on the performance of our algorithms. These parameters should be tuned well to trade spare network resource for energy saving.

Keywords: Heterogenous cellular networks (Hetnets); Pico base stations; Dynamic switching off algorithm; Energy saving; Blocking probability

1 Introduction

Nowadays, with the proliferation of smart devices, mobile data traffic is increasing exponentially. It is reported that the number of mobile-connected devices will exceed the world's population in this year. Furthermore, by 2018, mobile network connection speeds will increase twofold and mobile-connected tablets will generate nearly double the traffic generated by the entire global mobile network in 2013 [1]. To satisfy such incremental demand of the future communication, recently, a new framework called heterogenous cellular networks (Hetnets) [2-5] has emerged as a flexible and cost-effective solution. Different from typical macrocells of high power base stations (MBSs) serving as a coverage layer, in Hetnets, small cell access points such as relay nodes, picocell base stations (PBSs), femtocell base stations, and remote radio heads overlaid on macrocells serve as a capacity layer, which bring mobile networks closer to user equipments (UEs), thus enhancing network capacity [6-8].

The densification of small cells strives to excavate the spatial splitting gains, while also increases the energy consumption of the whole network simultaneously. Currently, it has been estimated that the overall energy consumed by information and communications technology (ICT) industry, which includes cellular networks, already constitutes about 2% of global carbon emissions and is projected to increase much further in the coming years [9]. Moreover, it has been revealed in [10] that up to 80% of the energy consumption in a cellular network is attributed to the operations and functionality of the base stations (BSs) in the radio access network while the remaining energy is expended in the switching and core networks.

Since the required network capacity and the number of BSs in a cellular network are typically dimensioned to serve the peak traffic, if all BSs remain active irrespective of traffic load, a tremendous amount of resource will

*Correspondence: jinshi@seu.edu.cn
[1]School of Information Science and Engineering, Southeast University, No. 2 Sipailou, Nanjing 210096, China
Full list of author information is available at the end of the article

be redundant during off-peak times, and energy is inefficiently consumed. Besides, even at the same time, traffic distribution of different network area can be no uniform. Therefore, methods that reduce energy consumption by adapting the network resource to the traffic demand are important research directions. Fortunately, such temporal and spatial traffic load fluctuation gives the opportunity to save energy significantly by switching off the underutilized MBSs or PBSs [11,12].

1.1 Contributions

We consider picocells which are low-power and low-cost cells covered by PBSs and designed to serve a small outdoor area such as hotspot or shaded region. The main objective of this paper is to switch off the redundant PBSs to reduce energy consumption of the network while simultaneously guaranteeing the quality of service (QoS) of UEs. Specifically, the fundamental limit of the achievable QoS of the macrocell UEs (MUEs) and the picocell UEs (PUEs) is given in terms of the required service rate. It is known that BS switching of finding the optimal operation mode of the network system is a difficult combinatorial problem. Since this problem is NP hard and requires high-computational complexity as well as large signaling overhead, there are some works [13-15] considering low complexity algorithms to tackle such problem. These algorithms are all sub-optimal, yet proved to be very useful in the given scenarios. Similarly, in our work, we propose practically implementable algorithms by considering the main characteristics of the Hetnets. The major contributions of this paper are summarized as follows:

1) We design a utility function of PBS considering some key factors particularly for PBSs in Hetnets. These factors within a certain picocell include the total rate of served PUEs, the PBS's traffic load, the number of served PUEs, the number of blocked PUEs, and the received interference signal strength from the nearby cells. The utility function assists the following proposed algorithms to select PBSs in a reasonable order to be tested whether they can be switched off.
2) We first propose a heuristic dynamic switching off (HDSO) algorithm, which tests PBSs to switch off one by one at each step, and is similar to the switching off algorithm proposed in [16]. It turns out that after this algorithm is executed, the number of active PBSs can track the network traffic profile very well. Simulations show that some parameters can significantly influence the performance of HDSO. Besides, its complexity can be prohibitive for large-size network with dense deployment of PBSs and peak traffic time.
3) To overcome the inefficiency of HDSO, we propose a progressive dynamic switching off (PDSO) algorithm (inspired by the works in [17]), by avoiding the unnecessary work of testing. Different from the HDSO algorithm, PDSO is carried out in a round by round manner. Specifically, in one round of testing, this algorithm is based on the utility function to classify the current active PBSs into two groups, then a switching-off process is carried out to test all PBSs in the group with relatively small utilities to be switched off. It intelligently decides whether to launch another round of testing according to the switching off result of the previous round. Simulations demonstrate that the PDSO algorithm is more efficient in switching off the redundant PBSs, for which the complexity of the algorithm is greatly reduced. Moreover, with properly tuned parameters, the PDSO algorithm is verified to have a better potential of energy saving.
4) We propose a PUE transferring algorithm to transfer the PUEs of the PBS which is tested to be switched off to the nearby BSs. As we aim at switching off the PBSs, this algorithm first attempts to transfer the PUE to the nearby MBSs. If this attempt fails, then the nearby PBSs will be considered as the acceptor BSs. Both the acceptor MBSs and PBSs need to reserve some resource blocks (RBs) for subsequent new UEs in order to reduce the network blocking probability. The principle to decide whether to switch off a PBS is that if all its served PUEs can be successfully transferred to the nearby BSs.

1.2 Related work

Different techniques of energy saving have been proposed. The challenge to these techniques is to maintain reliable service coverage and QoS, while simultaneously saving the most energy. In the context of Hetnets, a novel idle mode procedure has been proposed in [18], which allows the femto BS transmissions and associated processing to be switched off completely at all times when the femto BS does not need to support an active call. This is achieved by a low-power sniffer capability in the femto BS and a predetermined threshold of uplink (UL) received power. Besides, threshold-based approaches have also been used in [19,20] to apply sleep mode to PBSs. The basic idea is to switch off the PBSs when their traffic loads are below a certain threshold for a certain period of monitoring time, which means the PBSs' loads can be offloaded by their neighboring BSs. The threshold of either the UL received power or the PBS traffic load is an important metric in the performance of the algorithm, which greatly depends on automatic configuration and fine adjustment of the threshold during operation. In addition, these approaches are derived from the micro perspective (i.e., the cell level); however, from the macro perspective (i.e., the network level), there also exists a fraction threshold of the necessary PBSs needed to remain active. In [17], the authors developed a simple analytical model that allow optimal BS switch-off times to be identified as a function of the daily traffic pattern, in the case where several switch offs per day are permitted (progressively reducing the number of active base stations and the network energy). Some fractions of the BSs needed to be active at certain time instants per day are assumed. Inspired by this work, we propose our PDSO algorithm which adopts the threshold approach from the network perspective.

For the conventional cellular networks, dynamic switching off of BSs has been extensively studied in [11,16,21-26]. A solution proposed in [21] automatically switch off appropriate BSs or sectors, which become the compensation area needed to be covered through the tilting of antennas in the neighboring BSs. A novel energy-efficient cellular access network architecture based on the principle of ecological protocooperation was proposed in [22], which indicates that BSs can cooperatively and dynamically make intelligent decisions based on thresholds for switching between different power modes according to traffic conditions. Similarly, a transmission power increment is required for compensating the areas of switched off BSs. In contrast to these works, the use of coordinated multipoint for MBS switching off without transmission power adjustment from compensating neighboring BSs is presented in [23]. From a game theoretic perspective, the energy efficiency issues in multi-operator mobile networks was studied in [24], where cost-based functions are used to decide the best suitable BSs to remain active. This paper introduces the cost that has to be paid by an operator when its subscribers have to be served by another operator due to the fact that some BSs have been switched off. In [25], the authors proposed a practical switching on/off-based energy saving algorithm that can be realized distributively. The key principle of the algorithm was to switch off a BS one by one that minimally affect the network by using network impact, which takes into account the additional load increments brought to its neighboring BSs. Distance-based approaches have been developed in [11,16]. In [11], the authors used two real datasets (i.e., temporal and spatial) to estimate the energy savings, and they used a greedy algorithm by sequentially switching off the BSs with the minimum distance to its nearest active BSs if the coverage is met. A dynamic switching on/off algorithm where the number of active BSs adapts to the network condition was proposed in [16], which was based on BS traffic load and the position of the associated UEs. This algorithm preferentially tests the BS with larger average distance to its associated UEs to be switched off one by one and terminates when UEs of a BS cannot be accepted by the neighboring BSs. In realistic networks, the distances among BSs in homogeneous cellular network can be easily acquired, but the position of UEs may not be easily and accurately acquired. Besides, the policy that the algorithm just stops when one BS fails to transfer its UEs does not seem reasonable as other BSs with larger average distance may be lowly loaded, thus also have the potential to be switched off. It is important to remark that in the previous works regarding homogeneous cellular network, the decision to switch off BSs was based on either the distance factor or the traffic. However, since we focus on switching off PBSs in Hetnets, the distance between PBS and PUE cannot be accurately or easily obtained due to various reasons, such as technology limits and designing principles. As in some cases, the network side is not aware of the UEs' position, which is also the desire of the users. Moreover, in addition to the load, other factors such as UEs' service rate, the historical blocking probability, and the interference conditions should also be considered. Although BS switching off is primarily designed to reduce network energy consumption, it should be noted that the various approaches must ensure that the QoS in the coverage area is not compromised at all times. Emphasizing on this concern, a progressive BS switching on/off technique has been implemented through the coordination of multiple surrounding BSs in [26], where the main finding shows that the duration of BS sleeping and waking up transients is very short, with no significant reduction of the energy savings achievable with sleep mode approaches.

1.3 Organization

The remainder of the paper is organized as follows. Section 2 presents the system model for the Hetnets. In Section 3, we propose the algorithms and elaborate them in detail. In Section 4, extensive simulations have been performed to investigate the performance of these algorithms, and the simulation results are provided with detailed analysis. Finally, conclusions are drawn in Section 5.

2 System model

Figure 1 shows a system model of Hetnets which consists of a macrocell and multiple picocells and femtocells. The Hetnets considered in this paper consist of M MBSs and P PBSs. We focus on the downlink (DL) transmission scenario based on orthogonal frequency division multiple access (OFDMA), in which the total bandwidth B is divided into N_{RB} RBs with a set $\mathcal{N}$. Co-channel deployment is considered; hence, all RBs (i.e., the scheduled units) are simultaneously allocated by both MBSs and PBSs to their served UEs. Only large-scale fading of the channel model is considered, and small-scale fading is omitted. The subcarriers constituting a single RB are subjected to the same fading, and hence, the channel gain

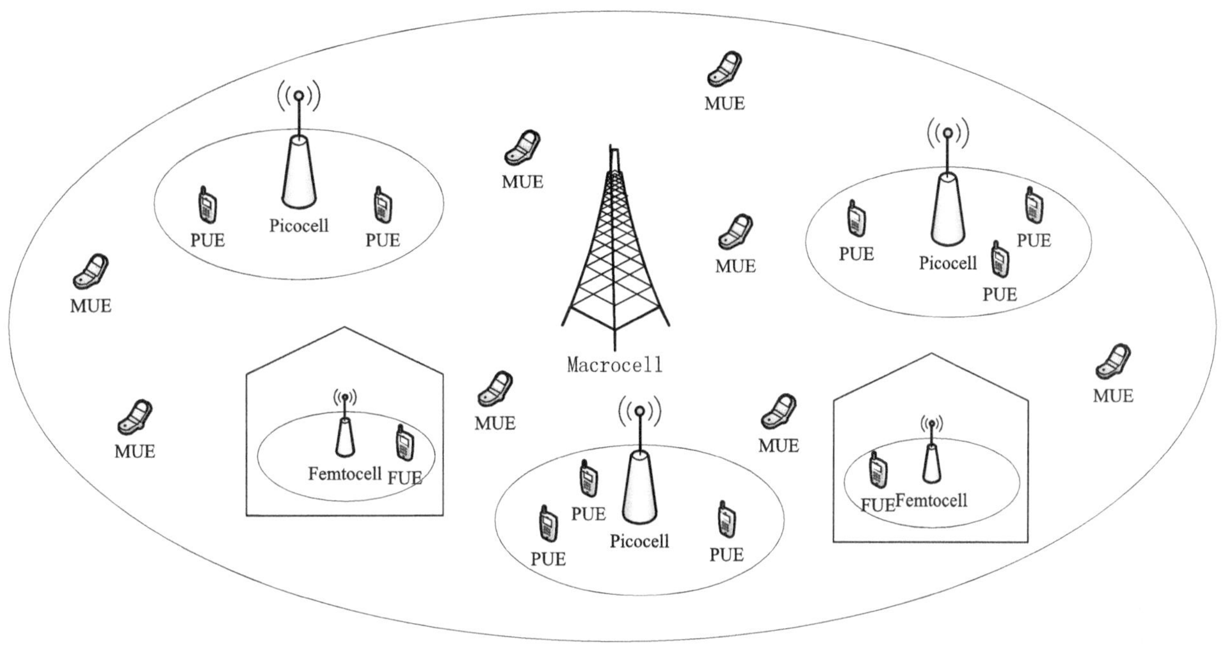

Figure 1 System model of the Hetnets deployments.

on the subcarriers of a single RB is assumed the same. Additionally, the fading is assumed to be independent identically distributed (i.i.d.) across RBs. We generally refer to a MBS or a PBS as a BS and a macrocell or a picocell as a cell. In order to guarantee seamless coverage, MBSs keep active, while PBSs can be switched off to save energy when traffic load is low.

For simplicity, power control is not considered in this paper. Assume that transmission power is uniformly allocated among all RBs for either MBS or PBS, namely, the power on each RB is calculated as $P_m = P_m^{\text{tot}}/N_{\text{RB}}$, $P_m = P_m^{\text{tot}}/N_{\text{RB}}$, where P_m^{tot} is the total transmission power of MBS, and P_p^{tot} is the total transmission power of PBS. We also assume that the unused portion of DL transmission power at the MBS and the PBS is not reallocated to other RBs in order to avoid additional interference toward normal UEs.

2.1 Achievable service rate calculation for UEs

In LTE, the different RBs allocated to a UE can have different modulation and coding schemes (MCSs). However, we do not take the different MCSs into account in this paper and just use the Shannon-Hartlay theorem to calculate the UE throughput for simplicity as in [21]. Assume that MUE k is served by the MBS m_i, $\mathcal{I}_{\text{RB},k}$ is the set of RBs allocated from MBS m_i to MUE k, then the achievable rate of MUE k can be expressed as:

$$R_k^{m_i} = \sum_{n\in\mathcal{I}_{\text{RB},k}} B_{\text{RB}}\log_2\left(1+\text{SINR}_{k,n}^{m_i}\right) \quad (1)$$

where B_{RB} is the bandwidth of one RB, $\text{SINR}_{k,n}^{m_i}$ is the signal-to-interference plus noise ratio (SINR) on the nth RB of MUE k, which can be expressed as:

$$\text{SINR}_{k,n}^{m_i} = \frac{P_m G_{k,n}^{m_i}}{I_{k,n}^{m_i}+\sigma_n^2} \quad (2)$$

where $G_{k,n}^{m_i}$ is the average channel gain from MBS m_i ('m_i' for the ith MBS, 'p_j' for the jth PBS) to MUE k on the nth RB, which includes path loss and shadowing. σ_n^2 is the power of additive white Gaussian noise on the nth RB, and $I_{k,n}^{m_i}$ is the interference MUE k receives on the nth RB, which is given by:

$$I_{k,n}^{m_i} = \sum_{i'=1,i'\neq i}^{M}\left(\sum_{k'\in\mathcal{U}_{m_{i'}}}\alpha_{k',n}^{m_{i'}}\right)\cdot P_m G_{k,n}^{m_{i'}} + \sum_{j=1}^{P}\left(\sum_{l\in\mathcal{U}_{p_j}}\alpha_{l,n}^{p_j}\right)\cdot P_p G_{k,n}^{p_j} \quad (3)$$

Note that the first term in the right hand side of (3) is the intra-layer interference from the other MBSs, where $\mathcal{U}_{m_{i'}}$ and $\mathcal{U}_{p_j}$ are the UE sets of the according MBS and PBS, respectively, $\alpha_{k',n}^{m_{i'}} \in \{0,1\}$ is a scheduling indicator which denotes that when MUE k' is associated with MBS $m_{i'}$, if the nth RB is allocated to MUE k', then $\alpha_{k',n}^{m_{i'}}$ is 1, otherwise 0. Since in LTE, every RB in a certain transmission time interval can only allocated to one UE, as for MBS $m_{i'}$, we have $\sum_{k'\in\mathcal{U}_{m_{i'}}} \alpha_{k',n}^{m_{i'}} \leq 1$. When the nth RB of MBS $m_{i'}$ is occupied by UE, MBS $m_{i'}$ will cause interference to MUE k, otherwise no interference. The second term in the right hand side of (3) is the inter-layer interference from all PBSs and detailed explanation is similar as the abovementioned.

Similarly, assume that PUE l is served by PBS p_j, and $\mathcal{I}_{\mathrm{RB},l}$ is the set of RBs allocated from PBS p_j to PUE l, then the rate of PUE l can be expressed as:

$$R_l^{p_j} = \sum_{n \in \mathcal{I}_{\mathrm{RB},l}} B_{\mathrm{RB}} \log_2 \left(1 + \mathrm{SINR}_{l,n}^{p_j}\right) \tag{4}$$

where $\mathrm{SINR}_{l,n}^{p_j}$ can be expressed as:

$$\mathrm{SINR}_{l,n}^{p_j} = \frac{P_p G_{l,n}^{p_j}}{I_{l,n}^{p_j} + \sigma_n^2} \tag{5}$$

and I_{i,k_p} can be expressed as:

$$I_{l,n}^{p_j} = \sum_{i=1}^{M} \left(\sum_{k \in \mathcal{U}_{m_i}} \alpha_{k,n}^{m_i} \right) \cdot P_m G_{l,n}^{m_i} + \sum_{j'=1, j' \neq j}^{P} \left(\sum_{l' \in \mathcal{U}_{p_{j'}}} \alpha_{l',n}^{p_{j'}} \right) \cdot P_p G_{l,n}^{p_{j'}} \tag{6}$$

2.2 Daily network traffic profile

We assume that the daily traffic profile of the whole Hetnets is the same and repeats periodically, which can be approximated by a sinusoidal-like periodic behavior as follows [27]:

$$\lambda(t) = \frac{A}{2^b} \left[1 + \sin\left(\frac{\pi t}{12} + B\pi \right) \right]^b + C \tag{7}$$

where A denotes the parameter that controls the amplitude of the traffic profile, k denotes the minimum traffic intensity in the network, $b \in \{1, 3\}$ is used to modulate the gradient of traffic profile curve (note that with $b = 3$ the curve has steeper slope, and the average traffic is lower), and B means the phase of traffic curve which regulate the position of the peak. Most of the traffic can be simulated through this formula. In this paper, the service arrival is modeled as a Poisson process with intensity $\lambda(t)$. In (7), we take $A = 19, b = 1, B = -\frac{11}{12}, C = 1$ and draw the traffic curve as shown in Figure 2, assuming that the network has statistic traffic information. This periodic sinusoidal traffic profile in realistic scenario has been provided in [11], which proves to be persuasive in investigating the performance of our following algorithms. Once a service arrives, a UE with a minimum rate requirement is located with certain probability in the network region, and this will be explained later. All UEs remain stationary until the transmission terminates. The transmission duration of each UE follows exponential distribution with mean $1/\mu = 180$s. According to the Little law [28], during the peak time, there are about $\lambda/\mu = 3,600$ UEs in the network.

3 Dynamic PBS switching off algorithms

In this section, we propose two dynamic PBS switching off algorithms for energy saving, which make use of the spatial and temporal traffic load fluctuation in Hetnets. The two proposed algorithms are both centralized which can guarantee a considerable energy saving performance; hence, we assume that there is a centralized controller unit (CCU) that carries out the algorithms. Since the traffic load fluctuates over time, the number of active PBSs should track this fluctuation to make a trade-off between saving energy and satisfying the UEs' service requirement. We divide the traffic period T_{total} into Z equal time intervals T with $Z + 1$ time spots as showed in Figure 3. At each time spot $t^{(i)}$, $i = 0, \ldots, Z$, the CCU carries out the proposed PBS switching off algorithm to determine all PBSs' working mode based on the current network information (the execution time of the algorithm is ignored), and all the reconfigured PBSs' mode keeps constant during the following time interval T. Note that our proposed algorithms should reckon for UEs' service requirement in the current time and the subsequent arriving UEs' service requirement (i.e., guarantee a tolerable blocking probability). The PBSs which can be switched off adjust their mode to sleeping state; the other PBSs remain active to provide service to the current and subsequent arriving UEs in the Hetnets during the time interval T. Note that time interval T which keeps equal just simplifies the resource management of the network. While in realistic scenario, the length of T can vary with $\lambda(t)$ so as to better track the network traffic variation.

At each time spot $t^{(i)}$, $i = 0, \ldots, Z$, the CCU collects the network information about UE-BS association and RB occupation, based on which the utility of each PBS can then be calculated. The utility is used as a metric which determines whether a certain PBS should be switched off or kept on. The utilities of the current active PBSs are iteratively calculated to keep up to date during the implementation of the proposed algorithms.

3.1 UE-BS association procedure and resource allocation scheme

On the expanded region (ER) of picocells in Hetnets, the power unbalance between DL and UL leads to a mismatch between the DL and UL handover boundaries. Therefore, associating a UE to the BS which provides the strongest DL reference signal receiving power (DL RSRP) may not always be the best strategy and is not an efficient way of network resource usage. Cell range expansion (CRE) is an alternative cell association that has been widely discussed in literature [29]. In CRE, generally, a positive bias is added to the DL RSRPs of PBSs pilot signals at UEs to increase PBSs' DL coverage footprints, thus compensating for the DL/UL mismatch. Although CRE allows more UEs to be associated with PBSs, the UEs in the expended region of

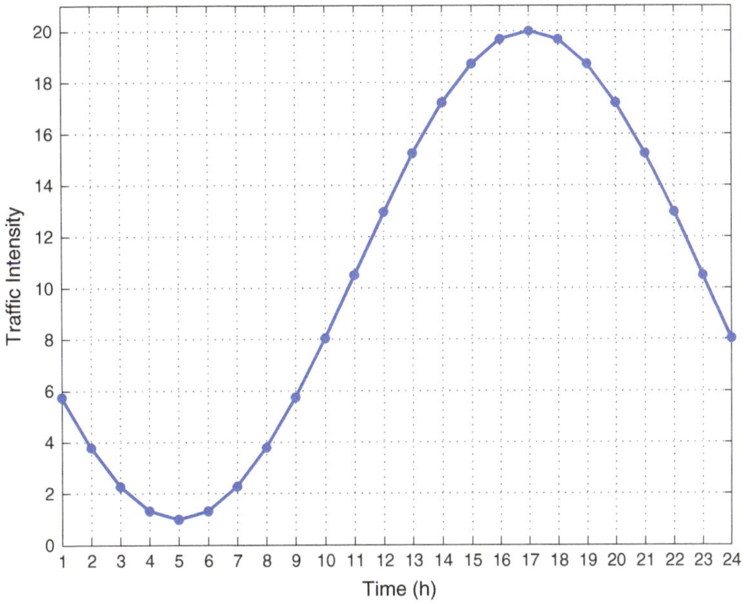

Figure 2 Average daily network traffic variations of UE service arrival rate.

PBSs without using any intercell interference coordination (ICIC) scheme will suffer severe DL cross-layer interference. Therefore, such victim CRE UEs are classified into the protected UE group. In order to improve the service quality of these UEs, a level of radio resource management coordination between the MBSs and the overlaid PBSs is needed. When a new UE enters the Hetnets, the UE-BS association process strives to balance the load of the network, aiming at lowering the network blocking probability. Therefore, the new UE is prone to choose these BSs with relatively lower load and better channel condition (i.e., larger RSRP). We then combine the conventional RSRP association method with CRE and ICIC technique. The path loss-based BS selection procedure is adopted in this paper to realize CRE, namely, associating UEs to BSs with the lowest path loss. If the new UE cannot find a proper MBS or PBS to serve its required rate, then it is blocked. Particularly, a new UE chooses a candidate MBS and a candidate PBS (if there exists one) from the potential serving BSs with the strongest RSRP, respectively, which are denoted as $P_{r,m}$ and $P_{r,p}$. Besides, the corresponding path losses are PL_m and PL_p. When $PL_m < PL_p$, the UE tries to be associated with the candidate MBS. While $PL_m \geq PL_p$, the UE tries to be associated with the candidate PBS. More particularly, if $P_{r,m} < P_{r,p}$, the UE is located in the picocell normal coverage area; otherwise, the UE is located in the picocell extended coverage area. Such association process iteratively continues until the association succeeds or all the candidate BSs cannot serve the UE. The candidate MBS and PBS which cannot afford the required rate will be eliminated from the potential serving cells in the next iteration.

In terms of ICIC technique, practically, the MBSs intelligently stop using or lower transmission power in the spectral and/or temporal resources allocated by PBSs to their CRE UEs [30]. The idea in [30] was that when a

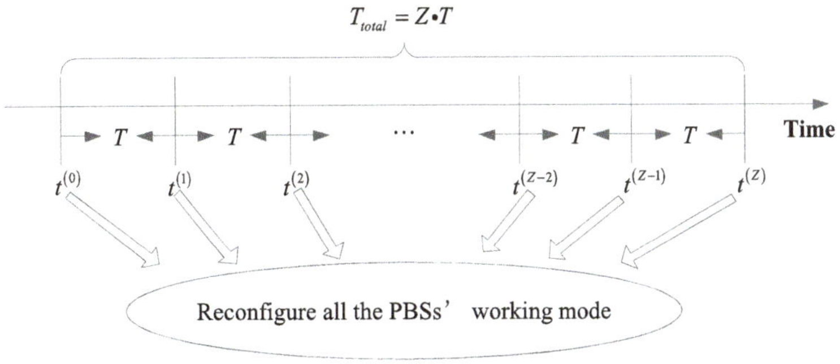

Figure 3 System operation over time.

UE enters or stays within the ER of a PBS, the PBS will inform the MBS about the set of RBs allocated to this CRE UE, and then the MBS will lower its transmission power in the specified RBs so that a desired DL QoS in terms of SINR is guaranteed to this CRE UE. Similarly, a new dynamic frequency distribution strategy was proposed in [31], where the MBSs restrict the transmission power on a part of frequency resources reserved only for CRE UEs operation. The ratio of the bandwidth of the reserved band to that of the whole band was established according to the number of UEs connected to the MBS before CRE and the recent number of CRE UEs. Noh, 2012 [32] proposed a distributed and dynamic cooperative scheme of silencing MBS transmission over part of the system bandwidth, where the silencing fraction of the whole bandwidth was determined online.

Provided that the DL transmission scenario in this paper is based on OFDMA, we adopt the MBS-PBS frequency domain ICIC technique in [30] to protect the victim CRE UEs through dynamic detection. The authors in [30] proposed a method to calculate the maximum power that the MBS can apply in each RB used by the CRE UEs to provide them with the desired QoS. However, in this paper, we focus on energy saving but not the research on ICIC techniques. Therefore, we employ the following reduced scheme for simplicity. When a UE enters the ER of a PBS, the PBS will inform the MBS about the set of RBs allocated to this CRE UE, and then the MBS will silence its transmission in the specified RBs so that a desired DL QoS in terms of required service rate is guaranteed to this CRE UE. When the service for this CRE UE ends, the occupied RBs of the PBS and silenced RBs of the MBS are released and can again be used for subsequent transmission. Cross-layer communication among MBSs and PBSs can be supported through the operator's backhaul and could be periodic or event triggered.

3.2 Utility function for PBS

Contrasting with MBSs, the traffic load of PBSs has more significant fluctuations in space and time due to a number of factors such as user mobility and behavior, as well as the fact that each PBS supports fewer simultaneous UEs. Therefore, when designing the rule of determining which PBS should first be tested to switch off, more factors should be included in the rule, instead of just the PBS traffic load as widely used in literature or the average distance between UEs and PBS as used in [16]. After having UEs in the Hetnets associated with their respective BSs and performing resource allocation as described in Section 3.1, a utility function for each PBS can then be computed. In this paper, the proposed utility function depends on the total rate of served PUEs, the PBS's traffic load, the number of served PUEs, the number of blocked PUEs, and the received interference signal strength from the nearby cells. We do not consider the average distance between a PBS and its associated PUEs, for the accurate positions of PUEs are hard to acquire in realistic scenario. The utility function is expressed as follows:

$$U_p = \frac{\alpha \cdot \frac{R_p}{R_{\max}} + \beta \cdot \frac{L_p}{L_{\max}} + \gamma \frac{W_p}{W_{\max}}}{I_p/I_{\max}} \tag{8}$$

with

$$L_p = \frac{\text{RBs utilized by the PUEs}}{\text{Total RBs available to the PBS}} \tag{9}$$

$$W_p = N_{\text{served},p} * \exp\left(P_{\text{blocked,th}} - \frac{N_{\text{blocked},p}}{N_{\text{served},p} + N_{\text{blocked},p}}\right) \tag{10}$$

where U_p is the utility of PBS p, α, β, γ are the weighting coefficients of the considered factors and satisfy $\alpha + \beta + \gamma = 1$, R_p is the total rate of all PUEs in PBS p, L_p is the traffic load on PBS p defined as [33] (the traffic load on MBS is defined likewise), W_p is the term that considers the number of served UEs $N_{\text{served},p}$ and the number of blocked UEs $N_{\text{blocked},p}$ in PBS p, which takes a similar form to the utility function adopted in [34]. In addition, $P_{\text{blocked},th} = 1\%$ is the blocking probability threshold, indicating the blocking probability that is tolerated in the network. The term I_p is the received interference signal strength from the nearby BSs. Specifically, I_p is once calculated over all RBs assuming that all RBs of nearby BSs are used to account for the worst interference case and then stored for the following utility calculation. The four terms with subscript max are the maximum values of the corresponding terms respectively and are used for normalization. The abovementioned parameters needed to calculate the utility can be obtained from the network information collected by the CCU.

In (8), the utility value is monotonously increasing with the three terms on the numerator, when the sleeping algorithm tries to switch off PBSs, we have to consider the load of the cell (i.e., RB occupation ratio and the number of served UEs) and the throughput of cell, in the interest of switching off those PBSs with relatively lower load and lower cell throughput, and thus trade-off between energy saving and network capacity ensuring. The explanation of W_p is similar to that in reference [34]. The number of the served PUEs, $N_{\text{served},p}$, determines the increase of the PBS utility as long as blocking probability threshold is not exceeded. When the number of blocked PUEs increases, the exponential term in W_p decreases. When the blocking probability threshold is exceeded, the term in the exponential becomes negative, and W_p also decreases. Besides, the interference term on the denominator is used to give more priority to those PBSs which receive relatively strong interference signal from nearby MBSs and other PBSs to be switched off.

It should be noted that the utility in (8) is selected as such because it satisfies the description presented above and integrates the characteristics of Hetnets. Nevertheless, other utility functions and metrics might be used to achieve the same purpose. The algorithms proposed in the next two sections are independent from the utility selected and can be implemented with any appropriate utility.

3.3 Proposed dynamic switching off PBS algorithms

In this section, we provide both heuristic and progressive solutions to dynamically switch off the redundant PBSs for energy saving, which uses the utility function provided in Section 3.2. The effect of the energy-saving algorithms is represented in terms of the number of active PBSs, yet the quantity of the energy saved considering practical power consumption model is an interesting topic for future study. Similar to the switching off algorithm proposed in [16], we first propose the HDSO algorithm, which tests all PBSs in the network one by one to see whether they can be switched off. Unlike the algorithm in [16] which just stops when one BS fails to transfer its UEs due to the unavailable resource the nearby BSs can provide, we consider that there still exists the potential of switching off other active PBSs when just one PBS fails to be switched off. The detailed description of the HDSO algorithm is summarized in Algorithm 1. When the algorithm terminates, the PBSs of set $\mathcal{B}_{\text{off}}$ are switched off, and the remaining PBSs remain active until the next time spot of reconfiguration.

Algorithm 1 Heuristic dynamic switching off PBS.

Initialize the set of all PBSs as $\mathcal{B}$, and the set of switched-off PBSs as $\mathcal{B}_{\text{off}} = \emptyset$
while $\mathcal{B} \neq \emptyset$ **do**
  Calculate the utility function values of PBSs in set $\mathcal{B}$ based on service state information collected by the CCU, assume p_{off} as the PBS with the minimum utility value
  Try to transfer the PUEs of PBS p_{off} to the nearby MBSs or PBSs
  if all PUEs of PBS p_{off} can be transferred **then**
    $\mathcal{B}_{\text{off}} = \mathcal{B}_{\text{off}} + \{p_{\text{off}}\}$
    Transfer PUEs of PBS p_{off} to the nearby MBSs and PBSs, update the UE association information of the network and the SINR of UEs
  else
    PBS p_{off} remains on-state and the PUE association state remains the same as before
  end if
  $\mathcal{B} = \mathcal{B} - \{p_{\mathit{off}}\}$
end while

The above HDSO algorithm greedily tests all PBSs to be switched off to see the overall energy energy potential and can commendably achieve the energy saving target. In addition, such brute-force approach is straightforward and easy to implement. However, this algorithm needs to traverse all PBSs and all PUEs in the network at any traffic condition, and obviously, such brute-force approach can be inefficient. Note that some PUEs can be transferred several times like playing a football till they finally settle. Besides, the complexity of this algorithm is proportional to the number of the PBSs and the PUEs, which can be very high and causes considerable network signaling overhead. It was mentioned in [20] that the sleep mode mechanism can be used to improve Hetnets energy efficiency at low and medium traffic load only, which indicates that most of PBSs cannot be switched off during peak traffic (this is also verified by our simulation); hence, most part of the traversal is unnecessary. Based on the utility given in (8), the probability that a PBS with relatively large utility can be switched off is smaller than that of a PBS with relatively small utility, which is intuitive and can be easily testified through simulation.

To overcome the shortcoming of the heuristic algorithm, we propose the following expeditious and intelligent PDSO algorithm, which is carried out in a round by round manner and presented in Algorithm 2. This algorithm is based on the utility function to classify the current active PBSs into two groups (one group of PBSs with smaller utility values than the other group), then a switching-off testing process is carried out to test all PBSs in the group with relatively small utilities to be switched off in one round. Upon the termination of this round, the algorithm determines whether to continue to classify the remaining active PBSs and launch another round of testing based on the switching-off result of the previous round. In the same way, the PDSO algorithm tests PBSs as far as possible, and does't stop when one PBS cannot transfer its PUEs successfully. Novelly, PDSO terminates when the predefined condition is met, and the condition can reflect the remaining energy saving potential in the current network. When the algorithm terminates, the PBSs of set $\mathcal{D}$ are switched off, and the remaining PBSs remain active until the next time spot of reconfiguration.

3.4 UE-BS re-association and RB re-allocation during the transferring process

When new UEs arrive in the Hetnets or current PUEs need to be transferred to other BSs, these UEs will select their serving BSs. In Section 3.3, at each time spot $t^{(i)}, i = 0, \ldots, Z$, the principle to decide whether to switch off a PBS is that if all the PUEs associated with this PBS can be successfully transferred to the nearby BSs. The detailed transferring process is elaborated in this section.

Algorithm 2 Progressive dynamic switching off PBS.

1: Allow the PBSs without any PUE association to be switched off, initialize the set of switched-off PBSs as $\mathcal{D}$, the remaining PBSs constitute the set of candidate switched-on PBSs
2: According to the network state information of the current active PBSs collected by the CCU, calculate the utility function values of these PBSs and sort the values incrementally
3: Denote P percentage of the PBSs with relatively small utility values as can testing set $\mathcal{T}$ (the number of elements is N), the remaining PBSs with relatively large utility values constitute the candidate switched-on set $\mathcal{S}$, and initialize the number of switched-off PBSs in this testing iteration as $\text{off}_{\text{count}} = 0$
4: **while** $\mathcal{T} \neq \emptyset$ **do**
5: Try to transfer the PUEs of PBS with the maximum utility in set $\mathcal{T}$ (denoted as M) to the nearby MBSs or PBSs
6: **if** all PUEs of PBS M can be transferred **then**
7: $\mathcal{T} = \mathcal{T} - \{M\}, \mathcal{D} = \mathcal{D} + \{M\}, \text{off}_{\text{count}} = \text{off}_{\text{count}} + 1$
8: PBS M can be switched off, transfer PUEs of PBS M to the nearby MBSs or PBSs, update the UE association information of the network and the SINR of UEs
9: **else**
10: $\mathcal{T} = \mathcal{T} - \{M\}, \mathcal{S} = \mathcal{S} + \{M\}$
11: PBS M remains on-state and PUE association state remains the same as before, update the elements of sets
12: **end if**
13: **end while**
14: **if** $\text{off}_{\text{count}}/\text{N} \geq \delta$ **then**
15: ruturn to step 2
16: **end if**

Assuming the to be transferred PUE as u in PBS p , we define the set of potential acceptor MBSs as $\mathcal{S}_m$, and the set of potential acceptor PBSs as $\mathcal{S}_p$. Particularly, we first consider to transfer PUE u to the potential acceptor MBSs, if these MBSs cannot accept the PUE, then the potential acceptor PBSs will attempt to accept it. We initialize the transmission rate as $R_s = 0$ and the allocated RB set as $\mathcal{R} = \emptyset$. During the transferring process, PUE u chooses the MBS with better channel condition $\bar{G}_u^{m_i}$ (i.e., the average channel gain between MBS m_i and PUE u) and lower load L_{m_i} as well, aiming at balancing the network load. In order to reduce the blocking probability of the subsequent new UEs, the acceptor MBS needs to reserve $\varepsilon_m \cdot N_{\text{RB}}$ RBs for the new UEs, where $\varepsilon_m \in (0, 1]$ is the bandwidth protection margin of each MBS. This leaves some spare bandwidth at each MBS to decrease the blocking probability of subsequent UEs entering the network and larger ε_m means less PBSs being switched off. Therefore, the number of available RBs of each acceptor MBS is restricted to $(1 - \varepsilon_m) \cdot N_{RB}$ in the execution phase of the PUE transfering algorithm. If all the potential acceptor MBSs cannot accept the PUE, the potential acceptor PBSs will be considered. PUE u chooses the PBS with better channel condition $\bar{G}_u^{p_j}$ and higher load L_{p_j}, which aims at switching off as many PBSs as possible and avoiding the possible repeated transferring of a single PUE. For the same purpose, the acceptor PBS needs to reserve $\varepsilon_p \in (0, 1]$ bandwidth for the subsequent new UEs. The bandwidth protection margin is consistent with the idea of system load threshold used in reference [25]. The detailed transferring process is presented in Algorithm 3. When the process ends, if the candidate BS set $\mathcal{S}_m \neq \emptyset$ or $\mathcal{S}_p \neq \emptyset$, PUE u successfully reselects a proper serving BS and the RBs in $\mathcal{R}$ are allocated to it. Otherwise, there is no BS that has enough resource for PUE u. That is PUE u can't be transferred to other BSs, and the PBS p will stay active in the following time interval.

Algorithm 3 Transferring PUE.

Initialize the to be transferred PUE as u, the set of potential acceptor MBSs as $\mathcal{S}_m$, and the set of potential acceptor PBSs as $\mathcal{S}_p$
while $\mathcal{S}_m \neq \emptyset$ **do**
$i^* = \arg \max_{m_i \in \mathcal{S}_m} \bar{G}_u^{m_i} / L_{m_i}, R_s = 0, \mathcal{R} = \emptyset$
while $R_s < R_{\min}$ and $\sum_{k \in \mathcal{U}_{m_{i^*}}} \sum_{n \in \mathcal{N}} \alpha_{k,n}^{m_{i^*}} < N_{\text{RB}}$ **do**
$n^* = \arg \max_{n \in \mathcal{N}} \left(1 - \sum_{k \in \mathcal{U}_{m_{i^*}}} \alpha_{k,n}^{m_{i^*}} \right) G_{u,n}^{m_{i^*}}$
$\mathcal{R} = \mathcal{R} + \{n^*\}, R_s = R_s + B_{\text{RB}} \log_2 \left(1 + \text{SINR}_{u,n^*}^{m_{i^*}}\right)$
end while
if $R_s >= R_{\min}$ and $L_{m_{i^*}} \leq (1 - \varepsilon_m)$ **then**
break
else
$\mathcal{S}_m = \mathcal{S}_m - \{m_{i^*}\}$
end if
end while
if $\mathcal{S}_m = \emptyset$ **then**
while $\mathcal{S}_p \neq \emptyset$ **do**
$j^* = \arg \max_{p_j \in \mathcal{S}_p} \bar{G}_u^{p_j} / L_{p_j}, R_s = 0, \mathcal{R} = \emptyset$
while $\sum_{l \in \mathcal{U}_{p_{j^*}}} \sum_{n \in \mathcal{N}} \alpha_{l,n}^{p_{j^*}} < N_{RB}$ **do**
$n^* = \arg \max_{n \in \mathcal{N}} \left(1 - \sum_{l \in \mathcal{U}_{p_{j^*}}} \alpha_{l,n}^{p_{j^*}} \right) G_{u,n}^{p_{j^*}}$
$\mathcal{R} = \mathcal{R} + \{n^*\}, R_s = R_s + B_{\text{RB}} \log_2 \left(1 + \text{SINR}_{u,n^*}^{p_{j^*}}\right)$
end while
if $R_s >= R_{\min}$ and $L_{p_{j^*}} \leq (1 - \varepsilon_p)$ **then**
break
else
$\mathcal{S}_p = \mathcal{S}_p - \{p_{j^*}\}$
end if
end while
end if

After the working modes of all PBSs are determined, all MBSs and active PBSs need not to reserve any frequency resource when serving the newly arriving UEs during the time interval, the UE association algorithm is similar to the UE transferring algorithm expect two points. Firstly, UEs choose the serving BSs with higher channel quality and lower traffic load as well, so as to balance the traffic load of the network to reduce system blocking probability. Secondly, there is no protection margin in UE association algorithm (i.e., $\varepsilon_m = 0$, $\varepsilon_p = 0$).

4 Performance evaluation

In this section, extensive numeric simulation is done to validate our analysis and evaluate the performance of our proposed algorithms in terms of energy saving and blocking probability. Next, we present the simulation scenario, the results and analysis of our algorithms.

4.1 Simulation scenario

The evaluation is performed through numerical simulations, and Table 1 gives the major simulation parameters [35]. We employ a 19-hexagonal macrocell model with three sectors per MBS, denoting the antenna downtilt of MBS, the horizontal and vertical lobe width of MBS as φ, $\phi_{3\text{ dB}}$ and $\theta_{3\text{ dB}}$, respectively. We assume that four PBSs are randomly deployed within each sector with a uniform distribution. We consider the cluster distribution for UE deployment configuration, where two thirds of the UEs are randomly located within 40 m from the PBSs with a uniform distribution (i.e., forming a hotspot area), and the location of the remaining one third of the UEs are assigned randomly across the network with a uniform distribution. It should be noted that the UE-BS association is decided in a centralized manner in our simulation. However, in realistic network, this process happens in a distributed way and involves the UE and the potential serving BSs in the vicinity of this UE. Wraparound is applied to eliminate the network edge effect and generate accurate simulation results.

As for PDSO, we generally take the threshold and the percentage as $\delta = 0.5, P = 0.5$ for simplicity, which is referred to as fixed PDSO (F-PDSO). Considering that adaptive thresholds and percentages matching the traffic variation can further enhance the efficiency of the algorithm in the realistic scenario; hence, we can have the adaptive version of PDSO (A-PDSO). In the following simulation, except for the one with respect to (w.r.t.) the traffic profile (Figures 4, 5, 6), we take PDSO as F-PDSO. For the adaptive version in the following simulation, we correspondingly take both the thresholds and the percentages in the same manner as the the ratios of switched-off PBSs when the F-PDSO algorithm is applied w.r.t. all traffic intensities. Note that these thresholds and percentages can be obtained by other methods in reality, such as the historical traffic trace about the network resource utilization as in [11] and the anticipative necessary network resource based on some traffic prediction techniques. We take the weighting coefficients in utility function as $\alpha = 0.3, \beta = 0.6, \gamma = 0.1$ by considering the importance of the corresponding factors.

4.2 Simulation results

In the following figures, we provide the results of our algorithms and the analysis of the performance using a realistic daily traffic profile.

First, we investigate the performance of the proposed algorithms w.r.t. varying traffic intensities of a single day. Figure 4 shows that the number of active PBSs tracks the variation of the statistic traffic profile very well. Figure 5 shows the average blocking probability of the proposed

Table 1 Simulation parameters

Parameters	Macrocell setting	Picocell setting
Cellular layout	19 cell sites, 3 sectors per site	4 picocells randomly distributed per sector
System Bandwidth	10 MHz	10 MHz
Number of RBs	50	50
Path loss model	$L = 128.1 + 37.6 \log 10(R)$, R in km	$L = 140.7 + 36.7 \log 10(R)$, R in km
Shadowing standard deviation	8 dB	10 dB
Antenna pattern	3D pattern	Omni-directional
Antenna gain	14 dBi	5 dBi
$\phi_{3\text{ dB}}, \theta_{3\text{ dB}}$	70°, 10°	N/A
Thermal noise density	-174 dBm/Hz	-174 dBm/Hz
BS transmission power	46 dBm	30 dBm
UE distribution radius	289 m	40 m
UE's rate requirement	400 Kbps	400 Kbps
Minimum distance requirement	Macro-pico: 75 m, Macro-UE: 35 m	Pico-pico: 40 m, Pico-UE: 10 m

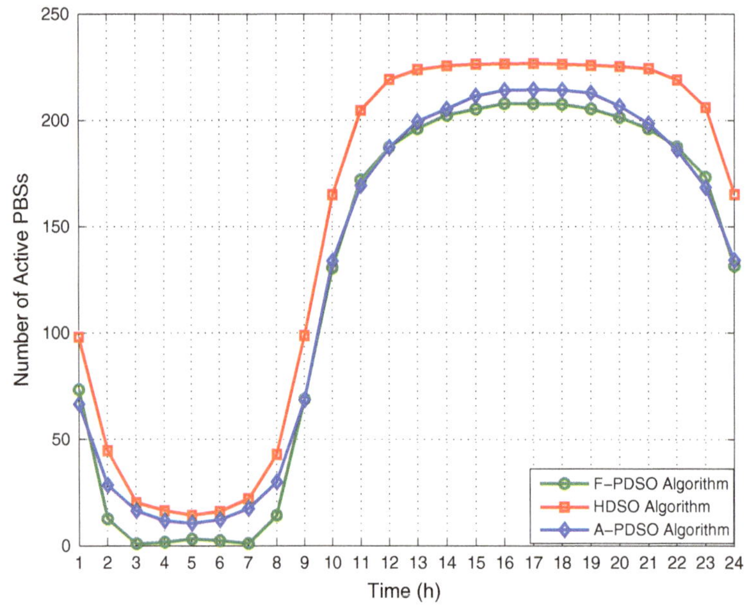

Figure 4 The number of active PBSs versus time, $\varepsilon_m = 0.6$, $\varepsilon_p = 0.9$, $\varphi = 13°$, $T = 3$ min.

algorithms. The F-PDSO algorithm can switch off averagely 25 more PBSs than the HDSO algorithm, while having a slight impact on the network blocking probability which can be observed from Figure 5. The more significant energy-saving effect is explained through the following analysis. Since the F-PDSO algorithm classifies the current active PBSs into two different groups and handles each group on its own characteristics, which guarantees that PBSs with similar characteristics can be handled (i.e., switched on or off) fairly, and gives more opportunity to some PBSs with relatively small utility to be switched off. However, the HDSO algorithm only tests PBSs one by one, there can be close-by PBSs all with low load, some PBSs act as acceptor PBSs to take over the transferred UEs which can be far from the respective PBSs and occupy more RBs than before. This phenomenon can increase the originally low-loaded acceptors' load and make them unable to be switched off.

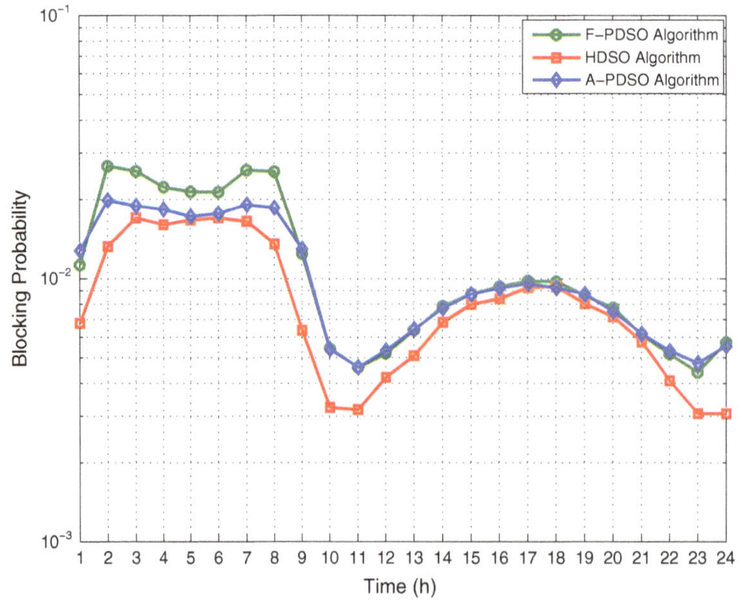

Figure 5 The blocking probability versus time, $\varepsilon_m = 0.6$, $\varepsilon_p = 0.9$, $\varphi = 13°$, $T = 3$ min.

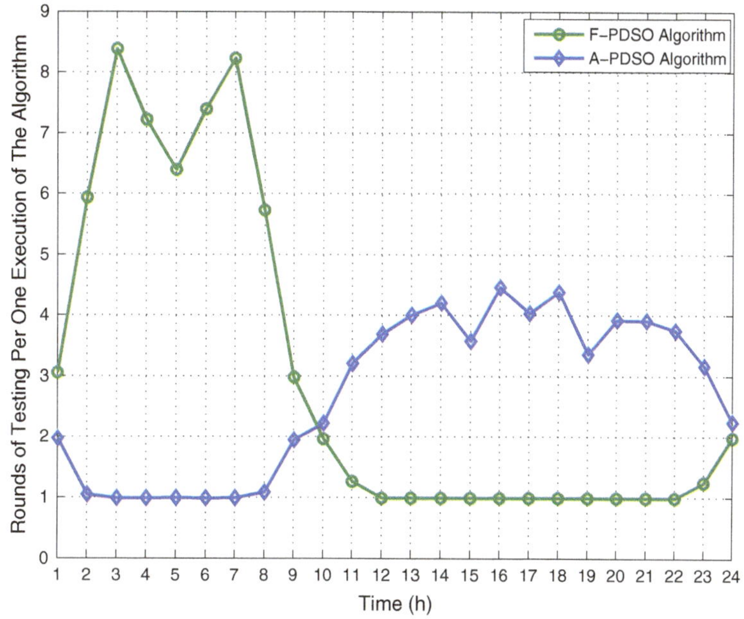

Figure 6 The rounds of testing per one execution of the proposed algorithms versus time, $\varepsilon_m = 0.6$, $\varepsilon_p = 0.9$, $\varphi = 13°$, $T = 3$ min.

Besides, the A-PDSO algorithm can keep more PBSs active than the fixed version during the bottom times and peak times. In order to explain this, we present the rounds of testing per one execution of the algorithms in Figure 6. According to the simulation result of F-PDSO algorithm, the percentages of switched-off PBSs during these two times are about 99% and 9%, respectively. In the first case, the A-PDSO tests 99% of the current active PBSs in one round aggressively, while the result of switching off may not be that enough to launch another round of testing (i.e., the termination condition of the A-PDSO algorithm can be more easily met, compared to that of F-PDSO). This is verified in Figure 6, as we can observe that during low traffic times, the A-PDSO only needs one round of testing. Therefore, more PBSs are left active at the end of the A-PDSO algorithm. However, the F-PDSO algorithm launches about seven rounds of testing to further switch off more PBSs. In the second case, the A-PDSO algorithm tests 9% of the current active PBSs in one round of testing conservatively. From Figure 6, we can see that about four rounds of testing are executed; hence, the A-PDSO algorithm can test at most 40% of the total PBSs. However, the F-PDSO algorithm just launches one round of testing and test 50% of the total PBSs, which is prone to switch off more PBSs. Therefore, as the process of the A-PDSO algorithm moves on, one round of testing only considers a small percentage of the current active PBSs, which gives more opportunity for the remaining PBSs to keep active. Now we can conclude that in extreme cases of the network traffic (i.e., bottom times and peak times), the A-PDSO algorithm is prone to keep more PBSs active so as to guarantee the QoS of the network. Moreover, the thresholds and the percentages used in PDSO have a significant influence on the performance of the algorithm.

In Figure 5, from 5:00 to 17:00, the same variation trend of blocking probability can be observed about the proposed algorithms, and it is not monotonous with the traffic variation. The average blocking probability is around 1%, which can satisfy the blocking probability constrain of general communication service. It can be observed that from 5:00 to 10:00, the blocking probability is decreasing with the incremental traffic, and the rationale can be explained as follows. When the traffic increases from the trough to the moderate intensity, more PBSs need to be active to provide service; thus, the resource is sufficient to serve the UEs, and the probability that an UE cannot find an appropriate BS to serve is relatively small. Therefore, the overall blocking probability is decreasing. But when the traffic continues to increase to the peak, the blocking probability is increasing contrarily even with more active PBSs. This is due to that the interference among UEs is becoming significant with more UEs served in the network, which results in decreasing the UEs' spectrum efficiency and makes BSs provide more RBs to UEs to satisfy their rate requirement. However, the frequency resource is limited, some newly arriving UEs needing more RBs can be blocked, and the increasing active PBSs cannot compensate the decreasing QoS of the network due to the decreasing UEs' spectrum efficiency. Hence, the blocking probability is increasing with the traffic curve until the peak. The explanation to the variation trend of the other two periods is the same as above owing to the

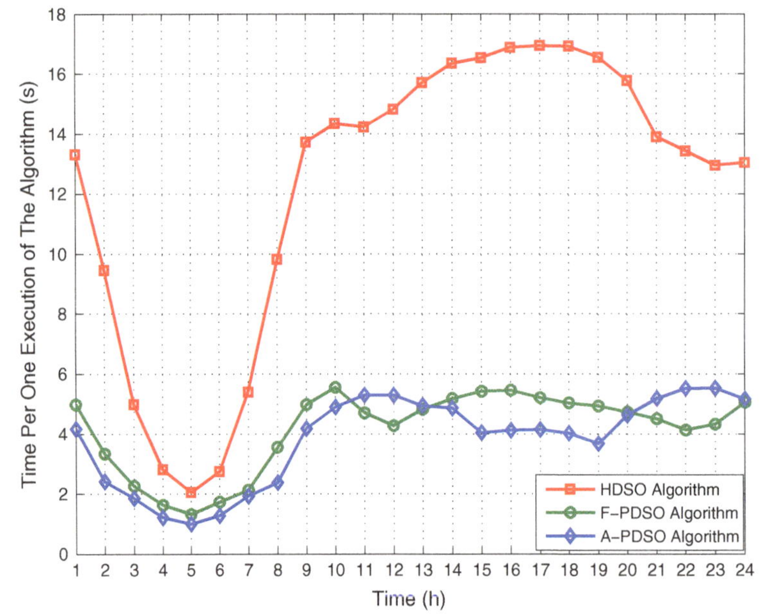

Figure 7 The time per one execution of the proposed algorithms versus time, $\varepsilon_m = 0.6$, $\varepsilon_p = 0.9$, $\varphi = 13°$, $T = 3$ min.

symmetrical variation of the sinusoidal-like traffic curve. Moreover, the average blocking probability during the peak traffic is lower than that during low traffic due to the densification of active PBSs, which provides sufficient network capacity. However, during low traffic, some UEs enter the macrocells' edge where no PBS exists cannot obtain the minimum service rate, and thus become easily blocked.

Figure 7 depicts the remarkable difference of the time per one execution of the proposed algorithms. Averagely, the PDSO algorithm takes at most one third time of the HDSO algorithm, which indicates that PDSO can reduce the time for the CCU making decision to a great extent and may also involve less network signaling overhead. Therefore, PDSO can quickly reconfigure the working mode of all PBSs to accommodate the traffic. Since

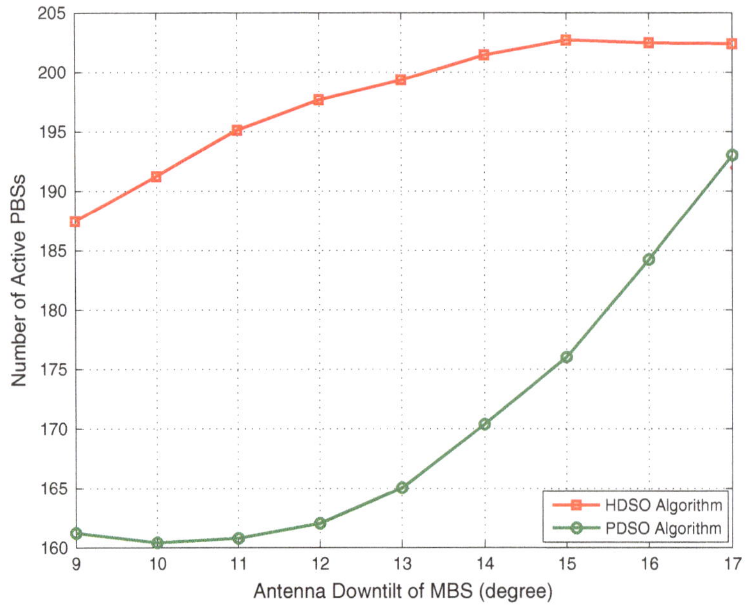

Figure 8 The number of active PBSs versus antenna downtilt of the MBS, $\varepsilon_m = 0.6$, $\varepsilon_p = 0.9$, $\lambda(t) = 10$, $T = 3$ min.

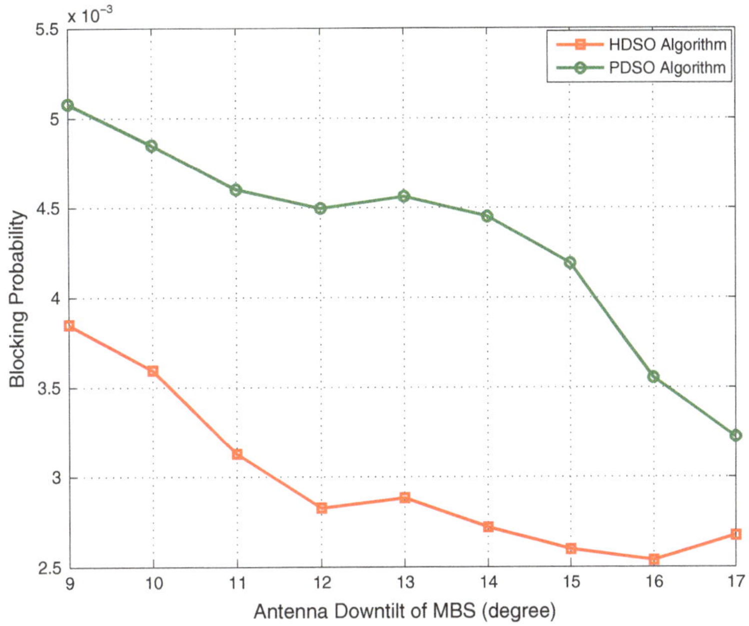

Figure 9 The blocking probability versus antenna downtilt of the MBS, $\varepsilon_{\mathrm{m}} = 0.6$, $\varepsilon_{\mathrm{p}} = 0.9$, $\lambda(t) = 10$, $T = 3$ min.

the PDSO algorithm avoids the unnecessary traversal of some PBSs and can adaptively decide when to stop the switching off process, a large amount of time can thus be saved. The phenomenon that the execution time of A-PDSO is slightly less than that of F-PDSO during most time reveals the adaptive version of PDSO algorithm can further quicken the decision-making of the CCU. As most PBSs without any association or being lightly loaded can be quickly judged whether they can be switched off or not during trough traffic, the difference between the PDSO and HDSO becomes the least. Note that the execution time is a simulation result and can vary among different simulation environments; however, the relative difference between the proposed two algorithms

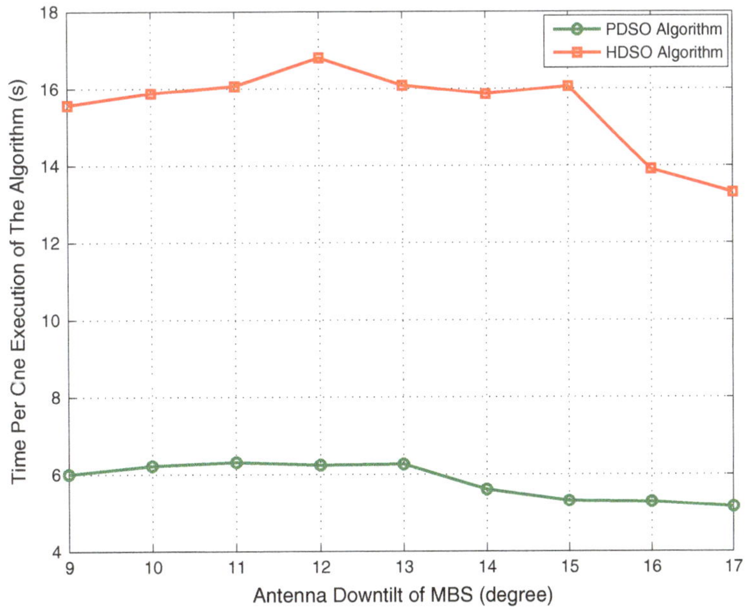

Figure 10 The time per one execution of the proposed algorithms versus antenna downtilt of the MBS, $\varepsilon_{\mathrm{m}} = 0.6$, $\varepsilon_{\mathrm{p}} = 0.9$, $\lambda(t) = 10$, $T = 3$ min.

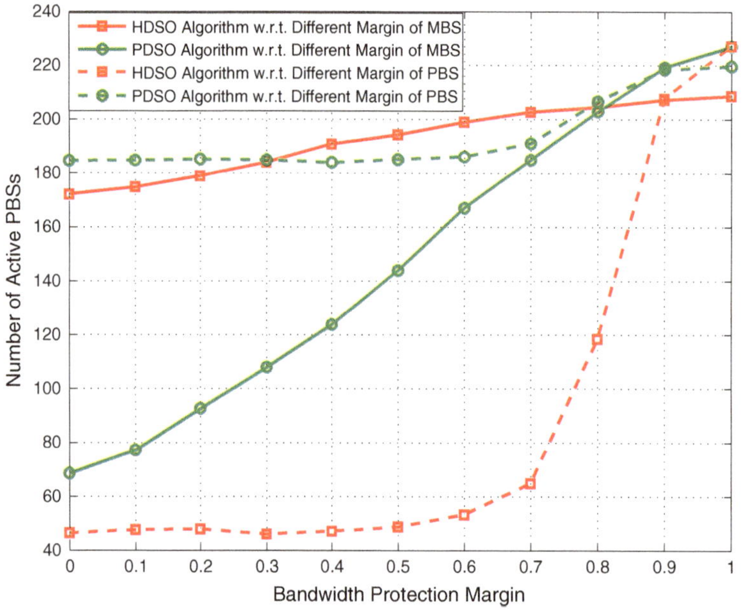

Figure 11 The number of active PBSs versus bandwidth protection margin, $\lambda(t) = 10, \varphi = 13°, T = 3$ min.

remains the same when they are simulated in the same environment.

Figures 8, 9, and 10 show the impact of antenna downtilt of the MBS on the performance of the proposed algorithms. It can be observed that the number of active PBSs increases with the downtilt, which is due to that when the MBSs reduce their coverage area, the emerging coverage holes need to be compensated by the nearby PBSs to provide constant QoS of the network. These PBSs' load increases, therefore they have less opportunity to be switched off. The difference should be noted is that the impact on the PDSO algorithm is relatively more significant than the HDSO algorithm, indicating that an appropriate smaller downtilt can lead to a better performance enhancement of PDSO algorithm compared to HDSO algorithm. Figure 9 shows the influence of the MBS antenna downtilt on the network blocking probability. From Figure 10, we can also observe that PDSO

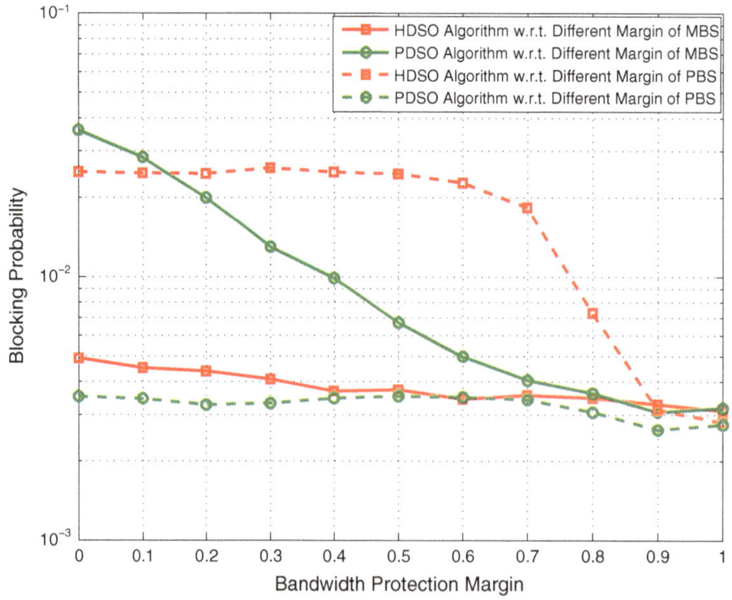

Figure 12 The blocking probability versus bandwidth protection margin, $\lambda(t) = 10, \varphi = 13°, T = 3$ min.

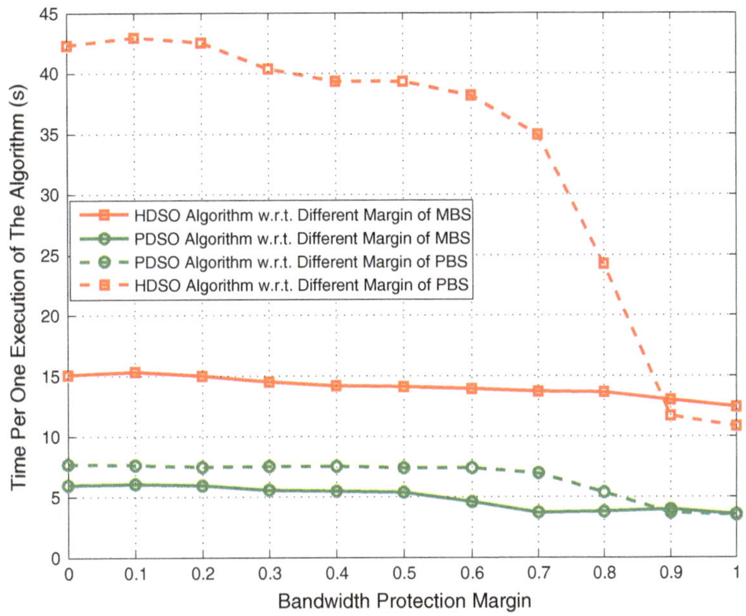

Figure 13 The time per one execution of the proposed algorithms versus bandwidth protection margin, $\lambda(t) = 10, \varphi = 13°, T = 3$ min.

algorithm exceeds HDSO algorithm in terms of algorithm complexity and that the downtilt has indistinctive impact on the execution time of each algorithm. As the downtilt increases, more PBSs are determined that their PUEs cannot be accepted by MBSs, which accelerates the execution of the two algorithms to some degree.

Figures 11, 12, and 13 show the impact of bandwidth protection margin of MBS and PBS on the performance of the proposed algorithms, respectively. As for the margin of MBS, the number of active PBSs of the two proposed algorithms both increases with the margin, and this is obvious. When the margin is low, the PDSO algorithm can switch off more PBSs and show more potential of saving energy while resulting in slight QoS deterioration observed from Figure 12. When $\varepsilon_{ma} \geq 0.4$, the PDSO algorithm can achieve the target that the blocking probability is maintained below 1%. In Figure 11, as the margin increases, the energy-saving difference

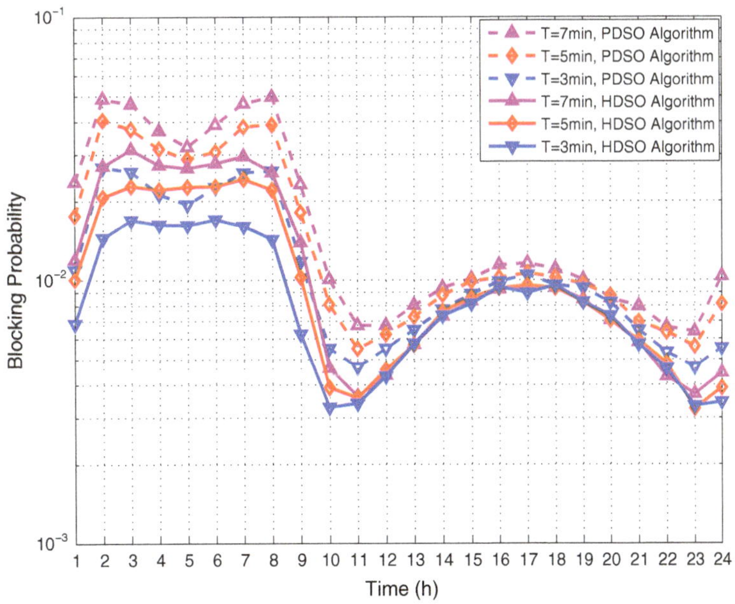

Figure 14 The blocking probability of the proposed algorithms under different time interval $T, \varepsilon_m = 0.6, \varepsilon_p = 0.9, \varphi = 13°$.

between the two algorithms reduces, and in extreme case, the energy-saving effects of the two algorithms become approximately the same. As for the margin of PBS, similar trends can be observed. However, the PDSO algorithm is more robust than the HDSO algorithm, and in extreme case, the effects of the two algorithms become approximately the same. An interesting phenomenon is that when $\varepsilon_{\mathrm{pi}} < 0.9$, the HDSO algorithm can switch off more PBSs than the PDSO algorithm. This is due to the termination condition of PDSO algorithm which is $\mathrm{off}_{\mathrm{count}}/N < \delta$. For the minimum distance restriction among PBSs, the PUEs of the switched-off PBSs are more likely transfered to the nearby MBSs than the nearby PBSs. However,the available RBs that MBSs can provide are limited, hence, it's likely that the PUEs of the switched-off PBSs can't be transferred, and these PBSs need to keep active. The termination condition can be easily met as we set δ=0.5, and the PDSO algorithm terminates to test the remaining PBSs which may have the potential to be switched off. However, the HDSO algorithm tests all PBSs which exploit the full potential of energy saving, which is at the expense of needing much more execution time. When $\varepsilon_{\mathrm{pi}} \geq 0.9$, the two algorithms show similar performance. The blocking probability is greatly related to the number of active PBSs, which can be observed from Figure 12. Figure 13 presents the time per one execution of the proposed algorithms. It can be observed that the time is decreasing with all bandwidth protection margins to some degree. Specifically, the time of the HDSO algorithm is much more sensitive to a different margin of PBS than the other three cases. From these results, we can see that the bandwidth protection margins have significant impacts on the performance of the proposed algorithms and should be tuned well in realistic scenarios.

Figure 14 shows the blocking probability of our proposed algorithms under different time interval T. We can see that the blocking probability varies in the same trend under different time intervals but increases with increasing time interval. The difference among different intervals is significant in the night zone of traffic profile, while unobvious during the peak traffic time. This is because most of the PBSs are switched off during the night zone leading to inadequate resource. During peak traffic time, what is important is that this result indicates that the sufficient resource for UEs benefiting from the densification of active PBSs provides more time for some PBSs to be switched off, while has slight influence on the blocking probability. The result implies that the proposed algorithms can track the variation of the traffic better when shortening the time interval of carrying out the algorithms, which yet increases the overhead of signaling due to the frequent reconfiguration. Hence, a tradeoff can again be made between the network signaling overhead and the QoS.

5 Conclusions

In this paper, two PBS switching off algorithms have been proposed to dynamically reconfigure PBSs' working mode for energy saving in Hetnets. With our algorithms, network energy consumption can be significantly reduced. The simulation results show that the number of active PBSs tracks the variation of traffic well while ensuring an acceptable blocking probability. Particularly, the PDSO algorithm can lead to a better effect of energy saving and slightly deteriorates the blocking probability. What is important is that the complexity of the PDSO algorithm is much lower than that of the HDSO algorithm, enabling the network respond to the traffic variation more promptly, which is demonstrated by the simulations that the execution time of the PDSO algorithm is at most one third of the HDSO algorithm. Additionally, simulations imply that the antenna downtilt of the MBS, the bandwidth protection margin of MBS and PBS, and the time interval T all have nonnegligible influence on the performance of our algorithms. As the downtilt increases, more PBSs remain active to guarantee the QoS of the network, which accelerates the execution of the algorithms to some degree. In terms of the bandwidth protection margin, the performance of the PDSO algorithm is more sensitive to the margin of MBS, and the performance of the HDSO algorithm is more sensitive to the margin of PBS, compared to the other two cases. The number of active PBSs of the algorithms increases with the two margins due to less available capacity to accept the transferred PUEs. The blocking probability increases with the time interval more significantly during low traffic times, indicating that the active PBSs set should change frequently in these times. The unobvious difference during peak traffic times indicates that sufficient resource for UEs benefiting from the densification of active PBSs provides more time for some PBSs to be switched off. Hence there exist tradeoffs that can be made to achieve a preferable system performance and energy saving.

Competing interests

The authors declare that they have no competing interests.

Acknowledgements

This work was supported by the National Natural Science Foundation of China under Grant 61222102, the National Basic Research Program of China (973 Program 2012CB316004), Research Project of Jiangsu Province under Grant BK2012021, the International Science & Technology Cooperation Program of China under Grant 2014DFT10300, the National Science and Technology Major Project of China under Grant 2013ZX03001032-004, and NEC Research Fund.

Author details

[1]School of Information Science and Engineering, Southeast University, No. 2 Sipailou, Nanjing 210096, China. [2]NEC Laboratories China, 11F, Bldg A, Innovation Plaza, Tsinghua Science Park, No. 1 Zhongguancun East Road, Haidian District, Beijing 100084, China.

References

1. Cisco: Visual networking index. White paper at Cisco.com (2014)
2. A Damnjanovic, J Montojo, Y Wei, T Ji, T Luo, M Vajapeyam, T Yoo, O Song, D Malladi, A survey on 3GPP heterogeneous networks. IEEE Wireless Commun. **18**(3), 10–21 (2011)
3. V Chandrasekhar, JG Andrews, A Gatherer, Femtocell networks: a survey. IEEE Commun. Mag. **46**(9), 59–67 (2008)
4. M Yavuz, F Meshkati, S Nanda, A Pokhariyal, N Johnson, B Raghothaman, A Richardson, Interference management and performance analysis of UMTS/HSPA+ femtocells. IEEE Commun. Mag. **47**(9), 102–109 (2009)
5. S-M Cheng, S-Y Lien, F-S Chu, K-C Chen, On exploiting cognitive radio to mitigate interference in macro/femto heterogeneous networks. IEEE Wireless Commun. **18**(3), 40–47 (2011)
6. D López-Pérez, I Guvenc, G De La Roche, M Kountouris, TQ Quek, J Zhang, Enhanced intercell interference coordination challenges in heterogeneous networks. IEEE Wireless Commun. **18**(3), 22–30 (2011)
7. J Sangiamwong, Y Saito, N Miki, T Abe, S Nagata, Y Okumura, in *11th European Wireless Conference 2011-Sustainable Wireless Technologies (European Wireless)*. Investigation on cell selection methods associated with inter-cell interference coordination in heterogeneous networks for LTE-advanced downlink (VDE, 2011)
8. R Madan, J Borran, A Sampath, N Bhushan, A Khandekar, T Ji, Cell association and interference coordination in heterogeneous LTE-A cellular networks. IEEE J. Selected Areas Commun. **28**(9), 1479–1489 (2010)
9. G Rittenhouse, S Goyal, DT Neilson, S Samuel, in *IEEE Technical Symposium at ITU Telecom World (ITU TW)*. Sustainable telecommunications, (2011)
10. F Richter, AJ Fehske, GP Fettweis, in *IEEE 70th Vehicular Technology Conference Fall (VTC 2009-Fall)*. Energy efficiency aspects of base station deployment strategies for cellular networks, (2009)
11. E Oh, B Krishnamachari, X Liu, Z Niu, Toward dynamic energy-efficient operation of cellular network infrastructure. IEEE Commun. Mag. **49**(6), 56–61 (2011)
12. I Ashraf, F Boccardi, L Ho, Sleep mode techniques for small cell deployments. IEEE Commun. Mag. **49**(8), 72–79 (2011)
13. J Gong, Z Niu, A dynamic programming approach for base station sleeping in cellular networks. IEICE Trans. Commun. **95**(2), 551–562 (2012)
14. S Cai, L Xiao, H Yang, J Wang, S Zhou, in *22nd Wireless and Optical Communication Conference (WOCC)*. A cross-layer optimization of the joint macro- and picocell deployment with sleep mode for green communications, (2013)
15. Y Shi, J Zhang, KB Letaief, Group sparse beamforming for green Cloud-RAN. IEEE Trans. Wireless Commun. **13**(5), 2809–2823 (2014)
16. A Bousia, E Kartsakli, L Alonso, C Verikoukis, in *IEEE Global Communications Conference (GLOBECOM)*. Dynamic energy efficient distance-aware base station switch on/off scheme for LTE-advanced, (2012)
17. MA Marsan, L Chiaraviglio, D Ciullo, M Meo, in *IEEE 4th International Conference on Communications and Electronics (ICCE)*. Multiple daily base station switch-offs in cellular networks, (2012)
18. H Claussen, I Ashraf, LT Ho, Dynamic idle mode procedures for femtocells. Bell Labs Tech. J. **15**(2), 95–116 (2010)
19. L Falconetti, P Frenger, H Kallin, T Rimhagen, in *IEEE Online Conference on Green Communications (GreenCom)*. Energy efficiency in heterogeneous networks, (2012)
20. AA Abdulkafi, TS Kiong, D Chieng, A Ting, J Koh, Energy efficiency improvements in heterogeneous network through traffic load balancing and sleep mode mechanisms. Wireless Pers. Commun. **75**(4), 2151–2164 (2014)
21. O Blume, H Eckhardt, S Klein, E Kuehn, WM Wajda, Energy savings in mobile networks based on adaptation to traffic statistics. Bell Labs Tech. J. **15**(2), 77–94 (2010)
22. MF Hossain, KS Munasinghe, A Jamalipour, in *IEEE Wireless Communications and Networking Conference (WCNC)*. An eco-inspired energy efficient access network architecture for next generation cellular systems, (2011)
23. G Cili, H Yanikomeroglu, FR Yu, in *IEEE International Conference on Communications (ICC)*. Cell switch off technique combined with coordinated multi-point (CoMP) transmission for energy efficiency in beyond-LTE cellular networks, (2012)
24. A Bousia, E Kartsakli, A Antonopoulos, L Alonso, C Verikoukis, in *IEEE International Conference on Communications (ICC)*. Game theoretic approach for switching off base stations in multi-operator environments, (2013)
25. E Oh, K Son, B Krishnamachari, Dynamic base station switching-on/off strategies for green cellular networks. IEEE Trans. Wireless Commun. **12**(5), 2126–2136 (2013)
26. A Conte, A Feki, L Chiaraviglio, D Ciullo, M Meo, MA Marsan, Cell wilting and blossoming for energy efficiency. IEEE Wireless Commun. **18**(5), 50–57 (2011)
27. MA Marsan, M Meos, in *Proc. of GreenMetrics Workshop in Conjunction with ACM SIGMETRICS Conf.* Energy efficient management of two cellular access networks, (2009)
28. JD Little, A proof for the queuing formula: $L = \lambda$ w. Oper. Res. **9**(3), 383–387 (1961)
29. D López-Pérez, X Chu, I Guvenc, On the expanded region of picocells in heterogeneous networks. IEEE J. Selected Topics Signal Process. **6**(3), 281–294 (2012)
30. D López-Pérez, X Chu, in *IEEE 20th International Conference on Computer Communications and Networks (ICCCN)*. Inter-cell interference coordination for expanded region picocells in heterogeneous networks, (2011)
31. Y Li, X Xu, J Han, W Liu, in *IEEE 78th Vehicular Technology Conference (VTC Fall)*. Coordinated inter-cell interference management for expanded region picocells in heterogeneous network, (2013)
32. W Noh, W Shin, C Shin, K Jang, H-h Choi, in *IEEE Wireless Communications and Networking Conference (WCNC)*. Distributed frequency resource control for intercell interference control in heterogeneous networks, (2012)
33. MW Arshad, A Vastberg, T Edler, in *IEEE Wireless Communications and Networking Conference (WCNC)*. Energy efficiency improvement through pico base stations for a green field operator, (2012)
34. E Yaacoub, in *19th International Conference on Telecommunications (ICT)*. Performance study of the implementation of green communications in LTE networks, (2012)
35. Evolved universal terrestrial radio access (E-UTRA); Further advancements for E-UTRA physical layer aspects. in *3GPP Tech. Rep., http://www.3gpp.org/DynaReport/36814.htm?device=desktop*, (2010)

5

Femtocell power control methods based on users' context information in two-tier heterogeneous networks

Reben Kurda[1*], Lila Boukhatem[1*] and Megumi Kaneko[2*]

Abstract

In this paper, we consider a two-tier macrocell/ femtocell overlaid heterogeneous network based on orthogonal frequency division multiple access (OFDMA) technology. Although the co-channel spectrum allocation provides larger bandwidth for both macrocell and femtocells, the resulting cross-tier interference may prevent macrocell users in the vicinity of femtocells to achieve their minimum required signal-to-interference plus noise ratio (SINR) in downlink. Therefore, we propose femtocell power control strategies for mitigating the interference experienced by macrocell users while preventing the femtocell throughput degradation. In particular, the proposed power control schemes make use of femto and macro users' context information in terms of positioning for setting the appropriate prioritization weights among the current victim macro users and the femto users in outage. System-level simulations show that our schemes enhance the throughput of macrocell users while maintaining a high performance for femtocell users compared to a conventional power allocation. Moreover, we show that the proposed prioritization weights allow to achieve the required level of macrocell/femtocell throughput trade-off.

Keywords: Heterogeneous networks; Macrocell; Femtocell; Power control; Interference management

1 Introduction

One of the main goals of the 4G long-term evolution advanced (LTE-A) system is to enhance the capacity of mobile radio networks and to improve the quality of service (QoS) levels experienced by prospective users [1]. Towards this end, LTE-A introduces the deployment of heterogeneous networks (HetNets), whereby low power nodes such as femtocell base stations (FBS)s are overlaid within a large macrocell served by a macrocell base station (MBS). Generally, FBSs provide several advantages such as offering high performance to their users due to the proximity between transmitters and receivers, and all the benefits of a backhaul connection (such as fiber optic or digital subscriber line DSL) which can be used to connect to the operator's macrocell. Moreover, FBSs, which are mainly deployed indoor, provide capacity offload to outdoor users or macrocell user equipments (MUEs), and some studies foresee that in future communication scenarios, almost 50% of voice and 70% of data traffic are likely to be generated indoors [2]. However, the cross-tier interference from the transmitting FBSs becomes a crucial issue for MUEs in the vicinity of femtocell coverage areas, as they may observe a severe degradation of their signal-to-interference plus noise ratio (SINR) level [3].

Several research works have addressed the cross-tier interference problem in two-tier HetNets based on orthogonal frequency division multiple access (OFDMA) transmission. This major issue can be addressed through adapted resource block (RB) allocation schemes at the scheduling level and/or evolved time, frequency, or power control techniques [1]. All these approaches propose to mitigate the cross-tier-interference on victim MUEs and can be classified as centralized, assistance-based, or decentralized solutions.

The centralized approaches rely on a central entity and a global knowledge of the parameters related to the different link gains whose transmission may induce extensive

*Correspondence: Reben.Kurda@lri.fr; Lila.Boukhatem@lri.fr; meg@i.kyoto-u.ac.jp
[1] Laboratoire de Recherche en Informatique (LRI), University of Paris-Sud 11 - CNRS, 15, rue Georges Clémenceau, 91405 Orsay Cedex, France
[2] Graduate School of Informatics, Kyoto University, Yoshida-Honmachi, Sakyo-ku, Kyoto 606–8501, Japan

signaling and generate long delays. In the assistance-based solutions, the MBS supports the FBSs in making their resource usage decisions using some global information and/or the reported measurements from MUEs/FUEs. In this category, the proposed power control solutions focus on adjusting dynamically the transmission power of FBSs given the evolving interference situation. In [4], the interfered victim MUEs report their reference signal received power (RSRP) with the corresponding cell's physical identity and channel quality indicator (CQI) to the macrocell who identifies the interfering (aggressor) femtocells and informs them about the interference situation. By adjusting the FBSs' transmit power, macrocells could ensure a low interference to their MUEs at the expense of a throughput degradation of femtocell users.

In [5], we proposed an MBS-assisted power adjustment strategy for interference mitigation in a mobility environment where macrocells and femtocells are co-located. The key benefits of this scheme were to achieve higher performance for nonclosed-subscriber group (CSG) macrocell users, as well as avoiding potential service degradation of femtocell users by taking into account mobile user locations and SINR at both macro- and femtocell sides. An extended approach was proposed in [6] to define two power adjustment strategies with different degrees of awareness of femtocell users' performance.

Another set of approaches consider the concept of clustering of femtocells for the power adjustment. In [7], the femtocell's initial power setting is correlated to the number of active femtocells in the cluster. Both MBS-assisted and distributed sensing are proposed to derive the number of active small cells.

Other assistance-based approaches address the interference issue from a frequency dimension basis. In [8], the frequency resources of the macrocell are reused by the femtocells while relying on an adapted spectrum scheduling. The key idea of the proposed scheme is that femtocells avoid using the same frequency resources of the nearby MUEs according to the sensing process performed independently by them and the scheduling scheme carried out by the MBS.

An intra-cell handover (IHO) scheme applied to both macrocell and femtocells is proposed in [9] to mitigate the cross-tier interference. The scheme relies on a received signal strength -based power control mechanism when the IHO scheme fails in getting free sub-channels in the interfering femtocell. [10] proposed a hybrid resource coordination for a mixed macrocell and femtocell deployment, where the MBS can notify the femtocell to release the interfered allocated resources relieving the victim MUE from interference. Even if the proposals in [9] and [10] improve the performance of MUEs, they cause nonnegligible throughput deteriorations to FUEs. It is shown that a proper combination of IHO and power control techniques reduces the outage probability for nonsubscribers compared to that of closed and open access.

Another category of approaches consider decentralized solutions for interference mitigation in OFDMA femtocell deployment. Most of the solutions involve a joint optimization of the power allocation and the channel allocation or spectrum assignment. The authors in [11] proposed a self-organized framework of femtocells which autonomously perform the spectrum assignment and transmission power based on users' reported measurements. The study evaluated the gain in spectral efficiency and power consumption for the cases with and without macrocell interference.

In [12-15], decentralized power allocation approaches are proposed under the assumption of limited MUE channel information. In the neighbor-friendly scheme of [12], FBSs allocate the total adjusted power on sub-channels in order to maintain targeted FUEs' sum rates while mitigating the interference on MUEs. However, some assumptions about channel stability over several frames may not be valid especially in mobility scenarios. In [13], the effect of different macrocell sizes on the power control is analyzed when assuming femtocells' overhearing ability of MUEs' RSRP measurements. Downlink (DL) power allocation methods are proposed in [14,15], where no control information exchange is assumed between the MBS and FBSs. Instead, each FBS predicts the RB allocation to its close-by MUEs based on their overheard channel state information (CSI) feedback from MUEs to MBS. Based on this MUE allocation prediction, each FBS can perform resource allocation to FUEs while mitigating interference towards close-by MUEs. However, interference mitigation among FBSs is not considered.

Finally, [16] showed that the use of radio environment maps (REM) information increases the efficiency of practical power control mechanisms of femtocells. The authors showed that under unknown shadowing terms, the use of statistical information retrieved from the REM database leads to a performance close to the ideal case.

In this paper, we mainly focus on power control approaches that take benefit from the radio resource control (RRC) signaling messages, and especially the RSRP signal [5,6,15], to mitigate the downlink interference using an MBS-assisted power adjustment strategy. Our key objective is to improve the spectral efficiency of MUEs, while preserving a high throughput to FUEs. More specifically, we propose two context-based adjustment schemes whose common objective is to mitigate the interference on victim MUEs while implementing two different degrees of awareness of FUEs' throughput degradation. Our two adaptive power control schemes, called respectively global power adjustment (GPA) and selective power adjustment

(SPA), differ in their selection strategy of the set of FBSs executing the power adjustment as well as the amount of power to reduce. The main advantage of our proposed approach over traditional ones is that it is based on power adjustment parameters whose values are dynamically adapted to the interference impact of each femtocell on the global interference situation. To estimate such femtocell's liability on the global interference, we use a *Score Function* which quantifies this value based not only on the RSS feedback returned by MUEs but also on some context information (such as the number and location of MUEs/FUEs) retrieved from a REM database. Exploiting this context information about users' number and localization enriches the decision policy and brings added knowledge value to the power allocation decision process. Finally, we discuss and analyze the effects of the Score Function weighting parameters on the global system performance in terms of macrocell and femtocell throughput.

The remainder of this paper is organized as follows. Section 2 describes the system model and main assumptions. Section 3 presents the proposed power adjustment schemes based on both users' context factors and dynamic power adjustment parameters. In Section 4, the simulation environment and results are presented and discussed. Finally, Section 5 concludes the paper.

2 System model and assumptions

We consider a macrocell/femtocell overlay system in which L femtocells are geographically positioned in streets located at the edge of the coverage area of one MBS referred as M (see Figure 1). We assume that I MUEs are uniformly distributed within the MBS coverage area and are not allowed to access a neighboring femtocell F_l, ($l = 1, \ldots, L$) when they approach its coverage area. In other words, all the outdoor users are considered non-CSG users for F_l's operator. Each F_l serves J FUEs uniformly distributed in its coverage area. The transmissions are OFDMA-based and all MUEs and FUEs share the same frequency band divided in K RBs. We also assume that MBS M is fully loaded and cannot execute any intra-channel allocation to mitigate the potential interference caused to MUEs. To perform the resource allocation of their RBs, both MBS and FBSs require their user equipments (UEs) to periodically report their signal quality expressed by the CQI in terms of SINR. The received DL SINR of a MUE_i associated with MBS M on RB k ($k = 1, \ldots, K$) can be expressed as

$$\text{SINR}^k_{M,\text{MUE}_i} = \frac{\frac{p^k_M}{\text{PL}_{M,\text{MUE}_i}} \left| G^k_{M,\text{MUE}_i} \right|^2}{\sum_{F_l \in \mathcal{Z}_{i,k}} \frac{p^k_{F_l}}{PL_{F_l,\text{MUE}_i}} \left| G^k_{F_l,\text{MUE}_i} \right|^2 + N_0}, \quad (1)$$

where p^k_M is the current transmit power allocated on RB k by the serving cell MBS M and $\left| G^k_{M,\text{MUE}_i} \right|$ is the channel fast fading gain between MBS M and MUE_i on RB k. Similarly, $p^k_{F_l}$ is the transmit power of neighboring femtocell F_l on RB k, while $\mathcal{Z}_{i,k}$ represents the set of all interfering FBSs on user MUE_i on RB k. Similarly, $\left| G^k_{F_l,\text{MUE}_i} \right|$ is the channel fast fading gain between MUE_i and FBS F_l on RB k. $\text{PL}_{M,\text{MUE}_i}$ and $\text{PL}_{F_l,\text{MUE}_i}$ represent the pathloss from MUE_i's serving cell MBS M and FBS F_l respectively, and N_0 is the power of the additive white Gaussian noise.

Similarly, the DL $\text{SINR}^k_{F_l,\text{FUE}_j}$ of a given FUE_j associated with a femtocell F_l on RB k can be expressed as:

$$\text{SINR}^k_{F_l,\text{FUE}_j} = \frac{\frac{p^k_{F_l}}{\text{PL}_{F_l,\text{FUE}_j}} \left| G^k_{F_l,\text{FUE}_j} \right|^2}{N_0 + A^k_j + \sum_{\substack{F_m \in \mathcal{V}_{j,k} \\ m \neq l}} \frac{p^k_{F_m}}{\text{PL}_{F_m,\text{FUE}_j}} \left| G^k_{F_m,\text{FUE}_j} \right|^2}, \quad (2)$$

where $p^k_{F_l}$ is the current transmit power allocated on RB k by the serving femtocell F_l and $|G^k_{F_l,\text{FUE}_j}|$ is the channel fast fading gain between FUE_j and its serving femtocell F_l. $p^k_{F_m}$ is the transmit power of a neighboring FBS F_m, while $\mathcal{V}_{j,k}$ represents the set of all interfering FBSs on user FUE_j on RB k. $|G^k_{F_m,\text{FUE}_j}|$ is the channel fast fading gain between FUE_j and its neighboring FBS F_m. $\text{PL}_{F_l,\text{FUE}_j}$ stands for the pathloss between FUE_j and FBS F_l while A^k_j is the interference generated by MBS M on resource block k, namely,

$$A^k_j = \frac{p^k_M}{\text{PL}_{M,\text{FUE}_j}} \left| G^k_{M,\text{FUE}_j} \right|^2. \quad (3)$$

For our system model needs, we define the following parameters:

- $\text{SINR}^{\text{target}}_{\text{MUE}}$: the predefined SINR requirement threshold of each MUE_i and each allocated RB k.
- $\text{SINR}^{\text{target}}_{\text{FUE}}$: the predefined SINR requirement threshold for each FUE_j and each allocated RB k.
- $\text{RSS}^k_{M,\text{MUE}_i}$ and $\text{RSS}^k_{F_l,\text{MUE}_i}$: the received signal strength by MUE_i from MBS M and FBS F_l on RB k, respectively [17]. They are expressed as:

$$\text{RSS}^k_{M,\text{MUE}_i} = \frac{p^k_M}{\text{PL}_{M,\text{MUE}_i}}, \quad (4)$$

$$\text{RSS}^k_{F_l,\text{MUE}_i} = \frac{p^k_{F_l}}{\text{PL}_{F_l,\text{MUE}_i}}. \quad (5)$$

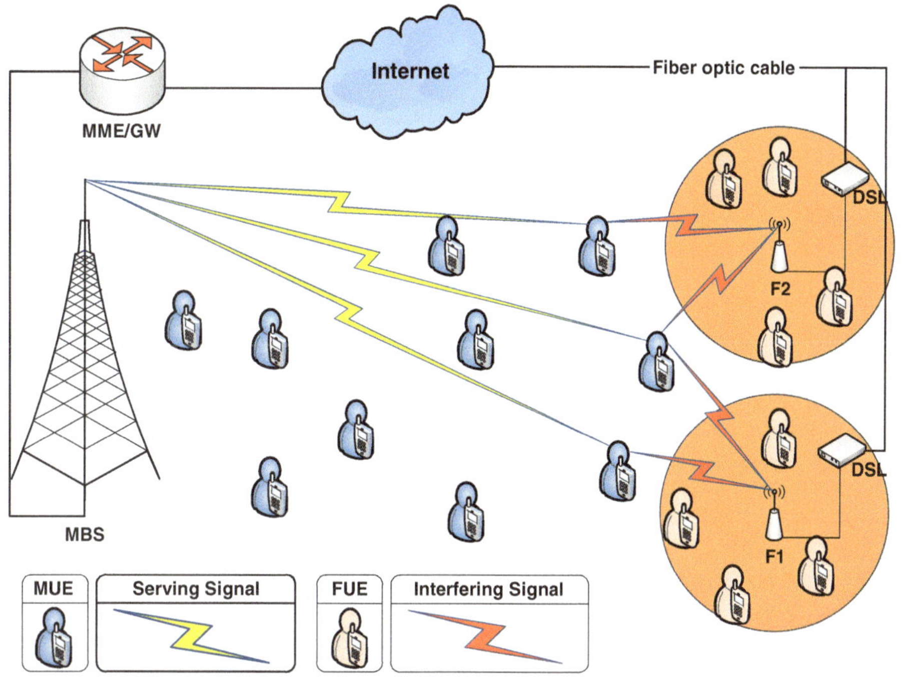

Figure 1 Two-tier heterogeneous network.

- ΔSINR: a positive protection margin, expressed in dBs, used together with $\text{SINR}_{\text{MUE}}^{\text{target}}$ to determine if MUE_i experiences a bad serving channel. This is confirmed if the following condition is verified:

$$\text{SINR}_{M,\text{MUE}_i}^{k} < \text{SINR}_{\text{MUE}}^{\text{target}} + \Delta\text{SINR}. \quad (6)$$

- ΔP: a positive power protection margin, measured in dBs and used with the reported $\text{RSS}_{F_l,\text{MUE}_i}^{k}$ to check whether $\text{RSS}_{M,\text{MUE}_i}^{k}$ from the serving MBS M is indicating a pending connection loss. This is verified if the following condition is true:

$$\text{RSS}_{M,\text{MUE}_i}^{k} < \text{RSS}_{F_l,\text{MUE}_i}^{k} + \Delta P. \quad (7)$$

Note that ΔP and ΔSINR values should be adequately tuned to avoid a potential oscillatory behavior. For example, if ΔP is too low, MUE_i may suffer too long from the interfering F_l before MBS M initiates the interference mitigation strategy. Inversely, if ΔP is too high, MBS M activates the interference mitigation procedure even if the interference from neighboring F_l is insubstantial, which leads to an increase in the signaling overhead.

Moreover, in this paper, we take benefit from the REM concept as a support to our interference mitigation techniques. The term REM is generally used to refer to a database (local and/or global) which gathers location-dependent information about the environment where the radio system operates [18]. The REM concept is useful to develop optimized resource allocation and enhanced inter-cell interference coordination (eICIC) solutions that require to capture the evolution of random processes such as the channel quality, user mobility, and traffic dynamics. Therefore, we propose in this paper to make use of local REM databases at both MBS and femtocell sides to store local information. More specifically, the MBS REM registers all the RSSs recently fed back by the MUEs to inform about the interference condition caused by the neighboring FBSs on their respective allocated RBs when conditions (6) and (7) are fulfilled. Moreover, the context parameters (to be defined in the next section) will also be recorded in each local REM to help the power adjustment decisions. The REM databases are also used to store other less dynamic parameters such as $\text{SINR}_{\text{MUE}}^{\text{target}}$, $\text{SINR}_{\text{FUE}}{}^{\text{target}}$, ΔSINR and ΔP values.

3 Femtocell power adjustment problem

3.1 Formulation of the optimization problem

Since the femtocells are deployed in the macrocell coverage area with full frequency reuse, the transmission signal from FBSs will generate a strong interference to neighboring MUEs, causing a large degradation of their received $\text{SINR}^k_{M,\text{MUE}_i}$. The goal of the considered femtocell power adjustment is to maximize the sum of the macrocell and femtocells' throughputs, while satisfying simultaneously MUEs' and FUEs' SINR requirements.

Since it is assumed that all RBs are fully loaded, i.e., they are all allocated to MUEs beforehand, the considered throughput optimization problem can be formulated as:

$$\max_{p^k_{F_l}} \sum_{k=1}^{K} \sum_{F_l} \tau^k_{F_l}\left(p^k_{F_l}\right) + \tau^k_M\left(p^k_{F_l}\right), \tag{8}$$

$$\text{s.t.} \quad \text{SINR}^k_{M,\text{MUE}_i} \geq \text{SINR}^{\text{target}}_{\text{MUE}} + \Delta\text{SINR} \quad \forall i \in \{1,\ldots,I\}, \forall k \in \mathcal{K}_i,$$

$$\text{SINR}^k_{F_l,\text{FUE}_j} \geq \text{SINR}^{\text{target}}_{\text{FUE}} \quad \forall j \in \{1,\ldots,J\}, \forall l \in \{1,\ldots,L\}, \forall k \in \mathcal{K}_j,$$

$$p^k_{F_l} \geq 0 \quad \forall F_l \in \{1,\ldots,L\}, \forall k \in \{1,\ldots,K\}$$

$$\sum_{k=1}^{K} p^k_{F_l} \leq P^{\text{Max}}_{Fl} \quad \forall k \in \{1,\ldots,K\},$$

where $\mathcal{K}_i$ denotes the set of RBs allocated to MUE_i, $\mathcal{K}_j$ the set of RBs allocated to FUE_j, and $P^{\text{Max}}_{F_l}$ denotes the maximum transmit power of femtocell F_l. $\tau^k_{F_l}$ and τ^k_M are the throughput of the femtocell and macrocell user, respectively, in RB k, and which can be expressed as:

$$\tau^k_M(p^k_{F_l}) = \text{B}^k_M.\left(1 - \text{BLER}\left(\text{SINR}^k_{M,\text{MUE}_i}\left(p^k_{F_l}\right)\right)\right), \tag{9}$$

$$\tau^k_{F_l}\left(p^k_{F_l}\right) = \text{B}^k_{F_l}.\left(1 - \text{BLER}\left(\text{SINR}^k_{F_l,\text{FUE}_j}(p^k_{F_l})\right)\right), \tag{10}$$

$$\text{B}^k_u = \Omega.b^k_u.\frac{N_{\text{RB}}S_{\text{RB}}}{t_{\text{RB}}}, u \in \{M, F_l\}, \tag{11}$$

where Ω is a fixed parameter that depends on the network configuration, b^k_u is the number of bits/symbol depending on the modulation and coding scheme selected by MUE_i for $u = M$ and by FUE_j for $u = F_l$. N_{RB} and S_{RB} are the number of data subcarriers and symbols per RB, respectively. Finally, t_{RB} is the RB time duration [19].

Note that similar optimization problems were studied in [12,20] and [21]. While [20,21] considered only power minimization problem, maximization of the weighted sum of femtocell and macrocell throughput was investigated in [12]. However, only one femtocell was considered and no QoS constraints such as minimum SINR were introduced. Even for that simplified problem, it was shown that the optimization problem is nonconvex with respect to the femtocell power allocation parameters $p^k_{F_l}$, making it difficult to find the global optimum. Thus, the global optimization of our problem by a centralized approach may be even more difficult and intractable. Therefore, we propose a heuristic approach for semi-distributed femtocell power control to achieve a good balance between the macro- and femtocells' performances. Note that, as the optimal solution cannot be derived under our dynamic environment assumptions, we will derive the upper bounds of the macro- and femtocells' throughputs to get a better idea about the performance of our sub-optimal solution.

3.2 Proposed power control strategies

In our model, various MUEs are randomly positioned within MBS M's coverage and each active MUE_i periodically sends to MBS M a CQI in terms of $\text{SINR}^k_{M,\text{MUE}_i}$ as expressed in Equation 1 and experienced on each allocated RB k. Before performing the resource allocation procedure based on the adopted scheduling strategy, MBS M first checks the quality of MUEs' communication channels based on the returned CQI [3-5].

For all MUEs and for all their allocated RBs for which condition (6) is fulfilled, MBS M asks the concerned users to send a measurement report (MR) in terms of RSSs of the received pilot signals from their current serving cell and all neighboring femtocells. The RSS reports help MBS M to identify the source of interference and to check whether it is caused by neighboring cells or if it is a consequence of any other effect such as fast fading or noise. For this purpose, condition (7) is checked to determine if for a given MUE_i, the signal received from the neighboring femtocell F_l covers the signal of its serving macrocell. If so, MBS M identifies F_l as an interfering cell for victim MUE_i and F_l is added to the set of interfering FBSs on MUE_i, $\mathcal{Z}_{i,k} : \mathcal{Z}_{i,k} \leftarrow \mathcal{Z}_{i,k} \cup \{F_l\}$. To keep track of the $\text{RSS}^k_{F_l,\text{MUE}_i}$ evolution perceived from each interfering FBS, MBS makes use of a local REM table to record the latest reported $\text{RSS}^k_{F_l,\text{MUE}_i}$ values.

In order to mitigate the interference generated by each interfering FBS in its local REM table, MBS M initiates the power adjustment decision process as follows. First, for each victim MUE_i, MBS identifies all the RBs k on which MUE_i suffers from interference (i.e., those which fulfill condition (7)). Then, it retrieves from the REM table the set of femtocells causing this interference on MUE_i on RB k (i.e., the set $\mathcal{Z}_{i,k}$).

The main objective of our approach is to express, for each interfering femtocell, its impact level on the interference situation. To do so, we define the Score Function

$f(F_l)$ for each interfering femtocell F_l in the REM table as follows:

$$f(F_l) = \theta_1 R_{F_l} + \theta_2 N_{F_l} - \theta_3 Q_{F_l}, \tag{12}$$

where R_{F_l}, N_{F_l} and Q_{F_l} are the context parameters defined as follows:

- $\mathcal{R}_{F_l}$: the set of MUEs impacted by femtocell F_l only (i.e., which verify condition (7)) and R_{F_l} denotes its cardinality,

$$\mathcal{R}_{F_l} = \{\text{MUE}_i, i = 1, \ldots, I; \exists k \in \mathcal{K}_i, \mathcal{Z}_{i,k} = \{F_l\}\},$$
$$R_{F_l} = |\mathcal{R}_{F_l}|$$

- $\mathcal{N}_{F_l}$: the set of MUEs located in the overlapping areas between F_l and any neighboring F_m, and N_{F_l} denotes its cardinality,

$$\mathcal{N}_{F_l} = \{\text{MUE}_i, i = 1, \ldots, I; \exists F_m \in \mathcal{L}, \exists k \in \mathcal{K}_i,$$
$$\mathcal{Z}_{i,k} \supseteq \{F_l\} \bigcup_{m \neq l} \{F_m\}, | \mathcal{Z}_{i,k} | \geqslant 2\},$$
$$N_{F_l} = | \mathcal{N}_{F_l} | .$$

- $\mathcal{Q}_{F_l}$: the set of FUEs observing an outage after decreasing the power of their serving cell F_l and Q_{F_l} denotes its cardinality,

$$\mathcal{Q}_{F_l} = \left\{\text{FUE}_j, j = 1, \ldots, J; \text{SINR}^k_{F_l,\text{FUE}_j} < \text{SINR}^{\text{target}}_{\text{FUE}}\right\},$$
$$Q_{F_l} = |\mathcal{Q}_{F_l}|.$$

Finally, θ_i, $(i = 1, \ldots, 3)$ are the positive weighting factors summing up to 1 and reflecting the priorities of the context parameters.

The context parameters refer to the number of victim MUEs and FUEs as well as their respective locations within the interference area. By assigning different weights to the MUEs affected by only one FBS and those affected by multiple FBSs in Equation 12, we differentiate the impact of those two types of interference situations in the Score Functions. Thus, these context parameters will help in identifying the set of femtocells which have the greater and shared impact on a given interference situation and hence allow for an adapted and coordinated power adjustment proportional to this impact. Figure 2 illustrates a simple scenario in which both interference situations are observed.

The Score Function $f(F_l)$ defined in Equation 12 can be either positive or negative. A positive $f(F_l)$ score means that the associated femtocell F_l impacts more importantly the MUE than the FUEs under its coverage and thus has the ability to tolerate more reduction of its transmission power. Conversely, a negative score value reflects a serious outage situation of F_l's FUE. Note that $f(F_l)$ is evaluated at each frame as long as the interference lasts. Its value will vary according to the power adjustment strategy to be applied in the current frame as well as the mobility of MUEs.

The main objective now is to estimate the exact amount of power to be used by each interfering femtocell for the adjustment purpose. To resolve this problem, we propose two power adjustment strategies baptized global power adjustment (GPA) and selective power adjustment (SPA). GPA and SPA offer a different degree of awareness of FUE performance and consequently impacts the MUE performance as well due to the tradeoff between their conflicting requirements in terms of spectral efficiency. More specifically, GPA and SPA implement a different selection mechanism of the interfering femtocells which are to apply the power adjustment and the amount of the power adjustment (increase or decrease). Besides, the associated power adjustment parameters to be introduced hereafter offer an additional leverage in managing the compromise between FUE and MUE performances.

For each victim MUE_i on RB k, MBS M retrieves from the REM table the set of femtocells responsible for this interference and determines their respective Score Function values. Then, MBS M finds the maximum interference $\text{MI}^k_{\text{MUE}_i}$ which can be allowed by a victim MUE_i on RB k as:

$$\text{MI}^k_{\text{MUE}_i} = \frac{\text{RSS}^k_{M,\text{MUE}_i}}{\text{SINR}^{\text{target}}_{\text{MUE}}} - N_0. \tag{13}$$

Afterwards, MBS M computes for each interfering F_l in the REM table the new value of its transmission power $p'^k_{F_l}$ in RB k as:

$$p'^k_{F_l} = \alpha^k_{F_l} . \beta^k_{F_l} . p^k_{F_l} \tag{14}$$

where we have

$$\beta^k_{F_l} = \frac{\text{MI}^k_{\text{MUE}_i}}{\sum_{l=1}^{L} \text{RSS}^k_{F_l,\text{MUE}_i}}. \tag{15}$$

$p^k_{F_l}$ represents the latest transmitted power of the interfering F_l in RB k, and $\text{RSS}^k_{F_l,\text{MUE}_i}$ is the measured received signal strength from F_l by MUE_i for the allocated RB k. First, the previous power level $p^k_{F_l}$ is adjusted by the parameter $\beta^k_{F_l}$ which represents the allowed per-femtocell interference level, where $\text{MI}^k_{\text{MUE}_i}$ has been normalized by the sum of all RSSs of interfering femtocells. Hence, $\beta^k_{F_l}$ is the same for all F_l and the received RSS by MUE_i after this adjustment becomes $\frac{\text{RSS}^k_{F_l,\text{MUE}_i}}{\sum_{l=1}^{L} \text{RSS}^k_{F_l,\text{MUE}_i}}$.

Figure 2 Macrocell/femtocells interference scenario.

Next, the resulting power is further adjusted by the boosting factor $\alpha^k_{F_l}$ whose value is proportional to $f(F_l)$ score. Using the boosting factor $\alpha^k_{F_l}$ enables to adapt the power allocation proportionally to the femtocells' role in the interference situation rather than uniformly as usually performed in traditional power adjustment approaches. However, the amount of power boosting will be limited by a tunable system parameter x, $(0 < x \leq 1)$ which represents the maximum allowed power reduction/increase ratio of the nominal power $\beta^k_{F_l} p^k_{F_l}$, i.e., $p'^k_{F_l} \in \left[\beta^k_{F_l} p^k_{F_l}(1-x), \beta^k_{F_l} p^k_{F_l}(1+x)\right]$. Therefore, we have:

$$\begin{cases} 1 - x < \alpha^k_{F_l} \leq 1 & if\, f(F_l) \geqslant 0 \\ 1 < \alpha^k_{F_l} \leq \min\left(1 + x, \frac{p^{\text{Max}}_{F_l}}{\beta^k_{F_l} p^k_{F_l}}\right) & \text{otherwise,} \end{cases}$$

Moreover, the $\alpha^k_{F_l}$ value for each F_l should be inversely proportional to its corresponding score $f(F_l)$. Indeed, the higher the $f(F_l)$ score, the greater the impact of F_l on the MUE interference so the more it should reduce its interference by setting a lower $\alpha^k_{F_l}$, with $1 - x \leq \alpha^k_{F_l} \leq 1$. On the contrary, a lower $f(F_l)$ value implies a larger $\alpha^k_{F_l}$. In particular, the case $f(F_l) < 0$ means that the proportion of unsatisfied FUEs is greater than that of victim MUEs, and hence $\alpha^k_{F_l}$ should be increased within the range $\left[1, \min\left(1 + x, \frac{p^{\text{Max}}_{F_l}}{\beta^k_{F_l} p^k_{F_l}}\right)\right]$. Thus, we propose the following setting of $\alpha^k_{F_l}$ in each case:

$$\begin{cases} 1 - \min\left(x, \frac{f(F_l)}{\max\limits_{\mathcal{Z}_{i,k}} f(F_l)}\right) & if\, f(Fl) \geqslant 0 \\ \min\left(1 + \min\left(x, \frac{f(F_l)}{\min\limits_{\mathcal{Z}_{i,k}} f(F_l)}\right), \frac{p^{\text{Max}}_{F_l}}{\beta^k_{F_l} p^k_{F_l}}\right) & \text{otherwise,} \end{cases} \tag{16}$$

where $\frac{f(F_l)}{\max\limits_{\mathcal{Z}_{i,k}} f(F_l)}$ and $\frac{f(F_l)}{\min\limits_{\mathcal{Z}_{i,k}} f(F_l)}$ act as the weights for FBS F_l when $f(F_l)$ is positive or negative, respectively.

At this step, the selection of the set of the interfering femtocells that are to apply the boosting factor $\alpha^k_{F_l}$ defines the power adjustment strategy. As mentioned above, we propose two adjustment strategies: GPA and SPA.

3.2.1 Global power adjustment

In the GPA strategy, we propose to apply the power adjustment using the boosting factors to all the interfering femtocells. It is worth noticing here, that GPA strategy acts on all femtocells participating in the interference whatever their Score Function values. Therefore, GPA will create a balance between the offered throughput to victim MUEs and the performance of FUEs which may suffer from the power adjustment (see Algorithm 1).

Algorithm 1 GPA algorithm

Let $\mathcal{I}$ be the set of MUEs of MBS M,
Let $\mathcal{L}$ be the set of FBSs,
Let $\mathcal{Z}_{i,k}$ be the set of interfering FBSs on MUE_i in a RB k,
Let $\mathcal{K}_i$ be the set of RBs allocated to MUE_i, $\mathcal{K}_i \neq \emptyset$,
initiate $\text{SINR}_{\text{MUE}}^{\text{target}}$
add $\mathcal{Z}_{i,k}$ **in** REM-Table
for all $\text{MUE}_i \in \mathcal{I}$ **and for all RB k** $\in \mathcal{K}_i$**do**
 initiate $\mathcal{Z}_{i,k} \leftarrow \phi$
if $\text{SINR}_{M,\text{MUE}_i}^{k} < \text{SINR}_{\text{MUE}}^{\text{target}} + \Delta\text{SINR}$**then**
  calculate $\text{RSS}_{M,\text{MUE}_i}^{k}$
   for all $F_l \in \mathcal{L}$ **do**
    calculate $\text{RSS}_{F_l,\text{MUE}_i}^{k}$
    if $\text{RSS}_{M,\text{MUE}_i}^{k} < \text{RSS}_{F_l,\text{MUE}_i}^{k} + \Delta P$**then**
     $\mathcal{Z}_{i,k} \leftarrow \mathcal{Z}_{i,k} \cup \{F_l\}$
    end if
   end for
  add $\mathcal{Z}_{i,k}$ **in** REM-Table
 end if
end for
for all $\text{MUE}_i \in \mathcal{I}$ **and for all RB k** $\in \mathcal{K}_i$ **do**
 get $\mathcal{Z}_{i,k}$ **from** REM-Table
 if $\mathcal{Z}_{i,k} \neq \phi$ **then**
  for all $F_l \in \mathcal{Z}_{i,k}$ **do**
   calculate R_{F_l}, N_{F_l}, **and** Q_{F_l} **from** REM-Table
   calculate $f(F_l) \leftarrow (\theta_1 R_{F_l}+\theta_2 N_{F_l}-\theta_3 Q_{F_l})$, //where $0\leq \theta_i \leq 1$, $i = 1,2,3$
   case $f(Fl) \geq 0$
     calculate $p_{F_l}^{'k} \leftarrow \alpha_{F_l}^{k} \cdot \beta_{F_l}^{k} . p_{F_l}^{k}$, //power adjustment at F_l
     where $\beta_{F_l}^{k} \leftarrow \frac{\text{MI}_{\text{MUE}_i}^{k}}{\sum_{l=1}^{L}\text{RSS}_{F_l,\text{MUE}_i}^{k}}$,
     $\alpha_{F_l}^{k} \leftarrow 1 - \min\left(x, \frac{f(F_l)}{\max\limits_{\mathcal{Z}_{i,k}} f(F_l)}\right)$
   case $f(Fl) < 0$
     calculate $p_{F_l}^{'k} \leftarrow \alpha_{F_l}^{k} \cdot \beta_{F_l}^{k} . p_{F_l}^{k}$, //power adjustment at F_l
     where $\beta_{F_l}^{k} \leftarrow \frac{\text{MI}_{\text{MUE}_i}^{k}}{\sum_{l=1}^{L}\text{RSS}_{F_l,\text{MUE}_i}^{k}}$,
     $\alpha_{F_l}^{k} \leftarrow \min\left(1+\min\left(x, \frac{f(F_l)}{\min\limits_{\mathcal{Z}_{i,k}} f(F_l)}\right), \frac{P_{F_l}^{\text{Max}}}{\beta_{F_l}^{k} p_{F_l}^{k}}\right)$
  end for
 end if
end for
update REM-Table

3.2.2 Selective power adjustment

The aim of this second strategy is to apply $\alpha_{F_l}^{k}$ adjustment parameters only for those femtocells which show a positive Score Function ($f(F_l) \geq 0$). The rest of interfering femtocells ($f(F_l) < 0$) are allocated a fixed and neutral parameter value $\alpha_{F_l}^{k}$=1. Such a strategy will give more priority to MUEs in terms of throughput since femtocells suffering from a severe FUE outage situation will not benefit from an adapted power reduction proportional to their 'suffering' level. As a consequence, in SPA strategy, $\alpha_{F_l}^{k}$ are derived as follows:

$$\begin{cases} 1 - \min\left(x, \frac{f(F_l)}{\max\limits_{\mathcal{Z}_{i,k}} f(F_l)}\right) & if\, f(F_l) \geqslant 0 \\ \alpha_{F_l}^{k} = 1 & \text{otherwise} \end{cases}$$

Similarly to GPA, the adjustment process in SPA is repeated recursively at each frame until the interference caused to MUE_i on RB k fades.

3.3 Required signaling information exchange between cells

The required signaling information between the macrocell and femtocells may be exchanged through two main possibilities: i) the backhaul connection using the core network infrastructure (optic fiber), through the S1 interface and femtocell gateway [22,23] and ii) the direct X2 interface between macrocells and femtocells [3]. These two communication means will impact differently the performance mainly in terms of delay and signaling overhead.

The X2 interface enables the support of resource management functionalities operating in short timescales such as interference, mobility, and handover [23]. Therefore, the X2 interface is more suitable to support the signaling exchange in our power adjustment strategy as the S1 interface will enduce longer delays. In fact, power adjustments and several context information (i.e., RSS, SINR, ..., etc) have to be periodically exchanged between the MBS and femtocell local REMs, and the delay imposed by the interface should be at a shorter timescale compared to this periodicity in order to allow a real-time response of the operations.

As regards to the signaling load, the amount of exchanged information becomes more stringent as the number of femtocells increases. In this respect, the backhaul connection may provide enough bandwidth for the signaling overhead but introduces harsh delay constraints, making it difficult to satisfy the real-time requirements [24]. It is worth noticing that the coordination between cells in our proposed scheme requires a limited signaling load for each updating cycle, namely:

- the measurement report Q_{F_l} which represents the number of F_l 's FUEs in outage situation. This message is triggered by F_l and sent to the corresponding MBS.
- the values of $p_{F_l}^{'k}$ for each RB k and each F_l, to be sent by the MBS to the corresponding FBS F_l.

4 Performance evaluation

4.1 Simulation environment

To investigate the performance of our proposed power allocation mechanisms, we implemented various simulation scenarios using LTE-Sim simulator [19]. The network topology consists of one MBS M and several neighboring FBSs positioned in a street with a reasonable distance from M. Each FBS is placed in a building and serves a maximum of four users registered as CSG users and moving at a low speed compared to the outdoor users (3 vs. 30 km/h). MUEs are randomly positioned within the macrocell and move according to a random walk mobility model constrained by the street layout of 30 m. The main simulation parameters are detailed in Table 1.

We used path loss models as specified in 3GPP TSG WG4 [17] with an external wall attenuation of 20 dBm. In a first set of experiments, the Score Function's weighting parameters θ_1, θ_2, and θ_3 have a fixed value of 0.3, 0.3, and 0.4, respectively. Note that the effect of these weighting parameters will be analyzed in Section 4.3.

4.2 Performance of the power adjustment strategies

In this first set of experiments, we evaluate the performance of our power allocation schemes using extensive simulations and gain further insight on the influence of the involved parameters on the macrocell/femtocell throughput performance. Regarding the scheduling module, we used the reference modified largest weighted delay first (MLWDF) algorithm. All MUEs/FUEs in the cell support three different types of application flows: CBR, VoIP, and Video. CBR traffic is generated at a rate of 5-byte packets/s, whereas the VoIP flows generate 20-byte packets/s using an ON/OFF model [25]. The Video service is a 242 kbps data source with H.264 coding [26].

Table 1 Simulation parameters

Parameters	Values
Number of macrocells	1
Number of femtocells	30
Number of MUEs	10 to 60
Number of FUEs	4
Bandwidth	10 MHz
Total number of RBs	50
Macro Tx power	43 dBm
Femto Tx power	23 dBm
Thermal noise density	-174 dBm/Hz
Carrier frequency	3.5 GHz
UE noise figure	2.5 dB
Macrocell radius	1 km
FUEs' average speed	3 km/h
MUEs' average speed	30 km/h
$SINR_{MUE}^{target}$	5.7 dB
$SINR_{FUE}^{target}$	12.6 dB
ΔP	0.1 dB

Quality of service in terms of throughput is measured for our proposed power allocation strategies GPA and SPA and compared with the Basic scheme where no power adjustment strategy is considered, as well as the Reference scheme proposed in [9], and which can be viewed as a classical femtocell power adjustment scheme that is unaware of the FUEs' performance. Note that in the Basic scheme, the femtocells transmit at their maximum transmit power on each RB. In other words, the Basic scheme corresponds to the upper bound of the FUEs' achievable throughput. The Basic and Reference schemes are noted 'Basic' and 'Ref.' in the figures, respectively. In order to better estimate the quality of our obtained suboptimal solution, we have also evaluated the MUEs' throughput upper bound denoted UP-Bound, which indicates the maximum achievable throughput for MUEs in the absence of femtocells, i.e., without any interference. Finally, we have compared our algorithms to the eICIC method based on almost blank subframes (ABS), denoted ABS in the legends. In this solution, the femtocells mute their transmission (of control and data signals) to enable the scheduling of victim MUEs by the macrocell, hence avoiding the cross-tier interference. We applied one of the ABS patterns considered in 3GPP with a duty cycle of 1/8 where femtocells remain silent during 1/8 of each subframe [27].

Figure 3 shows the average per-user throughput of MUEs for the supported CBR flows. We can easily observe that for all the schemes, the throughput degrades with the number of active MUEs as this number is correlated to the generated traffic. Moreover, we can also see that the gain in throughput for our proposed strategies is higher than for the Basic and Ref. strategies especially when the number of users is increased from 30 to 60. Our GPA and SPA strategies offer a gain of almost 12% and 16%, compared to the Basic scheme, and 9% and 14% compared to the Reference scheme, respectively. The same behavior is observed for VoIP flows (Figure 4) which shows almost a similar performance gain. This increase in throughput for higher user density is mainly due to an increased probability of interference situations when MUEs approach the femtocells' coverage. Figure 3 also shows that SPA strategy performs better than GPA since it gives a higher priority to MUEs, providing them a higher average throughput. We can also observe that SPA strategy achieves the closest performance to the MUEs' throughput upper bound. Both proposed schemes outperform all reference schemes in terms of MUEs' throughput.

Figures 5 and 6 illustrate FUEs' average per-user throughput for CBR and VoIP flows, when varying the

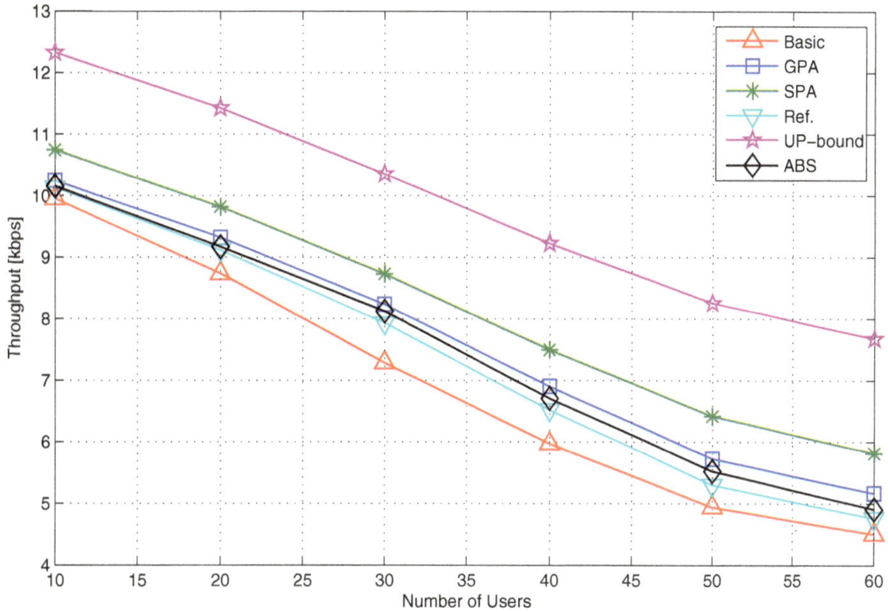

Figure 3 Average per-user throughput of MUEs for CBR flows.

number of active MUEs. As the power adjustment in general degrades the transmission power of femtocells to mitigate the interference effect on victim MUEs, the GPA scheme offers as expected a better average throughput for FUEs. However, we observe a similar FUE throughput performance for both SPA and Reference schemes, while ABS has the worst performance. Figures 5 and 6 show that the throughput degradation compared to the Basic scheme is only 5% and 7% as regards to GPA and SPA, respectively. Note that in the Basic scheme, the femtocells transmit at their maximum transmit power on each RB.

Figure 7 shows the average per-user throughput of MUEs for the supported Video flows. We have an improvement of 13% to 18% for SPA and 11% to 15% for

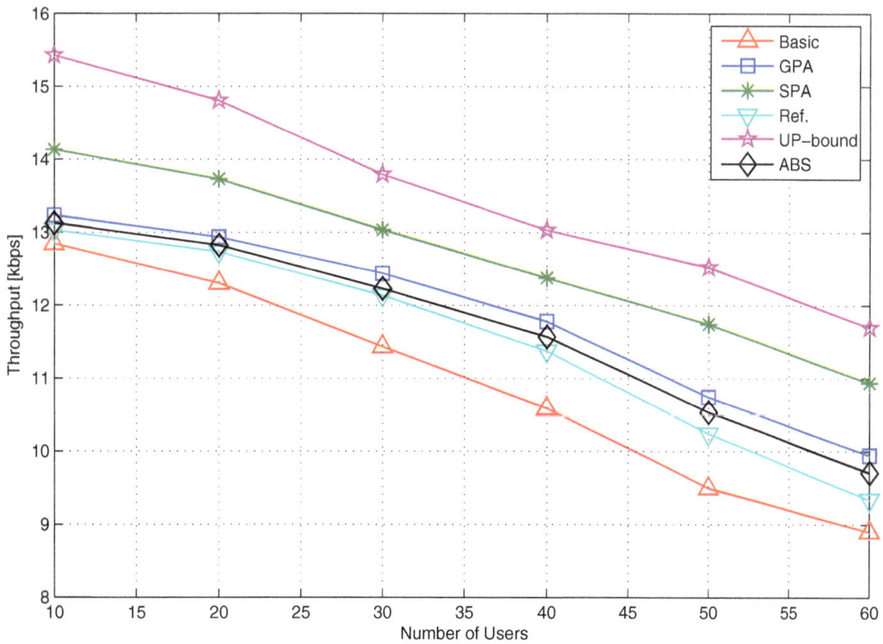

Figure 4 Average per-user throughput of MUEs for VoIP flows.

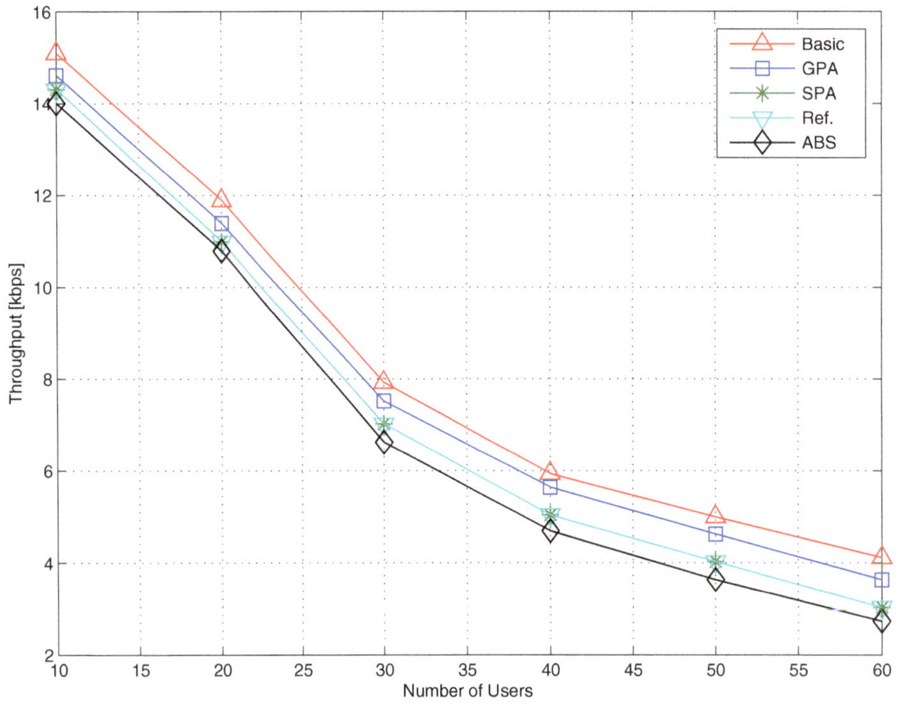

Figure 5 Average per-user throughput of FUEs for CBR flows as a function of the number of MUEs.

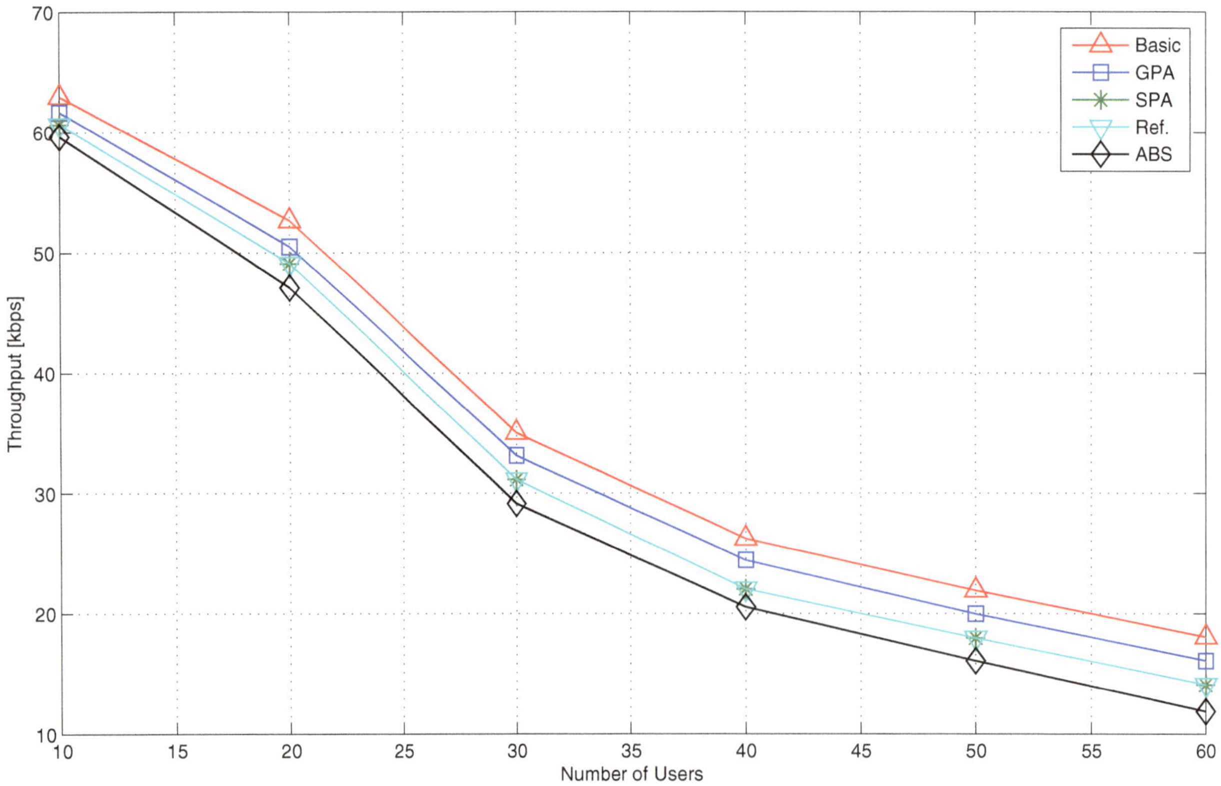

Figure 6 Average per-user throughput of FUEs for VoIP flows as a function of the number of MUEs.

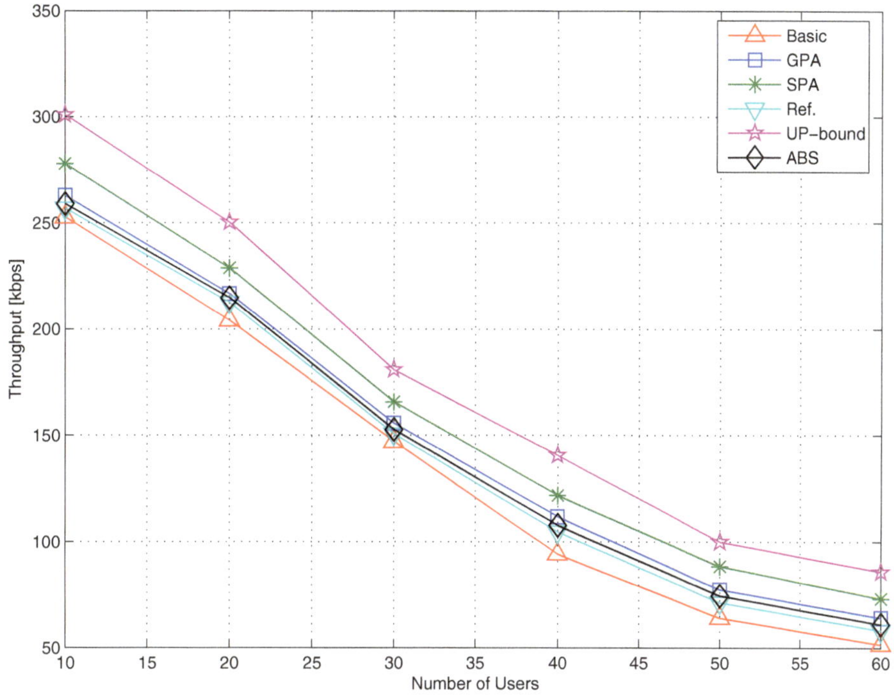

Figure 7 Average per-user throughput of MUEs for video flows.

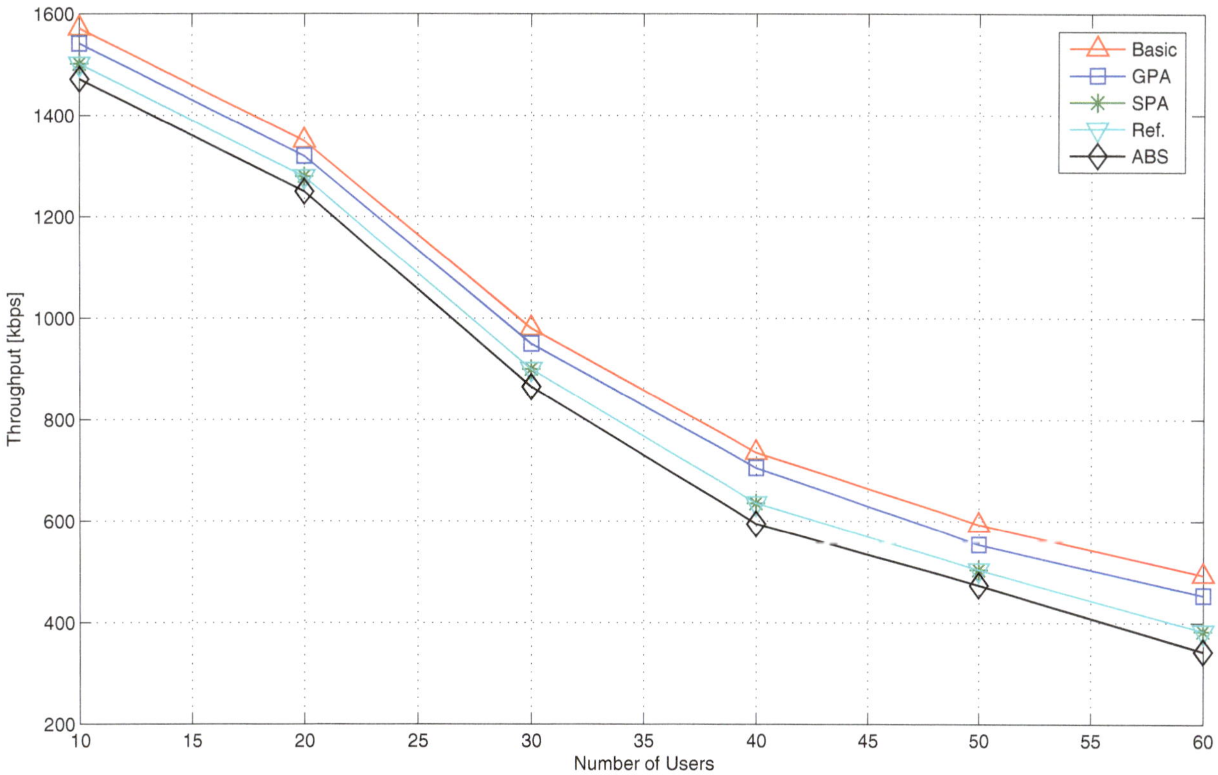

Figure 8 Average per-user throughput of FUEs for video flows as a function of the number of MUEs.

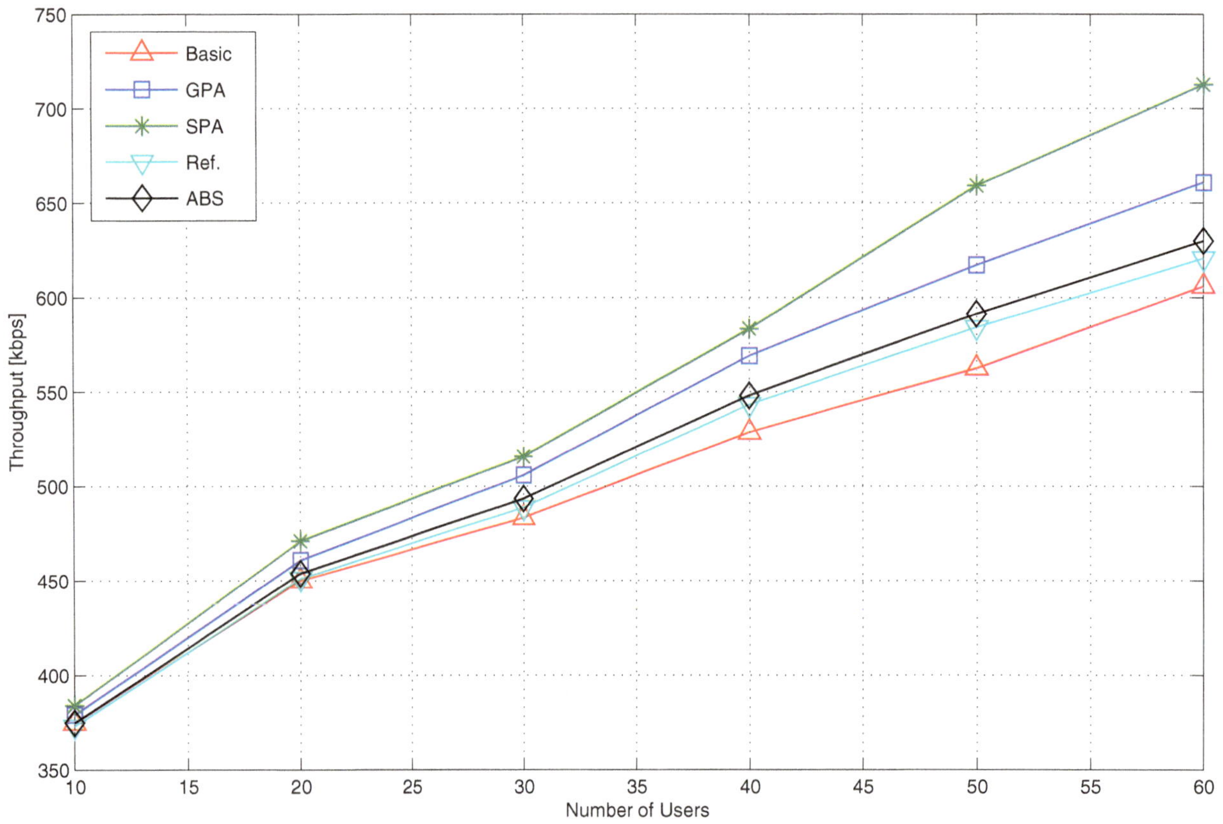

Figure 9 The overall system throughput for Voip flows as a function of the number of MUEs.

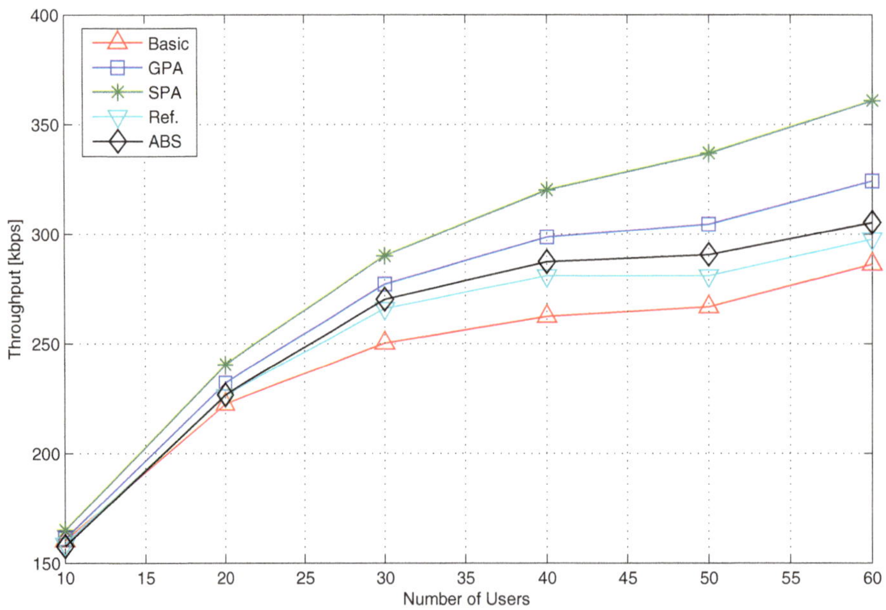

Figure 10 The overall system throughput for CBR flows as a function of the number of MUEs.

Table 2 Conducted Test Ids and corresponding θ_i values

	Test Ids	θ_i values
Case A (Priority to FUEs)	1	(0.1, 0.1, 0.8)
	2	(0.2, 0.2, 0.6)
Case B (MUEs/FUEs balanced)	3	(0.25, 0.25, 0.5)
	4	(0.4, 0.1, 0.5)
	5	(0.1, 0.4, 0.5)
Case C (Priority to MUEs)	6	(0.3, 0.3, 0.4)
	7	(0.4, 0.4, 0.2)

GPA compared to the Basic and Ref. schemes, respectively. Moreover, we observe that the Video traffic is more impacted by the gain in throughput, which is because Video flows usually require additional RBs compared to CBR or VoIP flows given the lower data rate and packet size. The gap between SPA and the MUEs' throughput upper bound is even smaller in the case of Video flows, indicating the validity of the proposed suboptimal solutions. Figure 8 depicts the average per-user throughput with Video flows for FUEs. Here again, our GPA strategy outperforms SPA and Reference schemes.

Finally, Figures 9 and 10 show the overall macrocell and femtocells' system throughput for VoIP and CBR traffics, respectively. In both cases, SPA outperforms GPA, the Reference and ABS schemes, while the Basic scheme offers the lowest performance. SPA strategy achieves the highest performance since it improves the spectral efficiency of MUEs while being agnostic to FUEs' throughput degradation. On the contrary, GPA improves the MUEs' throughput at the expense of a reduced system throughput.

Finally, note that the throughput gain achieved by our proposed techniques is more significant when the density of active UEs increases. Hence, these power adjustment strategies fit perfectly the dense and mobile scenarios of urban areas.

4.3 Analysis of the impact of the weighting parameters

In this section, we investigate and analyze the effect of the Score Function weighting parameters θ_i, $(i = 1, 2, 3)$ on the performance of our GPA and SPA schemes. As defined in Equation 12, the score of a given interfering FBS F_l depends on the values of θ_i parameters which rule the balance between the three context parameters. For a

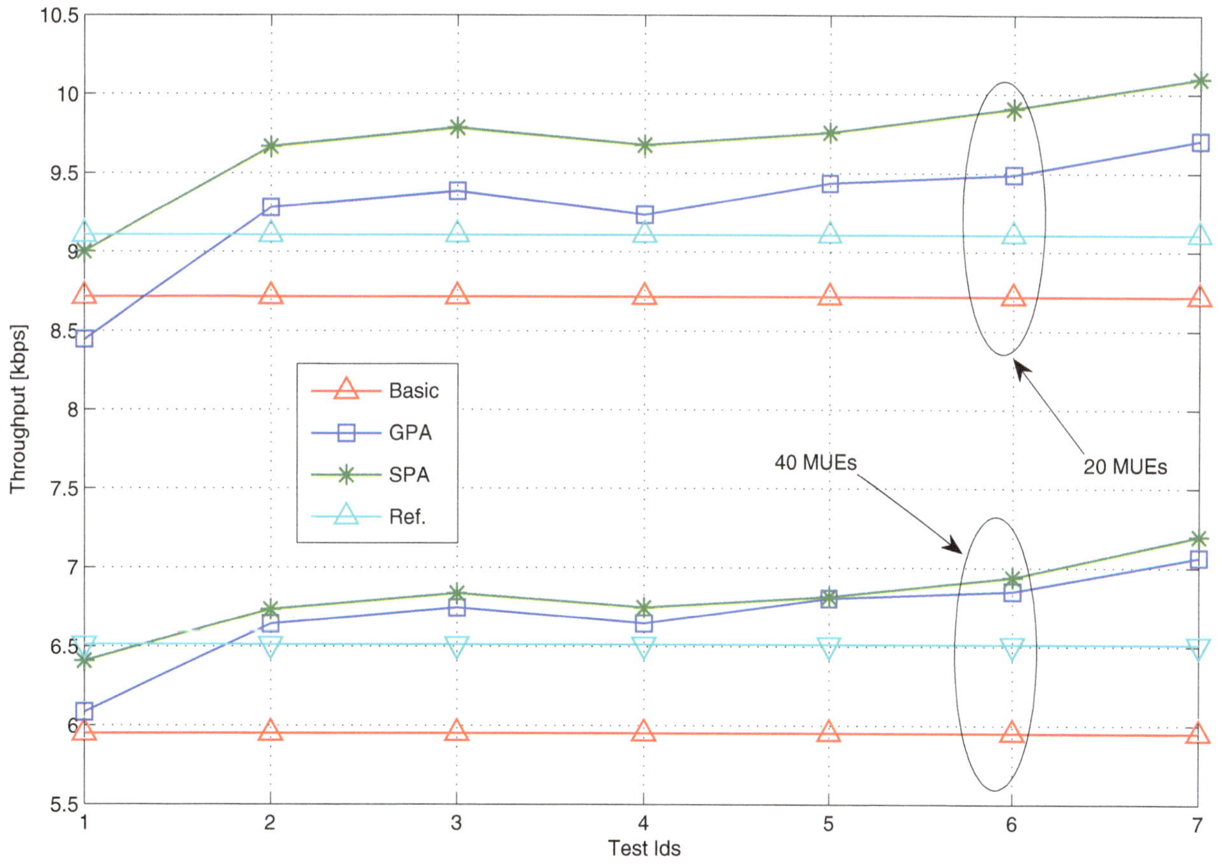

Figure 11 Average per-user throughput of MUEs (20 or 40 MUEs) for CBR flows in different scenarios.

better understanding of the weighting parameters' effect, we recall in the following their respective role in the Score Function:

- θ_1 : rules the weight of the 'single interference source' situation as it is associated to the context parameter R_{F_l}. The single interference situation refers here to the case where the victim MUEs suffer from the interference occurring only from the current FBS F_l.
- θ_2 : acts on the weight of the 'multiple interference sources' situation which indicates that the victim MUEs of sub-set $\mathcal{N}_{F_l}$ are suffering from the interference of other FBSs besides F_l's interference.
- θ_3 : is used to determine the weight of the outage FUEs on the global F_l score. In other words, θ_3 manages the degree of awareness of the FUEs' performance within the power adjustment strategy.

It is worth mentioning here that distinguishing the single and multiple interference situations provides a good means to take advanced and adapted power adjustment decisions. For example, the power adjustment strategy may demand more power decrease from FBSs involved in multiple interferences than those creating single interference to rapidly heal the global interference situation. Moreover, note that the association of the θ_i weighting factors and the adopted power adjustment strategy (GPA or SPA) provides two complementary means to realize the best trade-off between MUEs' and FUEs' performance.

To simplify the analysis and discussion, we organize the experiments in three different and significant cases (see Table 2):

1. *Case A:* corresponds to the scenarios which give a higher priority to FUEs over MUEs while maintaining the same weight between multiple and single interference situations. This case meets the following conditions, $\theta_1 = \theta_2$ and $\theta_3 > \theta_1 + \theta_2$.
2. *Case B:* corresponds to the balanced mode where FUEs and MUEs share the same priority. In this case, we have $\theta_1 + \theta_2 = \theta_3$.
3. *Case C:* represents the opposite situation of case A since we give more priority to MUEs by setting $\theta_3 < \theta_1 + \theta_2$.

We evaluate in the following the performance of our power allocation schemes in each of the cases mentioned

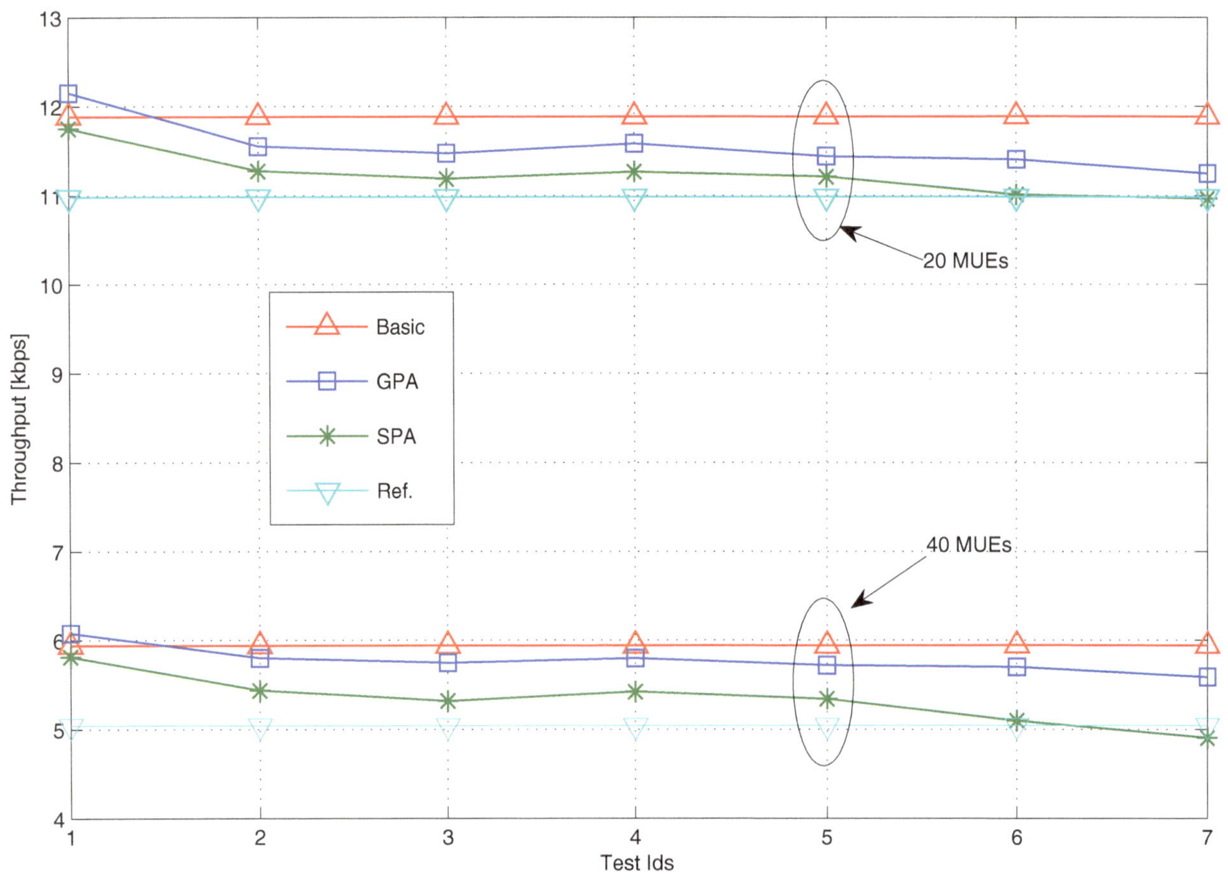

Figure 12 Average per-user throughput of FUEs (20 or 40 MUEs) for CBR flows in different scenarios.

above. To do so, we assume here CBR traffic flows with 20 and 40 active MUEs using the same simulation parameters provided in Table 1. The three different cases described above are conducted using the θ_i values provided in Table 2. The average throughput for MUEs and FUEs achieved for each test are presented in Figures 11 and 12, respectively. Note that the Basic and Reference schemes are also depicted in Figures 11 and 12, in order to provide a baseline performance and to have a better insight on how the parameters affect the results compared to the reference schemes. Except for the MUE throughput with Test Id 1, the proposed schemes always outperform the Reference one. Therefore, the following discussion will mainly focus on the weighting parameters' effect on our GPA and SPA strategies.

First of all, comparing the cases where $\theta_1 + \theta_2 < \theta_3$, i.e., tests 1 and 2, to the cases where $\theta_1 + \theta_2 > \theta_3$, i.e., tests 6 and 7, we observe that FUEs are prioritized against MUEs in the first cases and vice versa in the latter ones. Thus, the achieved throughputs are in concordance with the corresponding set of θ_i values, as higher values of $\theta_1 + \theta_2$ enable higher MUE throughput but lower FUE throughput given the constraint $\theta_1+\theta_2+\theta_3 = 1$. We observe that the value of $\theta_1 + \theta_2$ has more significant effects on the MUE throughput than the value of θ_3 on the FUE throughput, given the higher vulnerability of MUEs under FBS interference in our target scenario. Although smaller θ_3 implies a lower FUE throughput, Figure 12 shows that there is only a slight decrease, even with the SPA strategy where FBSs are not allowed to increase their power. We can also confirm that a very good trade-off between MUE and FUE throughput is achieved in the balanced case (case B, Test Ids 3,4,5) where $\theta_1 + \theta_2 = \theta_3$.

Moreover, we study the effect of the discrimination of θ_1 and θ_2 values, corresponding to different prioritization weights given to MUEs affected by only one femtocell and those affected by several femtocells. These scenarios are represented by Test Ids 4 and 5 in Table 2, assuming $\theta_1 + \theta_2 = \theta_3$. Clearly, choosing $\theta_2 > \theta_1$ (Test Id 5) provides a higher throughput to MUEs. Thus, giving a larger weight on the interference caused by multiple femtocells allows to decrease their power simultaneously and hence to improve the average MUE throughput. However, the power decrease on those femtocells does not decrease much the FUE average throughput, as shown in Figure 12, validating the efficiency of the proposed power adjustment strategies. In addition, note that compared to the completely balanced scenario (Test Id 3), the achieved MUE throughput is further enhanced by selecting $\theta_2 > \theta_1$ (Test Id 5), especially for the GPA strategy for both numbers of MUEs, while keeping the FUE throughput at the same level. This shows the utility of defining distinct θ_1 and θ_2 values depending on the number of interference sources.

Finally, the results obtained for different sets of θ_i values show that the selected power adjustment strategy, namely, GPA or SPA, along with the appropriate tuning of the θ_i values enable to achieve the required MUE/FUE prioritization level in terms of throughput in an efficient manner.

5 Conclusions

This paper has addressed the interference issue in a collocated macro/femtocell HetNets within an urban mobility environment. We proposed femtocell power adjustment methods whose key objective was to increase the average throughput of non-CSG MUEs by limiting the amount of interference caused by femtocells. The main contribution of our approach was to introduce power adjustment parameters whose values are dynamically adapted to the weight of each interfering femtocell through their Score Functions on the global interference situation. Given these weights, the two proposed power adjustment strategies have the common objective of mitigating the interference on victim MUEs while implementing two different degrees of awareness of FUEs' throughput degradation. The simulation results have shown the effectiveness of both proposed strategies in terms of macrocell and femtocell average throughput for different types of traffic. Moreover, it was shown that the proposed priority weights used in the femtocells' Score Functions provide an efficient means for achieving the desired level of macrocell/femtocell throughput trade-off.

Competing interests

The authors declare that they have no competing interests.

Acknowledgements

This work was jointly supported by the Grants-in-Aid for bilateral research between France and Japan, co-funded by JSPS and MAEE (PHC Sakura program), the European FP7 Newcom# NoE, and by the Grants-in-Aid for Scientific Research (KAKENHI) no. 23760334 and no. 26820143 from the Ministry of Education, Science, Sports, and Culture of Japan.

References

1. E Dahlman, S Parkvall, J Sköld, *3G Evolution: HSPA and LTE for Mobile Broadband*, second edition. (Elsevier Academic Press, London, UK, 2008)
2. V Chandrasekhar, JG Andrews, A Gatherer, Femtocell networks: a survey. IEEE Commun. Mag. **46**(9), 9-67 (2008)
3. D Lopez-Pérez, A Valcarce, J Zhang, OFDMA femtocells: a roadmap on interference avoidance. IEEE Commun. Mag. **47**, 41–48 (2009)
4. Z Wang, W Xiong, C Dong, J Wang, S Li, in *International Conference on Computational Problem-Solving (ICCP)*. A novel downlink power control scheme in LTE heterogeneous network (Leshan, China, 2011), pp. 241-245
5. R Kurda, L Boukhatem, TA Yahiya, in *IEEE on Wireless Communications and Networking Systems (WCNC)*. Interference mitigation in mobile environment through power adjustment in macro and femto cell systems (Istanbul, Turkey, 2014)
6. R Kurda, L Boukhatem, TA Yahia, M Kaneko, in *IEEE Symposium on Computers and Communications (ISCC)*. Power adjustment mechanism using context information for interference mitigation in two-tier heterogeneous networks (Madeira, Portugal, 2014)

7. MS Jin, S Chae, DI Kim, in *Proc. IEEE VTC 11- Spring*. Per cluster based opportunistic power control for heterogeneous networks (Budapest, Hungary, 2011)
8. P Gao, D Chen, M Feng, D Qu, T Jiang, in *IEEE on Wireless Communications and Networking Conference (WCNC)*. On the interference avoidance method in two-tier LTE networks with femtocells (Shanghai, China, 2013), pp. 3585-3590
9. D Lopez-Pérez, A Valcarce, A Ladanyi, J Zhang, Intracell handover for interference and handover mitigation in OFDMA two-tier macrocell-femtocell networks. EURASIP J Wireless Commun. Netw. **10** (2009)
10. Y Bai, J Zhou, L Liu, L Chen, H Otsuka, in *IEEE International Symposium on Personal, Indoor and Mobile Radio Communications (PIMRC)*. Resource coordination and interference mitigation between macrocell and femtocell (Tokyo, Japan, 2009), pp. 1401-1405
11. F Bernardo, R Agustí, J Cordero, C Crespo, in *Sixth Advanced International Conference on Date of Conference: Telecommunications (AICT)*. Self-optimization of spectrum assignment and transmission power in OFDMA femtocells (Barcelona, Spain, 2010), pp. 404-409
12. M Kim, H Won, FA Je Tobagi, in *IEEE Global Telecommunications Conference (GLOBECOM)*. Cross-tier interference mitigation for two-tier OFDMA femtocell networks with limited macrocell information (Miami, Florida, USA, 2010), pp. 1-5
13. M Bennis, L Giupponi, EM Diaz, M Lalam, M Maqbool, EC Strinati, A De Domenico, M Latva-aho, in *IEEE Wireless Communication, Vehicular Technology, Information Theory and Aerospace & Electronic Systems Technology (Wireless VITAE)*. Interference management in self-organized femtocell networks: the BeFEMTO approach (Chennai, India, 2011), pp. 1-6
14. T Nakano, M Kaneko, K Hayashi, H Sakai, in *Proc. of IEEE VTC-Spring*. Interference mitigation based on partial CSI feedback and overhearing in an OFDMA heterogeneous system (Dresden, Germany, 2013), pp. 1-5
15. T Nakano, M Kaneko, K Hayashi, H Sakai, in *IEEE International Symposium on Personal, Indoor and Mobile Radio Communications (PIMRC)*. Downlink power allocation with CSI overhearing in an OFDMA macrocell/femtocell coexisting system (Sydney, Australia, 2012), pp. 454-459
16. A Zalonis, N Dimitriou, A Polydoros, J Nasreddine, P Mähönen, in *IEEE Wireless Communications and Networking Conference (WCNC)*. Femtocell downlink power control based on radio environment maps (Paris, France, 2012), pp. 390–394
17. 3GPP, Simulation assumptions and parameters for FDD HeNB RF requirements, 3GPP TSG RAN WG4 R4-092042 (May 2008). http://www.3gpp.org/ftp/Specs
18. Beek van de J, T Cai, S Grimoud, B Sayrac, P Mähönen, J Nasreddine, J Riihijärvi, How a layered REM architecture brings cognition to today's mobile networks. IEEE Wireless Commun. **19**(4), 17-24 (2012)
19. G Piro, LA Grieco, G Boggia, F Capozzi, P Camarda, Simulating LTE cellular systems: an open source framework. IEEE Trans. Vehic. Technol. **60**, 498-513 (2010)
20. X Li, L Qian, D Kataria, in *Proc. of the Conference on Information Sciences and Systems (CISS)*. Downlink power control in co-channel macrocell femtocell overlay (Baltimore, USA, 2009)
21. G Cao, D Yang, X Ye, X Zhang, in *IEEE on Wireless Communications and Networking Conference (WCNC)*. A downlink joint power control and resource allocation scheme for co-channel macrocell-femtocell networks (Beijing, China, 2011), pp. 281-286
22. R Madan, J Borran, A Sampath, N Bhushan, A Khandekar, T Ji, Cell association and interference coordination in heterogeneous LTE-A cellular networks. IEEE J. Selected Areas Commun. **28**(9), 1479-1489 (2010)
23. B Sayrac, *D2.4: Final system architecture, deliverable of FARAMIR project*, (December 2011). available at http://www.ict-faramir.eu/
24. L Wang, Y Zhang, Z Wei, in *Proceedings of the 69th IEEE Vehicular Technology Conference (VTC 09)*. Mobility management schemes at radio network layer for LTE femtocells (Barcelona Spain, 2009)
25. C Chen-Nee, RH Katz, in *IEEE International Communication (ICC)*. Characterizing packet audio streams from Internet multimedia applications, vol. 2 (New York City, USA, 2002), pp. 1199-1203
26. Video trace library. http://trace.eas.asu.edu/
27. CATT, *Evaluations of RSRP/RSRQ measurement (R4-110284)*, (Austin, TX, 2011)

A low-complexity peak cancellation scheme and its FPGA implementation for peak-to-average power ratio reduction

Jiajia Song* and Hideki Ochiai

Abstract

The power amplifier (PA) is the most power-hungry component in a wireless base station transmitter, and reducing the peak-to-average power ratio (PAPR) of wireless signals is an important issue for its effective use. In this paper, we focus on a field-programmable gate array (FPGA) implementation of the peak cancellation (PC) technique, which is known as the simplest method for PAPR reduction. The design issue of effective peak-cancelling pulses under the constraint on the out-of-band emission is addressed. In order to reduce its hardware complexity, a novel approach for generating peak-cancelling pulses is also presented. The experimental results based on long-term evolution (LTE)/LTE-Advanced and multi-band Wideband Code Division Multiple Access (WCDMA) signals demonstrate the validity of the proposed scheme. It has been shown that the proposed PC scheme can achieve lower in-band distortion than the conventional PC with an acceptable loss in out-of-band performance. Our study also includes mapping the signal processing methods onto a Xilinx virtex-7 FPGA device running at 245.76 MHz and addresses the resource utilization and the hardware design in detail.

Keywords: PAPR reduction; Peak cancellation; FPGA; OFDM; LTE; LTE-Advanced; Multi-band; WCDMA

1 Introduction

Signals such as orthogonal frequency division multiplexing (OFDM) [1] and direct-sequence spread spectrum (DS-SS) for code division multiple access (CDMA) [2] are widely adopted in modern wireless communication systems due to their remarkable advantages such as flexible allocation of resources and high-spectrum utilization. As these signals are essentially a sum of multiple subcarriers/codes of multiple users in either frequency or time domain, the probability density functions (PDFs) of their signals tend to approach Gaussian, and thus, they exhibit high peak-to-average power ratio (PAPR) [3]. This poses strict demands on the dynamic range of data converters and especially limits efficient operation of the power amplifiers (PAs). Reducing the PAPR is hence important for boosting the PA efficiency by allowing higher average input power before saturation occurs. To mitigate this issue, extensive studies have been performed [4-6].

Some of the PAPR-reduction techniques for Wideband Code Division Multiple Access (WCDMA) system can be found in, for example, [7,8]. On the other hand, those for multi-carrier and OFDM signals are much more abundant and appear in many forms including selected mapping (SLM) [9] and partial transmit sequence (PTS) [10,11], just to mention a few. The aforementioned techniques do not incur distortion, but they either have a large computational complexity or have to modify the signal, which makes their implementation in high-speed real-time systems challenging or hinders standard-compliant operations. It should be noted that, as the digital techniques in a transmitter continue to scale, the power consumed by digital circuits also takes up a large portion of the total power consumption. Implementing the PAPR reduction with a high-speed and power-hungry digital signal processor (DSP) is obviously detrimental to the cost and power efficiency of the entire system.

*Correspondence: song-jiajia-kr@ynu.jp
Department of Electrical and Computer Engineering, Yokohama National University, 79-5 Tokiwadai, Hodogaya, Yokohama 240-8501, Japan

Motivated by the above observations, simple techniques such as clipping and filtering (CAF) [12-14], peak windowing (PW) [7,15,16] and peak cancellation (PC) [17-19], which have much lower complexity, can be considered as more realistic approaches from the viewpoint of practical implementation. These techniques essentially introduce nonlinear operations so that distortions are inevitable. Given that some degree of distortion is generally allowed for the transmitted signals, such techniques are very attractive. The major drawback of CAF is the peak regrowth caused by the filtering effect, and the amount of regrowth is generally intractable. This is undesirable for a transmitter with digital predistorter (DPD), as the DPD needs to strictly keep the peak below a predefined value to ensure that no signal sweeps into the saturation region of the PA. Although PAPR regrowth can be somewhat alleviated by iterative use of CAF [20,21], the resulting complexity will be increased several fold because of the duplicated functional blocks. Furthermore, the latency issue becomes also prohibitive.

In contrast, PC is a much overlooked technique that has advantages in several aspects. PC simply generates independent cancelling pulses to cancel the peak values to a given threshold. It allows more cost-effective hardware implementation than the CAF as no filtering operation, which involves either a large number of multipliers or a bank of fast Fourier transform (FFT) blocks, is required. Moreover, PC can be easily configured to make compliant operation to signals of different communication standards. This is because the cancelling pulses can be updated to support a variety of carrier configurations and bandwidths. The concept of PC can be also used to facilitate generating cancelling pulses in ACE [22] and tone reservation (TR) [23,24]. In [24], the cancelling pulse is generated by performing inverse fast Fourier transform (IFFT) of the distorted signal after clipping and filtering located in the peak reduction tones (PTRs). This method thus generates neither in-band distortion nor out-of-band (OOB) emission but at a cost of data rate loss. In [19], the cancelling pulse generated using the PRTs is repetitively employed without FFT and IFFT for low-complexity implementation. Nevertheless, given the high computational complexity and high latency, the above-mentioned approaches may not be suitable for practical applications.

Until now, only a few papers address hardware implementation of the PC, and even fewer of them have mentioned its application to actual signals observed in commercial transmitters. Therefore, it is meaningful to investigate the applicability of PC in practical settings through field-programmable gate array (FPGA)-based experiments, and this is our main contribution in this work. Specifically, we investigate the feasibility and realizability of PC through elaboration of hardware design issues upon FPGA implementation. Furthermore, a novel PC scheme with notably low-hardware complexity and improved error vector magnitude (EVM) performance is proposed with its effectiveness experimentally demonstrated.

This paper is organized as follows. Section 2 begins with the introduction of the basic model of PC considered throughout the paper, where the design of cancelling pulse is described. In Section 3, two conventional approaches with their respective advantages and drawbacks for implementing the PC are presented. Furthermore, a novel PC approach with much reduced hardware complexity is proposed and its implementation issues are discussed. Experiments using various signals are performed in Section 4 to demonstrate the benefits that can be achieved with the proposed PC scheme. Finally, our conclusion is given in Section 5.

2 PAPR reduction by peak cancellation

A basic diagram of the PC process considered in this work is sketched in Figure 1. Its principle is to generate cancelling pulses at the time instants where the peaks higher than the predetermined threshold are found. The generated pulses are linearly scaled and rotated with appropriate phase shift such that after their addition the original signals have the peaks reduced to the threshold [17].

2.1 Peak-cancelling process

To perform peak cancellation on the complex baseband signal, the target signal should be oversampled as the Nyquist-rate-sampled signals cannot correctly represent the actual amplitude of peaks of the continuous-time signals [3]. The discrete complex baseband signal s_n has the general form of:

$$s_n = r_n e^{j\theta_n}, \tag{1}$$

where r_n and θ_n represent the amplitude and phase, respectively, at the nth time instant. Suppose that there are N_p peaks that are larger than the predefined threshold A_{th} within a given time period T, and let $\rho_1, \rho_2, \cdots, \rho_{N_p}$ denote the corresponding successive peaks observed at the time instants $n_1, n_2 \cdots, n_{N_p}$, respectively. Let g_n denote the impulse response of the cancelling pulse centred at $n = 0$, i.e. g_0 representing its maximum value. Then, the ith peak cancelling pulse at the time instant n_i, where $i \in \{1, \cdots, N_p\}$, is expressed as:

$$p_n^{(i)} = \left(r_{n_i} - A_{th}\right) g_{n-n_i} e^{j\theta_{n_i}}, \tag{2}$$

where the phase is rotated by $e^{j\theta_{n_i}}$ to match the phase of the corresponding complex-valued peak sample, and the amplitude is scaled by $|r_{n_i} - A_{th}|$ such that the peak value at $n = n_i$ is equal to A_{th} after peak cancellation. Then, the overall signal after cancellation of the entire peaks is expressed as:

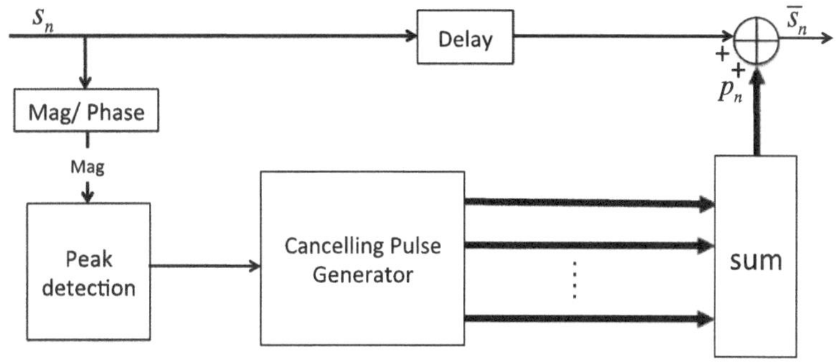

Figure 1 A simplified block diagram of the PC process.

$$\begin{aligned}\bar{s}_n &= s_n - \sum_{i=1}^{N_p} p_n^{(i)} \\ &= s_n - \underbrace{\left[\sum_{i=1}^{N_p} \left(r_{n_i} - A_{th}\right) e^{j\theta_{n_i}} g_{n-n_i}\right]}_{=p_n},\end{aligned} \tag{3}$$

where p_n is all the combined cancelling pulses located at the time instant n_i. If we ignore the change of the amplitude and average power due to the addition of all the cancelling pulses, A_{th} is the maximum amplitude after peak cancellation. In what follows, we refer to the corresponding PAPR determined by A_{th} as a *target PAPR.*

2.2 Effect of the cancelling pulse

The impulse response of the cancelling pulse g_n determines the resulting OOB radiation. In general, g_n should be compliant to the spectral mask of a given target standard. Suppose that s_n is the oversampled version of the band-limited signal and let J denote the oversampling factor such that the $\tilde{s}_{kJ}$ represents the samples at the Nyquist rate for an integer k.

The resulting signal s_n is then expressed as:

$$s_n = \sum_k \tilde{s}_{kJ} h_{n-kJ}, \tag{4}$$

where h_n is the corresponding impulse response of the pulse-shaping filter, and the summation is over the range of k where the impulse response has a non-negligible effect. It is worth mentioning that even though the OFDM signal is not explicitly shaped by a filter, we can still find an equivalent form [25] to represent the virtual pulse-shaping filter. Therefore, Equation 4 also applies to the conventional OFDM signals.

Now we consider the scenario where one peak ρ_i is detected and subtracted by a cancelling pulse $p_n^{(i)}$. It then follows that:

$$\begin{aligned}\bar{s}_n &= s_n - p_n^{(i)} \\ &= \sum_k \tilde{s}_{kJ} h_{n-kJ} - \left[\left(r_{n_i} - A_{th}\right) e^{j\theta_{n_i}} g_{n-n_i}\right].\end{aligned} \tag{5}$$

Suppose that the peak position is precisely given at $n_i = k_i J + b_i$, where k_i and b_i are some integers. Then, Equation 5 is rewritten as:

$$\begin{aligned}\bar{s}_n &= \sum_k \tilde{s}_{kJ} h_{n-kJ} - \left[\left(r_{n_i} - A_{th}\right) e^{j\theta_{n_i}} g_{n-(k_i J + b_i)}\right] \\ &= \sum_{k, k \neq k_i} \tilde{s}_{kJ} h_{n-kJ} + \left\{\tilde{s}_{k_i J} - \left(r_{n_i} - A_{th}\right) e^{j\theta_{n_i}}\right\} h_{n-k_i J} \\ &\quad + \underbrace{\left(r_{n_i} - A_{th}\right) e^{j\theta_{n_i}} \left\{h_{n-k_i J} - g_{n-k_i J - b_i}\right\}}_{d_n^{(i)}},\end{aligned} \tag{6}$$

which indicates that a proper design of g_n will avoid the out-of-band emission, but we still observe that distortion component $d_n^{(i)}$ will affect all the other sampling instants. This distortion component can be nullified only if g_n is set equal to h_n and the peak position occurs at the Nyquist point, i.e. $h_{n-k_i J} = g_{n-k_i J - b_i}$.

In other words, if the cancelling pulse g_n is identical to h_n, then no out-of-band regrowth will occur as the signal power is confined inside the pass-band of h_n. In fact, the clipping and filtering approach presented in [13] corresponds to this special case where g_n is the periodic sinc function [25]. Since the periodic sinc function has non-negligible impulse response over entire OFDM symbol, it causes considerable peak regrowth. Therefore, in practice, we wish to choose g_n such that its side lobe (in time domain) vanishes rapidly, and yet, its frequency response has acceptable out-of-band emission in terms of adjacent channel leakage ratio (ACLR).

2.3 Design of the cancelling pulse

As we have seen, the impulse response of the peak-cancelling pulse g_n, which is essentially a finite impulse response (FIR) filter, serves as a trade-off between the out-of-band radiation and in-band distortion. Specifically,

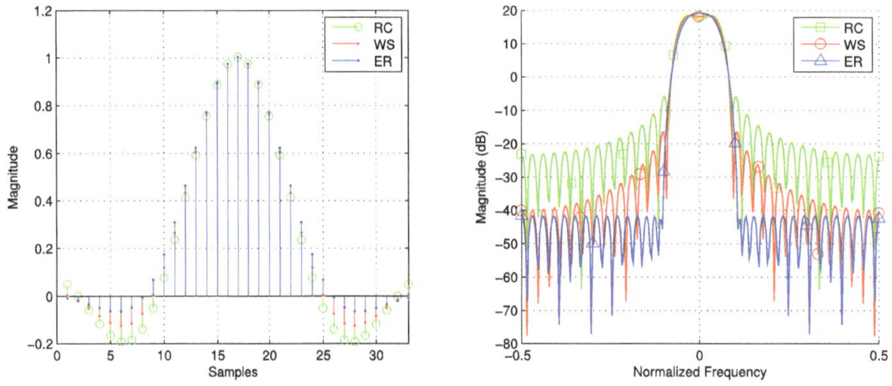

Figure 2 Performance comparison of the three different cancelling pulses designed based on different FIRs. The left hand side figure shows their impulse responses, whereas the right hand side figure shows the corresponding frequency responses. RC, raised cosine; WS, windowed sinc; ER, equal ripple.

shorter impulse response results in lower in-band distortion, but it will cause an increasing amount of out-of-band radiation that may violate the specified spectral mask. Therefore, careful design of the cancelling pulse is essential.

However, there exists no solid algorithm or closed-form deviation for finding the best cancelling pulse, and thus, exhaustive attempts are necessary to find the suitable one for a specified signal and to satisfy the design requirements. For instance, three different filters (cancelling pulses) of the same length are illustrated in Figure 2 with their respective impulse response and frequency response. Here, the windowed sinc (WS) is obtained by multiplying Kaiser window to sinc function. The raised cosine (RC) and sinc are both Nyquist filters as can be seen from the left hand of the figure. The equal ripple (ER) filter is obtained by the well-established Parks-McClellan algorithm [26] which minimizes the error in pass and stop bands by employing Chebyshev approximation. The performance of the peak cancellation based on the three cancelling pulses is demonstrated in Figure 3 using numerical simulation, where a WCDMA signal is used as its test signal. In this figure, as a practical measure for PAPR, we adopt the complementary cumulative distribution function (CCDF) of the instantaneous power normalized by its average power.

From the left hand side of Figure 3, we observe that similar PAPR performance is achieved. However, comparison of the power spectra in Figure 3 with their corresponding frequency responses in Figure 2 reveals that the frequency response of the cancelling pulses has the dominant effect on the resulting spectrum after peak cancellation.

The effects of pulse length on the in-band distortion (i.e. EVM) and out-of-band distortion (i.e. ACLR) are reported in Figure 4a,b, respectively. The measurement of EVM and ACLR follows the 3rd Generation Partnership Project (3GPP) frequency division duplexing (FDD) WCDMA downlink specification [27]. We have concluded

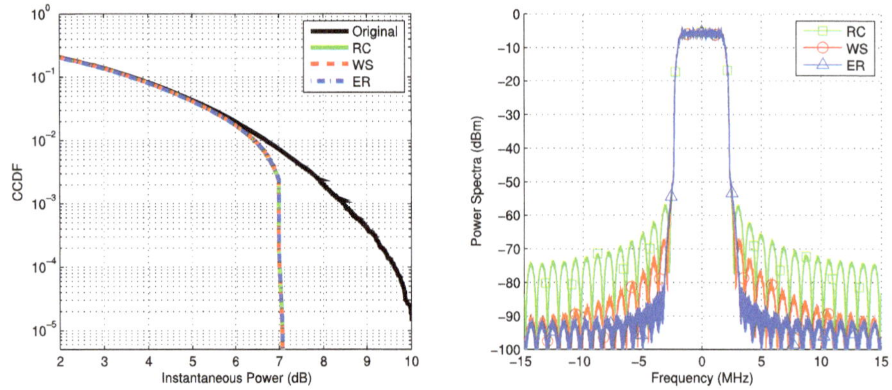

Figure 3 Peak-cancelled performance due to the different peak-cancelling pulses introduced in Figure 2. The left hand side figure shows the PAPR performance in terms of CCDF, whereas the right hand side figure shows the corresponding power spectra. CCDF, complementary cumulative distribution function; RC, raised cosine; WS, windowed sinc; ER, equal ripple.

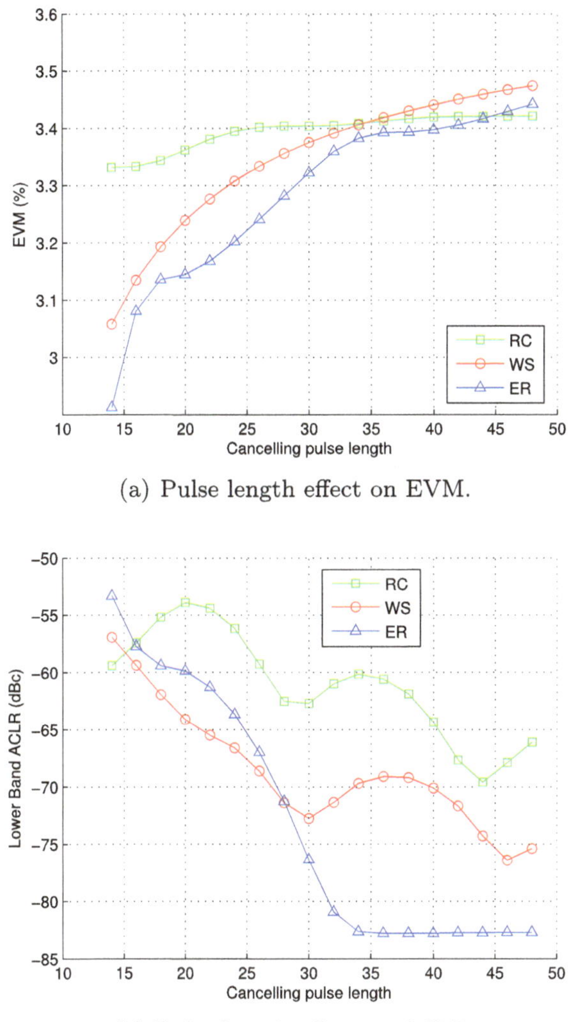

(a) Pulse length effect on EVM.

(b) Pulse length effect on ACLR.

Figure 4 Effect of different pulse length. (a) Pulse length effect on EVM. **(b)** Pulse length effect on ACLR. RC, raised cosine; WS, windowed sinc; ER, equal ripple; EVM, error vector magnitude; ACLR, adjacent channel leakage ratio.

in the last subsection that a longer pulse introduces more symbol errors to the target signal, and this is validated by Figure 4a. On the other hand, it can be easily grasped by inspection of Figure 4b that ACLR reduces with longer cancelling pulse length. However, the curves shown in the figure have some fluctuations. This can be understood by inspecting Figure 5, where the impulse response for a typical low-pass filter is shown. Truncation length of the impulse response also affects the shape of the resulting frequency response. For instance, L_1 has worse out-of-band attenuation than L_3 but may be better than L_2, because L_2 has non-zeroes at the head and end. This heuristic observation reveals a basic clue for choosing the cancelling pulse length: the cancelling pulse should be as short as possible to minimize the distortion but should be long enough to give admissible ACLR.

It can also be observed from Figure 4a,b that the cancelling pulse generated by an equal ripple filter gives the best performance both in terms of EVM and ACLR. Better EVM performance is due to the lower side lobe in time domain, and lower ACLR is due to higher out-of-band attenuation of the cancelling pulse, as can be seen from Figure 2. It can also be seen from Figure 4a that even though the raised-cosine filter completely conforms to the pulse-shaping filter for WCDMA signal, it shows worst performance when the cancelling pulse length is short. This is mainly due to the distortion effect and the high sidelobe of the raised-cosine filter. In general, designing cancelling pulse with Parks-McClellan algorithm leads to better performance than other filter types that are frequently found in the literature.

3 Hardware implementation of peak cancellation

From this section and later, we focus on efficient implementation of PC by hardware. We first introduce the conventional approach for the implementation of PC as well as its alternative method, which is followed by the description of the proposed PC as well as their detailed implementation issues.

3.1 Conventional implementation: scheme 1

A conventional scheme for implementing the PC [28], which we refer to as a *scheme 1* in what follows, is shown in Figure 6. The instantaneous magnitude and phase of the complex signal are computed with coordinate rotation digital computer (CORDIC) algorithm. The 'Peak Detect' block, which contains some registers and comparators, identifies the peak magnitude higher than the threshold. When a peak is detected, the corresponding phase and the magnitude amount higher than the threshold are latched. These values are used to scale and rotate the normalized real cancelling pulses, and these operations are accomplished by multiple CORDIC cores.

The cancelling-pulse generator block contains one counter, which is directly connected to the address port of the read only memory (ROM). The pre-determined cancelling pulse, which is compliant with the target signal spectrum, is stored in the ROM. The counter will be triggered as soon as one peak is found, and it is reset when it counts to the length of cancelling pulse L.

To deal with the occurrence of intensive peaks, multiple cancelling-pulse generators are necessary to generate independent cancelling pulses, and Figure 7 illustrates such an operation where the signal is propagating from left to right. When the first peak in the left is detected, the first generator will be turned on for L clocks to generate the first cancelling pulse. In the event that the second and third peaks are detected when the first generator is

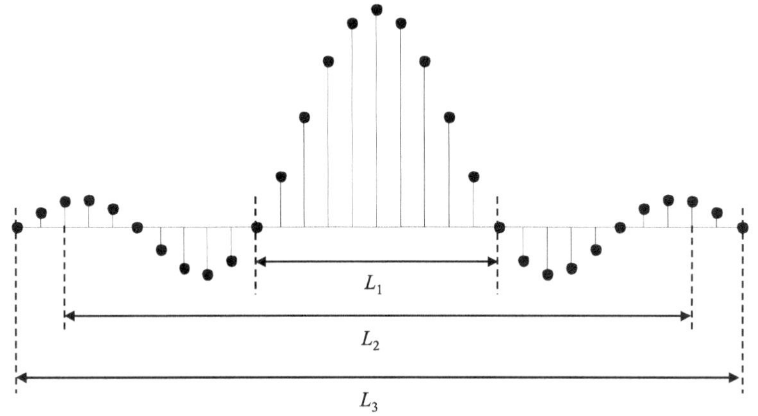

Figure 5 **An example of typical impulse response of cancelling pulse with various truncation lengths.**

still on, the second and third generators will be triggered to generate the second and third cancelling pulses. If the fourth peak is detected when ROM 1 is free, the first generator will be reused to generate the fourth cancelling pulse. In summary, when the previous cancelling-pulse generator is triggered and the next peak is found, the next cancelling-pulse generator will be triggered in sequel. This successive process continues until no peak higher than the threshold is found. All the outputs of the cancelling-pulse generators are summed and finally subtracted from the delayed original signal.

It becomes clear that the resource complexity of this scheme is bounded by the number of available cancelling-pulse generators. Note that whether this scheme can cancel all the peaks in one pass or not depends on the specific parameters such as the required pulse length, targeted PAPR and the number of cancelling-pulse generators. It is easy to see that the generators in the upper side of Figure 6 have a higher probability to be used while the cancelling-pulse generators in the bottom may be idle most of the time. Therefore, some compromise is necessary to determine how many cancelling-pulse generators are used. As a rule of thumb, five or six cancelling-pulse generators are enough when a reasonable threshold value is assumed, as the average number of the peaks above the threshold monotonically decreases as the threshold increases [3]. However, since the number of peaks itself is a random variable, with the fixed number of pulse generators, a failure to peak cancellation may occasionally happen. In this case, some iterative processing structure should be introduced, which may lead to increasing latency.

3.2 Conventional implementation: scheme 2

To overcome the issue of a peak cancellation failure, an alternative implementation scheme [29], which we refer to as *scheme 2*, can be applied. Let us rewrite Equation 3 as:

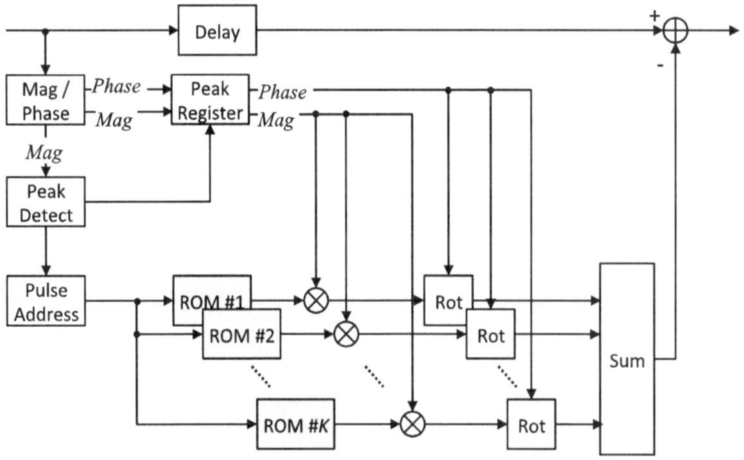

Figure 6 **Architecture of conventional PC hardware implementation - scheme 1.** ROM, read only memory.

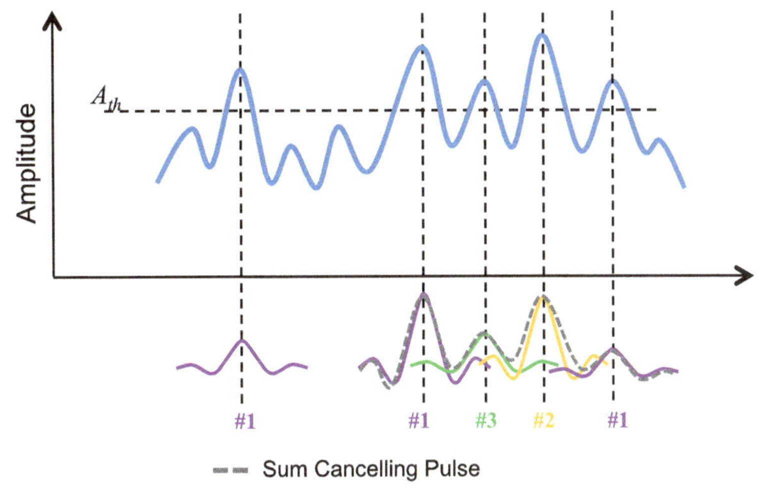

Figure 7 Principle of multiple cancelling pulse generation.

$$\bar{s}_n = s_n - \sum_{i=1}^{N_p} \left[\left(r_{n_i} - A_{th} \right) \cdot \delta \left(n - n_i \right) e^{j\theta_{n_i}} \right] * g_n, \quad (7)$$

where g_n serves as the coefficients (impulse response) of an FIR filter and $*$ denotes the convolution operation. The cancelling pulses can be thus generated by propagating the delta function train to this filter, and this principle is illustrated in Figure 8 and its hardware implementation is given in Figure 9.

Similar to the first scheme, a CORDIC core is used to compute the magnitude and phase of the signal. The 'Pulse Gen' block outputs a sample which is properly scaled and rotated when a peak is detected. The resulting delta function train is complex-valued. It is easy to see from Figure 9 that the resource complexity relies primarily on the filter, while the complexity of its counterpart in Figure 6 depends on the number of CORDIC cores. Implementing an FIR filter requires a large number of multipliers, which makes the scheme 1 preferable as it consumes less resources.

In summary, the first scheme has lower complexity in terms of fewer multipliers but has the problem of peak generation failure. The second scheme is easier to implement but has higher resource complexity due to the use of FIR filter. In order to cope with the peak generation failure, however, the first scheme may require more iterations which in turn increase the complexity several fold. In this sense, the second scheme that has a fixed hardware overhead is preferred.

3.3 The proposed peak cancellation

As can be observed from Figure 7, a sum of multiple overlapped pulses forms the final pulse when intensive peaks occur. The tails of the previous pulses may happen to be added in-phase to the peaks in sequel, resulting in less effective peak reduction. Inspired by these observations, we proposed a novel approach of peak cancellation. The principle of the proposed approach is illustrated in Figure 10. Instead of generating complete cancelling pulses stored in the ROMs as the one shown in Figure 6, the proposed scheme generates truncated

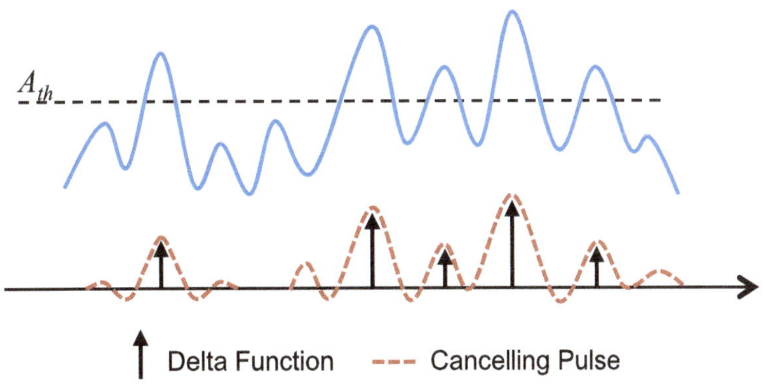

Figure 8 Principle of cancelling pulse generated by delta function trains.

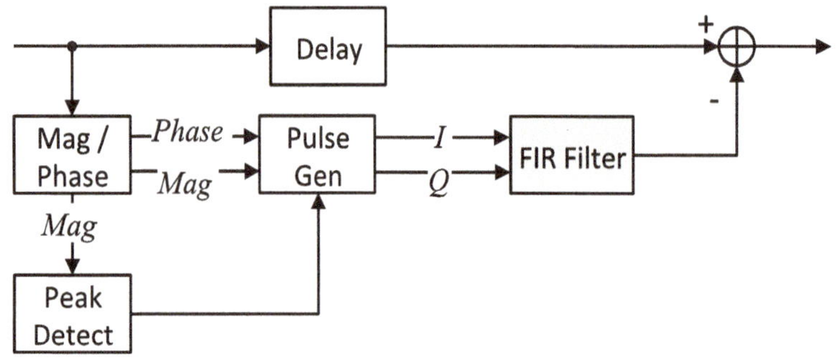

Figure 9 Architecture of conventional hardware implementation of peak cancellation - scheme 2. FIR, finite impulse response.

cancelling pulses when they are overlapping with each other. More specifically, when the interval of two contiguous peaks is detected to be less than the predefined cancelling-pulse length, the generation of the first cancelling pulse terminates in the middle of the interval and immediately triggers the second cancelling pulse. Since this scheme reduces the length for overlapping cancelling pulses, lower in-band error can be expected in view of the observation given in the Section 2.3.

However, as can be seen from Figure 10, the cancelling pulses show obvious discontinuity which would produce substantial out-of-band emission. To mitigate the out-of-band emission, a filter can be applied to smooth the discontinuities, and the resulting waveform of the smoothed cancelling pulse is also denoted by the dashed curve in Figure 10. In our method, we use a simple moving average filter as a smoothing filter, with which the out-of-band spurious level caused by non-continuous cancelling pulses can be reduced. Note that the smoothing filter is actually optional as we can still obtain a moderate out-of-band emission without it, considering that the occurrence of high peaks is a rare event [3]. The smoothing filter is only necessary in the ACLR-prior circumstance. Normally,

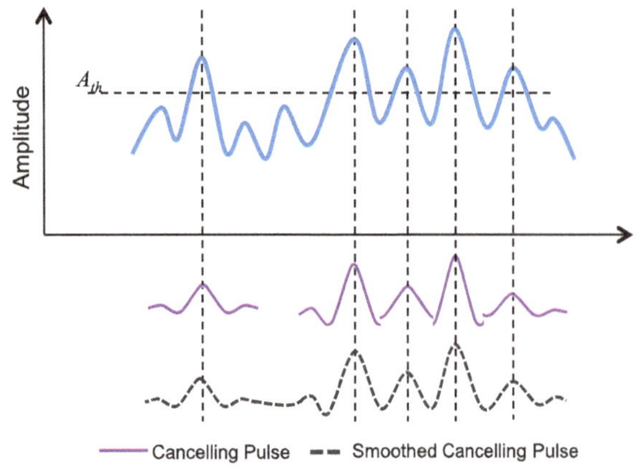

Figure 10 Principle of the proposed peak cancellation approach.

a short-length smoothing filter is adequate to further improve the ACLR performance. As the filter length is very short, it gives negligible effect on EVM.

3.4 Implementation of the proposed peak cancellation

We now describe the hardware implementation aspects of the proposed scheme. A detailed block diagram of the proposed scheme is given in Figure 11. Some example waveforms of the internal signals labelled in Figure 11 are plotted in Figure 12.

Similar to the previous schemes, a CORDIC core is used to compute the instantaneous magnitude and phase of the signal. The 'Peak Detect' block, which contains some registers and comparators, outputs a '1' when a magnitude peak is found, as shown in the second plot in Figure 12. The output of the 'Peak Detect' block is connected to the enable ports of the two latches which store the magnitude and phase of the corresponding peak. The 'Interval Locator' block generates a '1' in the middle of two peaks when the interval of these peaks is less than the cancelling-pulse length. The outputs of 'Interval Locator' and 'Peak Detect' (I) and (II) are combined with an OR gate. A 'Delay' block is used to align these two signals as the 'Interval Locator' has a fixed delay. The 'Cancelling Pulse Duration" block produces enable signal (III) for the counter which outputs the address of the ROM. The counting direction of the counter is controlled by a latch output which is reversed when triggered by the output of the OR gate. Operation of these signals can be easily seen from the second and third plots in Figure 12.

The ROM output (V) is then scaled and rotated by the latched magnitude and phase to form the cancelling pulses and fed to the FIR filter. The smoothed cancelling pulses (VI) are subtracted from the delayed original signal to form the PAPR-reduced signal. It should be noted that the four 'Delay' blocks shown in Figure 12 do not necessarily indicate that the delay values for these blocks are identical.

The moving average filter used here for smoothing the truncated pulses is much shorter than the one used in

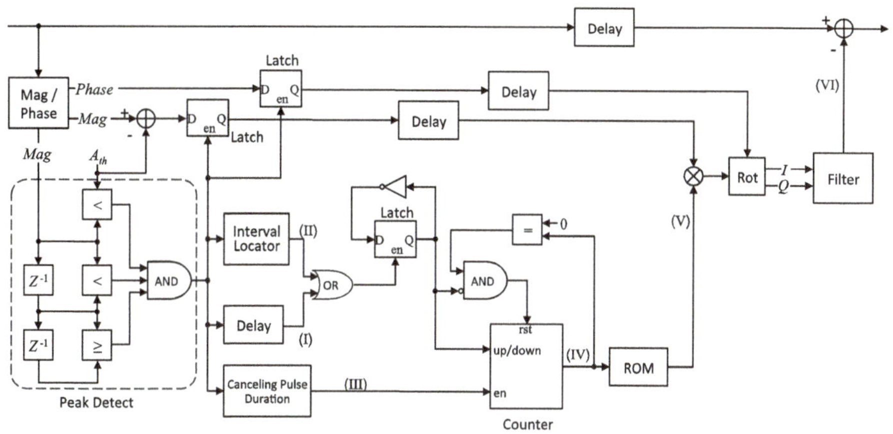

Figure 11 Hardware circuit of the proposed PC scheme using FPGA. ROM, read only memory.

Figure 9. Thus, it has less hardware complexity, even though both of them contain two CORDIC cores.

4 Performance evaluation of the proposed peak cancellation

In this section, the experimental results for implementation with FPGA are reported. We will compare the hardware complexity (in terms of resource utilization) of the proposed PC and conventional PC first. This is followed by the experimental results using a standard-conforming long-term evolution (LTE) signal as well as multi-standard signals.

4.1 Experiment description

The implementation is carried out using an FPGA evaluation board VC707, which contains a Xilinx Virtex-7 XC7VX485T-2FFG1761C device. The test WCDMA/LTE signal (baseband IQ) is generated by Matlab on the computer and is stored in a bank of RAMs in the FPGA as the signal source. The output of peak cancellation is captured by a series of integrated logic analyzers (ILAs) in parallel, which is arranged in a time-interleaved manner so as to receive the signal of long length. This signal is then transferred to computer through USB port and circularly shifted to align with the original signal and is analysed by Matlab.

A 245.76-MHz clock is synthesized by the on-chip mixed-mode clock manager (MMCM), which uses the on-board 200-MHz oscillator as the reference. The clock is set to integer times (64 times in our example implementation) of 3.84 MHz, targeting the specification of the 3GPP WCDMA and LTE signals.

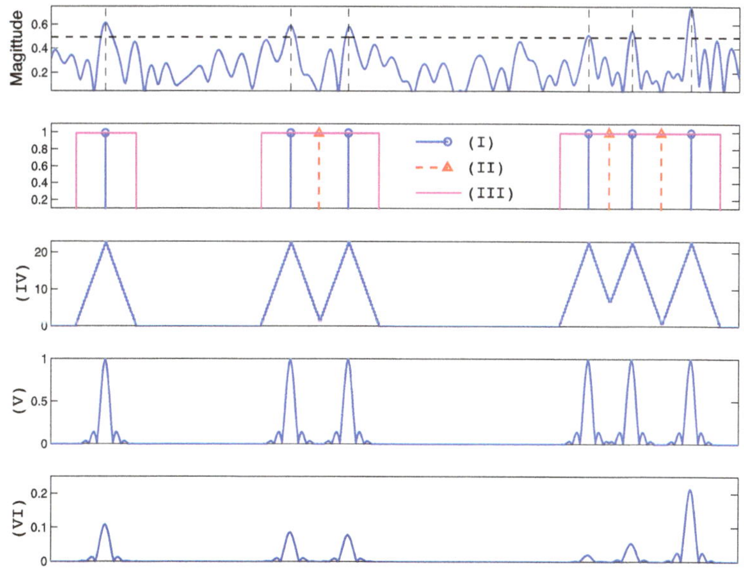

Figure 12 Internal signal illustration of the proposed PC circuit.

The cancelling-pulse length is set to 115, and the order of the moving averaging filter is set to 6 for the proposed PC. The complex cancelling pulse is assumed to support the asymmetric signal spectrum. As the smoothing filter is symmetric and one complex-variant multiplication uses three multipliers, the filter consumes $6 \div 2 \times 3$ multipliers. Furthermore, to optimize the speed, three multipliers are used to implement a CORDIC algorithm and the total of six multipliers by the two CORDIC cores. Another two multipliers are used to scale the cancelling pulse (real magnitude multiplying the complex cancelling pulse in I/Q) so that the total consumption of multipliers is 17.

4.2 Resource utilization

The FPGA resource utilization, which is evaluated with Resource Estimator using post-mapping, for the three PC approaches introduced are summarized in Table 1 where the scheme 1 is operated either without iteration or with one iteration. Only primary resources such as the slices and look-up tables (LUTs) are listed here, since they are generic for any FPGA, while the other resources used for our test (primarily block RAMs and IOs for ChipScope) are specific to the FPGA board and thus are not taken into consideration.

In scheme 1, the PC is comprised of four cancelling-pulse generators, and therefore, four RAMs are used to store the cancelling pulse and accordingly four CORDIC cores are needed. One can see that with a single iteration, the hardware resource of scheme 1 is roughly doubled. It is obvious that all of the three schemes consume only a small portion of the resources such as flip flops (FFs) and slice LUTs. The conventional scheme 2 costs more multipliers. This is because it requires implementing the FIR filter to generate the cancelling pulses of equivalent length. For circuit integration, the multipliers are most concerned as they generally cost more power and take up larger area than other simple arithmetic elements. In this sense, the proposed PC is rather cost effective as it consumes fewer multipliers. Although the conventional scheme 1 without iteration requires even fewer multipliers, the peak-missing problem will make it unworkable as will be demonstrated in the subsequent subsections. In view of this, the proposed PC yields the lowest complexity.

The signal has 16-bit precision, which gives a noise floor of approximately −80 dB. Note that since the systems such as LTE and WCDMA require an ACLR of −55 dB at most, actual implementation needs less bits.

Table 1 FPGA resource utilization table

	FFs	Slice LUTs	DSP48 (Multipliers)	RAM
Proposed	3,920	3,439	17	1
Conventional scheme1 (no iteration)	3,354	3,117	15	4
Conventional scheme1 (one iteration)	6,785	6,341	30	8
Conventional scheme2	5,749	5,383	171	0
Available in FPGA	607,200	303,600	2,800	1,030

FF, flip flop; LUT, look-up table.

4.3 Test results using LTE signals

The rest of the paper is devoted to the comparison of the three approaches in terms of in-band distortion, out-of-band emission and realizable PAPR.

The applied LTE signal is a 16-QAM OFDM signal with 1,200 subcarriers within 18.015-MHz occupied bandwidth and 20-MHz channel bandwidth. The basic sampling rate for such signal is 30.72 MHz since the FFT size is 2,048. The PC for LTE signal is operated at 245.76 MHz which represents an oversampling rate of 8. The signal spectrum and cancelling-pulse spectrum are demonstrated in Figure 13. The cancelling pulse used here was designed by Chebyshev approximation, as described in Section 2.3.

4.3.1 Peak power reduction capability comparison for different PC schemes

In what follows, we evaluate the CCDF of the normalized instantaneous power after employing the PC schemes considered in this work.

The CCDF plots of the signal with the proposed PC and the two conventional PC schemes are demonstrated in Figure 14 where the target PAPR values are set as 6 and 8 dB. As can be seen from the figure, the conventional scheme 1 without iteration exhibits high peaks due to

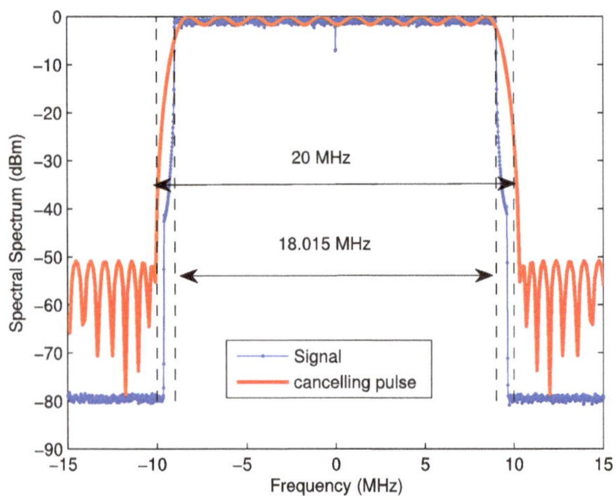

Figure 13 Demonstration of the spectra of LTE signal and its corresponding cancelling pulse.

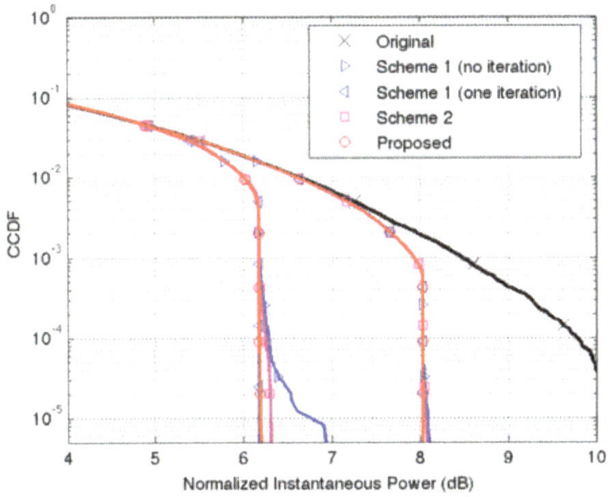

Figure 14 CCDF comparison in terms of normalized instantaneous power for different PC schemes with the target PAPR of 6 and 8 dB. CCDF, complementary cumulative distribution function.

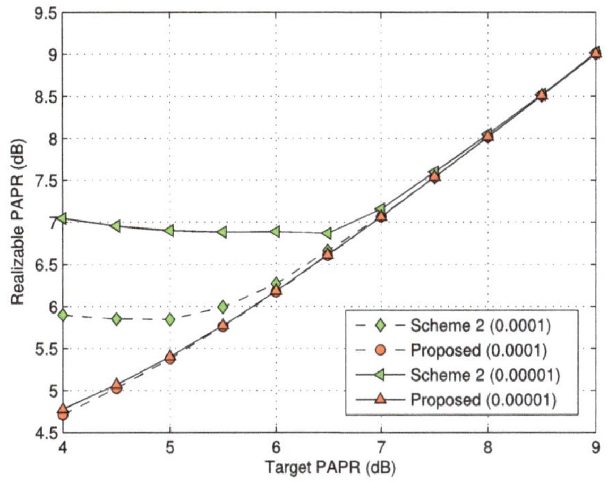

Figure 15 Realizable PAPR versus target PAPR for LTE signal. The realizable PAPR is defined as the corresponding threshold-normalized instantaneous power values measured at the CCDF of 10^{-4} and 10^{-5}. PAPR, peak-to-average power ratio.

the cancelling peak generation failure under the condition that only four ROMs are made available for this purpose. This problem can be solved by adding a single iteration, but this doubles the required resources as can be seen from Table 1. The conventional scheme 2 outperforms the scheme 1 without iteration, at a cost of increasing hardware complexity. The proposed PC, on the other hand, achieves performance comparable to the scheme 1 with iteration, even with lower hardware complexity.

Considering the fact that the scheme 1 may not necessarily achieve a target peak power reduction performance without iteration, the experiments hereinafter will be focused exclusively on the proposed scheme and conventional scheme 2.

An important factor for evaluating the performance of a PAPR-reduction technique is the actually realizable PAPR. It is well known that the CAF causes an unavoidable peak power regrowth, and this will make the precise control of PAPR challenging without resorting to increasing complexity such as iterative use of CAF [20], which may also introduce a prohibitive amount of latency. In contrast, the peak power achieved by PC schemes can reduce the effect of the peak power regrowth and thus achieve the PAPR close to the target PAPR. Figure 15 compares the peak power reduction capabilities of the schemes 2 and the proposed scheme. In this figure, in response to the target PAPR, the actual threshold values of the normalized instantaneous power at given specific CCDF of 10^{-4} and 10^{-5} are plotted as a realizable PAPR. It is observed that the proposed scheme outperforms the scheme 2 from the viewpoint of PAPR reduction as well.

4.3.2 In-band and out-of-band distortion comparison for different PC schemes

The constellation plots of the user data, which are 16-QAM signal, with target PAPR of 5 and 7 dB, are delineated in Figure 16. The pilot points (reference signal) in red, which is QPSK, are used to rotate, rescale and equalize the user data. It can be inspected from the figure that, for a target PAPR of 5 dB, the constellation of the user data is rather dispersive.

The measured EVM and ACLR with different target PAPRs are demonstrated in Figures 17 and 18, respectively. It can be inspected from the figures that the proposed scheme yields lower in-band distortion (in terms of lower EVM) and higher out-of-band distortion (in terms of higher ACLR). This stems from the fact that since the proposed method generates shorter cancelling pulses, it can essentially reduce the in-band distortion in time domain but with the broader side lobes in frequency domain, as discussed in Section 2. Also, even though the proposed scheme has higher out-of-band emission, the observed ACLR with this parameter setting is still satisfactory for practical applications.

4.4 Test results using various signals

Finally, multiple tests have been performed to validate the proposed scheme for a more general framework with multi-carrier and multi-standard signals.

4.4.1 Carrier-aggregated LTE signals

The first test here assumes an LTE-Advanced signal with carrier aggregation of three 20-MHz carrier components (CCs). The spacing of the two carriers are set to 20.1 MHz, which is an integer multiple of 15 KHz in order to maintain the OFDM subcarrier spacing. The target

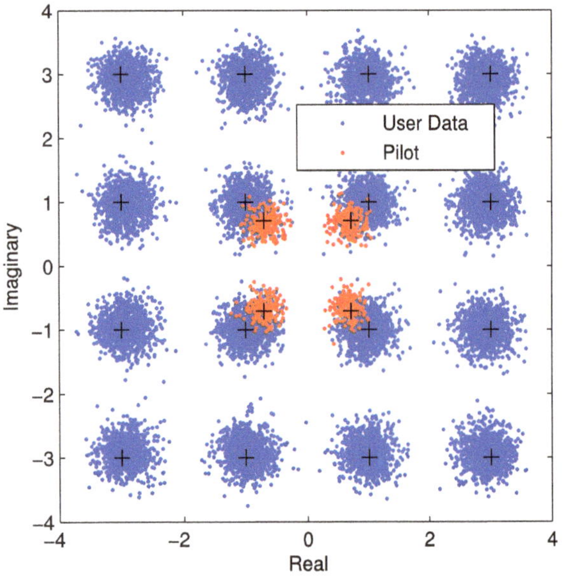

(a) when target PAPR=5 dB.

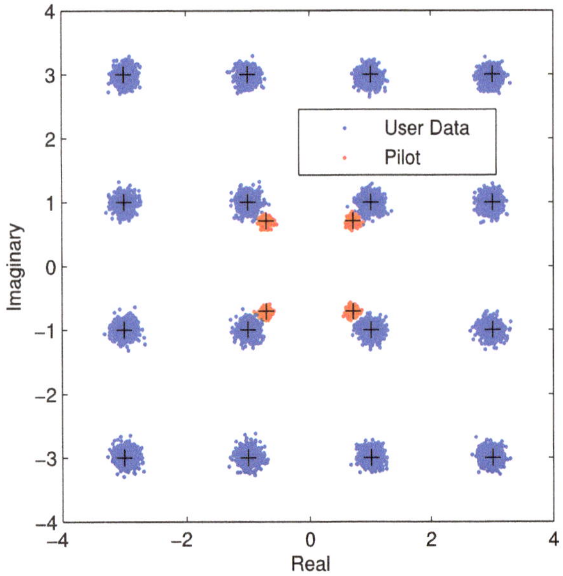

(b) when target PAPR=7 dB.

Figure 16 Constellation plots of user data. (a) When target PAPR = 5 dB. **(b)** When target PAPR = 7 dB. PAPR, peak-to-average power ratio.

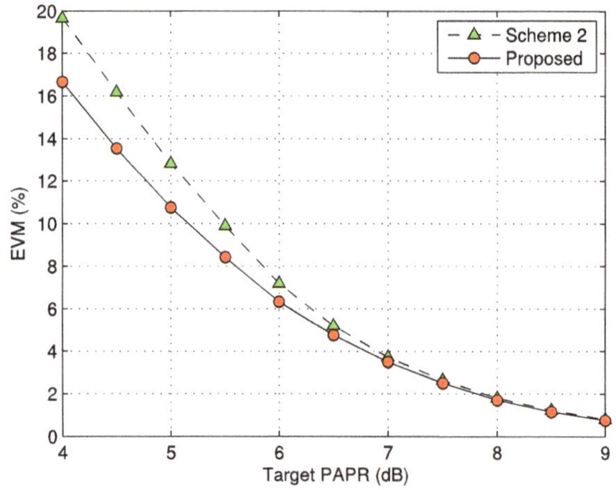

Figure 17 EVM versus target PAPR for LTE signal. EVM, error vector magnitude; PAPR, peak-to-average power ratio.

PAPR is set to 7 dB which can yield reasonable EVM. The resulting CCDF is plotted in Figure 19a to show the effectiveness of the proposed scheme. The spectral density plots provided in Figure 19b show that, with the given cancelling pulse, the proposed scheme yields very limited spectral regrowth. The EVMs for the respective CCs under the condition of different realizable PAPRs are plotted in Figure 20. We observe that the EVM curves for different CCs almost overlap with each other, indicating the effectiveness of the proposed PC under the system operated with multiple carrier frequencies.

4.4.2 Multi-standard signals

The second test uses a multi-standard signal, which contains three WCDMA carriers and a 20-MHz LTE carrier, where the two standard signals are spaced by 40 MHz. The WCDMA signal used in this paper is generated following the specification defined by 3GPP test model 3, for which 64 users (16-QAM) are multiplexed in dedicated physical data channel (DPDCH) to form the 3.84 Mc/s chip rate. The WCDMA chip is oversampled 64 times, and the PC is operated at a corresponding sampling frequency of 245.76 MHz.

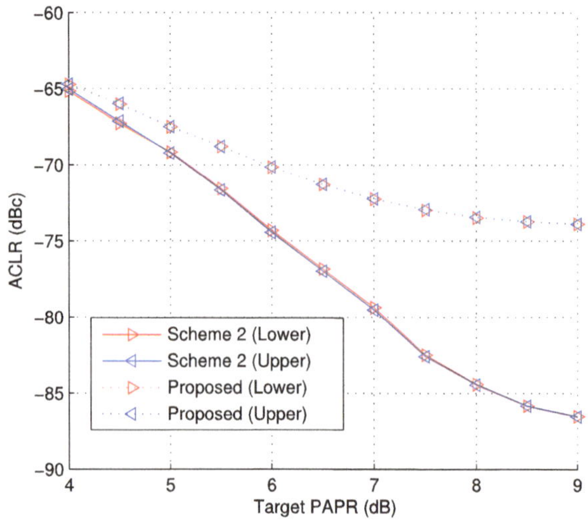

Figure 18 ACLRs (both lower and upper side bands) versus target PAPR for LTE signal. ACLR, adjacent channel leakage ratio; PAPR, peak-to-average power ratio.

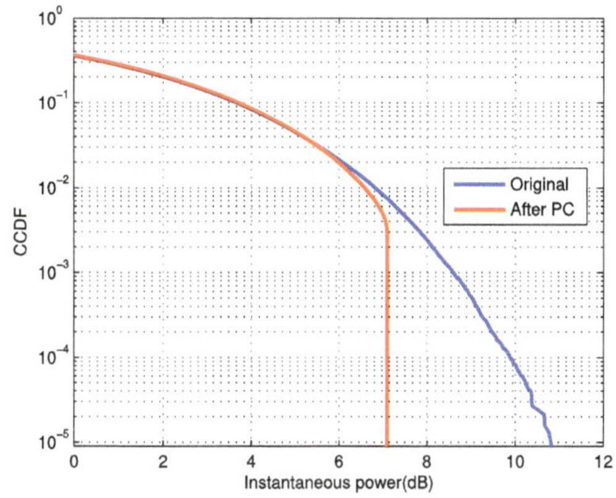

(a) CCDF when target PAPR is set to 7dB.

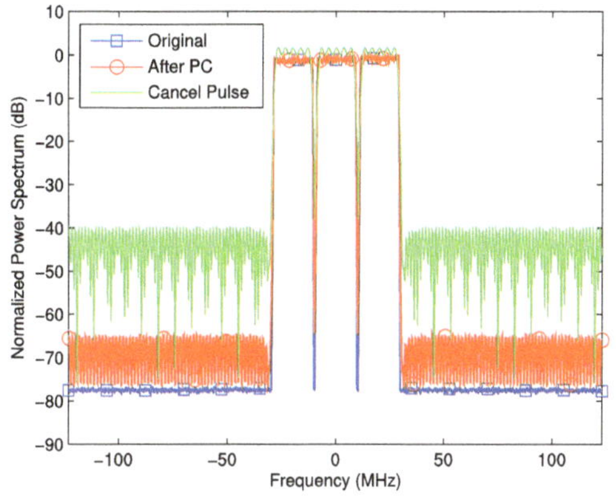

(b) The power spectral density plots when target PAPR is set to 7dB.

Figure 19 The performance comparison of PC for three-carrier LTE-Advanced signal. (a) CCDF when target PAPR is set to 7 dB. **(b)** The power spectral density plots when target PAPR is set to 7 dB. CCDF, complementary cumulative distribution function; PC, peak cancellation.

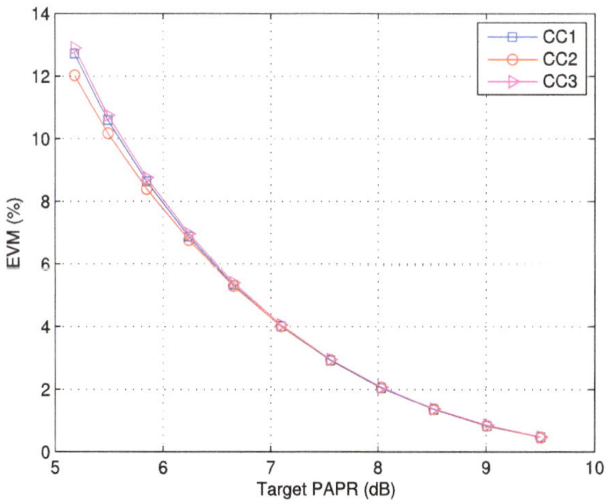

Figure 20 EVM of each LTE-Advanced component carrier. EVM, error vector magnitude; PAPR, peak-to-average power ratio.

The resulting CCDF and power spectral plots are given in Figure 21a,b, which demonstrate the adaptability of the proposed PC to the signal with non-contiguous spectrum. The proposed scheme, with a realizable PAPR of around 7 dB, can meet the imposed requirement on power spectrum well as can be inspected from Figure 21b.

5 Conclusions

In this paper, the peak cancellation technique as a general purpose PAPR reduction has been addressed. The design issues of cancelling pulses that determine the performance of PC have been also discussed. Our main focus was on their FPGA implementation with a special consideration on the hardware complexity. A novel PC scheme with notably low-hardware complexity has been also presented. The experimental results using various signals have demonstrated the validity of the proposed approach.

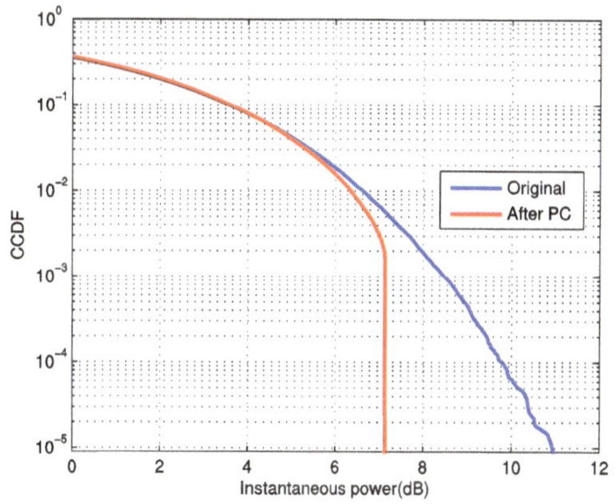

(a) CCDF performance when target PAPR is set to 7dB.

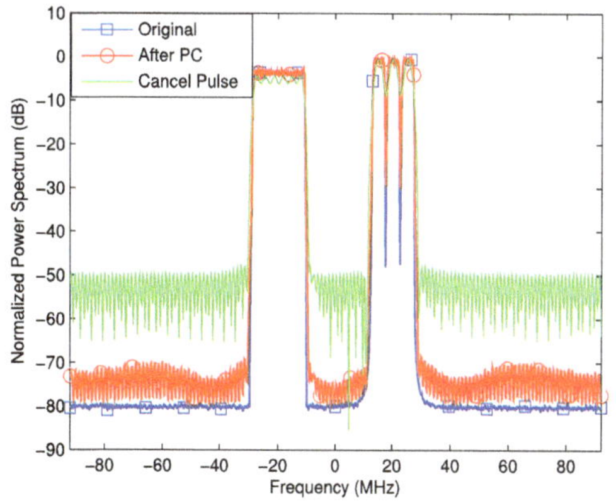

(b) The power spectral density plots when target PAPR is set to 7dB.

Figure 21 The performance comparison of PC for the aggregated signal of three-carrier WCDMA and LTE signal. (a) CCDF performance when target PAPR is set to 7 dB. **(b)** The power spectral density plots when target PAPR is set to 7 dB. CCDF, complementary cumulative distribution function; PC, peak cancellation.

Competing interests

The authors declare that they have no competing interests.

References

1. R van Nee, R Prasad, *OFDM for Wireless Multimedia Communications*. (Arthech House, Inc., Norwood, MA, USA, 2000)
2. H Honkasalo, K Pehkonen, MT Niemi, AT Leino, WCDMA and WLAN for 3G and beyond. IEEE Wireless Commun. **9**(2), 14–18 (2002)
3. H Ochiai, H Imai, On the distribution of the peak-to-average power ratio in OFDM signals. Commun. IEEE Trans. **49**(2), 282–289 (2001)
4. SH Han, JH Lee, An overview of peak-to-average power ratio reduction techniques for multicarrier transmission. Wireless Commun IEEE. **12**(2), 56–65 (2005)
5. T Jiang, Y Wu, An overview: peak-to-average power ratio reduction techniques for OFDM signals. Broadcasting IEEE Trans. **54**(2), 257–268 (2008)
6. Y Rahmatallah, S Mohan, Peak-to-average power ratio reduction in OFDM systems: a survey and taxonomy. Commun. Surveys Tutorials IEEE. **15**(4), 1567–1592 (2013)
7. O Väänänen, J Vankka, K Halonen, Simple algorithm for peak windowing and its application in GSM, EDGE and WCDMA systems. IEE Proc. Commun. **152**(3), 357–362 (2005)
8. W-J Kim, K-J Cho, SP Stapleton, J-H Kim, Doherty feed-forward amplifier performance using a novel crest factor reduction technique. Microwave Wireless Components Lett. IEEE. **17**(1), 82–84 (2007)
9. RW Bauml, RFH Fischer, JB Huber, Reducing the peak-to-average power ratio of multicarrier modulation by selected mapping. Electron. Lett. **32**(22), 2056–2057 (1996)
10. SH Müller, RW Bäuml, RFH Fischer, JB Huber, OFDM with reduced peak-to-average power ratio by multiple signal representation. Ann. Des Té, lécommunications. **52**(1-2), 58–67 (1997)
11. RJ Baxley, GT Zhou, Comparing selected mapping and partial transmit sequence for PAR reduction. Broadcasting IEEE Trans. **53**(4), 797–803 (2007)
12. X Li, LJ Cimini, Effects of clipping and filtering on the performance of OFDM. Commun. Lett. IEEE. **2**(5), 131–133 (1998)
13. H Ochiai, H Imai, Performance analysis of deliberately clipped OFDM signals. Commun. IEEE Trans. **50**(1), 89–101 (2002)
14. R O'Neill, LB Lopes, in *IEEE International Symposium on Personal, Indoor and Mobile Radio Communications (PIMRC'95). Wireless: Merging onto the Information Superhighway*. Envelope variations and spectral splatter in clipped multicarrier signals. vol. 1 IEEE (Toronto, 1995) 71–75
15. HN Mistry, Implementation of a peak windowing algorithm for crest factor reduction in WCDMA Master's thesis, School of Engineering Science-Simon Fraser. University, Burnaby, BC, Canada, (2006)
16. M Pauli, H-P Kuchenbecker, Minimization of the intermodulation distortion of a nonlinearly amplified OFDM signal. Wireless Personal Commun. **4**(1), 93–101 (1997)
17. T May, H Rohling, in *IEEE Vehicular Technology Conference (VTC98)*. Reducing the peak-to-average power ratio in OFDM radio transmission systems, vol. 3 (IEEE Ottawa, 1998), pp. 2474–2478
18. L Dan, Y Xiao, W Ni, S Li, Improved peak cancellation for PAPR reduction in OFDM systems. Commun. IEICE Trans. **E93-B**(1), 198–202 (2011)
19. H-B Jeon, J-S No, D-J Shin, A new PAPR reduction scheme using efficient peak cancellation for OFDM systems. Broadcasting IEEE Trans. **58**(4), 619–628 (2012)
20. J Armstrong, Peak-to-average power reduction for OFDM by repeated clipping and frequency domain filtering. Electron. Lett. **38**(5), 246–247 (2002)
21. Y Wang, Z Luo, Optimized iterative clipping and filtering for PAPR reduction of OFDM signals. Commun. IEEE Trans. **59**(1), 33–37 (2011)
22. BS Krongold, DL Jones, in *Proceedings of 2003 IEEE International Conference on Acoustics, Speech, and Signal Processing (ICASSP '03)*. PAR reduction in OFDM via active constellation extension, vol. 4, (April 2003), pp. IV–525–8
23. J Tellado, JM Cioffi, PAR reduction in multicarrier transmission systems. ANSI Document, T1E1.4 Tech. Subcommittee. **4**, 1–14 (1998)
24. L Wang, C Tellambura, Analysis of clipping noise and tone-reservation algorithms for peak reduction in OFDM systems. Vehicular Technol. IEEE Trans. **57**(3), 1675–1694 (2008)
25. H Ochiai, On instantaneous power distributions of single-carrier FDMA signals. Wireless Commun. Lett. IEEE. **1**(2), 73–76 (2012)
26. IW Selesnick, CS Burrus, Exchange algorithms that complement the Parks-Mcclellan algorithm for linear-phase FIR filter design. Circuits Syst. II: Analog Digital Signal Process. IEEE Trans. **44**(2), 137–143 (1997)
27. ETSI 3rd Generation Partnership Project (3GPP). Base station conformance testing (FDD). TS 25.141, the European Telecommunications Standards Institute, September 2010
28. Inc Xilinx, *LogiCORE IP Peak Cancellation Crest Factor Reduction v2.0*. (Xilinx, Inc., 1–28, 2009)
29. CA Azurdia-Meza, K Lee, K Lee, PAPR reduction by pulse shaping using Nyquist linear combination pulses. IEICE Electron. Express. **9**(19), 1534–1541 (2012)

Power minimization of cooperative beamforming networks with spectrum sharing

Cen Ling[1], Xuefeng Yin[1*], Silvia Ruiz Boqué[2] and Mario García-Lozano[2]

Abstract

Cooperative communication with spectrum sharing has been proved to be an efficient way to reduce energy consumption. In this work, an optimal power allocation algorithm is proposed for mobile cooperative beamforming networks, where several secondary users (SUs) collaboratively act as relay nodes to assist the transmission of a primary user (PU) in reward of transmitting their own information. The optimal power allocation scheme is obtained by solving a convex optimization problem aiming at minimizing network energy consumption with guaranteed quality of service (QoS) for both PU and SUs. Simulation results show that with the consideration of overheads, circuit power, and energy costs for message sharing, a cooperative beamforming network augmented with our proposed power allocation algorithm performs better in scenarios of more SUs in terms of higher energy efficiency and less power consumption per unit throughput. Additionally, the proposed scheme outperforms the conventional best-relay selection approach in the aspects of power consumption per unit throughput, energy efficiency, and network power consumption. Furthermore, simulations also demonstrate that the derived cooperative beamforming algorithm exhibits superior performance in energy saving than conventional methods in high-speed scenarios.

Keywords: Cooperative beamforming; Power minimization; Spectrum sharing; Mobile network

1 Introduction

Towards the fifth generation (5G) of wireless/mobile broadband communication, many advanced key technologies are going to be implemented, including spectrum sharing management, exploiting the Cloud-RAN (CRAN) and mobile clouds concepts as well as heterogeneous network coordination, in order to meet the demand for larger bandwidth, higher data rate up to Gbit/s and seamless connection access [1-3]. Furthermore, reduction of unnecessary energy consumption becomes one of the major concerns in 5G communications, which not only reduces environmental impact but also cuts overall network costs and helps make communication more practical and affordable. These studies belong to the research of 'green networking' which aims in increasing energy savings while maintaining satisfactory users' quality of service (QoS) [4-6]. Besides, cooperative techniques are promising in improving the reliability of the communications against channel impairments and achieved throughput through aggregating resources offered by different collaborative entities and realizing significant energy savings, targeting a so-called green and soft 5G system [7].

Apart from 'cell breathing' of base station coordination and clustered cooperative schemes [8,9], user cooperation, as a new form of spatial diversity, enables single-antenna mobiles in a multiuser environment to share their antennas and generate a virtual multiple-antenna transmitter, providing higher throughput, robustness to channel variations and ubiquitous mobile access for media-rich mobile devices. Recently, being extended with the concept of cognitive radio, user cooperation with spectrum sharing capability, i.e., secondary users (SUs) share the same spectrum with the primary user (PU), can further improve spectral efficiency. User-cooperative diversity protocols that combat fading induced by multipath propagation in wireless networks and the outage capacity characterization of various relaying protocols, i.e., fixed relaying, selective relaying, and incremental relaying, are characterized [10]. In literature, numerous criteria have been proposed for relay node selection based on channel conditions [11], the distances between nodes [12],

*Correspondence: yinxuefeng@tongji.edu.cn
[1]College of Electronics and Information Engineering, Tongji University, 4800 Caoan Road, Shanghai, 201804, China
Full list of author information is available at the end of the article

and the transmission contention [13], etc. It has been demonstrated in [14,15] that when more relay nodes, also called as cooperators, are involved in transmission, less energy costs and higher network throughput can be achieved simultaneously, compared with the single-relay transmission. This motivated the research on methodologies of selecting multiple entities to act as cooperators, also called as 'cooperative beamforming'. However, several issues should be considered from a media access control (MAC) layer perspective when multiple cooperating entities are involved. The first issue is that more control signaling overhead should be considered for selecting and coordinating multiple cooperating entities [16]. The second issue is related to users' willingness to cooperate. Furthermore, incentive mechanisms are of necessity for promoting user cooperation in order to prevent unwillingness and selfishness of the SUs via reputation evaluation, cost reduction, or increase of transmission time [17-20].

As the energy expenditure becomes a major concern for the operators of communication networks, constructing cooperative networks with the energy consumption minimized has become an important research field [21]. Multiple solutions of resource allocation were derived based on convex optimization or game theory via lowering power consumption under certain constraints, such as minimizing the outage probability [22-24], maximizing energy efficiency [25-28], maximizing overall throughput [29,30], maintaining service quality, or maximizing the sum of SUs' capacities and signal-to-noise power ratio (SNR) [31]. Joint relay selection and power allocation schemes are proposed by factoring the cost of acquiring channel state information (CSI) to minimize the total energy consumption [32]. The allocation schemes obtained are applicable to adapting the transmission power, bandwidth, time, and accessibility for the nodes involved in the cooperative network. In addition, circuit power consumption model has been taken into account in calculating the energy efficiency of cooperative heterogeneous and beamforming networks in [25,33].

The aforementioned resource allocation methods share some common drawbacks which limit their applicabilities in cooperative spectrum sharing networks in cases where a PU and multiple SUs coexist. First, the energy consumption minimization strategies adopted in these methods usually consider a single relay node, which is typical merely for the best relay selection scheme. The adaptation of the strategies in the situation with multiple SUs selected simultaneously as relay nodes, as in the cooperative beamforming scenario, has not been investigated thoroughly so far. Furthermore, SUs may be reluctant to relay the traffic for the PU, if SUs do not forsee an immediate benefit or reward from their cooperation or they have some security and privacy concerns or encounter resource limitation, e.g., battery shortage [14]. This issue concerning the incentive mechanisms so as to prevent unwillingness and selfishness of the SUs has not been paid great attention in recent researches on cooperative beamforming networks. Moreover, the existing power allocation schemes are derived under an ideal channel assumption that channel coefficients for multiple links among the communication nodes are independent random variables following identical distributions. This leads to nonrealistic conclusions for performance evaluation. It is necessary to re-design the resource allocation methods by taking into account some new constraints posed by the existence of multiple relays, the energy costs of the communication coordination overheads, the influence of the rewards for the relay nodes, and realistic propagation scenarios.

In this paper, we propose an energy-saving strategy for cooperative beamforming by finding the optimal power allocation scheme. This method is particularly useful in cooperative beamforming for mobile spectrum sharing scenarios where a PU and multiple SUs co-exist. The proposed power allocation scheme is obtained by solving a convex optimization problem with the boundary conditions specifying the minimum QoS for the PU and SUs, similar to the approach described in [19]. A model of constant circuit power consumption originally introduced in [25] is considered for calculating the network power consumption. Simulations under the setting of realistic fast fading channels generated by using an infinite impulse response (IIR) filter demonstrate that the energy efficiency which is defined as the ratio of outage capacity over total energy consumed, i.e., a necessary extension from single-relay scenarios [22,34] to multi-relay cooperative beamforming cases, can be significantly improved by implementing the proposed optimization scheme. Furthermore, the proposed scheme exhibits superior performance compared with some conventional methods including the best relay selection approach.

The rest of the paper proceeds as follows. In Section 2, the considered cooperative beamforming scenario in a cognitive network is introduced. In Section 3, the optimization method of power allocation and the derivation of energy efficiency for cooperative beamforming are presented. Simulation results are elaborated in Section 4 for performance assessment of the proposed scheme. Eventually, conclusive remarks are given in Section 5.

2 System model

Let us consider a wireless network where totally M SUs act simultaneously as relay nodes for one PU, as illustrated in Figure 1. A direct propagation path exists between the primary transmitter (PT) and primary receiver (PR). Each node is equipped with a single antenna for signal transmission and reception. A dual-hop decode-and-forward (DF) cooperative communication mode with

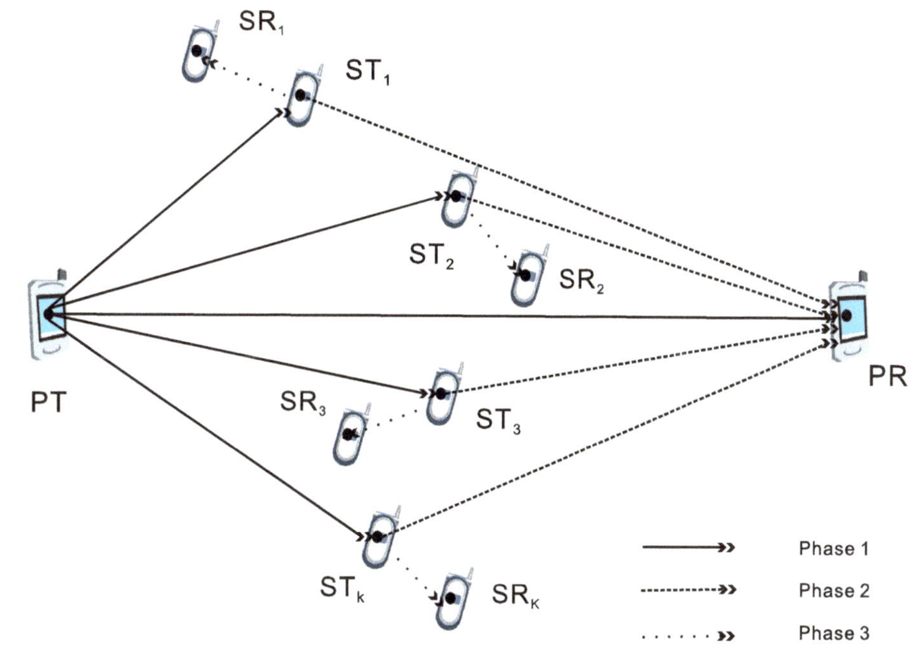

Figure 1 Cooperative beamforming model with three phases.

synchronization and diversity combining at the PR is considered. In return, the SU selected as a relay is rewarded with getting access to the channel for its own communication with a secondary receiver (SR) under the PU's authority.

Resource allocation can be performed with a frame or a slot structure according to the system design in a code division multiple access (CDMA) system, similar to Bluetooth, Wi-Fi, or sensor networks. The cooperative communication with all SUs and PU involved can be performed in three phases, as illustrated in Figure 2, where T and t_k denote the transmission time assigned to the PU and the SUs, respectively [35]. In the first phase, the PT broadcasts the data to the PR and all M SUs that could act as relays. In the second phase, the SUs decode the received signals individually and forward to the PR, acting as secondary transmitters (STs). At the PR, the signals from the PT and all STs are soft-combined and decoded. An ST's own transmission to its SR is performed during a short period t_k in the third phase. The overall time for completing these phases is then calculated as $T + t_k = \beta_k T$, where $\beta_k \geq 1$ can be determined optimally as a balance between the ST's throughput and power consumption, as will be discussed later in this paper.

Broadcast	Cooperative Beamforming	SUs' own transmission
$\frac{T}{2}$	$\frac{T}{2}$	t

Figure 2 Time slot allocation for primary and secondary user.

Noted that before the foregoing three phases, a short time period is usually allocated for the transmission of training signals and the estimation of CSI. The signal received for a link between any pair of nodes i and j, can be written as:

$$y_j = h_{ij}\sqrt{P_{ij}}x_i + n_j, \tag{1}$$

where h_{ij} is the complex channel coefficient from the node i to the node j, P_i denotes the transmitting power of the node i to the node j over the transmission bandwidth W, and n_j is the thermal noise generated by the node j and is modeled as an additive white Gaussian random variable with zero mean and spectral density height of N_o (W/Hz). The complex channel coefficient h_{ij} is assumed to be a complex circularly symmetric Gaussian random variable with magnitude $|h_{ij}|$ following a Rayleigh distribution. The mean of the Rayleigh distribution is dependent on the distance d_{ij} between the ith node and the jth node according to ${d_{ij}}^{\alpha}$, where α is an exponent constant describing the attenuation caused by the wave propagation. It is easy to show that $|h_{ij}|^2$ follows an exponential distribution with its rate parameter being $\lambda_{ij} = d_{ij}^{\alpha}$ [22]. From Equation 1, the SNR for the signals received at the node j when the node i transmits can be calculated as:

$$\mathrm{SNR}_{ij} = \frac{P_{ij}|h_{ij}|^2}{N_0}, \quad \gamma_{ij} \triangleq \frac{P_{ij}}{N_0}, \tag{2}$$

where γ_{ij} is defined as the ratio between the transmission signal power over the noise variance.

In order to successfully decode the information received by ST_k, the SNR of the link between PT and ST_k, i.e., SNR_{ps_k} is usually required to be higher than a threshold θ_1, i.e.,

$$SNR_{ps_k} = \frac{P_{ps_k}\left|h_{ps_k}\right|^2}{N_0} = \frac{P_{ps_k} d_{ps_k}^{-\alpha}}{N_0} > \theta_1, \quad (3)$$

where d_{ps_k} represents the distance from the PT to ST_k, P_{ps_k} denotes the transmission power of the PT, and h_{ps_k} is the channel coefficient of the link between the PT and ST_k. Applying the triangular principle for d_{ps_k}, the distance d_p from the PT to the PR, and the distance $d_{s_k d}$ from ST_k to the PR, it is easy to show that d_{ps_k} should be confined within the range described by the following inequalities:

$$\left|d_p - d_{s_k d}\right| < d_{ps_k} < \min\left\{\left(d_p + d_{s_k d}\right), \left(\frac{P_{\max}}{N_0\theta_1}\right)^{\frac{1}{\alpha}}\right\}, \quad (4)$$

where $P_{\max}$ represents the maximum transmission power of the network. Notice that Equation 4 can also be applied as a fundamental principle for relay selection when multiple STs are available in the network. In order to improve the understandings, explanations of the adopted symbols are demonstrated in Table 1.

In the following, an analytical expression is derived for the transmission rate of the cooperative communication system illustrated in Figure 1. In the first phase shown in Figure 2, the PT directly transmits to the PR without spectrum cooperation. The data rate per Hz R_d for the link between the PT and the PR can be calculated as:

$$R_d = \log_2\left(1 + \gamma_d |h_d|^2\right), \quad (5)$$

where γ_d denotes the ratio of the direct transmitting power P_d of the PT to the noise variance and h_d is the channel coefficient for the direct link from the PT to the PR.

In the second phase, cooperative beamforming is performed together with the direct transmission from the PT to the PR. The STs forward the signals to the PR after the decoding step, and the PR combines the signals originally transmitted by the PT and all STs using, e.g., the maximal ratio combining (MRC). The maximum data rate per second at the PR can be calculated by considering that all M STs transmit during the time $T/2$ to be:

$$R_c = \min_{k\in\kappa}\left\{R_{ps_k}, R_{MRC}\right\}, \quad (6)$$

with k being the indices of the STs, $\kappa \in \{1, 2, \ldots, M\}$, R_{ps_k} representing the data rate between the PT and ST_k, and R_{MRC} denoting the combination rate at the PR of the signals arriving from the PT and all STs. They can be calculated as respectively:

Table 1 Explanation of adopted symbols

Symbol	Explanation
y_j	Received signal at the *j*th node
h_{ij}	Channel coefficient from *i*th to *j*th node
P_{ij}	Transmitting power per bandwidth from the *i*th node to the *j*th node
x_i	Transmitting signal at the *i*th node
W	Transmission bandwidth
N_O	Spectral density height of thermal noise
n_j	Thermal noise
λ_{ij}	Rate parameter of exponential distribution
d_{ij}	Distance between the *i*th and *j*th nodes
R_d	Direct transmission data rate
γ_d	Transmitting SINR of direct transmission
h_d	Channel coefficient of direct transmission
R_c	Data rate of PR
R_{ps_k}	Data rate of the link of PT-SR_k
R_{MRC}	Data rate of MRC
γ_p	Ratio between the transmission signal power over the noise variance for PU
h_{ps_k}	Channel coefficient of the link PT-SR_k
$\gamma_{s_k d}$	Ratio between the transmission signal power over the noise variance between ST_k and PR
$h_{s_k d}$	Channel coefficient of the link ST_k-PR
R_{su_k}	Data rate of the *k*th SU
γ_{su_k}	Ratio between the transmission signal power over the noise variance for the *k*th SU
h_{su_k}	Channel coefficient of the link ST-SR
R_N	Cooperative network throughput
T	Time for the 1st and 2nd phase transmission
t_k	Time for the third phase transmission
P_p	Transmitting power of PT
P_d	Transmitting power of non-cooperative scheme
$P_{s_k d}$	Transmitting power of the ST_k
P_{su_k}	Transmitting power of the third phase
w_k	Beamforming weight of the ST_k
P_{OV}	Circuit power
P_{SR}	Overheads of cooperative beamforming
P_{TR}	Overheads of message sharing
P_c	Circuit power
M	Number of cooperative SUs
b_{RD}	Number of bits per symbol for cooperative beamforming
b_{SR}	Number of bits per symbol for message sharing
ρ	Amplifier efficiency
N_{TR}	Number of symbols of training period
N_{RD}	Number of symbols of cooperative beamforming
P_b	Power consumption per throughput

Table 1 Explanation of adopted symbols *(Continued)*

Symbol	Explanation
β_k	Time allocation ratio of the *k*th SU
Q_p	QoS of PU
Q_s	QoS of SU
P_d^{out}	Outage probability of direct transmission
C_d^{out}	Outage capacity of direct transmission
C_{su}^{out}	Outage capacity of SU
ε	Outage probability threshold
C_{cb}^{out}	Outage capacity of cooperative beamforming
η	Energy efficiency
θ_1	SNR threshold for decoding
θ_2	SNR of the worst link between ST_k and PR
d_p	Distance between PT and PR
d_{s_kd}	Distance between PT and PR
d_{su}	Distance between ST and SR
d_{ps_k}	Distance between PT and ST
P_{max}	Maximum transmitting power
γ_d	Ratio between the direct transmission signal power over the noise variance
ζ	Network energy consumption for unit bit of throughput
L_s	Number of SUs
v	Moving speed of users

$$R_{ps_k} = \frac{1}{2}\log_2\left(1+\gamma_p|h_{ps_k}|^2\right) \quad (7)$$

$$R_{\text{MRC}} = \frac{1}{2}\log_2\left(1+\gamma_p|h_d|^2+\sum_{k=1}^{M}\gamma_{s_kd}\left|h_{s_kd}\right|^2\right), \quad (8)$$

where γ_p denotes the ratio of the transmission power P_p of the PT to the noise variance N_0, γ_{s_kd} represents the ratio of the transmission power of the ST_k and N_0, and h_{s_kd} is the channel coefficient for the link between the ST_k and the PR.

In the last phase after the cooperative beamforming is executed, each ST gets the permission to use the channel between ST_k and SR_k. The data transmission rate can be calculated as:

$$R_{su_k} = \log_2\left(1+\gamma_{su_k}\left|h_{su_k}\right|^2\right), \quad (9)$$

where γ_{su_k} and h_{su_k} denote respectively the ratio of transmission power of the ST_k and the noise variance and the channel coefficient of the link between the ST_k and the SR_k.

The total network's throughput R_N can then be calculated as:

$$R_N = \frac{TR_c + \sum_{k=1}^{M} t_k R_{su_k}}{\bar{t}+T}, \quad (10)$$

where t_k is the time utilized in the third phase for the kth ST, and $\bar{t}$ represents the average of t_k, $k = 1, \ldots, M$. In the system considered here, every SU transmits its own information in its allocated channel, and the individual transmission time t_k is determined using an optimization algorithm described in the later sections of the paper.

The network's power consumption P_N, as a holistic and system-wide metric, which includes all the energy consumption such as transmission power, circuit power, and signaling overhead in the entire network, can be calculated as:

$$P_N = \frac{\frac{T}{2}\left(P_p + \sum_{k=1}^{M} w_k P_{s_kd}\right) + \sum_{k=1}^{M} w_k t_k P_{su_k}}{\bar{t}+T} + P_{\text{OV}}, \quad (11)$$

where the optimal beamforming weight w_k for the kth ST can be calculated as $\left|h_{s_kd}\right|/\|\boldsymbol{h}\|$ with $\|\cdot\|$ being the norm of given vector and $\boldsymbol{h} = \left[\left|h_{s_1d}\right|, \left|h_{s_2d}\right|, \ldots, \left|h_{s_Md}\right|\right]^T$ representing the vector consisting of the channel gains from the STs to the PR and P_{OV} denotes the power overheads required in the source message sharing and channel estimation phases [25].

The overhead consumption P_{OV} is obtained by using the constant circuit-power consumption model [25]:

$$P_{\text{OV}} = \left(\frac{P_{\text{SR}}}{\rho} + (1+M)P_c\right)\frac{b_{\text{RD}}}{b_{\text{SR}}} + \left(\frac{P_{\text{TR}}}{\rho} + (1+M)P_c\right)\frac{N_{\text{TR}}}{N_{\text{RD}}}, \quad (12)$$

where P_{TR} denotes the power used for the training signals, P_{SR} signifies the overhead power of message sharing period, ρ represents the amplifier efficiency, P_c is the circuit power, b_{RD} and b_{SR} denote respectively the number of bits per symbol during the cooperative beamforming period and during the message sharing period, W is the available transmission bandwidth, and finally, N_{TR} and N_{RD} represent the number of symbols transmitted during the two periods, respectively. P_{TR}, is selected such that the SNR of the worst link between the ST_k and PR is θ_2. In addition, P_{SR} can be calculated as:

$$\begin{aligned} P_{\text{SR}} &= \frac{\gamma_{\text{th}}(b_{\text{SR}})N_0W}{\min\left(\left|h_{ps_k}\right|^2\right)}, \\ \gamma_{\text{th}}(b_{\text{SR}}) &= -\frac{1}{c_2}\left(2^{b_{\text{SR}}}-1\right)\ln\left(\frac{p_{\text{th}}}{c_1}\right), \\ p_{\text{th}}(b) &= c_1\exp\left(-\frac{c_2 p_p}{\left(2^b-1\right)N_oW}\right), \end{aligned} \quad (13)$$

where $\gamma_{\text{th}}(b_{\text{SR}})$ is the SNR required to achieve the target M-ary quadrature amplitude modulation (M-QAM) bit error probability, p_{th} and $c_1 = 0.2$, $c_2 = 1.5$.

To compute the average transmission rates for the entire time period ($T + t_k = \beta_k T$), we consider that the cooperative communication is performed within $1/\beta_k$ of the whole time period, and the secondary communication utilizes the spectrum in the rest of time, i.e., $(\beta_k-1)/\beta_k$ of the whole period. The average transmission rates of PU in the first and second phases and of SUs in the third phase can then be calculated as, respectively:

$$\bar{R}_p = \frac{R_c}{\bar{\beta}_k}, \bar{R}_{su} = \sum_{k=1}^{M} \frac{\beta_k-1}{\beta_k} R_{su_k} \text{ with } k = 1, \ldots, M. \quad (14)$$

3 An algorithm for energy consumption minimization and cooperative beamforming

3.1 Optimal power allocation for cooperative beamforming

We now derive an algorithm for determining the optimal power allocation with minimal energy consumption for the considered cooperative beamforming system. It is reasonable to assume that within the channel coherence time, CSI, correctly detected by corresponding UEs, has been sent to the PT. Based on the CSI, the transmission powers, i.e., P_p, P_{s_kd}, and P_{su_k}, shall be determined with the objective to minimize the average energy consumption of the network while guaranteeing users' QoS. To be specific, P_{s_kd} denotes the transmission power of the kth ST to the PR, and P_{su_k} represents the transmission power of the SU$_k$'s own transmission in the third phase. The predefined QoS of the PU and SU denoted with Q_p and Q_s, respectively, are identified with the required transmission rate. The optimization problem can be formulated as:

$$\min_{P,\beta_k} T \cdot \left(\frac{1}{2\beta_k} \left(P_p + P_{s_kd}\right) + \frac{\beta_k - 1}{\beta_k} P_{su_k} \right), \quad (15)$$

$$\text{s.t. } \frac{1}{2\beta_k} \left(\log_2 \left(1 + \frac{P_p |h_p|^2}{N_o} \right) + \log_2 \left(1 + \frac{P_{s_kd} |h_{s_kd}|^2}{N_o} \right) \right) \geq Q_p, \quad (16)$$

$$\frac{\beta_k - 1}{\beta_k} \log_2 \left(1 + \frac{P_{su_k} |h_{su_k}|^2}{N_o} \right) \geq Q_s, \quad (17)$$

$$\beta_k > 1. \quad (18)$$

The inequalities Equations 16 and 17 represent respectively the QoS requirements of the PU and of the SU. The left-hand sides of Equations 16 and 17 separately denote the transmission rate for PU during the cooperative communication (phases 1 and 2) and SU's own transmission (phase 3).

This convex optimization problem in Equations 15 to 18 satisfies the Karush-Kuhn-Tucker conditions, and it can be solved using the Lagrangian equation [36,37]. Besides, T is normalized as 1. Detailed convexity analysis and the solution of the optimization problem are given in Appendix 1 and Appendix 2, respectively.

Despite the unawareness of β_k, the optimal powers can be computed as, respectively:

$$P_p = N_o \left(\frac{2^{\beta_k Q_p}}{|h_p| |h_{s_kd}|} - \frac{1}{|h_p|^2} \right),$$
$$P_{s_kd} = N_o \left(\frac{2^{\beta_k Q_P}}{|h_p| |h_{s_kd}|} - \frac{1}{|h_{s_kd}|^2} \right),$$
$$P_{su_k} = \frac{N_o}{|h_{su_k}|^2} \left(2^{\frac{\beta_k Q_s}{\beta_k - 1}} - 1 \right). \quad (19)$$

Once these optimal transmission powers are known, the optimal β_k can be calculated by solving the function:

$$\frac{2^{\beta_k Q_p}}{|h_{s_kd}| |h_p|} (Q_p \beta_k - 1) - \frac{\beta_k Q_s 2^{\frac{\beta_k Q_s}{\beta_k - 1}}}{|h_{su_k}|^2 (\beta_k - 1)} + \frac{2^{\frac{\beta_k Q_s}{\beta_k-1}} - 1}{|h_{su_k}|^2} + \frac{1}{2|h_p|^2} + \frac{1}{2|h_{s_kd}|^2} = 0. \quad (20)$$

It can be readily shown in Appendix 2 that the left-hand side of Equation 20 has the derivative with respect to β_k always larger than zero and, hence, increases monotonically along with β_k. Thus, the optimal β_k can be determined by applying the Newton's iterative method as described in [38].

Notice that when the cooperative beamforming is not executed, the optimization problem for the power allocation via minimizing the energy consumption reduces to;

$$\min \{P_d\} \text{ s.t. } \log_2 \left(1 + \frac{P_d |h_d|^2}{N_o} \right) \geq Q_p. \quad (21)$$

The optimal power for the direct transmission in this case can be calculated as:

$$P_d = \frac{1}{|h_d|^2} \left(e^{Q_p} - 1 \right). \quad (22)$$

This scheme is used as a reference for evaluating the performance of the proposed method in simulations.

3.2 Energy efficiency of the cooperative beamforming scheme

The expressions of the outage probability and outage capacity for the proposed cooperative beamforming scheme are derived in the following. It is easy to show that during the direct transmission from the PT to the PR, the outage probability between the PT and PR can be calculated as:

$$P_d^{\text{out}} = 1 - \exp \left(-d_d^{\alpha} \cdot \frac{2^R - 1}{\gamma_d} \right). \quad (23)$$

Hence, for a fixed outage probability ε, the outage capacity can be obtained as:

$$C_d^{out} = \log_2\left(1 + \frac{\gamma_d}{d_d^\alpha}\ln\frac{1}{(1-\varepsilon)}\right). \tag{24}$$

In the third phase, the outage capacity for an SU can be formulated similarly as:

$$C_{su}^{out} = \log_2\left(1 + \frac{\gamma_{su}}{d_{su}^\alpha}\ln\frac{1}{(1-\varepsilon)}\right), \tag{25}$$

where d_{su} represents the distance between an ST and the corresponding SR.

From the data rate expressions in the second phase as shown in Equations 6 to 7, the outage probability P_c^{out} of cooperative beamforming can be calculated as:

$$P_c^{out} = \Pr\left(\frac{1}{2}\min\left(R_{ps_1}, R_{ps_2}\cdot\cdot R_{ps_k}\cdot\cdot, R_{ps_M}, R_{MRC}\right) < R\right) = 1 - \prod_{k=1}^{M}\left(1 - \Pr\left(R_{ps_k} < 2R\right)\right)\left(1 - \Pr(R_{MRC} < 2R)\right). \tag{26}$$

The two cumulative density functions $\Pr\left(R_{ps_k} < 2R\right)$ and $\Pr(R_{MRC} < 2R)$ in the right-hand side of Equation 26 can be calculated as, respectively:

$$\Pr\left(R_{ps_k} < 2R\right) = 1 - \exp\left(-d_{ps_k}^\alpha \cdot \frac{2^R - 1}{\gamma_{ps_k}}\right), \tag{27}$$

$$\Pr(R_{MRC} < 2R) = F\left(2^{2R} - 1\right), \tag{28}$$

where $F(z)$ can be shown to be:

$$F(z) = \sum_{k=1}^{M+1}\frac{(1 - \exp(-\lambda_k z))\prod_{i=1,i\neq k}^{i=M+1}\lambda_i}{\prod_{i=1,i\neq k}^{i=M+1}(\lambda_k - \lambda_i)} \tag{29}$$

with $z = \sum_{k=1}^{M}\gamma_{s_kd}\left|h_{s_kd}\right|^2 + \gamma_p\left|h_p\right|^2$, $\lambda_k = d_{s_kd}^\alpha$, and $\lambda_{M+1} = d_p^\alpha/\gamma_p$.

Substituting Equations 27 to 29 into 26, the outage probability P_c^{out} is rewritten as:

$$P_c^{out} = 1 - \sum_{k=1}^{M}\left(g_{s_kd}\right)^{d_{s_kd}^\alpha}\left(1 - a\left(g_p\right)^{d_p^\alpha} + (1+a)\left(g_{s_kd}\right)^{d_{s_kd}^\alpha}\right), \tag{30}$$

with

$$a = \left(\frac{P_{s_kd}}{P_p}\left(\frac{d_p}{d_{s_kd}}\right)^\alpha - 1\right)^{-1},$$
$$g_{s_kd} = \exp\left(-\frac{N_0\left(2^{2R} - 1\right)}{P_{s_kd}}\right),$$
$$g_p = \exp\left(-\frac{N_0\left(2^{2R} - 1\right)}{P_p}\right). \tag{31}$$

From Equations 26 to 27, it is obvious that P_c^{out} increases along with R, while $g(x)$ is a decreasing function of R. Since $g_{s_kd} = g_p^{\frac{P_p}{P_{s_kd}}}$, by replacing g_{s_kd} in Equation 30 with g_p, it can be shown that P_c^{out} is a monotonically decreasing function of g_p. Thus, by equating Equation 30 with a threshold ε of outage probability, we can solve for a specific g_p, denoted with $g_{p|\varepsilon}$ by using the Newton method.

$$C_{cb}^{out} = \log_2\left(1 + \gamma_p\ln\frac{1}{g_{p|\varepsilon}}\right). \tag{32}$$

Eventually, the energy efficiency of the cooperative beamforming scheme, which can be used as a metric evaluating the amount of useful information bit per unitary energy, is obtained as:

$$\eta = \frac{C_{cb}^{out}}{P_N}, \tag{33}$$

where P_N can be obtained from Equation 11.

A widely adopted metric for evaluating the validity for saving network energy consumption is the value of 'bits-per-Joule' for the network, i.e., the network energy consumption for unit bit of throughput, denoted with ζ and calculated as $\zeta = P_N/R_N$. Energy efficiency, denoted with η, is defined as ratio of outage capacity over network power consumption.

4 Simulation results for performance evaluation

Monte Carlo simulations are conducted to evaluate the performance of the proposed minimum power allocation algorithm utilized in cooperative beamforming for a spectrum sharing system. The mobile wireless channel is generated using an IIR filter followed by linear interpolation based on the 'sum-of-sinusoids' principle proposed by Jakes in [39,40]. The distance d_p between the PT and the PR varies within the range from 0 to 1,000 m. The distances between the STs and the PR are fixed to be 300 m, and the distances between the STs and their corresponding SRs are set to be 200 m. In the simulations, a ST is selected to be a relay if the distance between the PT and the ST satisfies Equation 4 and is upper-bounded by 800 m. The parameter setting of the simulations is reported in Table 2.

Table 2 Parameter settings for simulation

Description	Value
Amplifier efficiency ρ	38%
Circuit power p_C	1 mw
Power for training signals P_{TR}	1 mw
Bits per symbol of MS b_{RD}	6
Bits per symbol of CB b_{SR}	4
Number of training symbols N_{TR}	1
Number of symbols for CB N_{RD}	99
Outage probability threshold ε	0.01
Required QoS of PU/SU	1/0.5 bps/Hz
Pathloss exponent α	3.5
PT-PR distance d_p	$0 < d_p \leq 1{,}000$ m
PT-ST distance d_{ps_k}	$d_{ps_k} \leq 800$ m
ST-PR distance d_{s_kd}	$d_{s_kd} = 300$ m
ST-SR distance d_{su}	$d_{su} = 200$ m
Maximum transmission power P_{max}	40 mw
Noise spectral density N_o	0.5
Number of snapshots per meter	20
SNR threshold for decoding θ_1	0.01
SNR of the worst link between ST_k and PR θ_2	0.0001

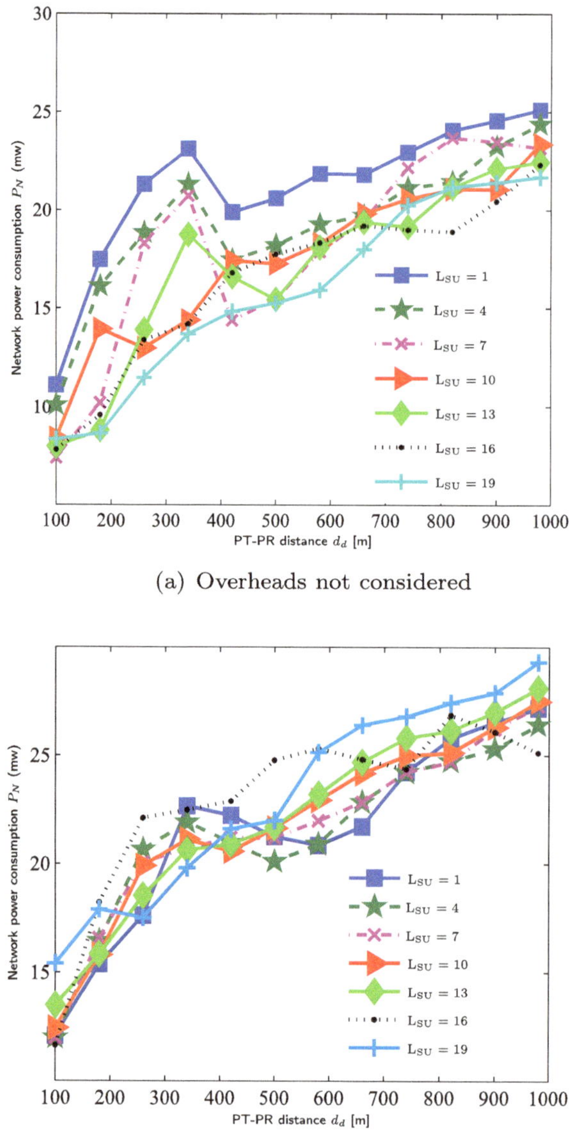

(a) Overheads not considered

(b) Overheads considered

Figure 3 Average minimum power consumption versus the PT-PR distance d_d for different SU numbers with/without overhead considered. (a) Overheads not considered. **(b)** Overheads considered.

4.1 Performance evaluation with the number L_s of SUs as a parameter

Figure 3a,b depicts the average minimum network power consumption P_N required for performing cooperative beamforming versus the PT-PR distance d_d for the cases with and without considering the overhead cost P_{OV}, respectively. The number L_s of SUs is considered as a parameter. Totally, 2,000 snapshots were simulated to calculate each point along the graph of average P_N versus d_d. It can be observed by comparing Figure 3a,b that in the case without considering P_{OV}, the average minimum network power consumption P_N decreases dramatically when L_s increases, while in the situation where P_{OV} is taken into account, the average P_N graphs appear to be less deviated from each other when L_s changes. This observation indicates that the energy costs for relay scheduling and information sharing can increase along with the number of the SUs involved. The advantage of cooperative beamforming is counteracted by the increasing of overheads to some extent.

Figure 4 illustrates the variation of ζ, i.e., the network energy consumption per unit throughput versus the PT-PR distance d_d with the overhead consumption P_{OV} considered. The number L_s of SUs is also taken as a parameter in the simulations. It can be observed that ζ decreases significantly with the growth of the SU number. This is due to the fact that the network throughput R_N increases enormously along with L_s, and meanwhile, no evident increase of network power consumption is observed when L_s grows when the overhead and scheduling consumption is considered.

The variation of the energy efficiency η for the cooperative beamforming scheme when L_s changes is illustrated in Figure 5. The overhead consumption has been considered in the simulations. Every scatter plot in Figure 5 is calculated by averaging the energy efficiency obtained with various d_d from 2,000 snapshots. It is obvious from Figure 5 that the average η grows linearly with the increasing number of SUs. A linear curve with a slope of 10

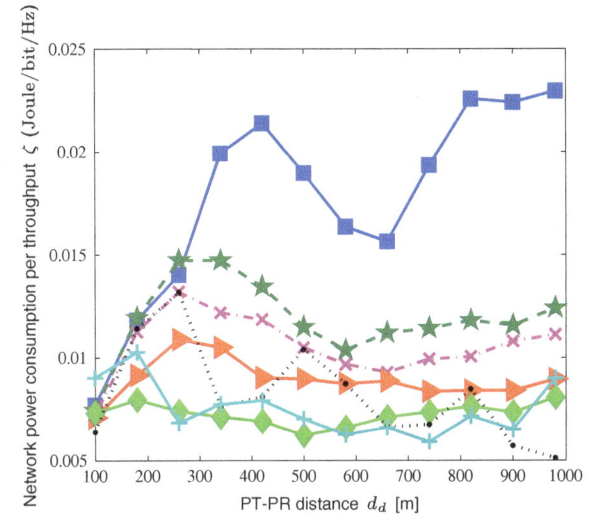

Figure 4 Network power consumption per unit throughput ζ versus the PT-PR distance d_d for different SU numbers with overhead considered. The legend shown in Figure 3 also applies here.

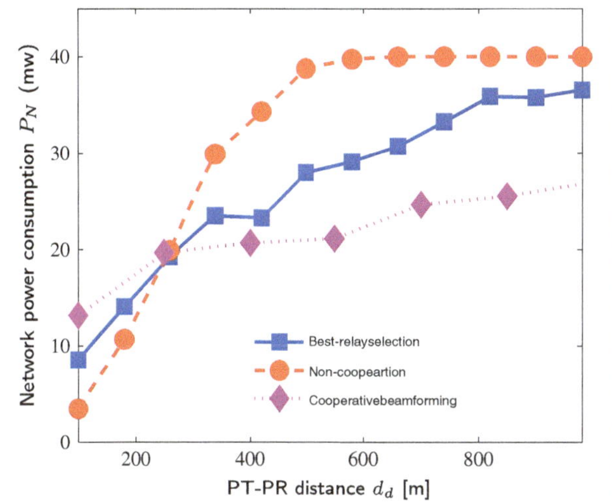

Figure 6 Average minimum power consumption versus the PT-PR distance d_d for three communication schemes with the number of SUs $L_s = 6$.

bit/s/J/SU is found fitting well with the simulation results. The observation of more bits being transmitted when L_s increases is due to the joint effect that the enhanced capacity of the cooperative beamforming system for larger L_s, and at the same time, the network power consumption in the cooperative beamforming phase is reduced with more SUs involved.

4.2 Comparison among three communication schemes

Figure 6 compares the network power consumption for three cases, i.e., best relay selection with the number of SUs $L_s = 6$, non-cooperation scheme with single relay, and the cooperative beamforming with $L_s = 6$.

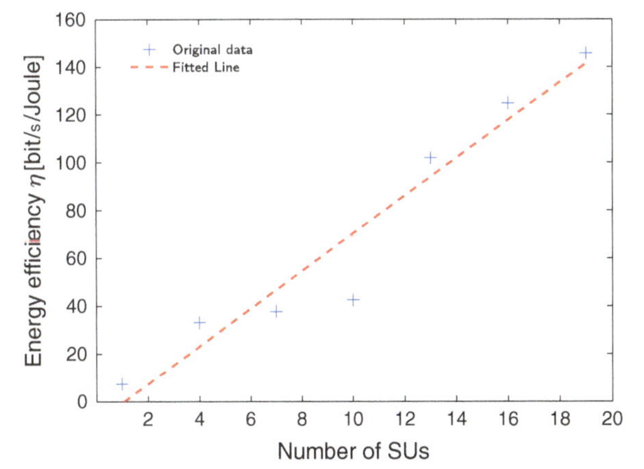

Figure 5 Energy efficiency versus SU number with overhead considered.

In the best relay selection algorithm, the optimal relay is selected with the criterium of maximum energy efficiency from M candidate relays [35]. The threshold of QoS requirement of the PU and the SUs are set as respectively, 1 and 0.5 bps/Hz in the simulated communication schemes. It can be observed that the proposed cooperative beamforming scheme outperforms the other two, and the best-relay selection algorithm exhibits the intermediate performance. This is the rationale as the implementation of the relay selection process may deteriorate the performance in some cases when compared with the cooperative beamforming scheme. Furthermore, for a fixed number of SUs, without significant benefits from the energy minimization, the best relay selection scheme consumes the similar amount of energy for scheduling and of circuit power as the cooperative beamforming does.

Figure 7a,b illustrates the throughput R_d of the direct transmission, the network throughput R_N, the average PU's throughput R_p in the first and second phases, and the average SU's throughput $\bar{R}_S = \frac{1}{M}\sum_{k=1}^{M} R_{su_k}$, versus the PT-PR distance d_d for the best-relay-selection scheme and the proposed cooperative beamforming scheme, respectively. The number L_s of SUs equals 6 in the simulations. It can be found from Figure 7a,b that the direct transmission provides less R_d for increasing d_d, and on the contrary, the PU's throughput R_P and the network throughput R_N with cooperative beamforming increase along with d_d. The throughput of SUs remains a constant level irrespective of d_d as expected. These results reveal that the relay and the cooperative beamforming systems can provide higher throughput than the direct transmission. It can be furthermore observed by comparing Figure 7a,b that

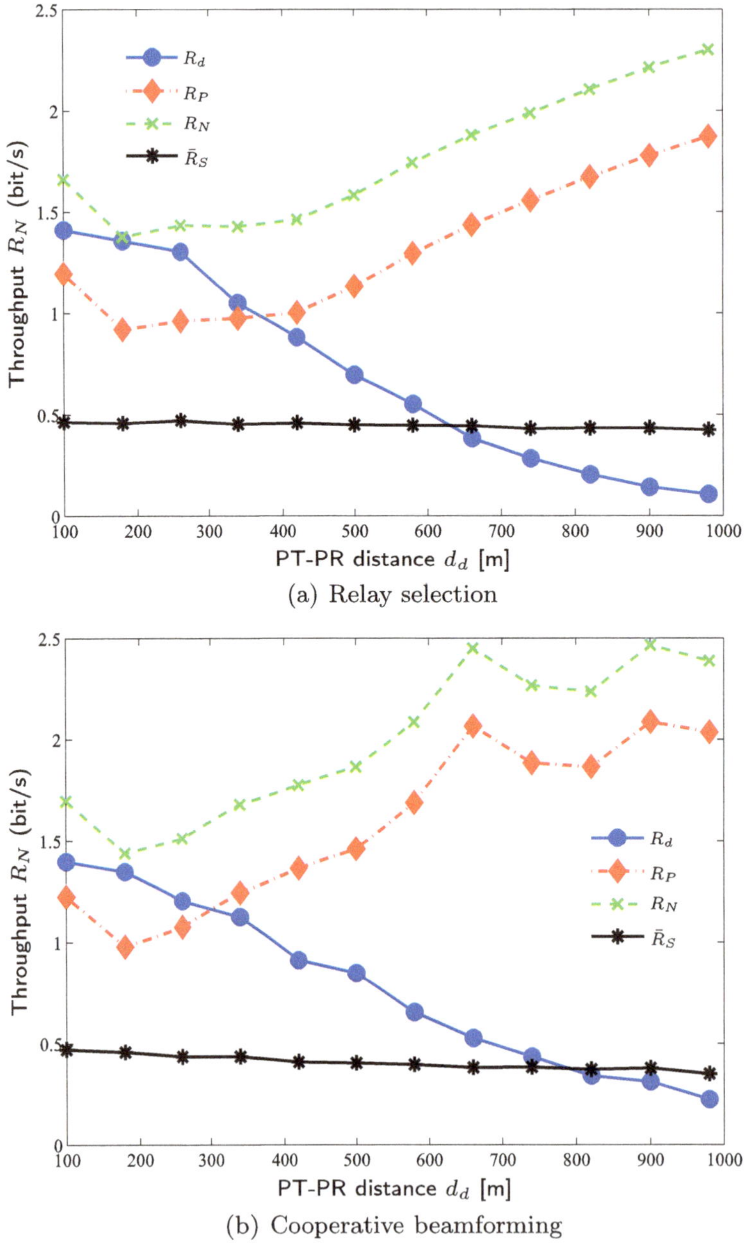

Figure 7 Throughput versus the PT-PR distance d_d for relay-selection and cooperative beamforming schemes with the number of SUs $L_s = 6$. (a) relay selection. **(b)** Cooperative beamforming.

the proposed cooperative beamforming scheme exhibits throughput graphs similar to those obtained for the best relay scheme. However, since the network consumption is less for the cooperative beamforming than for the best relay scheme as illustrated in Figure 6, it is a natural conjecture that the energy efficiency, i.e., the network power consumption per unit throughput would be less in the case of cooperative beamforming than of the best relay. This is confirmed by the results shown in Figure 8 where the 'Joule-per-bit' for the cooperative beamforming is nearly a half of that observed for best relay selection with large $d_{\rm d}$.

Figure 9 compares the network energy efficiency η versus the PT-PR distance d_d for the best relay selection, non-cooperative and cooperative beamforming schemes with $L_s = 6$. It can be observed that the network energy efficiency of cooperative beamforming remains constant for different values of d_d and is the highest among all three schemes considered. Furthermore, the graph of the energy efficiency of the best relay scheme exhibits

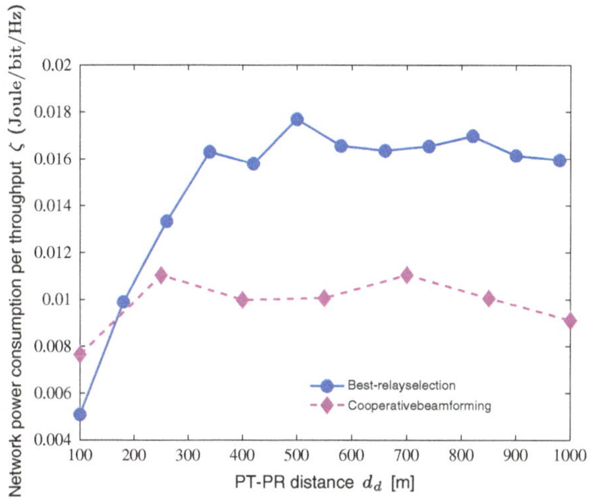

Figure 8 Network power consumption per unit throughput ζ versus the PT-PR distance d_d of two cases with $L_s = 6$.

a concave shape with the maximum obtained with the distance equal to 500 m. This is consistent with the analysis shown in [41,42] that the network energy efficiency is concave with respect to the transmission power which monotonically increases along with the distance. Furthermore, in the case with short distances, the outage capacity is small, and a certain level of total power consumption still remains. Consequently, the energy efficiency becomes extremely low. In cases where the distance is sufficiently large, the outage capacity becomes saturated. In such case, since the power consumption keeps increasing, the energy efficiency decreases and converges to zero for large d_d.

4.3 Performance evaluation in vehicular scenarios

The performance of the cooperative beamforming scheme with proposed power allocation strategy is also evaluated by simulations in time-variant cases where the nodes in the network move in high speeds. In the simulations, the speeds of the PU and of the SUs are set to be identical and represented with v. The channel coefficients among the nodes are generated by using a tap-delay-line (TDL) model with six discrete paths. The number L_s of SUs is fixed to be 5.

Figure 10 depicts the network power consumption P_N versus the PT-PR distance d_d for moving speed $v = 5$, 50, and 100 km/h. It can be observed from Figure 10 that the graph of P_N versus d_d observed for $v = 100$ km/h is only slightly above those obtained with less v. This indicates that the network power consumption for the cooperative beamforming is insensible to the speed of the nodes involved. Figures 11, 12, and 13 depict, respectively, the network throughput R_N, the energy consumption per unit throughput, and the energy efficiency η versus the PT-PR distance d_d for different v. From these figures, it can be observed that the performance of the cooperative beamforming scheme with proposed power allocation only slightly degrades when the speed of nodes increases. These results demonstrate the robustness of the proposed scheme against the time variability of the channels and the applicability of the scheme in vehicular scenarios.

5 Conclusions

In this paper, a novel power allocation scheme has been proposed for energy-efficient mobile cooperative communication with spectrum sharing. In the system, SUs act collaboratively as beamforming relays and help the PU in data transmission in reward of transmitting their own information. The optimal power allocation derived solves a convex optimization problem of minimizing network

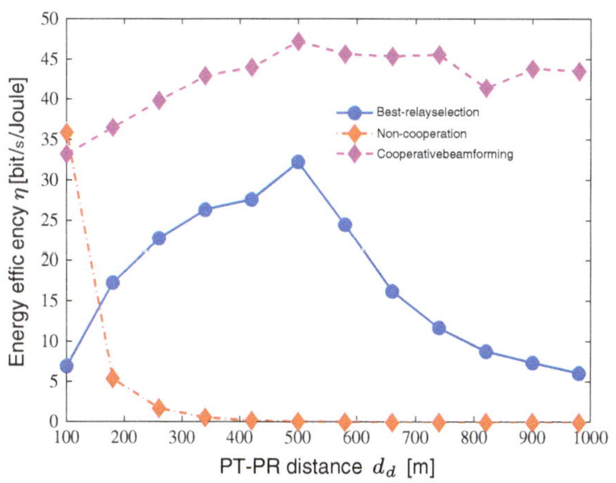

Figure 9 Network energy efficiency versus the PT-PR distance d_d of three cases with $L_s = 6$.

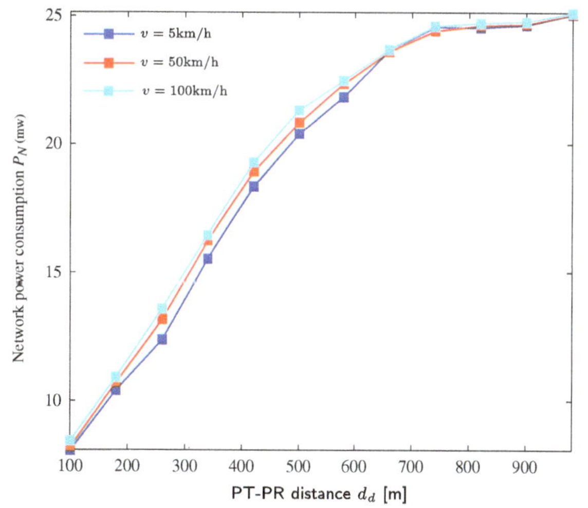

Figure 10 Network power consumption P_N versus the PT-PR distance d_d with UE speed as a parameter.

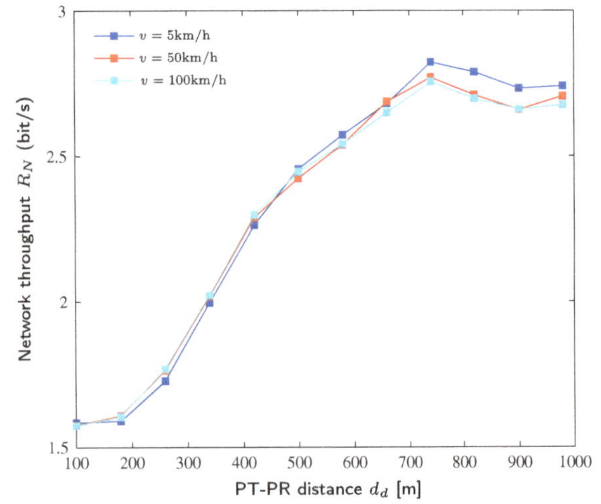

Figure 11 Network throughput R_N versus the PT-PR distance d_d with UE speed as a parameter.

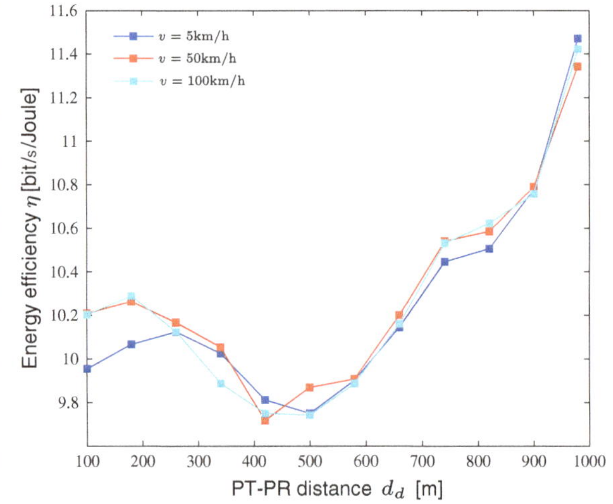

Figure 13 Network energy efficiency η versus the PT-PR distance d_d with UE speed as a parameter.

energy consumption under the constraint of the guaranteed quality of service for both PU and SUs. Analytical expressions were derived based on the resultant optimal power allocation for outage capacity of the system. Numerical simulations have been conducted for evaluating the performance of the proposed scheme. The results demonstrated that the network power consumption exhibited different behaviors depending on whether the overhead power consumption is considered. By taking into account the overhead power consumption, circuit power, and message sharing energy costs, the cooperative beamforming cognitive network together with the proposed power allocation method performs better when more SUs are involved, in terms of enhanced energy efficiency and reduced power consumption per unit throughput. Furthermore, simulations demonstrated that the proposed cooperative beamforming scheme is superior to the conventional best relay selection scheme in the aspects of network power consumption, power consumption per unit throughput, and energy efficiency. In addition, the cooperative beamforming scheme with the optimal power allocation proposed here exhibited stable performance in the time-variant scenarios where both the PU and SUs move in high speeds.

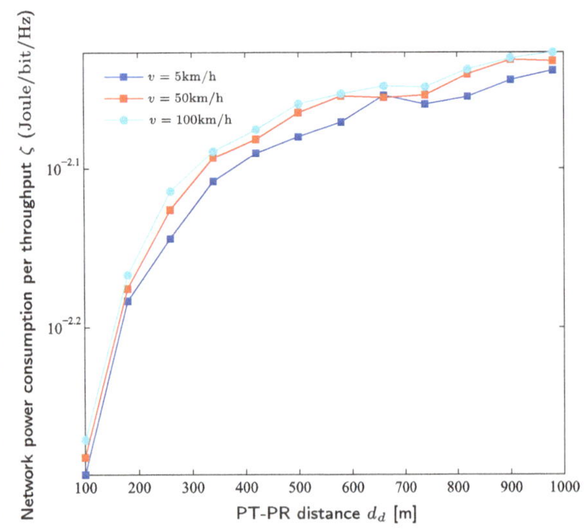

Figure 12 Network power consumption per unit throughput ζ versus the PT-PR distance d_d with UE speed as a parameter.

Appendix 1

Convexity analysis of the optimization problem in Equations 15 to 18

A convex optimization problem is one in which the objective and constraint functions are convex [37]. In general, a continuous and twice differentiable function is concave, if and only if its Hessian matrix is negative semi-definite. On the contrary, if the Hessian matrix is positively semi-definite, the function is convex. Now, we discuss the convexity of the two inequality constraints (16) and (17), i.e.,

$$\beta_k Q_p - \frac{1}{2}\log_2\left(1+\frac{P_p|h_p|^2}{N_o}\right) - \frac{1}{2}\log_2\left(1+\frac{P_{s_k d}|h_{s_k d}|^2}{N_o}\right) \leq 0 \quad \text{and}$$

$$Q_s - \frac{\beta_k - 1}{\beta_k}\log_2\left(1+\frac{P_{su_k}|h_{su_k}|^2}{N_0}\right) \leq 0.$$

Denoting $f\left(P_p, P_{s_k d}, \beta_k\right) = \beta_k Q_p - \frac{1}{2}\log_2\left(1+\frac{P_p|h_p|^2}{N_o}\right) - \frac{1}{2}\log_2\left(1+\frac{P_{s_k d}|h_{s_k d}|^2}{N_o}\right)$, we can compute the Hessian matrix of $f\left(P_p, P_{s_k d}, \beta_k\right)$ as:

$$H_f = \begin{pmatrix} \frac{\partial^2 f(P_p,P_{s_kd},\beta_k)}{\partial {P_p}^2} & \frac{\partial^2 f(P_p,P_{s_kd},\beta_k)}{\partial P_p \partial P_{s_kd}} & \frac{\partial^2 f(P_p,P_{s_kd},\beta_k)}{\partial P_p \partial \beta_k} \\ \frac{\partial^2 f(P_p,P_{s_kd},\beta_k)}{\partial P_{s_kd} \partial P_p} & \frac{\partial^2 f(P_p,P_{s_kd},\beta_k)}{\partial {P_{s_kd}}^2} & \frac{\partial^2 f(P_p,P_{s_kd},\beta_k)}{\partial P_{s_kd} \partial \beta_k} \\ \frac{\partial^2 f(P_p,P_{s_kd},\beta_k)}{\partial \beta_k \partial P_p} & \frac{\partial^2 f(P_p,P_{s_kd},\beta_k)}{\partial \beta_k \partial P_{s_kd}} & \frac{\partial^2 f(P_p,P_{s_kd},\beta_k)}{\partial {\beta_k}^2} \end{pmatrix}$$

$$= \begin{pmatrix} \frac{1}{2}\left(\frac{|h_p|^2 \ln 2}{N_0+P_p|h_p|^2}\right)^2 & 0 & 0 \\ 0 & \frac{1}{2}\left(\frac{\left|h_{s_kd}\right|^2 \ln 2}{N_0+P_{s_kd}\left|h_{s_kd}\right|^2}\right)^2 & 0 \\ 0 & 0 & 0 \end{pmatrix}.$$

Thus, the eigenvalues of H_f are all above or equal to zero, which shows that H_f is a positive semi-definite matrix, an implication that $f(P_p, P_{s_kd}, \beta_k)$ is a convex function and Equation 16 is a convex constraint.

Similarly, denoting $g(P_{su_k}, \beta_k) = Q_s - \frac{\beta_k - 1}{\beta_k} \log_2 \left(1 + \frac{P_{su_k}\left|h_{su_k}\right|^2}{N_0}\right)$, we can obtain the Hessian matrix of $g(P_{su_k}, \beta_k)$ as:

$$H_g = \begin{pmatrix} \frac{\beta_k - 1}{\beta_k} \frac{\left|h_{su_k}\right|^4 (\ln 2)^2}{\left[N_0+P_{su_k}\left|h_{su_k}\right|^2\right]^2} & \frac{1}{{\beta_k}^2} \frac{\left|h_{su_k}\right|^2 \ln 2}{\left[N_0+P_{su_k}\left|h_{su_k}\right|^2\right]} \\ \frac{1}{{\beta_k}^2} \frac{\left|h_{su_k}\right|^2 \ln 2}{\left[N_0+P_{su_k}\left|h_{su_k}\right|^2\right]} & \frac{2}{{\beta_k}^3} \log_2 \left(1 + \frac{P_{su_k}\left|h_{su_k}\right|^2}{N_0}\right) \end{pmatrix}.$$

Thus, the determinants of all leading principal minors of the H_g are given by:

$$\left|H_g^1\right| = \frac{\beta_k - 1}{\beta_k} \frac{\left|h_{su_k}\right|^4 (\ln 2)^2}{\left[N_0 + P_{su_k}\left|h_{su_k}\right|^2\right]^2} > 0$$

$$\left|H_g^2\right| = \left[\frac{1}{{\beta_k}^4} \frac{\left|h_{su_k}\right|^4 (\ln 2)^2}{\left[N_0 + P_{su_k}\left|h_{su_k}\right|^2\right]^2}\right] [2(\beta_k - 1)\log_2 \left(1 + \frac{P_{su_k}\left|h_{su_k}\right|^2}{N_0}\right) - 1].$$

From Equation 17, it can be shown that:

$$\left[2(\beta_k - 1)\log_2 \left(1 + \frac{P_{su_k}\left|h_{su_k}\right|^2}{N_0}\right) - 1\right] \geq 2\beta_k Q_s - 1.$$

Thus, as long as the requirement of $2\beta_k Q_s - 1 \geq 0$, i.e., $Q_s \geq 0.5$ is satisfied, it is easy to show that Equation 17 is a convex constraint.

Furthermore, denoting the objective function in Equation 15 $z(P_p, P_{s_kd}, P_{su_k}, \beta_k) = \frac{1}{2\beta_k}(P_p + P_{s_kd}) + \frac{\beta_k - 1}{\beta_k} P_{su_k}$, the Hessian matrix of $z(P_p, P_{s_kd}, P_{su_k}, \beta_k)$ can be shown as:

$$H_z = \begin{pmatrix} 0 & 0 & 0 & -\frac{1}{2{\beta_k}^2} \\ 0 & 0 & 0 & -\frac{1}{2{\beta_k}^2} \\ 0 & 0 & 0 & -\frac{1}{{\beta_k}^2} \\ -\frac{1}{2{\beta_k}^2} & -\frac{1}{2{\beta_k}^2} & -\frac{1}{{\beta_k}^2} & \frac{1}{{\beta_k}^3}\left[P_p + P_{s_kd} - 2P_{su_k}\right] \end{pmatrix}.$$

It can be seen that all leading principal minors of H_z are equal to zero matrixes. Thus, $z(P_p, P_{s_kd}, P_{su_k}, \beta_k)$ is a convex function in terms of $(P_p, P_{s_kd}, P_{su_k}, \beta_k)$ and the objective function in Equation 15 is convex.

In addition, the constraint (18) is a linear function concerning to β_k. Thus, the Hessian matrixes is zero matrix, indicating the convexity of the constraint (18).

Based on these analyses, we conclude that the formulated optimization problem in Equations 15 to 18 is convex, under the condition that QoS of SU is larger than or equal to 0.5.

Appendix 2

Solution of the optimization problem in Equations 15 to 18

As Equations 15 to 18 are a convex optimization problem with differentiable objective and constraint, the KKT conditions are necessary and sufficient for the optimality. Furthermore, T is normalized to be 1 without loss of generality. By adding Lagrange multipliers, i.e., λ, μ, ν, according to the constraints of Equations 16 to 18, the Lagrange dual function can be formulated as:

$$\begin{aligned} L(P, \beta_k, \lambda, \mu, \nu) = &\frac{1}{2\beta_k}(P_p + P_{s_kd}) + \frac{\beta_k - 1}{\beta_k} P_{su_k} \\ &+ \lambda \left(\beta_k Q_p - \frac{1}{2}\log_2\left(1 + \frac{P_p|h_p|^2}{N_o}\right) - \frac{1}{2}\log_2\left(1 + \frac{P_{s_kd}\left|h_{s_kd}\right|^2}{N_o}\right)\right) \\ &+ \mu \left(Q_s - \frac{\beta_k - 1}{\beta_k}\log_2\left(1 + \frac{P_{su_k}\left|h_{su_k}\right|^2}{N_0}\right)\right) + \nu(1 - \beta_k). \end{aligned}$$

By solving the following equations simultaneously:

$$\frac{\partial L(P, \beta_k, \lambda, \mu, \nu)}{\partial P_p} = 0,$$

$$\frac{\partial L(P, \beta_k, \lambda, \mu, \nu)}{\partial P_{s_kd}} = 0,$$

$$\frac{\partial L(P, \beta_k, \lambda, \mu, \nu)}{\partial P_{su_k}} = 0,$$

$$\frac{\partial L(P, \beta_k, \lambda, \mu, \nu)}{\partial \beta_k} = 0,$$

$$\lambda \left(\beta_k Q_p - \frac{1}{2}\log_2\left(1 + \frac{P_p|h_p|^2}{N_o}\right) - \frac{1}{2}\log_2\left(1 + \frac{P_{s_kd}\left|h_{s_kd}\right|^2}{N_o}\right)\right) = 0,$$

$$\mu \left(Q_s - \frac{\beta_k - 1}{\beta_k}\log_2\left(1 + \frac{P_{su_k}\left|h_{su_k}\right|^2}{N_0}\right)\right) = 0,$$

$$\nu(1 - \beta_k) = 0,$$

$$\beta_k Q_p - \frac{1}{2}\log_2\left(1 + \frac{P_p|h_p|^2}{N_o}\right) - \frac{1}{2}\log_2\left(1 + \frac{P_{s_kd}\left|h_{s_kd}\right|^2}{N_o}\right) \geq 0,$$

$$Q_s - \frac{\beta_k - 1}{\beta_k}\log_2\left(1 + \frac{P_{su_k}\left|h_{su_k}\right|^2}{N_0}\right) \geq 0,$$

$$1 - \beta_k \geq 0,$$

we obtain the optimal point in terms of the powers:

$$P_p = N_o \left(\frac{2^{\beta_k Q_p}}{|h_p| \, |h_{s_k d}|} - \frac{1}{|h_p|^2} \right),$$
$$P_{s_k d} = N_o \left(\frac{2^{\beta_k Q_P}}{|h_p| \, |h_{s_k d}|} - \frac{1}{|h_{s_k d}|^2} \right),$$
$$P_{su_k} = \frac{N_o}{|h_{su_k}|^2} \left(2^{\frac{\beta_k Q_s}{\beta_k - 1}} - 1 \right). \tag{34}$$

Once these optimal transmission powers are known, the optimal β_k can be calculated by solving the function:

$$\frac{2^{\beta_k Q_p}}{|h_{s_k d}| \, |h_p|} (Q_p \beta_k - 1) - \frac{\beta_k Q_s 2^{\frac{\beta_k Q_s}{\beta_k - 1}}}{|h_{su_k}|^2 (\beta_k - 1)} + \frac{2^{\frac{\beta_k Q_s}{\beta_k - 1}} - 1}{|h_{su_k}|^2} + \frac{1}{2|h_p|^2} + \frac{1}{2|h_{s_k d}|^2} = 0. \tag{35}$$

Denoting the left-hand side of Equation 35 as $f(\beta_k)$ and taking the derivative of $f(\beta_k)$ with respect to β_k as follows:

$$f'(\beta_k) = \frac{{Q_p}^2 \beta_k 2^{\beta_k Q_p}}{|h_{s_k d}| \, |h_p|} + \frac{{Q_s}^2 \beta_k 2^{\frac{\beta_k Q_s}{\beta_k - 1}}}{|h_{su_k}|^2 (\beta_k - 1)^3}, \tag{36}$$

it is clear that Equation 36 is always larger than zero, and hence, $f(\beta_k)$ increases monotonically along with β_k. Thus, the optimal β_k can be determined by applying the Newton's iterative method as described in [38]. Given an initial β_k as $\beta_k(0)$, the following process is repeated:

$$\beta_k(n+1) = \beta_k(n) - \frac{f(\beta_k(n))}{f'(\beta_k(n))} \tag{37}$$

until a sufficiently accurate value is reached.

Competing interests
The authors declare that they have no competing interests.

Authors' contributions
CL carried out the generic studies, performed the simulation platform, and drafted the manuscript. XY gave guidance in the implementation of the simulation software and modification of the contribution. SRB conceived of the study, participated in its design and coordination, and helped to draft the manuscript. MGL offered some help in the conceiving of the research. All authors read and approved the final manuscript.

Acknowledgements
This work has been done as a collaboration between research institutions members of European COST IC1004 Action on Cooperative Radio Communications for Green Smart Environments. It has been jointly supported by the Spanish Ministry of Science under the project TEC2011-27723-C02-01, the Sino-Spanish Campus between Tongji University and UPC, the project 'System design and demo-construction for cooperative networks of high-efficiency 4G wireless communications in urban hot-spot environments' granted by the Science and Technology Commission of Shanghai Municipality, China, and the NSFC general project with Grant No. 61471268. Thanks for the technical guidance of Prof. Joan Olmos from the University of Polytechnic in Catalunya.

Author details
[1]College of Electronics and Information Engineering, Tongji University, 4800 Caoan Road, Shanghai, 201804, China. [2]Department of Signal Theory and Communications, University of Polytechnic in Catalunya, Building C4 C. Esteve Terradas, 7 Castelldefels, 08860, Spain.

References
1. J Andrews, S Buzzi, W Choi, S Hanly, A Lozano, A Soong, J Zhang, What will 5G be. IEEE J. Sel. Areas Commun. **32**, 1065–1082 (2014)
2. I C-Lin, C Rowell, S Han, Z Xu, G Li, Z Pan, Toward green and soft: a 5G perspective. IEEE Commun. Mag. **52**, 66–73 (2014)
3. P Demestichas, A Georgakopoulos, D Karvounas, K Tsagkaris, V Stavroulaki, J Lu, C Xiong, J Yao, 5G on the horizon: key challenges for the radio-access network. IEEE Veh. Technol. Mag. **8**, 47–53 (2013)
4. S Chen, J Zhao, The requirements, challenges, and technologies for 5G of terrestrial mobile telecommunication. IEEE Commun. Mag. **52**, 36–43 (2014)
5. Y Wu, Y Chen, J Tang, D So, Z Xu, C-L I, P Ferrand, J-M Gorce, C-H Tang, P-R Li, K-T Feng, L-C Wang, K Borner, L Thiele, Green transmission technologies for balancing the energy efficiency and spectrum efficiency trade-off. IEEE Commun. Mag. **52**, 112–120 (2014)
6. X Hong, J Wang, C-X Wang, J Shi, Cognitive radio in 5G: a perspective on energy-spectral efficiency trade-off. IEEE Commun. Mag. **52**, 46–53 (2014)
7. W Zhuang, M Ismail, Cooperation in wireless communication networks. IEEE Wireless Commun. **19**, 10–20 (2012)
8. A Papadogiannis, D Gesbert, E Hardouin, in *Proceedings of IEEE International Conference on Communications (ICC)*. A dynamic clustering approach in wireless networks with multi-cell cooperative processing (IEEE Beijing, China, 2008), pp. 4033–4037
9. Z Niu, Y Wu, J Gong, Z Yang, Cell zooming for cost-efficient green cellular networks. IEEE Commun. Mag. **48**, 74–79 (2010)
10. J Laneman, D Tse, GW Wornell, Cooperative diversity in wireless networks: efficient protocols and outage behavior. IEEE Trans. Inf. Theory. **50**, 3062–3080 (2004)
11. A Bletsas, A Khisti, D Reed, A Lippman, A simple cooperative diversity method based on network path selection. IEEE J. Selec. Areas Commun. **24**, 659–672 (2006)
12. YJ, H Jafarkhani, Network beamforming using relays with perfect channel information. IEEE Trans. Inf. Theory. **55**, 2499–2517 (2009)
13. C Lo, S Vishwanath, R Heath, Relay subset selection in wireless networks using partial decode-and-forward transmission. IEEE Trans. Veh. Technol. **58**, 692–704 (2009)
14. B Hamdaoui, T Alshammari, M Guizani, Exploiting 4G mobile user cooperation for energy conservation: challenges and opportunities. IEEE Wireless Commun. **20**, 62–67 (2013)
15. Z Zhou, S Zhou, S Cui, J-H Cui, in *Proceedings of IEEE Military Communications Conference (MILCOM)*. Energy-efficient cooperative communication in clustered wireless sensor networks (IEEE Washington, USA, 2006), pp. 1–7
16. H Shan, W Zhuang, Z Wang, Distributed cooperative MAC for multihop wireless networks. IEEE Commun. Mag. **47**, 126–133 (2009)
17. D Yang, X Fang, G Xue, Game theory in cooperative communications. IEEE Wireless Commun. Mag. **19**, 44–49 (2012)
18. G R Bella, G Costantino, Evaluating the device reputation through full observation in MANETs. J. Inf. Ass. Sec. **4**, 458–465 (2009)
19. D Liu, W Wang, W Guo, 'Green' cooperative spectrum sharing communication. IEEE Commun. Letters. **17**, 459–462 (2013)
20. C Comaniciu, NB Mandayam, H Poor, J-M Gorce, An auctioning mechanism for green radio. J. Commun. Netw-s. Kor. **12**, 114–121 (2010)
21. J Furthmuller, O Waldhorst, in *Proceedings of Eighth International Conference on Wireless On-Demand Network Systems and Services (WONS)*. Energy-aware resource sharing with mobile devices (IEEE Bardonecchia Italy, 2011), pp. 52–59
22. K Woradit, T Quek, W Suwansantisuk, H Wymeersch, L Wuttisittikulkij, M Win, Outage behavior of selective relaying schemes. IEEE Trans. Wireless. Commun. **8**, 3890–3895 (2009)
23. E Larsson, Y Cao, Collaborative transmit diversity with adaptive radio resource and power allocation. IEEE Commun. Lett. **9**, 511–513 (2005)

24. Y Cao, B Vojcic, M Souryal, in *Proceedings of IEEE Vehicular Technology Conference (VTC2004-Fal)*. User-cooperative transmission with channel feedback in slow fading environmentA (IEEE Los Angeles, USA, 2004), pp. 2063–2067
25. G Lim, J Cimini, LJ, Energy-efficient cooperative beamforming in clustered wireless networks. IEEE Trans. Wireless. Commun. **12**, 1376–1385 (2013)
26. C Isheden, Z Chong, E Jorswieck, G Fettweis, Framework for link-level energy efficiency optimization with informed transmitter. IEEE Trans. Wireless. Commun. **11**, 2946–2957 (2012)
27. D To, T To, J Choi, in *Proceedings of IEEE Vehicular Technology Conference (VTC Spring)*. Energy efficient distributed beamforming with sensor selection in wireless sensor networks (IEEE Yokohama, Japan, 2012), pp. 1–5
28. Y Zhang, H Dai, Energy-efficiency and transmission strategy selection in cooperative wireless sensor networks. J. Commun. Netw-s. Kor. **9**, 473–481 (2007)
29. X Zhang, Z Zheng, J Liu, X Shen, L-L Xie, in *Proceedings of IEEE Global Communications Conference (GLOBECOM)*. Optimal power allocation and AP deployment in green wireless cooperative communications (IEEE Anaheim, USA, 2012), pp. 4000–4005
30. GB S Farshad, ML, in *Proceedings of International Union of Radio Science General Assembly and Scientific Symposium (URSI GASS)*. A Q-learning game-theory-based algorithm to improve the energy efficiency of a multiple relay-aided network (IEEE Beijing, China, 2014), pp. 1–4
31. L Zhang, Y Liang, Y Xin, Joint beamforming and power allocation for multiple access channels in cognitive radio networks. IEEE J. Sel. Areas Commun. **26**, 38–51 (2008)
32. R Madan, N Mehta, A Molisch, J Zhang, Energy-efficient cooperative relaying over fading channels with simple relay selection. IEEE Trans. Wireless Commun. **7**, 3013–3025 (2008)
33. G Lim, L Cimini, Energy-efficient cooperative relaying in heterogeneous radio access networks. IEEE Wireless Commun. Lett. **1**, 476–479 (2012)
34. Q Guan, F Yu, S Jiang, V Leung, H Mehrvar, Topology control in mobile ad hoc networks with cooperative communications. IEEE Wireless Commun. **19**, 74–79 (2012)
35. C Ling, X Yin, SR Boqule, M Garcia-Lozano, in *Proceedings of International Union of Radio Science General Assembly and Scientific Symposium (URSI GASS)*. Optimal power allocation and relay selection in spectrum sharing cooperative networks (IEEE Beijing, China, 2014), pp. 1–4
36. RT Rockafellar, *Convex Analysis*. (Princeton University Press, USA, 1996)
37. LV Stephen Boyd, *Convex Optimization*. (Cambridge University Press, UK, 2006)
38. R Kress, *Numerical analysis*. (World Book Publishing, China, 2003)
39. WC Jakes, *Microwave Mobile Communications*. (John Wiley & Sons, USA, 1974)
40. C Komninakis, in *Proceedings of IEEE Global Telecommunications Conference (GLOBECOM)*. A fast and accurate Rayleigh fading simulator (IEEE San Francisco, USA, 2003), pp. 3306–3310
41. H Yu, Y Li, M Kountouris, X Xu, J Wang, in *Proceedings of 12th International Symposium on Modeling and Optimization in Mobile, Ad Hoc, and Wireless Networks (WiOpt)*. Energy efficiency analysis of relay-assisted cellular networks using stochastic geometry (IEEE Hammamet, Tunis, 2014), pp. 667–671
42. J Chen, X Chen, T Liu, L Lei, in *Proceedings of IEEE international conference on Communications in China (ICCC)*. Energy-efficient power allocation for secure communications in large-scale MIMO relaying systems (IEEE Shanghai, China, 2014), pp. 385–390

8

METIS research advances towards the 5G mobile and wireless system definition

Jose F Monserrat[1*], Genevieve Mange[2], Volker Braun[2], Hugo Tullberg[3], Gerd Zimmermann[4] and Ömer Bulakci[5]

Abstract

The Mobile and wireless communications Enablers for the Twenty-twenty Information Society (METIS) project is laying the foundations of Fifth Generation (5G) mobile and wireless communication system putting together the point of view of vendors, operators, vertical players, and academia. METIS envisions a 5G system concept that efficiently integrates new applications developed in the METIS horizontal topics and evolved versions of existing services and systems. This article provides a first view on the METIS system concept, highlights the main features including architecture, and addresses the challenges while discussing perspectives for the further research work.

Keywords: 5G; Beyond 2020; METIS; System concept

1 Introduction

In the decade beyond 2020, it will be necessary to support 1,000 times [a] higher mobile data volume per area [1] together with new wireless broadband communication services coming from a plethora of different market segments. These requirements will go beyond the natural evolution of IMT-Advanced technologies, which show the need for a new mobile generation, with certain disruptive features with respect to legacy technologies. Although there is no unanimous agreement so far, this seems to be the birth of the Fifth Generation (5G) technologies.

Mobile and wireless communications Enablers for the Twenty-twenty Information Society (METIS) is an integrated research project partly funded by the European Commission under the Framework Programme 7 (FP7) research framework [2]. METIS aims at laying the foundation for the beyond 2020 wireless communication systems by providing the technical enablers needed to address the very challenging requirements foreseen for this time frame. Such a system has to (1) be significantly more efficient in terms of energy, cost, and resource utilization than today's systems, (2) more versatile to support a significant diversity of requirements (e.g., payload size, availability, mobility, and Quality-of-Service (QoS)) and new scenario use cases, and (3) provide better scalability in terms of number of connected devices, densely deployed access points, spectrum usage, energy, and cost. The technical goals derived from these main objectives for METIS [3] are:

- 1,000 times higher mobile data volume per area,
- 10 to 100 times higher typical user data rate,
- 10 to 100 times higher number of connected devices,
- 10 times longer battery life for low-power devices,
- 5 times reduced end-to-end (E2E) latency, reaching a target of 5 ms for road safety applications.

The key challenge is to meet these goals at a similar cost and energy consumption as today's networks.

From the point of view of METIS, a complete redesign of the Internet is discarded. Since the current Internet has become so large, the implementation of new architectural principles is impractical due to the commercial and operational difficulties it poses. Some concepts such as information-centric networking (ICN) [4] are considered unrealistic, although some of its fundamentals, as the universal caching, have to be taken into account in order to efficiently distribute the traffic load in the network. If 5G includes a revolution, it will come from the radio interface design, where some new paradigms under discussion represent a radical change in the current view we have on mobile networks in addition to some fundamental

* Correspondence: jomondel@iteam.upv.es
[1]Universitat Politecnica de Valencia, Camino de Vera, s/n, Valencia 46022, Spain
Full list of author information is available at the end of the article

changes in the mobile core functionalities, e.g., driven by new concepts of mobility.

In this challenging task, the METIS project has defined a set of five essential Horizontal Topics (HTs) to be consolidated into a single 5G system concept. These HTs are direct device-to-device communications (D2D), massive machine communications (MMC), moving networks (MN), ultra-dense networks (UDN), and ultra-reliable communications (URC) [5].

The present paper discusses a number of design choices and features that change the way we understand mobile networks. It is worth noting that this article represents the common vision of the set of network operators, industrial, and academic partners that work together in the METIS project. It does not provide the final solution, but highlights the most promising research lines, which combine efforts in common directions in the project.

In the next section, we present the METIS view on the key functional and technical features of the next generation mobile system, detailing the network architecture in Section 3. Section 4 highlights the technical enablers and research challenges that address the technical goals of the project for 5G. We finally draw the main conclusions of the paper in Section 5.

2 Main features of the 5G system

All mobile generation changes so far have been based on a new radio link concept and have provided an increase of the peak data rate of approximately two orders of magnitude. The 5G system must meet the requirements of increased rate and capacity needed beyond 2020 and requirements on reduced latency. However, the integration of new services and application areas is as important as increasing rate and reducing latency. The 5G system will be the wireless enabler of the Internet of things and, in addition to human users, must cater to different machine-type communications with widely different requirements. All together, the range of requirements will increase as compared with current mobile broadband (MBB) technologies. For instance, the data rates will range from very low for sensor data to very high rates for high-definition video. Latencies will range from extremely low for safety-critical applications to applications where latency is not really a constraint. Packet sizes will vary from small for, e.g., smartphone applications to large for, e.g., file transfers.

By using a system architecture that supports also D2D communication and UDN deployments, METIS foresees a multi-Radio Access Technologies (RAT) system that efficiently integrates fundamental building blocks as:

- *Evolved mobile broadband (eMBB)* will provide high data rates and low latency communications improving quality of experience (QoE) for the users. As depicted in Figure 1, eMBB will go much beyond current MBB solutions.
- *Massive machine communications (MMC)* will provide up- and down-scalable connectivity solutions for tens of billions of network-enabled devices, where scalable connectivity is vital to the future mobile and wireless communications systems.
- *Vehicle to Vehicle, Device and Infrastructure (V2X) and driver assistance services* require cooperation between vehicles and between vehicles and their environment (e.g., between vehicles and vulnerable road users over smartphones) in order to improve road safety and traffic efficiency in the future. Such V2X services for MNs require reliable communication links that enable the transmission of data packets with guaranteed maximum latencies even at high vehicle speeds.
- *Ultra-reliable communications (URC)* will enable high degrees of availability. It is required to provide scalable and cost-efficient solutions for networks supporting services with extreme requirements on availability and reliability. Reliable service decomposition provides mechanisms for graceful degradation of rate and increase of latency, instead of dropped connections, as the number of users increases.

To address these challenges, new flexible air interfaces, new possible waveforms, and new multiple access schemes, medium access control (MAC), and radio resource management (RRM) solutions (e.g., [6,7]), and signaling protocols must be investigated to discard the idea that physical layer improvements are already close to their upper limit. However, the foreseen steps towards the ubiquitous high bit-rate services pass mainly by managing multi-cell and multi-user

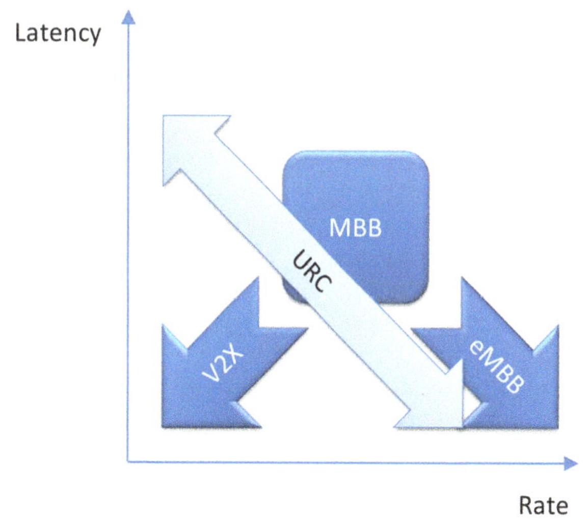

Figure 1 Extension of requirements as compared with current MBB technologies.

MIMO together with new paradigms of network deployment with multiple RATs and multilayer networks.

The key supporting enablers include:

- *Dense and dynamic RAN* providing a new generation of dynamic radio access networks (RANs). The term RAN 2.0 could be also used referring to this flexibility of the RAN.
- *The spectrum toolbox* contains a set of enablers (tools) to allow 5G systems to operate under different regulatory and spectrum sharing scenarios.
- *Flexible air interface* incorporating several radio interface technologies operating as a function of the user needs.
- *Massive MIMO,* in which the number of transmit and receive antennas increase over an order of magnitude.
- *New lean signaling/control information* is necessary to guarantee latency and reliability, support spectrum flexibility, allow separation of data and control information, support large variety of devices with very different capabilities, and ensure energy efficiency.
- *Localized contents/traffic flows* allow offloading, aggregation, and distribution of real-time and cached content.

2.1 Dense and dynamic RAN

There is a consensus that future traffic demand will require a high density of a variety of access nodes [8,9]. UDN refers to this new paradigm of wireless communication network, which includes network cooperation and ultra-dense availability of access points. Nomadic and moving nodes, mounted on a car, bus, or train, can provide connectivity to users in their proximity and increase data rates by reducing the radio distance to the nearest access node. D2D communications will guarantee the ubiquity of high-quality services and offload the infrastructure transport network [10].

Network densification has been used for wireless evolution during last years and will continue this trend in the next decade [11]. Even some authors have speculated about a 10 year-out situation in which the number of access points could surpass that of active users [12]. UDNs present serious challenges in terms of mobility support, interference management, and operation (cost, maintenance, and backhaul).

In the context of mobility in UDNs, the classical approach of user-initiated mobility management based on measurement reports poses a severe problem, provided the increased signaling and device measurement overhead. Therefore, current trends of self-configuration of handover parameters [13] must be revisited to adapt to the 5G scenarios with random layout and a variety of cell ranges. Moreover, mobility prediction will be fundamental to initiate the handover with enough time in advance.

On the other hand, from a radio planning point of view, the disparity in transmission powers in heterogeneous deployments invalidates traditional DL-based cell association mechanisms. Cell range expansion (RE) seems to be the solution to overcome this problem. An additional benefit is a more effective offloading from the macro layer to the small cells [14]. Whereas RE partially compensates UL interference issues, DL interference is increased at the new small-cell edge due to the DL transmissions from the macro cell to other users. For this reason, 3GPP has already introduced improvements in the inter-cell interference coordination mechanisms of LTE-A [15,16]. To sum up, higher densification is nowadays being pursued by applying patches to the old ideas, and this has reached its limit with the current densification values. Given this, it is clear that new system designs are needed for 5G networks to tackle the root of the problem, that is, the huge imbalance between the nature of mobile devices and cell infrastructure. This leads to the idea of asymmetric user association and the fact that UL and DL should be treated as independent networks. This way, the unity of the cell is broken, and UL and DL are totally decoupled and served by different access points [17].

This decoupling philosophy is also being proposed in the 3GPP for the splitting of control and user planes (C/U-plane) [18]. Given this, it is possible to transmit the C-plane, and consequently HO messages, from a stable macro cells whereas the U-plane is transmitted from the closest small cell (see Figure 2). The recent work by Li Yan et al. argues about the suitability of this approach for future 5G networks and provides interesting insights on practical aspects [19].

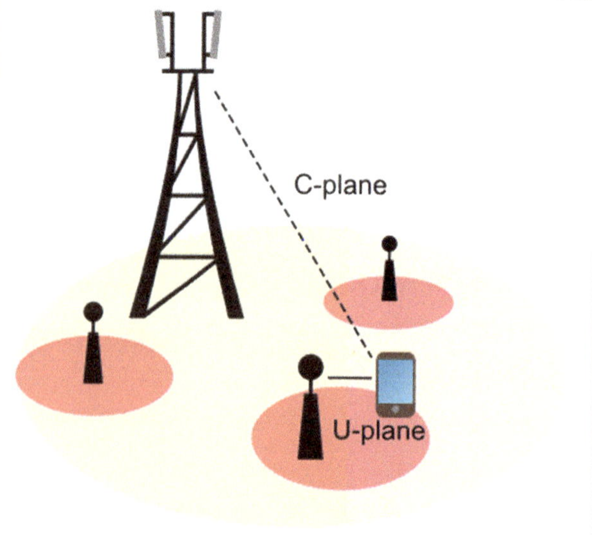

Figure 2 Decoupling of control and user plane in UDN.

Another important problem is the provision of high-quality services within vehicles on the move. One possible solution to address the problem in the future will be the deployment of a relay within the vehicle. The relay can be connected to the vehicular rooftop antenna, in order to establish a robust (relay/backhaul) link to the cellular network. On the other hand, relays allow for optimum coverage conditions inside the vehicle and a certain control of the consumed resources through the usage of closed subscriber groups. Since the use of a relay eliminates the vehicle penetration loss (VPL) and provides additional antenna gains due to better characteristics of the rooftop antenna, this solution improves significantly the link budget [20] and allows for a more efficient resource utilization of the cellular network. However, in current LTE networks, an additional overhead burden due to signaling exists (around 10%), and therefore, the benefit from the use of those relays is to be demonstrated.

Concerning the research on mobile relaying, most studies focus on network architecture aspects [21-23], mostly on the basis of its usage in LTE/LTE-Advanced (see Figure 3). In this case, the mobile Relay Node (RN) is connected as a regular user to a base station, known as donor e-Node B (DeNB). Moreover, regarding the system performance assessment, the expected gain derived from the usage of moving relays is investigated, for instance in [23]. Other studies evaluate the impact of deploying mobile femtocells from the point of view of spectral efficiency and signaling overhead considering opportunistic use of spectrum and scheduling [24,25]. Current research on the integration of moving networks into 5G focuses on improved procedures, channel prediction, reduced signaling, and optimum backhauling [26].

In order to reduce the link distance, the 3GPP initiated in Release 12 the study item on the use of device-to-device (D2D) communications under the name of proximity services (ProSe) [27]. The 3GPP defines D2D communication as a type of communication between two nearby UEs in which there is not any routing through the core network. ProSe in LTE-Advanced is also aimed at critical situations and must be functional when cellular networks are not operational.

In 3GPP, ProSe studies focus on two aspects, namely D2D discovery, also known as proximity detection, and D2D communication. D2D discovery aims at identifying other neighbor UEs for possible D2D communication, whereas D2D communication is the actual direct link for data transfer. For both operations, it is of paramount importance to design and efficient control plane to synchronize terminals, ensure radio link security, and keep track on the mobility during the communication phase. Figure 4 shows the current approach for this functional split in 3GPP. More details on this control plane design for D2D communications can be found in [28] and references therein.

D2D discovery is usually localized in static resources while D2D communications should use resources in an opportunistic manner, only when necessary. However, this communication mode may introduce severe interferences into the overlaying cellular network, which results in the need for dynamic and network-controlled radio resource allocation to the D2D link [29]. Several methods have been proposed to deal with this coexistence [30]. Feng et al. [31] proposed a resource allocation scheme for D2D communications operating in opportunistic manner focusing on the guarantee of QoS requirements in the heterogeneous network. On the other hand, Xu et al. in [32] designed an iterative auction for resource allocation whereas Lingyang et al. proposed a game for this resource allocation [33].

Beyond these concepts, METIS foresees a new generation of dynamic RAN, referred to as RAN 2.0. Following an analogy with the Web 2.0 concept, in RAN 2.0, the user devices can become active entities of the RAN, in addition to existing network nodes (see Figure 5). This user-centric network (UCN) is a new paradigm of wireless networking in which user devices will temporarily take over the role of access nodes for other users, e.g., to guarantee the ubiquity of high-quality services. This multi-hop cellular solution

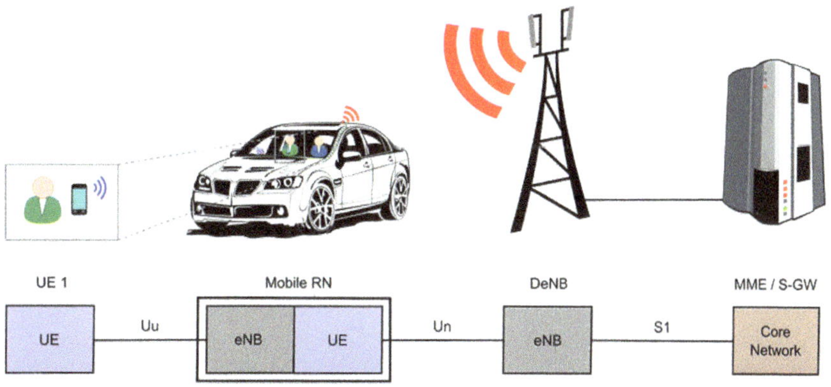

Figure 3 Current LTE-A architecture for moving networks.

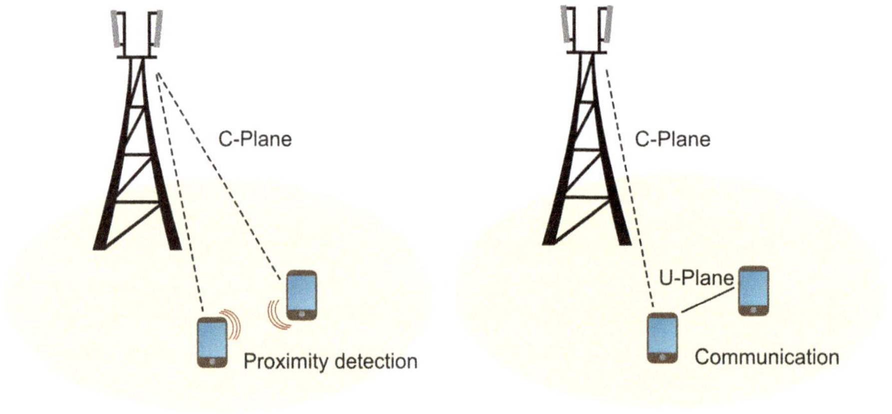

Figure 4 D2D approach in 3GPP.

opens the door for an interconnection of devices that collaborates within a group of users they trust searching for mutual benefit, similar to a social network. RAN 2.0 implies a flat architecture from service point of view resulting in low latency. It is also accompanied by an agile infrastructure support, since *ad hoc* and smart coordinated setup of networks is expected under this service/user-centric model. Finally, RAN 2.0 also entails a different distribution of intrinsic functions that can be executed in any node depending on its hardware and software capabilities.

This concept has been also discussed under the name of Spontaneous Smartphone Network (SSN) [34], since it depends heavily on the network functionalities provided by the smartphone operating system. All current operating systems have tethering capabilities to route packets from the cellular link to another short-range wireless interface like Wi-Fi, but operating at IP layer. However, this tethering characteristic is not working transparently to the end user, lacks adequate security control procedures for the pairing and ciphering of the information, cannot make use of cooperation and does not support self-organizing principles to manage the links, and ensures minimum interference and dynamic configuration to face unexpected changes in the channel conditions. Current research is focusing mainly on the dynamic configuration of the network [34] and on security [35], likely the main showstopper of this 5G enabler.

2.2 The spectrum toolbox

Of course, new spectrum bands must be identified in the International Telecommunication Union-Radio-communication (ITU-R) Radio Regulations to support the increase of traffic demand. In this direction, the worldwide allocation of a number of bands to International Mobile Telecommunications (IMT) technologies is on the agenda for the next World Radio Conference 2015 (WRC-15). Figure 6 shows

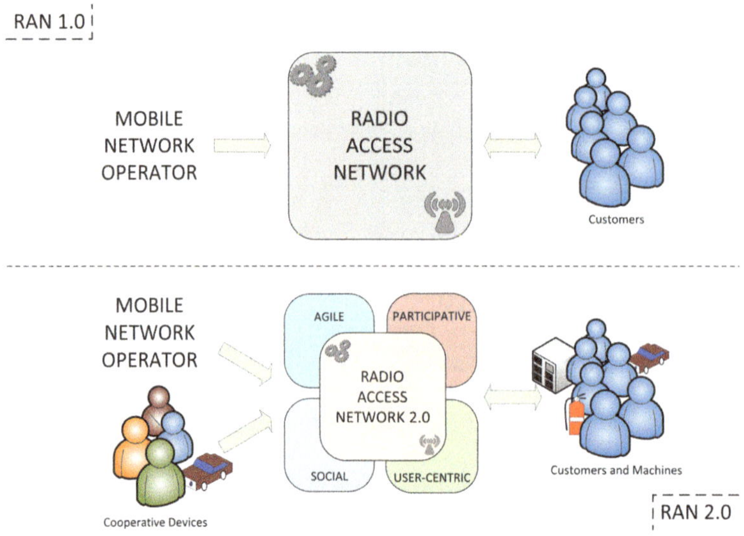

Figure 5 RAN 2.0 concept.

the possible candidate bands for IMT under WRC-15 agenda [36]. Despite the fact that the current practice of predominantly using dedicated licensed spectrum will remain the main stream, new regulatory tools and approaches of sharing the spectrum and optimizing its use must be devised [37]. On the other hand, the coverage of higher frequency bands may be limited, for example, to hot spots or dense areas. MMC and low latency on the other hand are required in wide area coverage.

For its special relevance, we need to mention here the use of millimeter wave (mmW) communications (30 to 300 GHz). Although conventionally beyond 6 GHz spectrum has been discarded due to its high attenuation, recent research points towards its potential use for backhaul, D2D and even cellular communications [38]. Moreover, semiconductor technologies are getting mature enough to reduce cost and power consumption of wireless communications in these new bands [9]. On the other hand, propagation problems can be tackled thanks to the use of massive MIMO and the resulting huge antenna array gains. Recent measurements at 28 and 73 GHz in New York University [39] have revealed that, with an accurate beam-selection, signals can be detected at a distance between 100 to 200 m even in the absence of line of sight (LoS). This result has motivated an increased interest of the wireless community on cellular mmW communications. This proposal still requires intense research to overcome its main challenges, that is, the support of mobility for the optimum beam selection and the design of transmission techniques that alleviate the need for feedback on channel state information (CSI) to the transmitter [40].

2.3 Flexible air interface

The variety of spectrum options leads to a system concept incorporating a number of air interfaces, each optimized for a specific set of use cases. The air interfaces including their protocol stacks will be configurable to a great extent in order to meet the diverging traffic demands of different applications. They will be integrated in smart and dynamic wireless software-defined networks (SDNs), [41], which means a generalization of the software-defined radio (SDR) concept [42]. In contrast to existing cellular systems, sophisticated technologies such as ultra-dense access node deployments and multi-node coordination will be supported to achieve optimum performance [43]. Frequency division duplex (FDD) and time division duplex (TDD) systems are expected to further coexist, while TDD-only operation is expected to become more widely used in higher frequency bands. Full duplex is under investigation, but its use will probably be restricted to low-power radio nodes, e.g., for indoor and outdoor small cell applications (including in-band wireless backhaul). Evolved versions of existing communication systems need to be efficiently integrated.

2.4 Massive MIMO

In massive MIMO, we see a very large antenna array at each base station and in general an order of magnitude more antenna elements as compared with conventional systems. Massive MIMO can be used for a more efficient backhaul wireless link or even for the access link, in which a large number of users are served simultaneously. Here, massive MIMO is interpreted as multiuser MIMO with lots of base station antennas. Although massive MIMO has been historically associated with the use of beamforming, new trends are arising towards the use of spatial multiplexing techniques in conjunction with beamforming to make the most of the richness of the channel [44]. With the incorporation of nomadic and moving nodes, sophisticated MIMO solutions can be implemented in a vehicle node, making advanced multi-user MIMO feasible.

Massive MIMO can increase system capacity by a factor of 10 and, at the same time, increase the energy efficiency on the order of 100 times [45]. This boost in capacity comes from the use of aggressive schemes of spatial multiplexing whereas the improvement in energy consumption is due to the concentration of power in very concrete space regions.

The main concern when the concept of massive MIMO comes to real deployments relies on the factor form of an antenna of hundreds of element. In this sense, the concept of massive MIMO is intrinsically related to its use in high frequencies, e.g., in the centimeter wave (cmW) and mmW bands, which will reduce the size of each of these elements [46]. Many different configurations for the real antenna arrays can be used, as depicted in Figure 7, ranging from distributed antenna systems, to planar and cylindrical arrays.

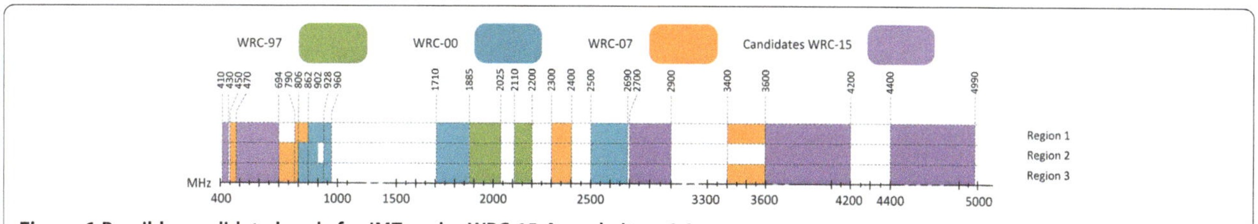

Figure 6 Possible candidate bands for IMT under WRC-15 Agenda Item 1.1.

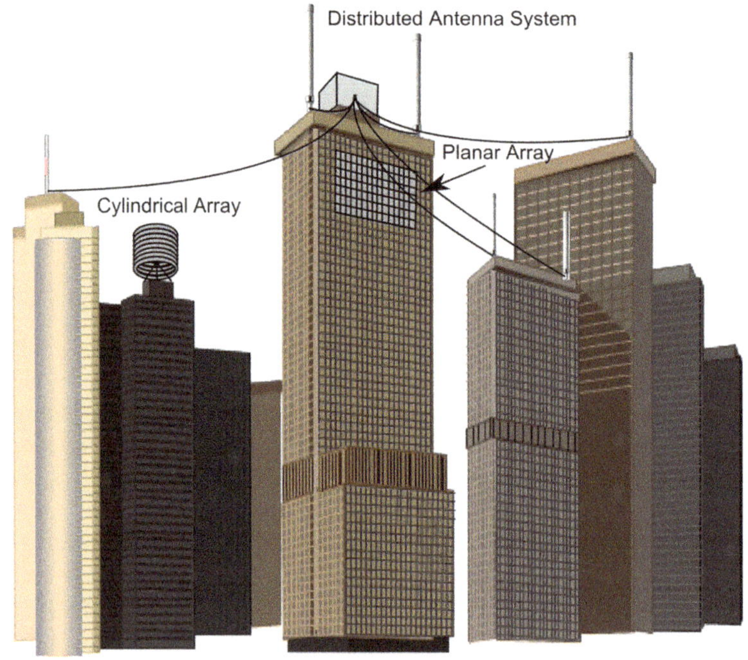

Figure 7 Different alternatives for the deployment of massive MIMO, including distributed antennas, planar arrays and cylindrical arrays.

Extensive use of beamforming will be an essential tool to improve link budgets. In addition, spatial multiplexing using massive MIMO techniques at mmW along with small cell geometries seems to be a symbiotic convergence for throughput boost [47]. The success of these techniques might be threatened by impairments in the transmission and reception chain, such as hardware limitations and lack of appropriate channel knowledge. This will be a matter of exhaustive research in the coming years. Another important problem with massive MIMO is pilot contamination. The number of orthogonal uplink pilot sequences is limited and depends on the coherence interval, which is very small at higher frequency bands. When reusing the same pilot from two transmitters, the negative consequences derived from interferences are referred to as pilot contamination. A beamforming scheme based on the channel estimates obtained from contaminated pilots results in interferences directed to those transmitters using the same pilot [48]. Therefore, new transmission schemes based on blind channel estimation are to be designed for the next generation mobile [49].

2.5 New signaling/control information

The control information/signaling needs to be fundamentally readdressed in 5G systems to accommodate the different needs of different services. In order to facilitate the spectrum-flexible multi-layer connectivity for eMBB services, a separation of control and data plane can be used. One example is mmW communication, where the control channel can be established at lower frequencies. Another example is network-controlled communication of content via D2D connection, offloading the cellular network. More generally, the user centricity of 5G systems shown in Figure 5 induces separation of control and data planes.

MMC, on the other hand, benefits from a closer coupling between the control and data plane, even integration of the control and data planes. MMC also requires optimized sleep mode solutions for battery-operated devices and mobility procedures with a minimum of signaling and measurements.

URC requires guaranteed latency and reliability. Here, it should be noted that the successful reception of control information is a prerequisite for communication of the data part. For very reliable, low-latency V2X connection, the device discovery can be assisted by the wide-area network. The increasing number of network nodes requires lean signaling for energy performance boost.

Another aspect that requires further attention is security. In the sensible case of D2D communications, the wide-area system shall provide security parameters to both communications links based on the internal credentials of the users. This network-assisted security establishment is a challenging issue in the development of D2D communications.

2.6 Localized contents and traffic flows

One of the main goals of METIS is to reduce current E2E latency performance to less than 5 ms. However, the delay budget analysis of legacy technologies reveals that most of the delay comes from the Internet and the core network parts of the E2E connection. Therefore, localized traffic flows, including data traffic offloading, aggregation, caching, and local routing, contribute to meeting this target.

The network operator improves the user experience by providing, e.g., authentication and security features while reducing the load on the data transport. In this framework, context information and network assistance for D2D discovery are of paramount importance to enable such direct communication.

In MMC, the use of concentrators acting as local gateways could allow direct communication among sensors located in capillary networks without the need to reach the core network gateway. For MMC, the localized traffic flows allow low-power access to the network. The network edge nodes can provide aggregation and information fusion of sensor data reducing the transport load and provide local added information value. Further, the necessary context information for MMC operations can be stored locally.

For delay-sensitive services, e.g., road safety applications, it is necessary to turnaround the traffic flow and perform critical computations close to the user to meet the latency constraints of less than 5 ms.

Moreover, the concept of caching could be shifted to the network edges, reaching access nodes or even the own devices that could act as proxies in case of having the requested content in the memory [50].

3 Flexible network architecture for 5G

The 5G architecture development is driven by three key aspects, namely, flexibility, scalability, and service-oriented management. All of these three aspects are complementary to each other such that the diverse set of 5G technologies which aim to fulfill broad range of service requirements can be efficiently supported. Flexibility is the core aspect to enable dynamic configuration of the necessary network functionalities for the efficient realization of a target service and use case. Furthermore, the future-proof architecture shall be flexible enough to handle diverse requirements of the use cases and services that we foresee today and the ones that may emerge in the future and cannot be anticipated now. On this basis, scalability will be assisted by flexibility to fulfill the requirements of extremely contradicting services, e.g., MMC verses multi-user ultra high definition (UHD) tele-presence.

As addressed above, the new generation of RAN networks envisaged for 5G needs to efficiently handle multiple layers and a variety of air interfaces in the access and the backhaul domains. They have to control and cope with the dynamics of traffic, user behavior, and active nodes involved and need to be able to differentiate a larger variety of QoS characteristics, such as ultra-low latency traffic, ultra-reliable communications, and broadcast traffic. SDN, Network Function Virtualization (NFV) [51], and advanced self-organizing network (SON) technologies will play an important role in the implementation and control of the network functionalities, in particular, to improve scalability and reliability as shown later in this paper with respect to achievement of METIS goals.

The generic METIS 5G network architecture, as sketched in Figure 8 from the topological point of view, must accommodate a number of technical enablers and communication paradigms while taking into account existing and emerging evolutionary and revolutionary architectural trends. The network topology will comprise various flavors of Cloud-RAN (C-RAN) [52], traditional access nodes as well as new virtual access nodes where the fixed-cell concept disappears in favor of device-centric communications. Traditional access nodes are the ones that we have today with the hardware and software components existing together, e.g., dedicated hardware implementations of base stations. The virtual

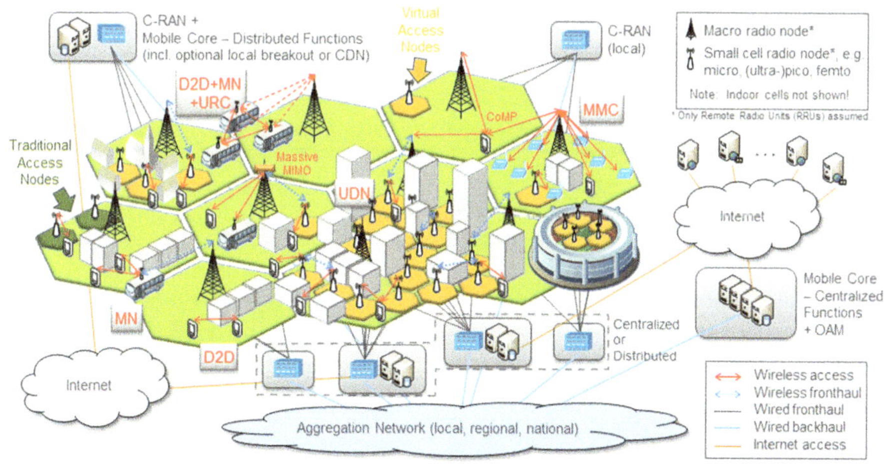

Figure 8 Generic 5G network architecture (high-level topological view).

access node is a new notion anticipated for 5G mobile and wireless communications networks, where the functionalities at an access node will be run by various virtual machines that can be modified or extended based on a need basis of the services and/or use cases to be supported.

Moreover, 5G network architecture will also be scenario and use-case specific, e.g., it may be different in areas with low user density compared to deployments in ultra-dense areas, such as Mega-Cities.

Implementation of radio network and service functions in C-RAN environments will simplify the mapping of SDN and NFV features (known from core function virtualization in data centers) to the RAN. It also increases flexibility with respect to integration of decentralized core functions in C-RAN processing units like local mobility management, local breakouts, and content delivery networks (CDNs) with caching capabilities [50]. Due to centralized processing and minimum delay among baseband processing units (BBUs), C-RAN environments allow simplified clustering of cells for joint RRM and interference coordination (including coordinated multipoint transmission/reception (CoMP)). Dependent on network infrastructure availability of the mobile network operator and delay limitations set on back-/fronthaul links [b], e.g., by CoMP schemes, C-RANs can be deployed in a distributed or more centralized way, which differentiate especially in terms of the number of contained BBUs (realizable via virtual machines on general purpose processors, possibly with support of hardware accelerators for dedicated physical layer computing tasks). Nevertheless, the BBU number of a local C-RAN can be also high in case of UDN, e.g., for a stadium scenario.

A first consideration of the main functional building blocks to appear in the METIS 5G RAN architecture from the logical point of view is shown in Figure 9. Flexibility shall enable a dynamic configuration of the necessary network functionalities in order to fulfill the requirements of a target service. This can necessitate new or tailored functionalities that will be made available on-demand based on the capabilities, such as processing and memory, of the network elements. On this basis, the functions may be updated or enabled on different time scales and spatial dimension. For example, a machine may be updated by cluster-head functionality when there are at least a minimum number of other machines in the proximity, and the candidate cluster-head machine has the capabilities of running that functionality. The Service Flow Management, as depicted in the logical view of architecture, has the task of analyzing the services and outlining the requirements for data flows through network, which will be orchestrated by the Function Coordinator. We note that to realize such flexibility end-user devices, e.g., nomadic nodes, D2D relays, and MMC gateways, are to be integrated as network elements. Function Coordinator has an interface with Function Agents by which it can configure the Base-Band Elements (BBEs) and core network elements (CNEs) based on the service requirements, also known as service orchestration, and forms the service chain. For example, addition of new functions and updates of the existing functions at BBEs and CNEs are performed by the Function Agent considering the configurations received from the Function Coordinator. Here, service requirements are determined by the Service Flow Management. Wireless SDN (WSDN) Controller sets up the service chain on the physical network infrastructure taking into account the configurations orchestrated by the Function Coordinator. WSDN Controller then constructs the data plane processing for the

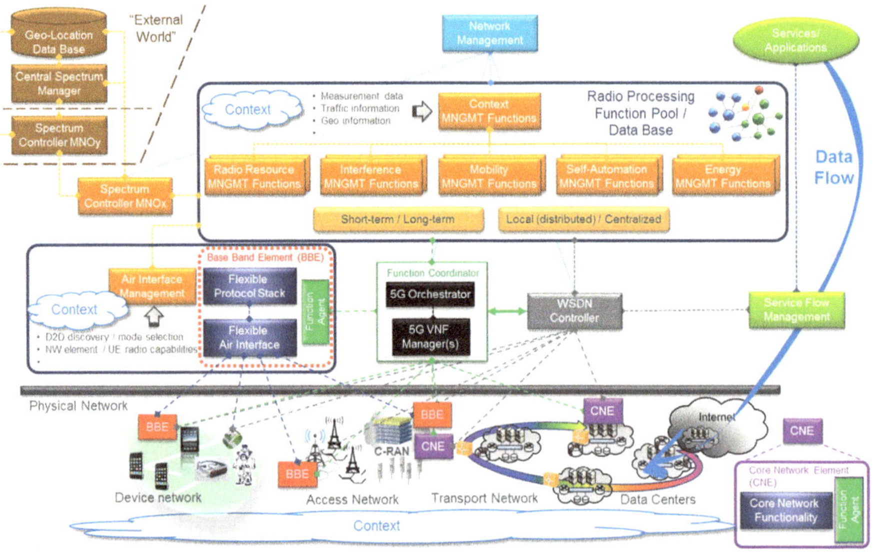

Figure 9 Generic 5G network architecture (high-level functional/logical view).

data flow, i.e., it builds up the connections for the service chain of CNEs and BBEs in the physical network through which the data flow passes. Accordingly, WSDN Controller configures the routers/switches (i.e., has interfaces to routers/switchers) also by factoring in radio-specific functionalities like resource scheduling and mobility management. It is worth noting that conventionally C-RAN infrastructure comprises BBUs. This corresponds to BBEs in the envisioned logical architecture. Nevertheless, thanks to the flexibility, which is anticipated for 5G, some core functionalities can also be moved to C-RAN. Therefore, C-RAN can compromise multiple BBEs as well as CNEs as depicted in Figure 5.

The envisioned architecture will provide the necessary flexibility to realize efficient integration and cooperation of functional blocks according to the individual service and network function needs of the HTs as well as functions for future evolution of existing cellular and wireless networks. The functions can be flexibly modified, tailored, and created by the Function Coordinator (consisting of 5G NFV Orchestrator and 5G virtual network function (VNF) Managers) according to the data flows and can be moved to the relevant network elements on-demand. In this regard, the context information, which is collected from the whole network, will be efficiently utilized by management functions. The architecture is based on WSDN approach to enable on-demand creation of customized virtual networks using shared resource pools and allowing effective service-adaptive decoupling of control and data plane in order to optimize routing and mobility management across the whole service transport chain. From the perspective of the end-to-end flow, even protocol stacks will be optimized and customized via software according to the required services and the network topology, i.e., the protocol stack becomes more and more a set of functional building blocks flexibly/dynamically combined on-demand to fulfill the very specific tasks for a certain service. On this basis, BBEs comprise flexible protocol stack, flexible air interface, and Function Agent. Moreover, it is worth noting that WSDN Controller can also be virtualized running as a VNF with its Element Management System (EMS) & VNF Manager [53].

In contrast to present 4G, certain functionalities of the user equipments (UEs) may be partially controlled by the operator, e.g., for D2D relaying. Furthermore, not only do the UEs connect to the network, but also a wide range of devices, such as, sensors and robots, will co-exist, and the operator may have further degrees of freedom to coordinate the functionalities of these devices. It has to be noted that flexibility may be limited by capabilities of the network elements involved, such as sensors, which may not be updated with all new functionalities, e.g., due to hardware limitations.

Especially for operators with both fixed and mobile network infrastructure, cost reduction in future deployment stages is of great importance. For reuse of common network infrastructure on transport and access layer (Fixed-Mobile Convergence (FMC)), WSDN and NFV are, as well, seen as the key enablers [54] allowing multi-operator network infrastructure and resource sharing in a cost-optimized way. The integration of joint core functionalities belongs to the next steps towards FMC.

An exemplary use case is described in the following. The motivation is to enhance the user experience of an extremely high density of active users and machines in an area where normally the mobile access network nodes are sparsely deployed. The challenge is then to fulfill the requirements of such a massive connectivity of a mixture of eMBB and MMC devices for a certain time period by the network that is highly under-dimensioned for such a use case. Within the framework of the envisioned flexible architecture, service flow management determines that there is an over-flow of mixture of services to be provisioned to massive number of devices and triggers the Function Coordinator regarding this situation. The Function Coordinator (here 5G Orchestrator) in turn updates the service chain to handle these requirements. One solution to tackle the signaling congestion is to perform clustering mechanisms where a cluster head coordinates the scheduling of a group of devices. In this regard, Function Coordinator (here VNF Manager) commands the Function Agents of the selected cluster heads (selection can be performed by the Context Management Entity according to the collected context, e.g., device capabilities reported by Function Agents) in the target region to install the cluster-head functionality in the BBEs of those selected cluster heads. The functionality definition is taken from the Radio Processing Function Pool via Function Coordinator (here 5G Orchestrator) and Function Agents. It is worth noting that the cluster head functionality is only utilized for a given time period when the service is required. Due to limited machine capabilities, e.g., small memory, such cluster head functionality, may be removed depending on the service and energy efficiency requirements. Based on the input from Function Coordinator (here 5G Orchestrator), the WSDN Controller can now perform the re-routing of the data flows according to service chain in line with the updated network topology.

Although in the envisioned flexible architecture, there are additional logical entities introduced, which are needed to enable the anticipated flexibility, compared to today's network architecture, there is a general consensus that adopting these architectural trends has various advantages in terms of cost savings (see, e.g., [51]). For instance, a virtualized network enables faster service provisioning, i.e., increased speed of time to market, and extension of the "classical" service portfolio to new offerings for, e.g., 3rd party connectivity providers (XaaS incl. network sharing), which implies revenue increase for the operators as the impact on the hardware deployments is minimized. As instead of dedicated hardware implementations, commercial-off-the-shelf hardware can be utilized, a better capital efficiency can be reached. Along with the cost savings, better optimization of the network resources can contribute to the energy savings in the network, which is in turn coupled with the cost savings, as well. It is worth noting that this trend is already ongoing on the fixed network side [55], and this vision for 5G mobile network environment can create synergy among fixed and mobile networks.

4 Meeting the METIS goals for 5G - Research challenges

In this section, we briefly discuss the primary methods envisaged for the main features of the 5G system concept that can contribute most beneficially to achieve the METIS goals stated at the beginning of this paper.

4.1 1,000 times higher mobile data volume per area

The primary methods to address this goal are to use more frequency spectrum, to use more network nodes, and to enhance network performance and spectral efficiency.

4.1.1 Use more frequency spectrum

Both higher data rates and higher data volume require access to more frequency spectrum for mobile communications. As stated before, in WRC-15, it is expected that clearly more spectrum will be released for mobile communications. This new spectrum lies in the frequency range between 300 MHz and 6 GHz. However, for the future 5G system, these new spectrum opportunities will not be sufficient. Additionally, the release of new spectrum in the mmW bands was postponed to WRC-18.

Moreover, new spectrum access schemes are being considered such as Licensed Shared Access (LSA) [56], which would allow using further spectrum for mobile communications especially in higher frequency bands in a more flexible way. Accordingly, research is conducted on further novel spectrum access schemes, e.g., for coordinated spectrum sharing between different UDNs, supporting network densification as addressed below.

Also, wireless local area network (WLAN)-type systems [57] operating in the unlicensed bands, such as the ISM and U-NII bands at 2.4 and 5 GHz, as well as the 60 GHz band, can be more tightly integrated.

4.1.2 Use more network nodes

Densification of networks by deploying more nodes is already applied today with femto or pico cells providing local connectivity for increased capacity in addition to the macro cells in the wide-area. This trend is expected to continue, where the small cells can make use of higher frequency bands, such as the 3.5 GHz bands or the mmW bands. The use of TDD with higher frequencies can become more prominent as compared with 3G/LTE today. The different network layers/cell hierarchies and RATs are to be integrated by means of smarter networks. Dedicated implementation solutions such as distributed RRUs can be developed for specific deployments such as for providing coverage of very crowded places, e.g., stadiums and events. To this end, intensive research is conducted to support MN and UDN, e.g., solutions for in-band backhaul, where the access and backhaul share the same spectrum, specific distributed RRM schemes, or dynamic nomadic node selection for backhaul optimization.

4.1.3 Enhance network performance and spectral efficiency

Enhanced network performance and spectral efficiency are to be achieved by applying more sophisticated protocol and air interface technologies as compared to LTE-A [58]. The primary air interface technologies in this regard are the use of multiple antennas, possibly involving massive antenna constellations, multi-stream transmission such as spatial multiplexing or non-orthogonal multiple access, and cooperation techniques such as coordinated multi-node transmission or solutions based on interference alignment.

4.2 10 times to 100 times higher typical user data rate

The primary methods to address this goal are to use spectrum aggregation and more dynamic spectrum access, usage of higher frequencies up to mmW frequencies and short-range communications.

4.2.1 Spectrum aggregation and more dynamic spectrum access

Spectrum aggregation is already defined by 3GPP LTE-A and enables peak data rates up to about 1 Gbps. This may evolve further to take into account the new bands released during WRC-15. Aggregation of multiple bands over a wide range of frequencies poses challenges concerning the cost and efficiency of radio frequency filter and transceiver solutions, in particular for the UEs. Global research must be conducted to propose new schemes for spectrum opportunity detection and assessment and spectrum management concept for LSA. Based on the need for spectrum, these technical enablers shall help to obtain spectrum by negotiating access to it with other entities and also assessing the use of spectrum.

4.2.2 Usage of higher frequencies

Usage of higher frequencies up to mmW frequencies will further increase the achievable data rates to 10 Gbps or beyond. To compensate for the path loss at such high frequencies, high antenna directivity is required if transmission range shall be beyond a few meters. Accordingly, flexible air interfaces for mmW frequencies must be developed. The identification of suitable frequency bands and the characterization of the propagation channel demand new investigations.

4.2.3 Short-range communications

Finally, reducing the transmission distance, approaching short-range communications, e.g., D2D or UDN, is also an efficient means to increase user data rate. Indeed, node densification and dynamic RAN (cf. RAN 2.0 described previously) can allow confining interference to very specific areas of use while reducing the signal attenuation. Both effects shall boost the signal to interference plus noise ratio (SINR) thus increasing the user data rate by a logarithmic law.

4.3 10 times to 100 times higher number of connected devices

The primary methods to address this goal are to reduce signaling overhead, to use dynamic profiles of users, and to implement traffic concentration, in particular for MMC applications.

4.3.1 Reduce signaling overhead

Signaling overhead can be reduced by designing more efficient protocols and by thinning out some protocol layers, e.g., for the simplification of end-to-end procedures. On the air interface, the signaling for random access, time adjustment, and/or resource assignment can be reduced, for example, by using contention-based access, waveforms that are more tolerant to timing mismatch as compared to orthogonal frequency division multiplexing (OFDM), and/or by resource reservation, making the air interface more flexible and suitable for small payload traffic. For instance, methods dedicated to contention-based access for massive number of devices without scheduled access and multi-user detection facilitated by compressive sensing are good candidates to reduce signaling.

4.3.2 Use dynamic profiles of users

NFV technologies can play a role to overcome current limitations of a rigid protocol stack by allowing flexible configuration of the network functionalities. The future air interface shall support this flexibility by using dynamic profiles of users, of which entries are configurable to pre-defined use cases. In this sense, some radio functions will be activated/deactivated on demand depending on the specific needs of the service and the status of the network. Again, the user-centric paradigm applies, since the network shall adapt its topology and operation mode according to the needs of each user.

4.3.3 Traffic concentration

Finally, traffic concentration can be achieved by means of gateways, for example, by linking a capillary MMC network to a wide area cellular network. In this framework, the RAN might configure one specific device to act as such concentrator (cluster head) by using D2D communications to collect all data from the surrounding machines. It is expected that this last hop link should work on a different radio interface to host a large number of mesh devices with severe energy constraints. For instance, techniques are developed to reduce the amount of traffic in gateway-based MMC solutions by context-based device grouping and signaling.

4.4 10 times longer battery life for low power MMC devices

The primary methods to address this goal are to improve air interface, procedures, and signaling and to reduce the distance between the MMC devices and the access node. Techniques to improve the air interface include the use of sleep modes, energy-efficient modulation, coding and multiple access schemes, for example, use of constrained modulation techniques such as Gaussian Minimum Shift Keying (GMSK), very robust coding and/or spreading allowing very low transmission powers, and simplified procedures such as those introduced for the previous goal.

The MMC devices can, in some scenarios, be brought closer to the access node, for example, by gateways connecting a capillary network to a wide area network or by denser deployment of small cells.

4.5 Times reduced End-to-End (E2E) latency

The primary methods to address this goal are to use more efficient network architectures, more efficient air interface designs, signaling and procedures, better QoS differentiation, and direct D2D communication.

4.5.1 Efficient network architectures

Low-latency network architectures are concerned with shortening the distances and number of hops between the user and the content, for example, by distributing some network functions that are centralized today or by applying local and universal caching as well as local breakouts to external networks like Internet, service provider, or enterprise networks.

4.5.2 Efficient air interface designs

The air interface may adapt the frame structures in order to reduce the Transmission Time Interval (TTI) and/or Hybrid Automatic Repeat reQuest (HARQ) round trip times. While this can efficiently be implemented at the higher frequencies where more bandwidth is available, this can become very inefficient for wide area coverage. Signaling and procedures can be thinned out, by simplifying their components.

Table 1 Main enablers for the 5G design

Technique	Impact	Pros	Cons
D2D communications	System capacity can be increased by a factor of 2 using the same bandwidth when some RRM mechanisms are used for opportunistic access and there exists full cooperation among devices. Latency is also reduced to the order of the TTI length (e.g., 1–2 ms).	With properly selected safety distance, D2D communication can use the spectrum allocated to small cells (mostly uplink band) without affecting the small cell performance.	How to motivate cooperation of users is still one open issue. Battery consumption is something critical in current systems and D2D communication model is only accepted by the end user for him/her own benefit. Specification is far from being ready for D2D integration. Opposition from some mobile network operators is also an important barrier to overcome.
New air interfaces	Access time can be reduced down to 1.5 ms if new air interfaces are coupled with efficient access procedures (as shown by FBMC evaluations [60]). Due to the very good frequency localization, the transmit power can be concentrated on only very few subcarriers to eventually enhance significantly the expected coverage or to reduce battery consumption. This is of special relevance for machine-type communications, in which payload is very small.	The well-localized signal energy in frequency domain of the multi-carrier signal also allows for efficient access to fragmented spectrum and efficient spectrum sharing, as a minimum amount of guard bands are needed for the signal separation in frequency.	Changing the waveform impacts the signal structures, implying a revolutionary step towards a new radio generation. Backward compatibility thus cannot be guaranteed, however, for a new radio generation this should not be crucial requirement.
Ultra-dense network	Capacity is directly proportional with the number of nodes, provided centralized interference coordination. For indoor cases, the coefficient of proportionality could be as high as 0.73 with ISD of 10 m.	Together with increasing the bandwidth, the use of more nodes is an easy means to achieve desired levels of capacity.	Cost is the main issue of this approach, together with the need for interference coordination. In case of certain isolation between cells, this need for coordination is relaxed.
Traffic concentration	The use of accumulators or concentrators for machine-type communications improves range of coverage and sensors' throughput. The improvement factor is between two and three thanks to those relays. For the same throughput and coverage needs, traffic concentration reduces battery consumption.	Concentrators also reduce signaling overhead thanks to the coupling of parallel signaling flows.	Performance heavily depends upon the appropriate relay selection.
Moving networks	In vehicles equipped with two access points, one for outside transmission/reception and another for inside users, their best reception chain improves the link budget for the end user up to 9 dB, in cases where the user is outside the car, and up to 24 dB, when the user is within the car. This results in better coverage or higher user throughput, mostly in the cell-edge.	Battery is not a big issue for vehicles, which opens the door for more active collaboration between cars and end-users. Moreover, the number of antennas integrated in vehicles can be much higher than in the handheld devices. This allows for massive-MIMO solutions.	Mobility of access points increase management complexity mostly due to the higher dynamicity of the network.
Localized traffic flows	With a dedicated bandwidth of 80 MHz, end-to-end latency is reduced to 60%, as compared with current LTE-A system. Moreover, half of the traffic can be offloaded from the cellular system.	Use of localized traffic flows is simple to implement and can be easily integrated into current networks.	Universal caching requires storage resources as well as the design of specific signaling mechanisms.
Massive MIMO	Spectral efficiency can be increased by a factor of 20 when using 256 antennas in the transmitter and receiver side as compared with four antenna systems. For the same spatial multiplexing capability as legacy systems (8 streams), beamforming gain reaches 15 dB.	Simple way of increasing cell efficiency mostly for small cell deployments. Fits together with the use of higher frequencies above 6 GHz due to the reduced antenna size.	Pilot contamination is one of the main showstoppers of the use of massive MIMO. Moreover, TDD mode seems to be a must to reduce signaling overhead thanks to the use of channel reciprocity principle. Form factor forces the use of centimeter wave or mmW to compact massive antennas. Finally, performance is very sensitive to mobility and computation burden could make multi-user solutions unaffordable.

Table 1 Main enablers for the 5G design *(Continued)*

Spectrum above 6 GHz	Meeting the 5G user expectations will require much larger bandwidths as today, in the order of 2 GHz, which can only be available at frequencies above 27 GHz (e.g., 27.5 to 29.5 GHz; 40.5 to 42.5 GHz; 47.2 to 50.2 GHz; 57 to 76 GHz; or 81to 86 GHz).	No restrictions on previous allocations of paired FDD spectrum allows for the use of TDD mode, which is much more efficient for channel estimation. Higher frequencies also reduce antenna size, thus permitting massive MIMO implementations.	Path losses are much higher at such high frequencies, which reduces the coverage to small distances or relies on the use of higher order beamforming to overcome such high attenuations.

4.5.3 Better QoS differentiation

Better QoS differentiation is important in this context. As the low-latency transmission may become inefficient, it should only be applied with the services that require such low latencies and should be avoided with delay-insensitive services.

4.5.4 D2D

D2D can be used between devices within transmission range of each other which saves a radio hop and the routing of traffic through the network. In some case, it can be applied also for local broadcasting, for example, where vehicles broadcast position/velocity information to enhance traffic safety.

5 Conclusions

This paper has presented the current view on the 5G system concept from the METIS project. Within the 5G system concept and its main features, a generic overview for the envisioned flexible 5G network architecture is provided. More details on this architecture proposal can be found in [59]. In contrast to current 3GPP work, D2D communications must be a cornerstone of the new 5G technology going towards the new RAN 2.0 paradigm. Flexible radio air interface can offer the best solutions targeted to different use cases. Moreover, the ultra density of access nodes, including vehicles, devices, and base stations, requires further research into the definition of more efficient mobility procedures and signaling simplification. On the other hand, a crucial difference of the future 5G networks in comparison to existing 4G networks is that the system must be from-start designed not only for human-centric, but also for machine-centric traffic in MMC with different service demands. Intensive research is required in the fields of overlaying multiple transmissions, new air interfaces, smart pre-allocation of resources, reduction of the synchronization requirements, and context/service-aware configuration of the radio access. Finally, special attention must be paid to the development of URC techniques, capable of supporting new services like tactile Internet or V2X communications for traffic safety.

Simulation results in METIS have proven the potential of some of the techniques discussed along this paper. Table 1 summarizes the impact of most significant techniques together with some comments concerning the pros and cons of their use in 5G systems. Note that the quantitative analysis provided in Table 1 is derived from several simulations in up to 12 different test cases. For the sake of simplicity, we are not going to include in this paper the details of all these test cases. Please refer to METIS deliverables for additional information [2,60].

Of significant relevance is the use of D2D communications, which could increase average system capacity by a factor of 2 (with high density of users and common interests in the consumed contents), and moving networks, in which cars can operate in relaying mode for other users, e.g., out-car users in case of nomadic node operation [61,62]. Thanks to relaying, vehicle penetration losses are avoided and the reception chain improves, due to the better characteristics of the antenna and the receptor in the vehicle. The mobile relay operates in a separate band in full duplex mode, that is, is totally transparent for the network point of view that treats the vehicle cell like a regular user. Moreover, the collaboration of cars allows for a feasible implementation of universal caching, by which cars forward cached content directly to interested user. Results have proven two positive effects of this universal caching. First, caching reduces end-to-end latencies of cached users. This reduction increases with smaller packets, that is, when the non-radio-transmission delay is more relevant over the whole end-to-end latency. The second positive effect of universal caching is the offloading of traffic. According to current shape of cellular network contents, one fourth of the users may be offloaded from the classical network. With the current traffic type distribution, about one half of the traffic load might be offloaded.

5.1 Endnotes

[a]In comparison with data traffic in 2010.

[b]Backhaul is denoted as the link between aggregation (alternatively core) network and BBU of the radio node. The fronthaul is denoted as the link between BBU and Radio Remote Unit (RRU). This denotation is valid also for the C-RAN case [52].

Competing interests

The authors declare that they have no competing interests.

Acknowledgements

Part of this work has been performed in the framework of the FP7 project ICT-317669 METIS, which is partly funded by the European Commission. The authors would like to acknowledge the contributions of their colleagues in METIS with special thanks to Petar Popovski, Peter Fertl, David Gozalvez-Serrano, Andreas Höglund, Zexian Li, and Krystian Pawlak. Also thanks to Josef Eichinger and Malte Schellmann for the fruitful discussions during the revision of this article.

Author details

[1]Universitat Politecnica de Valencia, Camino de Vera, s/n, Valencia 46022, Spain. [2]Alcatel-Lucent Bell Labs, Stuttgart, Germany. [3]Ericsson, Stockholm, Sweden. [4]Deutsche Telekom, Darmstadt, Germany. [5]Huawei ERC, Munich, Germany.

References

1. Cisco, in *Global Mobile Data Traffic Forecast Update*, 2014–2019 White Paper, February 2015. http://www.cisco.com/c/en/us/solutions/collateral/service-provider/visual-networking-index-vni/white_paper_c11-520862.pdf
2. METIS, in *Mobile and wireless communications Enablers for the Twenty-twenty Information Society*, EU 7th Framework Programme project, http://www.metis2020.com.
3. ICT-317669 METIS project, in *Scenarios, requirements and KPIs for 5G mobile and wireless system*, Deliverable D1.1, May 2013, https://www.metis2020.com/documents/deliverables/
4. B Ahlgren, C Dannewitz, C Imbrenda, D Kutscher, B Ohlman, A survey of information-centric networking. IEEE Commun Mag **50**(7), 26–36 (2012)
5. A Osseiran, F Boccardi, V Braun, K Kusume, P Marsch, M Maternia, O Queseth, M Schellmann, H Schotten, H Taoka, H Tullberg, MA Uusitalo, B Timus, M Fallgren, Scenarios for the 5G mobile and wireless communications: the vision of the METIS project. IEEE Commun Mag **52**(5), 26–35 (2014)
6. D Gomez-Barquero, D Calabuig, JF Monserrat, N Garcia and J Perez-Romero, Hopfield neural network - based approach for joint dynamic resource allocation in heterogeneous wireless networks, in *Proceedings 64th IEEE Vehicular Technology Conference (VTC)*, Montreal. 2006
7. JF Monserrat, P Sroka, G Auer, J Cabrejas, D Martin-Sacristan, A Mihovska, R Rossi, A. Saul, R. Schoenen, Advanced Radio Resource Management for IMT-Advanced in WINNER+ (II), in *Proc. Future Network and Mobile Summit*, pp.1-9, June 2010.
8. F Boccardi, RW Heath, A Lozano, TL Marzetta, P Popovski, Five disruptive technology directions for 5G. IEEE Commun Mag **52**(2), 74–80 (2014)
9. JG Andrews, S Buzzi, C Wan, SV Hanly, A Lozano, ACK Soong, JC Zhang, What will 5G be? IEEE J Sel Area Comm **32**(6), 1065–1082 (2014)
10. MN Tehrani, M Uysal, H Yanikomeroglu, Device-to-device communication in 5G cellular networks: challenges, solutions, and future directions. IEEE Commun Mag **52**(5), 86–92 (2014)
11. N Bhushan, L Junyi, D Malladi, R Gilmore, D Brenner, A Damnjanovic, R Sukhavasi, C Patel, S Geirhofer, Network densification: the dominant theme for wireless evolution into 5G. IEEE Commun Mag **52**(2), 82–89 (2014)
12. K Okino, T Nakayama, C Yamazaki, H Sato, Y Kusano, Pico Cell Range Expansion with Interference Mitigation toward LTE-Advanced Heterogeneous Networks, in *Proc. of IEEE International Conference on Communications (ICC)*, 2011.
13. P Mugen, L Dong, W Yao, L Jian Li, C Hsiao-Hwa, Self-configuration and self-optimization in LTE-advanced heterogeneous networks. IEEE Commun Mag **51**(5), 36–45 (2013)
14. I Siomina, D Yuan, Load Balancing in Heterogeneous LTE: Range Optimization via Offset and Load-coupling Characterization, in *Proc. of IEEE Int. Conference on Communications (ICC)*. June 2012.
15. KI Pedersen, Y Wang, B Soret, F Frederiksen, eICIC Functionality and Performance for LTE HetNet Co-Channel Deployments, in *Proc. of IEEE Vehicular Technology Conf*, Sep 2012
16. X Gu, X Deng, Q Li, L Zhang, W Li, Capacity Analysis and Optimization in Heterogeneous Network with Adaptive Cell Range Control, Int. J. Antennas. Propag. **2014**(215803), 10 (2014)
17. K Smiljkovikj, P Popovski, L Gavrilovska, Analysis of the Decoupled Access for DL and UL in Wireless Heterogeneous Networks, in *IEEE Wireless Communications Letters*, in press, doi:10.1109/LWC.2015.2388676.
18. P Agyapong, M. Iwamura, D. Staehle, W. Kiess, A. Benjebbour, Design considerations for a 5G network architecture. IEEE Commun Mag **52**(11), 65–75 (2014)
19. L Yan, X Fang, Reliability Evaluation of 5G C/U-plane Decoupled Architecture for High-speed Railway. EURASIP J Wirel Commun Netw **2014**, 127 (2014)
20. B Zafar, S Gherekhloo, M Haardt, Analysis of multihop relaying networks: communication between range-limited and cooperative nodes. IEEE Veh Technol Mag **7**(3), 40–47 (2012)
21. *Study on Mobile Relay for Evolved Universal Terrestrial Radio Access (E-UTRA)*, 3GPP TR 36.836, V2.0.2, July 2013.
22. A Krendzel, LTE-A Mobile Relay Handling: Architecture Aspects, in *Proc. of the 19th European Wireless Conference (EW)*, Guildford, UK, pp. 1–6, 2013.
23. M Khanfouci, Y Sui, A Papadogiannis, and M Färber, *Moving Relays and Mobility aspects*, ARTIST4G project deliverable D3.5c-v2.0, 2012.
24. F Haider, M Dianati, and R Tafazolli, A Simulation Based Study of Mobile Femtocell Assisted LTE Networks, in *Proc. Of the 7th International Wireless Communications and Mobile Computing Conference (IWCMC)*, Istanbul, Turkey, pp. 2198–2203, 2011
25. F Haider, W Haiming, H Haas, Y Dongfeng, W Haiming, G Xiqi, Y Xiao-Hu, E Hepsaydir, Spectral efficiency analysis of mobile Femtocell based cellular systems, in *Proc. of the 13th International Conference on Communication Technology (ICCT)*, Jinan, pp. 347–351, September 2011.
26. ICT-317669 METIS project, *Initial report on horizontal topics, first results and 5G system concept*, Deliverable D6.2, April 2014, https://www.metis2020.com/documents/deliverables/
27. *Study on LTE Device to Device Proximity Services*, 3GPP TR 36.843, 2014.
28. V Yazıcı, UC Kozat, M Oguz, Sunay, A new control plane for 5G network architecture with a case study on unified handoff, mobility, and routing management. IEEE Commun Mag **52**(11), 76–85 (2014)
29. F Malandrino, C Casetti, C-F Chiasserini, Toward D2D-enhanced heterogeneous networks. IEEE Commun Mag **52**(11), 94–100 (2014)
30. A Asadi, Q Wang, V Mancuso, A survey on device-to-Device communication in cellular networks. IEEE Commun Surv Tutor **16**(4), 1801–1819 (2014)
31. D Feng, L Lu, YY Wu, GY Li, G Feng, S Li, Device-to-device communications underlaying cellular networks. IEEE Trans Commun **61**(8), 3541–3551 (2013)
32. C Xu, L Song, Z Han, Q Zhao, X Wang, X Cheng, B Jiao, Efficiency resource allocation for device-to-device underlay communication systems: a reverse iterative combinatorial auction based approach. IEEE J Sel Area Comm **31**(9), 348–358 (2013)
33. S Lingyang, D Niyato, H Zhu, E Hossain, Game-theoretic resource allocation methods for device-to-device communication. IEEE Wireless Commun **21**(3), 136–144 (2014)
34. G Aloi, M Di Felice, V Loscrì, P Pace, G Ruggeri, Spontaneous smartphone networks as a user-centric solution for the future internet. IEEE Commun Mag **52**(12), 26–33 (2014)
35. PA Frangoudis, GC Polyzos, Security and performance challenges for user-centric wireless networking. IEEE Commun Mag **52**(12), 48–55 (2014)
36. ITU-R M.2079, in *Technical and operational information for identifying Spectrum for the terrestrial component of future development of IMT-2000 and IMT-Advanced*, 2006
37. AB MacKenzie, LA DaSilva, Application of signal processing to addressing wireless data demand [in the spotlight]. IEEE Signal Process Mag **29**(6), 168–166 (2012)
38. X Cheng, Y Koucheryavy, Y Li, F Zhao, T Znati (ed.), Dynamic Spectrum Access for Throughput, Delay, and Fairness Enhancement In Cognitive Radio Networks, EURASIP J Wirel Commun Netw, November 2014
39. MR Akdeniz, Y Liu, MK Samimi, S Sun, S Rangan, TS Rappaport, E Erkip, Millimeter wave channel modeling and cellular capacity evaluation. IEEE J Sel Area Comm **32**(6), 1164–1179 (2014)
40. A Adhikary, E Al Safadi, M Samimi, R Wang, G Caire, TS Rappaport, AF Molisch, Joint spatial division and multiplexing for mm-wave channels. IEEE J Sel Area Comm **32**(6), 1239–1255 (2014)
41. K Pentikousis, Y Wang, W Hu, Mobileflow: toward software-defined mobile networks. IEEE Commun Mag **51**(7), 44–53 (2013)
42. E3 D2.4. *Cognitive Function mapping to Networks Architectures, Standard Engineering and Software Technologies for Cognitive Radios*, E3 Project Deliverable 2.4, December 2009.
43. R Wang, H Hu, X Yang, Potentials and challenges of C-RAN supporting Multi-RATs toward 5G mobile networks. IEEE. Access. **2**(1187), 1195 (2014)
44. V Jungnickel, K Manolakis, W Zirwas, B Panzner, V Braun, M Lossow, M Sternad, R Apelfrojd, T Svensson, The role of small cells, coordinated multipoint, and massive MIMO in 5G. IEEE Commun Mag **52**(5), 44–51 (2014)

45. E Larsson, O Edfors, F Tufvesson, T Marzetta, Massive MIMO for next generation wireless systems. IEEE Commun Mag **52**(2), 186–195 (2014)
46. W Roh, S Ji-Yun, P Jeongho, L Byunghwan, L Jaekon, K Yungsoo, C Jaeweon, C Kyungwhoon, F Aryanfar, Millimeter-wave Beamforming as an Enabling Technology for 5G Cellular Communications: Theoretical Feasibility and Prototype Results. IEEE Commun Mag **2**(2), 106–113 (2014)
47. AL Swindlehust, E Ayanoglu, P Heydari, F Capolino, Millimeter-wave massive MIMO: the next wireless revolution? IEEE Commun Mag **52**(9), 56–62 (2014)
48. L Lu, GY Li, AL Swindlehurst, A Ashikhmin, Z Rui, An overview of massive MIMO: benefits and challenges. IEEE J Sel Top Signal Process **8**(5), 742–758 (2014)
49. S Roger, D Calabuig, J Cabrejas, JF Monserrat, Multi-user non-coherent detection for downlink MIMO communication. IEEE Signal Process Lett **21**(10), 1225–1229 (2014)
50. X Wang, M Chen, T Taleb, A Ksentini, V Leung, Cache in the air: exploiting content caching and delivery techniques for 5G systems. IEEE Commun Mag **52**(2), 131–139 (2014)
51. ETSI ISG NFV (Operator Group), *Network Functions Virtualisation – Network Operator Perspectives on Industry Progress*, Updated White Paper, October 2013
52. NGMN Alliance, in *Suggestions on potential solutions for C-RAN*, White Paper, January 2013
53. ETSI ISG NFV, *Network Functions Virtualisation (NFV); Virtual Network Functions Architecture*, v1.1.1, Dec 2014.
54. A Tzanakaki, MP Anastasopoulos, GS Zervas, BR Rofoee, R Nejabati, D Simeonidou, Virtualization of heterogeneous wireless-optical network and IT infrastructures in support of cloud and mobile cloud services. IEEE Commun Mag **51**(8), 155–161 (2013)
55. A Manzalini, R Saracco, C Buyukkoc, P Chemuouil, S Kukliński, A Gladisch, M Fukui, W Shen, M Fujiwara, K Shimano, E Dekel, D Soldani, M Ulema, W Cerroni, F Callegati, G Schembra, V Riccobene, C Mas Machuca, A Galis, J Mueller, Software-Defined Networks for Future Networks and Services: Main Technical Challenges and Business Implications, IEEE Workshop SDN4FNS, 1–16, 2013
56. CEPT ECC, in *Licensed Shared Access (LSA)*, ECC Report 205, February 2014
57. IEEE 802.11, in *IEEE 802.11-2012 Part11: Wireless LAN Medium Access Control (MAC) and Physical Layer (PHY) Specifications*, IEEE Standard, March 2012
58. D Martín-Sacristán, JF Monserrat, J Cabrejas-Peñuelas, D Calabuig, S Garrigas, N Cardona, On the way towards fourth-generation mobile: 3GPP LTE and LTE-Advanced. EURASIP J Wirel Commun Netw **2009**, 10 (2009)
59. ICT-317669 METIS, *Final report on architecture*, Deliverable D6.4, January 2015, https://www.metis2020.com/documents/deliverables/
60. ICT-317669 METIS, *Report on simulation results and evaluations*, Deliverable D6.5, February 2015, https://www.metis2020.com/documents/deliverables/
61. Ö Bulakci, Z Ren, C Zhou, J Eichinger, P Fertl, S Stanczak, *Dynamic Nomadic Node Selection for Performance Enhancement in Composite Fading/Shadowing Environments*, (IEEE VTC 2014-Spring, Seoul, South Korea)
62. ICT-317669 METIS, *Final report on network-level solutions*, Deliverable D4.3 Version 1, February 2015

9

Radio resource management for uplink carrier aggregation in LTE-Advanced

Hua Wang[1*], Claudio Rosa[2] and Klaus I Pedersen[1,2]

Abstract

This paper investigates the uplink resource allocation problem in the context of Long-Term Evolution (LTE)-Advanced systems with carrier aggregation (CA) and dual-cluster scheduling. On one hand, these Rel'10 functionalities can increase the available transmission bandwidth and scheduling flexibility in uplink. On the other hand, they will introduce additional power back-off for the power amplifier in the user equipment (UE) with non-contiguous resource allocation. Taking into account that the uplink is inherently limited by the maximum transmission power of the UE, the assignment of uplink CA and/or dual cluster transmission for LTE-Advanced UEs has to be careful. A pathloss-threshold-based component carrier (CC) and cluster configuration algorithm is proposed to determine whether an LTE-Advanced UE should be configured with multiple CCs and/or dual cluster scheduling. An extended bandwidth-expansion-based packet scheduling algorithm is proposed for dual-cluster transmission, which tightly couples the bandwidth allocation and packet scheduling together to exploit the frequency domain diversity with low complexity. Simulation results show that with proper differentiation between power-limited and non-power-limited UEs, Rel'10 CA with dual-cluster scheduling can maintain similar coverage performance as in Rel'8 case while achieving substantial gains in median and peak user throughputs. Moreover, the proposed CC assignment algorithm achieves higher user throughput as compared to blindly assigning all UEs on all available CCs.

1 Introduction

Carrier aggregation (CA) is one of the key features for Long-Term Evolution (LTE)-Advanced to support higher transmission bandwidth up to 100 MHz, enabling peak date rates requirements of up to 1 Gbps in downlink and 500 Mbps in uplink to be satisfied. This is achieved by aggregating two or more individual component carriers (CCs) of the same or different bandwidth belonging to contiguous or non-contiguous frequency bands, subject to spectrum availability and the user equipment (UE)'s capability [1-5]. In addition to bandwidth extension enabled by CA, dual-cluster transmission has also been introduced in Rel'10 to further improve the spectral efficiency in uplink. With dual-cluster transmission, within a CC, a UE can be allocated to a maximum of two non-contiguous clusters, each of which includes one or more consecutive radio resource blocks [6,7]. Dual-cluster scheduling has the advantage of higher scheduling flexibility as compared to single carrier frequency division multiple access (SC-FDMA), while keeping the peak-to-average power ratio (PAPR) or signal cubic metric (CM) relatively low as compared to orthogonal frequency division multiple access (OFDMA).

In the open literature, there are quite many studies on the topic of CA. The performance analysis of downlink CA is investigated in [8-10]. Enhancements of downlink control channel resource allocation and mobility management for CA are studied in [11-13]. Different carrier load balancing schemes and packet scheduling algorithms are analyzed in [14] by means of theoretical formulations as well as system level simulations. Enhanced frequency diversity schemes for CA are exploited in [15-17]. However, most of the existing work on CA for LTE-Advanced is mainly concentrated on the downlink. There are a few studies in the uplink. An overview of uplink multiple access transmission schemes in support of CA is provided in [18]. A dynamic uplink CA scheme is proposed in [19] to improve the energy efficiency of uplink communications. An edge-prioritized channel and traffic aware CC assignment resource and allocation algorithm is proposed in [20]. The performance

* Correspondence: huw@es.aau.dk
[1]Aalborg University, Fredrik Bajers Vej 7, DK-9220 Aalborg Ø, Denmark
Full list of author information is available at the end of the article

of uplink CA in LTE-Advanced systems with different CC allocation schemes is investigated in [21-24].

As different CCs may operate at different frequencies and bandwidths, questions arise as how to assign the CCs to each user, and how to multiplex the users within each CC. Different from the downlink, the UE is limited by the maximum transmission power in the uplink, especially for cell edge users since they usually suffer from unfavorable channel conditions. Furthermore, an additional power back-off is needed in the UE power amplifier (PA) with non-contiguous resource allocation in the uplink, which in practice means a reduction of the UE maximum transmission power [25]. The UE transmission power constraint together with the additional power back-off required with non-contiguous resource allocation might counterbalance the gain brought by multi-CC and/or dual-cluster transmission and even results in a performance loss as compared to the case without CA where the SC-FDMA properties of the transmitted signals are maintained (single-CC and single-cluster transmission). Therefore, the selection of UEs to operate with uplink CA has to be carefully considered. The CC selection for uplink CA has been studied in [22] for the case of intra-band contiguous CA and single-cluster transmission. In this paper, we extend the work of [22] to design efficient radio resource management (RRM) algorithms for uplink CA for cases of both intra-band contiguous CA and inter-band non-contiguous CA, as well as the support for multi-cluster transmission. Our main contributions are as follows:

1) We first focus on the derivation of a simple yet effective pathloss-threshold-based CC configuration algorithm which can be applied to both intra-band CA and inter-band CA. The proposed algorithm takes some of the key uplink issues into considerations, such as uplink power control and power back-off requirements in the UE PA with non-contiguous resource allocation.
2) To support multi-cluster transmission with in a CC, an extended bandwidth-expansion-based packet scheduling algorithm is proposed based on the work in [23,26]. It aims at optimizing the resource allocation with low computation complexity.
3) We present an extensive performance analysis under realistic conditions. Due to the complexity of the system model and the various RRM elements involved, the corresponding performance is best evaluated via advanced system-level simulations under realistic multi-cell, multi-user conditions with dynamic birth-death traffic, using commonly accepted stochastic models and accurate representation of the many mechanisms that influence on the performance.

The rest of the paper is organized as follows: Section 2 outlines the system model of RRM framework for LTE-Advanced, with special attention on multi-cluster transmission, UE power back-off model, and the uplink power control. The proposed CC configuration and packet scheduling algorithms are described in Section 3. Section 4 introduces the simulation methodology and main assumptions. Simulation results and performance analysis are presented in Section 5. Finally, some conclusions are drawn in Section 6.

2 System model

In this study, it is assumed that eNodeB (eNB) antennas are collocated and have the same beam patterns for the CCs (also known as CA scenario 1 in the 3rd Generation Partnership Project (3GPP)). This would be a typical scenario for intra-band CA when the CCs are within the same band or different bands but with modest frequency separation, providing nearly the same coverage on all CCs. The RRM framework for multi-CC LTE-Advanced system is illustrated in Figure 1. The layer-3 CC assignment module in the eNB configures one or multiple CCs for each user based on the quality-of-service (QoS) requirements, terminal capability, and CC deployment scenario, etc. Separate RRM blocks such as independent link adaption, packet scheduling, and hybrid automatic repeat request (HARQ) operate independently on each CC in coherence with LTE Rel'8 assumptions [5]. The channel state information (CSI) is extracted from sounding reference signals (SRS) sent by the UE and is used for packet scheduling and link adaptation. As a user may be assigned on multiple CCs, the scheduler in each CC may need to exchange the scheduling information across different CCs in order to achieve better performance [14]. Since the UEs are limited by the transmission power, uplink power control is also an important issue. It is worth mentioning that the algorithms for admission control, CC assignment, and packet scheduling are not part of the standard but are vendor specific.

2.1 Types of CA and UE transceiver architecture

Three types of CA have been defined by 3GPP [2]: 1) intra-band CA with contiguous CCs, 2) intra-band CA with non-contiguous CCs, and 3) inter-band CA with non contiguous CCs. For the radio frequency (RF) aspects of UEs supporting CA, two options are considered for the baseline UE transceiver architecture, illustrated in Figure 2 [27].

a) Single RF front-end with single wideband analog-to-digital converter (ADC) and dual base band (BB) processor
b) Dual RF front-ends with dual narrowband ADCs and dual BBs

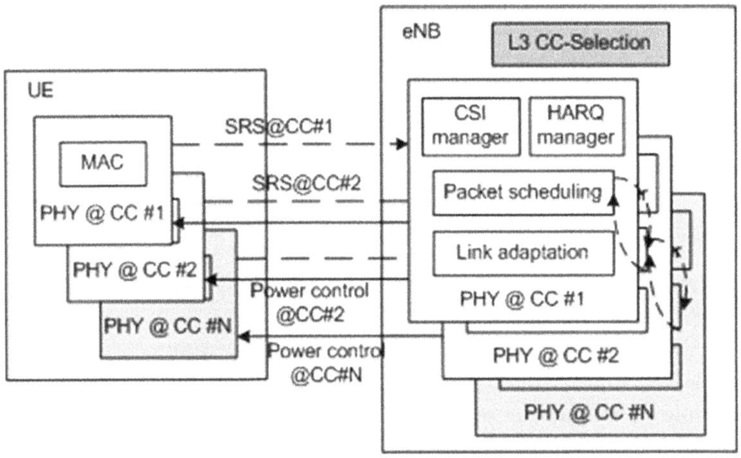

Figure 1 RRM framework of a multi-component carrier LTE-Advanced system.

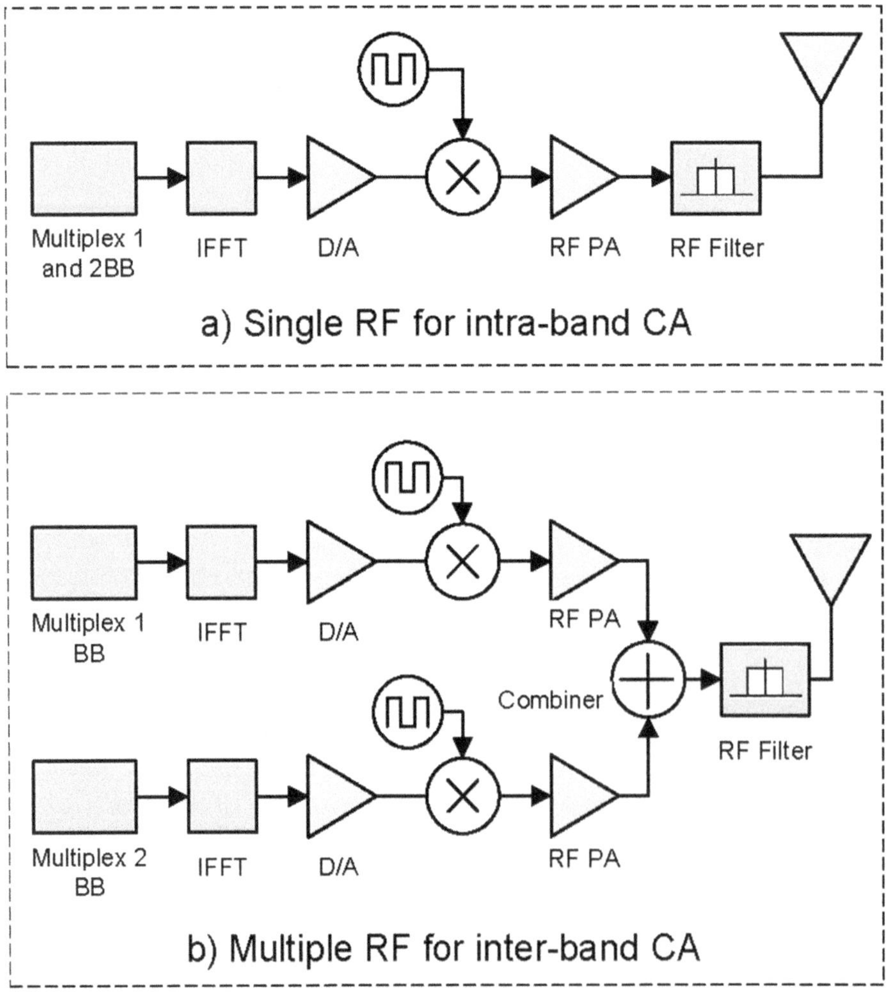

Figure 2 UE transceiver architecture for supporting CA. **(a)** Single RF for intra-band CA; **(b)** multiple RF for inter-band CA.

Option a) is only applicable for intra-band CA with contiguous CCs, because it cannot filter undesired frequency bands between the non-contiguous CCs. The advantage of option a) is keeping the UE transceiver complexity low. For cases of non-contiguous CA, only option b) is applicable at the expense of increased complexity. In this paper, we focus on intra-band CA with contiguous CCs and inter-band CA with non-contiguous CCs. The default UE transceiver architectures for the two considered CA scenarios are therefore option a) and option b), respectively.

2.2 Multi-cluster transmission in uplink

In SC-FDMA, UEs in principle can only be scheduled on one set of continuous sub-carriers, which allows SC-FDMA to reach very low signal CM but with less scheduling flexibility compared to OFDMA. Multi-cluster transmission has been introduced in Rel'10 to further improve the spectral efficiency by allowing UEs to be scheduled on non-contiguous sub-carriers in the uplink. The minimum resource allocation unit in multi-cluster scheduling is a sub-band, which consists of an integer number of physical resource blocks (PRBs[a]). Several contiguous sub-bands can be seen as a cluster, and a UE can be allocated to multiple clusters not adjacent to each other within one CC. Therefore, multi-cluster transmission has higher scheduling flexibility compared to SC-FDMA while keeping the power back-off requirements at the UE at a reasonable level. Though not considered in this study, it is worth mentioning that the scheduling flexibility of multi-cluster transmission is particular beneficial when combined with non-perfectly overlapping multi-user multiple-input and multiple-output (MU-MIMO) [28]. The three different multiple access schemes are illustrated in Figure 3. A maximum of two clusters within each CC are supported in Rel'10. The sub-band size is a fixed value and depends on the uplink system bandwidth [6]. In our study, the sub-band size corresponds to two PRBs.

2.3 Maximum power reduction for non-contiguous allocation

With a single RF front-end at the UE side (e.g., UEs supporting intra-band CA), the UE may be allocated non-contiguous resources when transmitting over multiple CCs, thus the single carrier property of SC-FDMA is no longer preserved. The increased CM of the transmitted signal with non-contiguous resource allocation, and the need to still fulfill the in-band and out-band emission masks [29] require the UE PA to operate with an additional power back-off [27-29]. Furthermore, the multi-cluster transmission within a CC imposes more stringent linearity requirements on the PA. The exact value of the required power back-off is hard to determine as it is UE implementation specific and is affected by many aspects of the specific uplink resource allocation such as modulation and coding scheme (MCS), number of clusters, size of clusters, frequency separation between clusters, distance from the edge of the first/last cluster to the left/right hand edge of the first/last CC, etc. [30,31]. In this study, we adopt the maximum power reduction (MPR) mask equation proposed by 3GPP [25], though in practice the proposed MPR mask only represents an upper bound of the required power back-off. The MPR value (in dB) is determined solely based on the ratio between the allocated and the available bandwidth aggregated from the assigned CCs. For multi-CC transmission, the estimated MPR mask is calculated as:

$$P_{\mathrm{MPR}}^{\mathrm{CA}} = \begin{cases} 8.2, & 0 < A < 0.025 \\ 9.2\text{–}40A, & 0.025 < A \leq 0.05 \\ 8\text{–}16A, & 0.05 < A \leq 0.25 \\ 4.83\text{–}3.33A, & 0.25 < A \leq 0.4 \\ 3.83\text{–}0.83A, & 0.4 < A \leq 1 \end{cases} \tag{1}$$

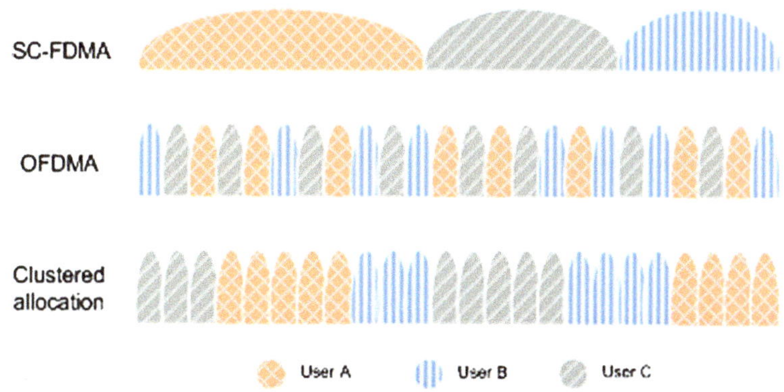

Figure 3 Different multiple access schemes in LTE-Advanced.

In case of single CC with multi-cluster transmission, the estimated MPR mask is given by:

$$P_{\text{MPR}}^{\text{SC}} = \begin{cases} 8-10.2A, & 0 < A \leq 0.33 \\ 5.67-3.07A, & 0.33 < A \leq 0.77 \\ 3.31, & 0.77 < A \leq 1 \end{cases} \quad (2)$$

where $A = N_{\text{PRB\_alloc}}/N_{\text{PRB\_agg}}$ is the ratio between the number of allocated PRBs ($N_{\text{PRB\_alloc}}$) and total available PRBs over all assigned CCs ($N_{\text{PRB\_agg}}$).

With dual RF front-ends at the UE side (e.g., UEs supporting inter-band CA), each carrier is amplified by a separate PA and the two carriers are added up in the combiner as illustrated in Figure 2. In this case, the power back-off issue is less important if two PA are sufficiently isolated from each other. Note that the use of a combiner results in insertion loss of at least 2 dB [27]. However, if multi-cluster transmission is enabled within each CC, power back-off is still needed in the UE PA with the MPR value specified in Equation 2. It is worth mentioning that the power back-off model for inter-band CA with multi-cluster transmission is built from existing models using simplified assumptions. Table 1 summarizes the required power back-off values for different CA types and uplink resource allocation schemes.

2.4 Uplink power control

As the UE is limited by the maximum transmission power, the uplink power control is an important issue. It consists of open loop power control and closed loop power control. The open loop power control aims at compensating for slow variations of channel conditions, i.e., pathloss and shadowing. The closed loop power control targets at further adjusting the UE's transmission power to optimize the system performance.

In Rel'10, independent power control is applied in each CC so that Rel'8 power control formula can also be reused in case of CA. The estimated transmission power on each assigned CC depends on the allocated bandwidth and on the CC specific power control parameters. When user i is scheduled for transmission on CC k, the estimated transmission power $\tilde{P}_{i,k}(t)$, expressed in dBm, is calculated as [32]:

$$\tilde{P}_{i,k}(t) = 10\log_{10}M_{i,k}(t) + P_{0,k}(t) + f\left(\Delta_{i,k}(t)\right) + \alpha_k \cdot L_{i,k} \quad (3)$$

where $M_{i,k}(t)$ is the number of PRBs allocated to user i on CC k, $P_{0,k}(t)$ is the normalized power density on CC k, α_k is the pathloss compensation factor on CC k, $L_{i,k}$ is the measured pathloss between user i and its serving base station on CC k, and $f(\Delta_{i,k}(t))$ is the closed loop power control correction for user i on CC k. More details on the uplink power control formula for LTE-Advanced with CA can be found in [7].

Previous studies [33,34] have shown that the system performance in terms of coverage (5th percentile user throughput) and cell throughput is highly dependent on the setting of $P_{0,k}$ and α_k. Load adaptive power control (LAPC) is proposed in [34] to dynamically adjust the users' power spectral density $P_{0,k}$. If the optimization goal is to maximize the coverage, the 5th percentile cell edge users should transmit with the maximum transmission power. Then, the power spectral density is updated periodically according to the following equation:

$$P_{0,k}(t) + f\left(\Delta_{i,k}(t)\right) = P_{\max} - 10\log_{10}M_k(t) - \alpha_k \cdot L_{95\%,k} \quad (4)$$

where $P_{\max}$ is the maximum UE transmission power in dBm, $M_k(t)$ is the estimated average number of allocated PRBs allocated to each user on CC k, and $L_{95\ \%,k}$ is the estimated 95th percentile user pathloss in the corresponding cell on CC k. For simplicity, $\bar{M}_k(t)$ is approximated as the total number of available PRBs divided by the number of UEs served by the corresponding cell.

Because power control is performed independently within each CC, the aggregated transmission power allocated over all CCs may exceed the maximum UE transmission power constraint even if the individual transmission power within each CC is below the limit. In this case, the UE needs to scale down the transmission power with different priorities according to the channel type [35]. In our study, the reduction of transmission power for user i on each assigned CC follows the same ratio. The actual transmission power $P_{i,k}(t)$ of user i on CC k then becomes:

$$P_{i,k}(t) = \tilde{P}_{i,k}(t) - \max\left\{0\ ,\ 10\log_{10}\left(\sum_{k=1}^{N}10^{\tilde{P}_{i,k}(t)/10}\right) - (P_{\max} - P_{\text{backoff}})\right\} \quad (5)$$

where N is the number of CCs assigned to user i, and P_{backoff} is the required power back-off with the values listed in Table 1.

Table 1 Required power back-off values for different types of CA and uplink resource allocation schemes

CA type	Resource allocation	Required power back-off (dB)
Intra-band CA with one RF front-end	Multi-CC transmission	$P_{\text{MPR}}^{\text{CA}}$
	Single-CC with multi-cluster	$P_{\text{MPR}}^{\text{SC}}$
	Single-CC with single-cluster	0
Inter-band CA with dual RF front-ends	Multi-CC with single-cluster	2
	Multi-CC with multi-cluster	$P_{\text{MPR}}^{\text{SC}} + 2$

3 Component carrier configuration and packet scheduling

The two main RRM functionalities that determine the radio resource allocation among the users in the system are the layer 3 CC configuration which selects a CC set for each user, and the layer 2 packet scheduling which allocates the resources among the admitted users on each CC. Hence, these two RRM functionalities are also the ones that impact the most on the performance of uplink CA. The proposed algorithms for these two sets of RRM algorithms are therefore motivated and outlined in the following.

3.1 Component carrier configuration

In a multi-CC LTE-Advanced system, both legacy and LTE-Advanced users may co-exist. The legacy Rel'8 users naturally can only be assigned on a single CC. With intra-band CA, the radio channel characteristics of different CCs are more or less the same. For optimal system performance, it is desirable to distribute the load equally on each CC. So, a simple yet effective load balancing scheme is applied for Rel'8 users, which assigns the CC with the least load (e.g. the number of users). With inter-band CA, the radio channel characteristics can be different at different frequency carriers. The study in [10] indicates that the CC selection should take both traffic load and radio channel characteristics into considerations. Therefore, with inter-band CA, cell-edge Rel'8 users (e.g., 5th percentile in pathloss distribution) are assigned to the CC with better coverage (i.e., low-frequency carrier) to improve the coverage performance, while other Rel'8 users are assigned to the CC with the least number of users to balance the load on each CC.

LTE-Advanced users can be assigned on multiple CCs. A CC is defined as its primary cell (PCell) and can only be changed via handover. Different users may not necessarily use the same CC as their PCell. If more than one CC is configured, the additional CCs are denoted as secondary cells (SCells). The SCells can be activated/deactivated by explicit signaling from the eNodeB, as well as based on appositely configured timers [36].

LTE-Advanced users configured with multiple CCs have the possibility to transmit on multiple CCs using the corresponding power control settings in each CC, therefore requiring higher transmission power as compared to the case with single CC configuration. Configuring LTE-Advanced users with multiple CCs and/or dual-cluster transmission in principle only makes sense if its total transmission power does not exceed the maximum UE power capability. Otherwise, they do not have sufficient power to exploit the increased transmission bandwidth and scheduling flexibility. For an LTE-Advanced user i assigned on N CCs, the total transmission power is calculated as:

$$P_i(t) = 10\log_{10}\left(\sum_{k=1}^{N} 10^{P_{i,k}(t)/10}\right) \le P_{\max} - P_{\text{backoff}} \quad (6)$$

Substituting Equations 3 and 4 into Equation 6 and assuming $M_{i,k}(t) = M_k(t)$, we obtain:

$$\sum_{k=1}^{N} 10^{\frac{\alpha_k\left(L_{i,k} - L_{95\%,k}\right)}{10}} \le 10^{-\frac{P_{\text{backoff}}}{10}} \quad (7)$$

where P_{backoff} is the expected power back-off with non-contiguous resource allocation. So with uplink CA, Equation 7 can be used to estimate whether the UE has enough transmission power to manage transmissions on multiple CCs simultaneously, assuming LAPC is applied. For the case of intra-band CA, it is often assumed that α_k and $L_{95\ \%,k}$ are the same on each CC k, then Equation 7 can be further simplified into:

$$L_i \le L_{95\%} - \frac{10\log_{10}N + P_{\text{backoff}}}{\alpha} \quad (8)$$

Based on the derived pathloss threshold, we propose a CC configuration algorithm for uplink CA. The idea is to distinguish between power-limited and non-power-limited users. LTE-Advanced users whose pathloss satisfies Equation 7 are considered to be non-power-limited and can be assigned on multiple CCs with dual-cluster transmission. Otherwise, they are considered to be power-limited and are only assigned on one CC with single-cluster transmission. The selection of the CC for power-limited LTE-Advanced users is the same as for Rel'8 users. By separating the power-limited and non-power-limited LTE-Advanced users, cell-edge LTE-Advanced users will not experience any performance loss with non-contiguous resource allocation, while non-power-limited LTE-Advanced users can benefit from increased transmission bandwidth and scheduling flexibility with CA and multi-cluster transmission. It is worth mentioning that with non-contiguous resource allocation, the required power back-off specified in Equation 1 is not dependent on the number of clusters, but rather depends on the ratio between the allocated and total available bandwidth [30]. Therefore, for non-power-limited LTE-Advanced users, it is always beneficial to use dual-cluster transmission in combination with CA in order to increase the scheduling flexibility. The proposed CC and multi-cluster configuration algorithm is illustrated in Figure 4.

3.2 Bandwidth allocation and packet scheduling

After the CC and cluster assignment for each user, the layer 2 time and frequency domain packet scheduler

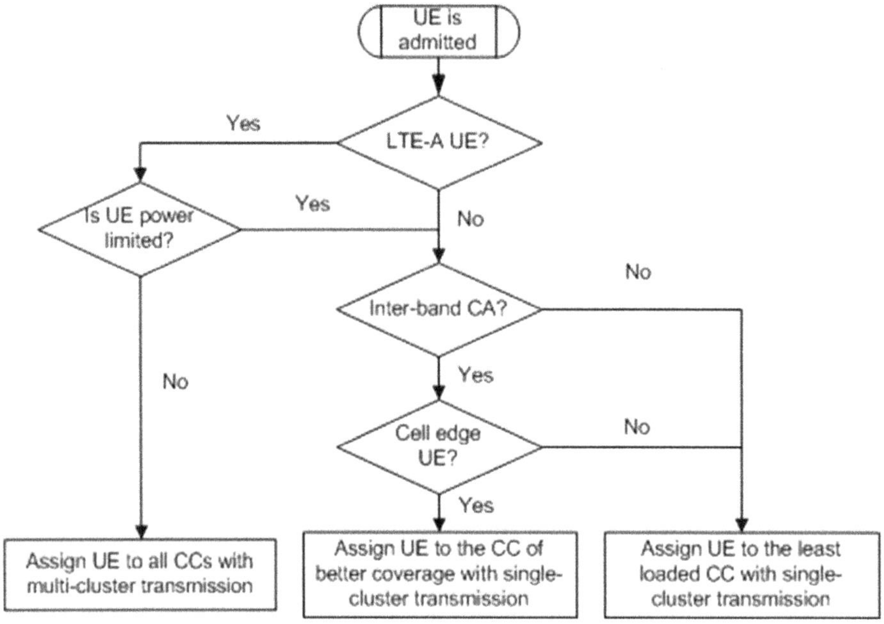

Figure 4 Proposed CC and multi-cluster configuration algorithm in uplink LTE-Advanced.

allocates the resources in terms of PRBs among the admitted users on each CC according to certain constraints with the objective to improve the spectral efficiency while maintaining fairness among scheduled users. The use of sounding reference signals (SRS) allows for uplink link adaptation and channel-aware frequency domain packet scheduling [7]. The eNB may configure the SRS patterns on each carrier depending on the accuracy of uplink SRS measurements.

In [26], the authors proposed an adaptive transmission bandwidth (ATB)-based packet scheduling algorithm for SC-FDMA with single-cluster transmission in uplink, which tightly couples the bandwidth allocation and the packet scheduling together to exploit the bandwidth flexibility. In this study, we extend the ATB algorithm so that it is catered to support multi-cluster transmission [23]. With multi-cluster transmission, the allocated PRBs for a user can be non-contiguous subject to the constraint of not exceeding the maximum number of configured clusters, which is determined according to the algorithm depicted in Figure 4. The basic idea behind the proposed scheduling algorithm is to produce an allocation table which closely follows the envelope of the UEs' scheduling metrics.

We assume that packet scheduling is performed in each CC. The well-known proportional fair (PF) scheduler is applied. The problem with baseline PF scheduler is that it will allocate more resources to the users connected to multiple CCs (with CA) than users without CA, resulting in an unfair resource allocation. Therefore, cross-carrier PF scheduling across multiple CCs is recommended when calculating the scheduling metric in each CC [14]. In cross-carrier PF scheduler, the denominator of the PF metric is calculated as the sum of the average scheduled throughput over all CCs where the user has been assigned and scheduled in the past. By applying this modification, the scheduler can distribute the radio resources among the users more fairly. It simply requires information exchange on the average scheduled throughput across different CCs. The scheduling metric of user i at sub-band j on CC k, denoted as $M_{i,j,k}$, is therefore calculated as:

$$M_{i,j,k}(t) = \frac{R_{i,j,k}(t)}{\sum_{k=1}^{N} R_{i,k}(t)} \tag{9}$$

where $R_{i,j,k}(t)$ is the estimated throughput of user i at sub-band j on CC k, and $R_{i,k}(t)$ is the exponentially filtered average throughput of user i on CC k.

At each scheduling instance, the scheduler in each CC first generates a scheduling matrix of each user on each sub-band. Then, the scheduler selects a UE and the corresponding sub-band with the highest scheduling metric and checks the power constraint of that UE with Equation 3. If the selected UE exceeds its maximum transmission power capability with the new resource allocation, the scheduler removes that UE from the candidate list and reselects the UE and the sub-band from the scheduling matrix. After the power constraint check, the scheduler checks the number of allocated clusters of the selected UE. If the new resource allocation causes the number of allocated clusters to exceed the maximum number of configured clusters (i.e., single-cluster for Rel'8 and power-limited LTE-A users, and dual-cluster for non-power-limited LTE-A users), the current sub-band is disabled for that UE and the algorithm restarts from the beginning. Otherwise, the scheduler allocates the current sub-band to that UE and expands its transmission bandwidth to the adjacent sub-bands on both sides until either another UE has a higher scheduling metric or the maximum transmission power of that UE is exceeded. Details of the bandwidth expansion procedure can be found in [26]. Then, the algorithm restarts from the beginning and continues the loop until either all UEs have been scheduled or there are no resources left. A detailed description of the proposed multi-cluster scheduling algorithm is illustrated in Figure 5 [23].

4 Simulation assumptions

The performance of the proposed algorithms is evaluated in a quasi-static uplink multi-cell system level simulator that follows the LTE specifications defined in [37], including detailed modeling of major RRM functionalities such as layer 3 CC configuration, layer 2 packet scheduling, HARQ, link adaptation, and uplink power control. The simulation scenario is 3GPP macro-cell case number 1 with seven sites and three sectors per site using the wrap-around technique. A directional 3D antenna pattern with default 15° down-tilt is modeled for the macro cells. The ITU defined geometrical channel model is applied, where UE to macro links follow the urban macro model (UMa), which includes separate models for line-of-sight (LOS) and non-LOS (NLOS) [38]. Selection between the LOS and NLOS model is random for each link, where the probabilities for selecting LOS or NLOS vary with the distance between the UE and eNB. Small-scale fading (also known as fast fading) is modeled according to the commonly accepted stochastic typical urban model. It is assumed that distance-dependent path loss and shadowing are maintained constant for each UE, but fast fading is updated every TTI independently on each CC based on the ITU typical urban power delay profile and UEs' speed. For intra-band CA, two CCs each with 20 MHz bandwidth are deployed at 2 GHz. For inter-band CA, two CCs each with 20 MHz bandwidth are deployed at 1.8 and 2.6 GHz, respectively. Up to two clusters are supported in each CC for uplink multi-cluster transmission. The SRS resolution is two PRBs and the SRS period is set to be ten subframes. Imperfect SRS measurements are assumed. The measurement error is a function of the experienced signal-to-interference-plus-noise ratio (SINR) and sounding bandwidth. The link to system level mapping is based on the actual value interface (AVI) method [39], including channel estimation error with uplink clustered transmission which is modeled by modifying the experienced SINR before it is used for AVI mapping. Both full buffer and bursty traffic models are considered. In full-buffer traffic model, we assume each sector has a fixed number of UEs with full buffer traffic. For bursty traffic model, a dynamic birth-death traffic model is applied for generating user calls, where call arrival is according to a Poisson process with arrival rate λ per sector. Each call has a finite payload size of $B = 2$ Mbits.

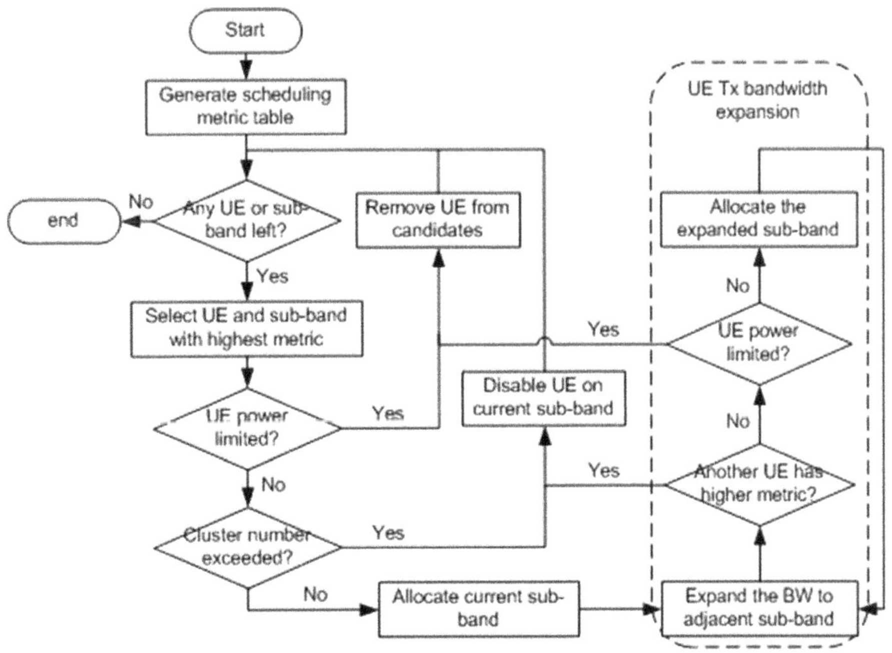

Figure 5 Proposed uplink multi-cluster scheduling algorithm in LTE-Advanced.

Once the payload has been successfully transmitted by the UE, the call is terminated and the UE is removed from the simulation. Thus, the average offered load per sector equals $\lambda \times B$. Joint proportional fair scheduling in frequency domain is used together with multi-cluster bandwidth allocation. Independent LAPC is enabled on each CC to dynamically adjust the UE power spectral density based on the variable load conditions. The optimization goal is to maximize the 5th percentile cell edge user throughput. The average UE power back-off P_{backoff} with non-contiguous resource allocation is set to be 6 dB (close to worst-case assumption considering the power back-off model in Equation 1). Table 2 summarizes the main parameter settings used in the system-level simulations.

5 Simulation results and analysis

5.1 Multi-cluster transmission gain in uplink

We start our analysis by first looking at the scenario with a fixed number of UEs per sector and full-buffer traffic model. Only one CC is configured with 10 MHz@2 GHz in this scenario. Figure 6 shows the dual-cluster scheduling gain over single-cluster scheduling and the average number of scheduled clusters per UE versus the different number of UEs per macro sector, respectively. It is shown that the dual-cluster scheduling gain is closely correlated to the average number of scheduled clusters per UE. Specifically, the dual-cluster scheduling gain increases as the number of UEs increases until reaching the maximum value, i.e., 16% gain at six UEs per macro sector in our case. Then, the gain gradually decreases as the number of UEs increases. That is because when the number of UEs per macro sector is low (e.g., less than six), dual-cluster scheduling users have more chance to exploit the increased frequency diversity order as compared to single-cluster transmission. However, when the number of UEs per macro sector is high (e.g., larger than six), the probability of a user being scheduled on multiple clusters decreases, therefore the gain brought by dual-cluster scheduling decreases.

5.2 Effectiveness of the proposed CC configuration algorithm

Next, we evaluate the effectiveness of the derived pathloss threshold in Equation 8 against other pathloss thresholds with bursty traffic model. Two contiguous CCs each with 20 MHz bandwidth are configured for intra-band CA in this scenario. Figure 7 shows the 5th percentile and 50th percentile user throughput versus the different pathloss thresholds under different traffic loads (5 and 20 Mbps offered loads correspond to approximately 20% and approximately 70% PRB utilization, respectively). Here, we assume 100% LTE-Advanced UEs. The 5th percentile user throughput stays almost constant when the pathloss threshold is smaller than a certain value (approximately 109 dB). After that point, the 5th percentile user throughput decreases as the pathloss threshold increases. For the 50th percentile user throughput, it increases as the threshold increases until the maximum value is reached also around 109 dB. Then, the throughput decreases with further increase of the pathloss threshold. The reasons are as follows: if the pathloss threshold is set to be low, most UEs, including power-limited and some non-power-limited UEs are only assigned on one CC with single-cluster transmission. As a result, those non-power-limited UEs cannot benefit from the advantages of transmission bandwidth expansion by using CA and increased scheduling flexibility with dual-cluster transmission. Therefore, small value of the threshold would decrease the 50th percentile user throughput. The 5th percentile user throughput remains steady as they are assigned only on one CC. On the other hand, if the pathloss threshold is set to be high, not only non-power-limited but also some power-limited UEs are assigned on both CCs with dual-cluster transmission. As a result, those power-limited cell edge UEs will experience performance loss from being scheduled over multiple CCs and clusters due to the effect of additional power back-off with non-contiguous resource allocation. Since cell edge users usually utilize the resources with low efficiency due to poor channel conditions, the increase of their activity time in the network worsens the resource utilization of other users. Therefore, higher value of the threshold would decrease both the 5th percentile and 50th percentile user throughput. As a conclusion, the value of pathloss threshold has to be carefully determined in order to optimize the system performance. It is shown from Figure 7 that our derived pathloss threshold works effectively to achieve good performance at both 5th percentile and 50th percentile user throughput under different traffic conditions. Thus, it is robust against variations in the traffic load.

5.3 Overall performance

Then, we evaluate the performance of intra-band and inter-band CA with the proposed CA algorithm as compared to the case without CA in a bursty traffic model. The performance with 'blind' CC configuration (i.e. all UEs are configured with uplink CA) is also included for comparisons. Figure 8 shows the 5th percentile user throughput versus the offered load with different CA and multi-cluster transmission schemes. As expected, the user throughput decreases as the load in the network increases. For intra-band CA, there is a performance loss if all UEs are configured with CA and/or multi-cluster transmission, as compared to the case without CA and single-cluster transmission. This is because at the cell edge, UEs usually experience large pathloss and are

Table 2 Summary of main simulation parameters

Parameters	Settings	
Simulation scenario	3GPP macro case number 1	
Network layout	Seven macro sites - three sectors/site - wrap around	
Channel model (UMa) [38][a]	LOS pathloss	$22\log_{10}(d) + 28.0 + 20.0\log_{10}(f)$, shadow fading std = 4 dB, $10\text{ m} < d < d'_{BP}$
		$40\log_{10}(d) + 7.8 - 18\log_{10}\left(h'_{BS}\right) - 18\log_{10}\left(h'_{UT}\right) + 2\log_{10}(f)$, shadow fading std = 4 dB, $d'_{BP} < d < 5{,}000\text{ m}$
	NLOS pathloss	$161.04 - 7.1\log_{10}(W) + 7.5\log_{10}(h) - (24.37 - 3.7(h/h_{BS})^2)\log_{10}(h_{BS}) + (43.42 - 3.1\log_{10}(h_{BS}))(\log_{10}(d) - 3) + 20.0\log_{10}(f) - (3.2(\log_{10}(11.75h_{UT}))^2 - 4.97)$, shadow fading std = 6 dB, $10\text{ m} < d < 5{,}000\text{ m}$
	LOS probability	$\min\left(\frac{18}{d}, 1\right)\left(1 - \exp\left(-\frac{d}{63}\right)\right) + \exp\left(-\frac{d}{63}\right)$
Inter-site distance	500 m	
Component carriers	Intra-band CA: 2 × 20 MHz @ 2 GHz, inter-band CA: 20 MHz@1.8 GHz + 20 MHz@2.6 GHz	
	96 available PRBs for PDSCH per CC	
Number of clusters	Single or dual clusters	
PRBs per sub-band	Two PRBs	
Sounding resolution	Two PRBs	
Sounding period	Ten subframes	
Sounding method	Imperfect SRS	
eNode-B receiver	4-Rx MMSE	
Max UE power	200 mW (23 dBm)	
UE transceiver	Single transceiver for intra-band CA	
	Two separate transceivers for inter-band CA	
UE Tx bandwidth	ATB for multi-cluster	
Packet scheduling	Joint proportional fair	
Bursty traffic model	Finite buffer with different Poisson arrival rate	
	Fixed payload size of 2 Mbits per UE	
Full-buf. traffic model	Full buffer with fixed number of UEs	
UE speed	3 km/h	
Available MCSs	BPSK (R = 1/5,1/3)	
	QPSK (R = 1/4,1/3,1/2,2/3,3/4)	
	16 QAM (R = 1/2,2/3,3/4,5/6)	
HARQ	Synchronous and adaptive with maximum of four transmissions	
First Tx BLER target	20%	
Link adaptation	Fast adaptive modulation and coding (AMC)	
Pathloss compensation	$\alpha = 0.8$	
95th percentile user pathloss	118 dB@1.8 GHz, 120 dB@2 GHz, 123 dB@2.6 GHz	
Power spectral density	Independent LAPC in each CC [34]	
Average power back-off	$P_{backoff} = 6$ dB	

[a] f is the center frequency (Hz), h_{BS} and h_{UT} are the actual antenna heights at the BS and the UT (m), h'_{BS} and h'_{UT} are the effective antenna heights at the BS and the UT (m), d is the distance (m), d'_{BP} is the break point distance $d'_{BP} = 4h'_{BS}h'_{UT}f/c$ where c is the propagation velocity in free space, W is the street width (m), and h is the average building height (m).

limited by the transmission power. In fact, with LAPC, cell edge UEs (worst 5th percentile UEs) are configured to transmit with the maximum transmission power on the assigned CC. Transmitting over multiple CCs and/or multiple clusters will cause further power reduction due to the effect of additional power back-off with non-contiguous resource allocation, therefore degrading the 5th percentile UE throughput performance. In the proposed CA algorithm, the power-limited cell edge UEs are only assigned on one CC with single-cluster transmission so that the SC-FDMA properties of the transmitted signals are maintained. Therefore, there is no performance loss as compared to Rel'8 operation. For inter-band CA with single-cluster transmission, the performance is slightly worse as compared to

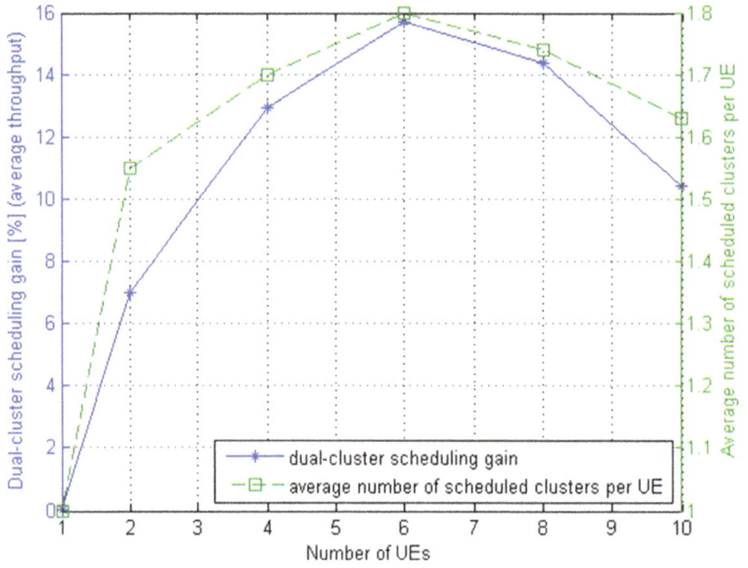

Figure 6 Multi-cluster scheduling gain with full buffer traffic.

the case of intra-band CA, due to the 2-dB insertion loss in the combiner with two RF front-ends and the higher frequency band used for inter-band CA. For inter-band CA with dual-cluster transmission, the dual-cluster transmission should be avoided for cell edge users as it will cause additional power back-off similar to the case of intra-band CA.

Figure 9 shows the 50th percentile user throughput versus the offered load with different CA and multi-cluster transmission schemes. For intra-band CA, the

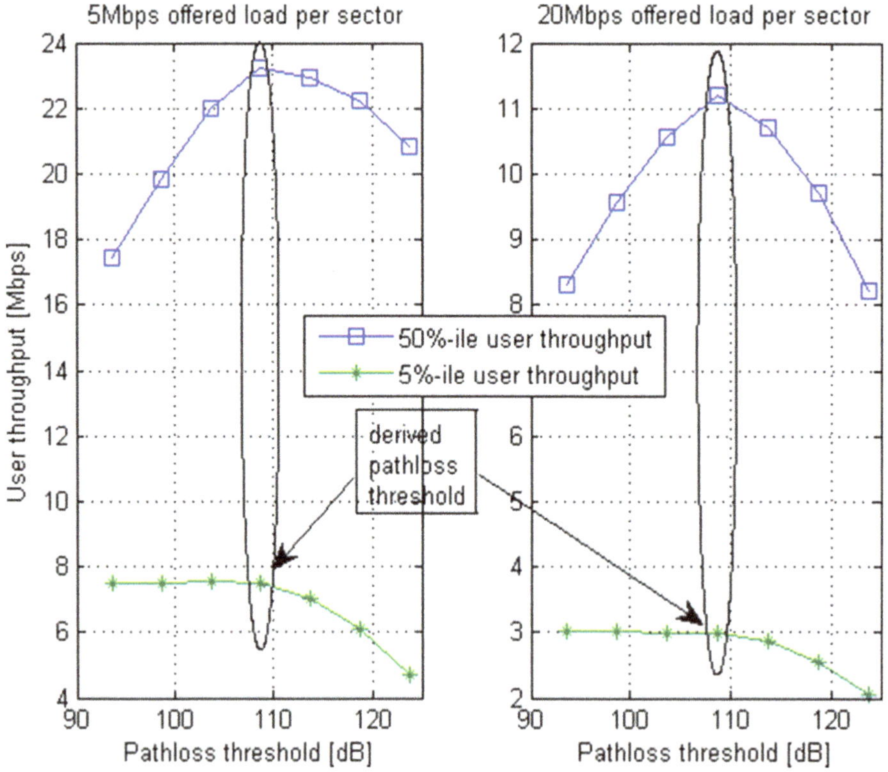

Figure 7 5th and 50th percentile user throughput with different pathloss thresholds, with 5 and 20 Mbps offered load.

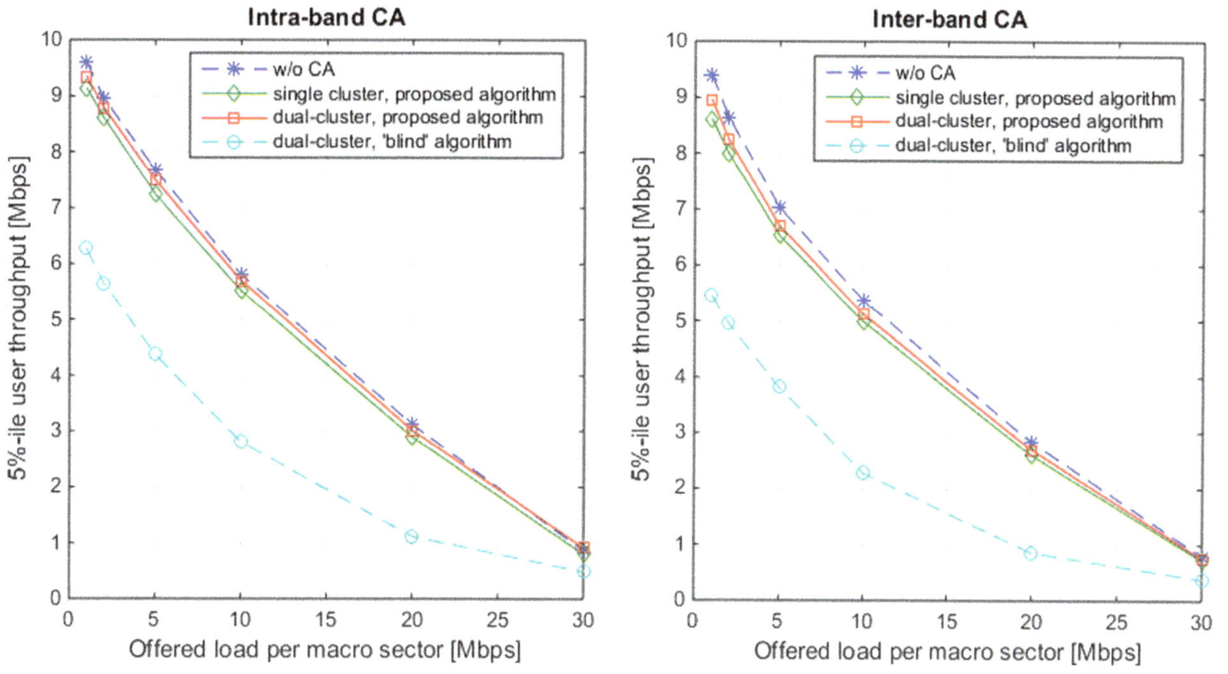

Figure 8 5th percentile user throughput under different traffic loads in different scenarios.

user throughput with the proposed CA algorithm is significantly higher than that with 'blind' CA configuration and the case without CA. Also, it is observed that dual-cluster transmission can further increase the user throughput. That is because with 'blind' CA configuration, cell edge UEs will stay in the system for longer time due to the poor performance as shown in Figure 8, therefore occupying more resources. In the proposed CA algorithm, only the non-power-limited UEs are assigned on multiple CCs with dual-cluster transmission so that they can benefit from the advantages of CA (i.e., increased transmission bandwidth with multiple CCs)

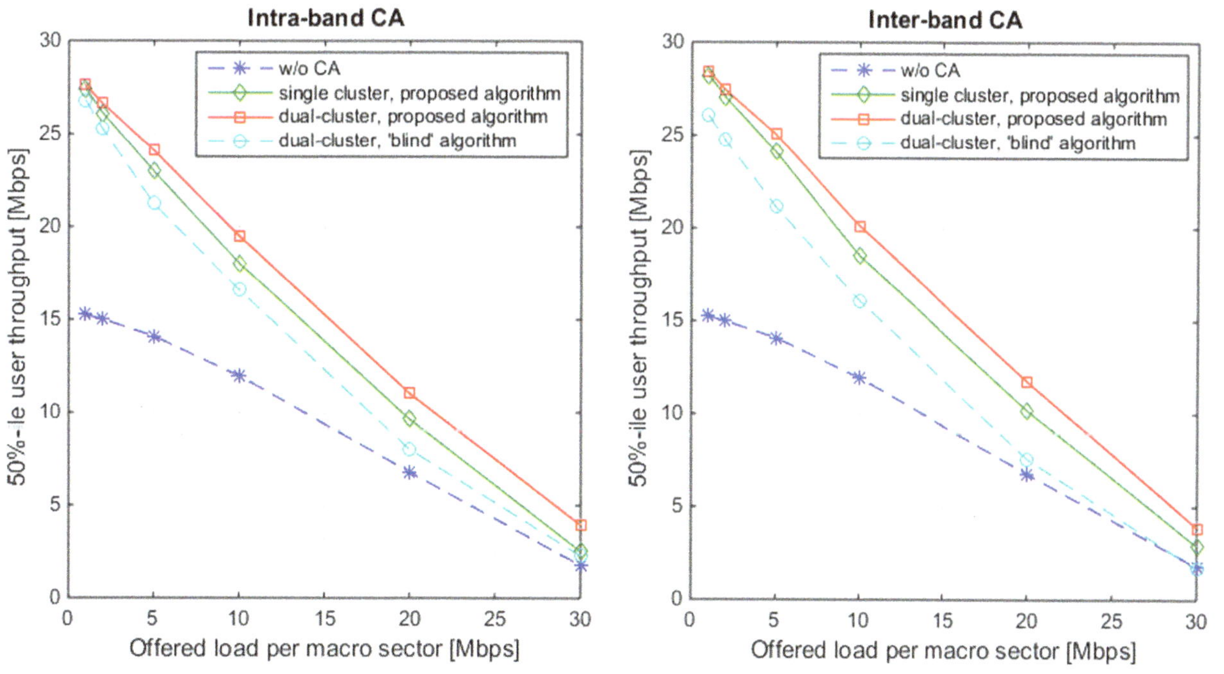

Figure 9 50th percentile user throughput under different traffic loads in different scenarios.

and dual-cluster scheduling (i.e., increased scheduling flexibility within one CC). The power-limited cell edge users are configured only on one CC with single-cluster transmission so that they will not suffer from performance loss. For the case of inter-band CA, similar phenomenon can be observed.

From Figure 8 and 9, it clearly suggests avoiding using uplink CA and dual-cluster transmission at cell edge but activating them when a user is identified to be in a non-power-limited region in order to benefit from increased transmission bandwidth and scheduling flexibility. At low load, the 50th percentile user throughput gain with CA as compared to the case without CA lies in the range of 70% to 80%. When the load increases, the probability of a user being scheduled on both CCs is reduced, and so is the gain.

Figure 10 shows the empirical cumulative distribution function (cdf) of the average UE transmission power with 10 Mbps offered load (approximately 40% PRB utilization) including the effect of power back-off, under different CA and multi-cluster transmission schemes. The curves with inter-band CA are not plotted as they are similar to the case with intra-band CA. The configuration of CA would increase the UE transmission power at the lower region of the cdf, which means that UEs not operating close to the maximum power can increase their transmission power to exploit the benefit of increased bandwidth with CA. At the higher region of the cdf, the UEs are operating in power limited conditions. With all UEs configured with CA, cell edge UEs cannot transmit with the maximum power (the maximum transmission power is 21 dBm for cases when all UEs are configured with CA) due to the power back-off requirement specified in Equation 1. As cell edge users usually experience bad channel conditions, lowering the transmission power would decrease the SINR, thus degrading the throughput performance of those users as seen in Figure 8. With the proposed CA algorithm, the power-limited cell edge UEs are not configured with CA and dual-cluster transmission, thus those users do not have to back off their transmission power (the maximum transmission power with the proposed algorithm is 23 dBm). Therefore the transmission power with the proposed algorithm increases at the lower region, but gets closer to the case without CA at the higher region. Though not shown in the figure, it is observed in the study that the UE transmission power distributions without CA and with the proposed CC configuration algorithm do not vary much at different loads, indicating good adaptation and robustness of the LAPC algorithm and the proposed CC configuration algorithm.

6 Conclusions

In this paper, we have investigated the resource allocation problem for uplink CA in LTE-Advanced. For layer 3 CC configuration, we have derived a pathloss threshold that separates the UEs into two categories: power-limited and non-power-limited UEs, and assign only one CC with single-cluster transmission to power-limited LTE-A UEs but assign multiple CCs with dual-cluster transmission to non-power-limited LTE-Advanced UEs. For layer 2 packet scheduling, we have proposed an extended bandwidth expansion resource allocation algorithm with joint PF

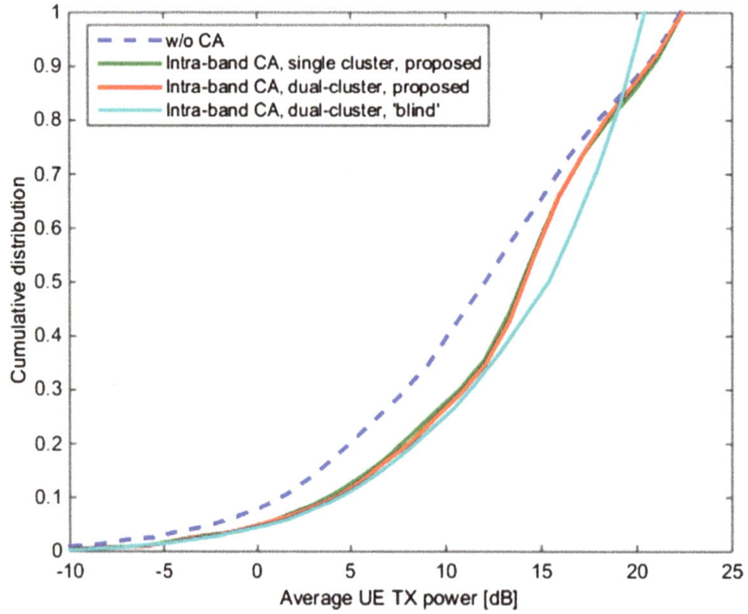

Figure 10 Average UE transmission power under different CA schemes and CC configuration algorithms, offered load = 10 Mbps.

scheduling that closely follows the envelope of the UEs' scheduling metrics. The performance of the proposed resource allocation algorithms has been evaluated for intra-band and inter-band CA via system level simulations. It is recommended that uplink CA and multi-cluster scheduling should not be configured for cell edge users, due to the limitation of UE's transmission power and the additional power back-off requirements when transmitting over non-contiguous resources. On the other hand, CA and multi-cluster scheduling should be configured for non-power-limited UEs so that they can benefit from increased transmission bandwidth and scheduling flexibility. The results show that intra-band and inter-band CA exhibit similar trend. Specifically, with the proposed CC configuration algorithm, there is no loss at 5th percentile user throughput and up to 70% to 80% gain can be achieved at 50th percentile user throughput, as compared to the case without CA and dual-cluster transmission. The throughput gain is especially visible at low load but decreases gradually as the load increases. Refinements of the power back-off model used for inter-band CA with dual-cluster transmission and other uplink power control methods could be interesting topics for future studies.

7 Endnotes

[a]In LTE, a PRB consists of 12 sub-carriers, each of which is 15 kHz, and is thus equal to 180 kHz.

Competing interests

The authors declare that they have no competing interests.

Acknowledgement

The authors would like to express their gratitude to Nokia Networks colleagues Huiyu Yuan, Yuyu Yan, Jens Steiner, and Mads Brix for their assistance in the development of the system-level simulator.

Author details

[1]Aalborg University, Fredrik Bajers Vej 7, DK-9220 Aalborg Ø, Denmark. [2]Nokia Networks, Alfred Nobel Vej 21, DK-9220 Aalborg, Ø, Denmark.

References

1. A Ghosh, R Ratasuk, B Mondal, N Mangalvedhe, T Thomas, LTE-advanced: next-generation wireless broadband technology. IEEE Wirel. Commun. **17**(3), 1536–1284 (2010)
2. G Yuan, X Zhang, W Wang, Y Yang, Carrier aggregation for LTE-advanced mobile communication systems. IEEE Wirel. Commun. **48**(2), 88–93 (2010)
3. M Iwamura, K Etemad, F Mo-Han, R Nory, R Love, Carrier aggregation framework in 3GPP LTE-Advanced. IEEE Commun. Mag. **48**(8), 60–7 (2010)
4. S Zukang, A Papasakellariou, J Montojo, D Gerstenberger, X Fangli, Overview of 3GPP LTE-advanced carrier aggregation for 4G wirelesscommunications. IEEE Commun. Mag. **50**(2), 122–30 (2012)
5. K Pedersen, F Frederiksen, C Rosa, H Nguyen, L Garcia, W Yuanye, Carrier aggregation for LTE-Advanced: functionality and performance aspects. IEEE Commun. Mag. **49**(6), 89–95 (2011)
6. 3GPP TS 36.213, v10.12.0, Release 10 physical layer procedures, March 2014. Available at www.3gpp.org/dynareport/36213.htm
7. H Holma, A Toskala, *LTE for UMTS: Evolution to LTE-Advanced*, 2nd edn. (John Wiley & Sons, Chichester, UK, 2011)
8. L Chen, W Chen, X Zhang, D Yang, Analysis and simulation for spectrum aggregation in LTE-Advanced system, Paper presented at the 2009 IEEE vehicular technology conference (VTC) Fall, IEEE, Anchorage, AK, 20–23 September 2009
9. Y Kakishima, T Kawamura, Y Kishiyama, H Taoka, T Nakamura, Experimental evaluation on throughput performance of asymmetric carrier aggregation in LTE-Advanced, Paper presented at the 2011 IEEE vehicular technology conference (VTC Spring),IEEE, Yokohama, 15–18 May 2011
10. H Wang, C Rosa, K Pedersen, Performance analysis of downlink inter-band carrier aggregation in LTE-Advanced, Paper presented at the 2011 IEEE vehicular technology conference (VTC Fall), IEEE, San Francisco, CA, 5–8 September 2011
11. Y Yan, A Li, A Harada, H Kayama, Enhanced downlink control channel resource allocation algorithm for cross-carrier scheduling in LTE-Advanced carrier aggregation system, Paper presented the 2011 IEEE vehicular technology conference (VTC Springer), IEEE, Budapest, 15–18 May 2011
12. H Nguyen, I Kovacs, Y Wang, K Pedersen, Feedback compression schemes for downlink carrier aggregation in LTE-Advanced, Paper presented at the 2011 IEEE vehicular technology conference (VTC Fall), IEEE, San Francisco, CA, 5–8 September 2011
13. K Yagyu, T Nakamori, H Ishii, M Iwamura, N Miki, T Asai, J Hagiwara, Investigation on mobility management for carrier aggregation in LTE-Advanced, Paper presented at the 2011 IEEE vehicular technology conference (VTC Fall), IEEE, San Francisco, CA, 5–8 September 2011
14. Y Wang, K Pedersen, P Mogensen, T Sorensen, Carrier load balancing and packet scheduling for multicarrier systems. IEEE Trans. on Wireless Communications **9**(5), 1780–9 (2010)
15. W Qiu, H Minn, C Chong, Enhanced frequency diversity exploitation in carrier aggregation for LTE-Advanced systems, Paper presented at the 2011 IEEE vehicular technology conference (VTC Fall), IEEE, San Francisco, CA, 5–8 September 2011
16. A Li, K Takeda, N Miki, Y Yuan, K Hidetoshi, Search space design for cross-carrier scheduling in carrier aggregation of LTE-Advanced system, Paper presented at the 2011 IEEE international conference on communications (ICC), IEEE, Kyoto, 5–9 June 2011
17. S Shi, C Feng, C Guo, A resource scheduling algorithm based on user grouping for LTE-Advanced system with carrier aggregation. Paper presented at the 2009 international symposium on computer network and multimedia technology, IEEE, Wuhan, 18–20 January 2009
18. Y Rui, P Cheng, M Li, Q Zhang, M Guizani, Carrier aggregation for LTE-advanced: uplink multiple access and transmission enhancement features. IEEE Wirel. Commun. **20**(4), 101–8 (2013)
19. F Liu, K Zheng, W Xiang, H Zhao, Design and performance analysis of an energy-efficient uplink carrier aggregation scheme. IEEE Journal on Selected Areas in Communications **32**(2), 197–207 (2014)
20. R Sivaraj, A Pande, K Zeng, K Govindan, P Mohapatra, *Edge-prioritized channel- and traffic-aware uplink carrier aggregation in LTE-advanced systems (IEEE International Symposium on World of Wireless, Mobile and Multimedia Networks (WoWMoM), 2012), p*, 2012
21. H Wang, C Rosa, K Pedersen, Performance of uplink carrier aggregation in LTE-Advanced systems, Paper presented at the 2010 IEEE vehicular technology conference (VTC fall)IEEE, Ottawa, ON, 6–9 September 2010
22. H Wang, C Rosa, K Pedersen, Uplink component carrier selection for LTE-Advanced systems with carrier aggregation. Paper presented at the 2011 IEEE international conference on communications (ICC), IEEE, Kyoto, 5–9 June 2011
23. H Wang, C Rosa, K Pedersen, Uplink Multi-Cluster Scheduling with MU-MIMO for LTE-Advanced with Carrier Aggregation", in Proc. IEEE Wireless Communications and Networking Conference (WCNC) 2012, pp. 1-5, April 2012
24. M Lema, M Garcia-Lozano, S Ruiz, D Gonzalez, Improved component carrier selection considering MPR information for LTE-A uplink systems, in Proc. IEEE 24th International Symposium on Personal Indoor and Mobile Radio Communications (PIMRC), pp. 2191-2196, Sep. 2013
25. 3GPP TS 36.101, v11.11.0, User equipment (UE) radio transmission and reception (Release 11), Jan. 2015. Available at www.3gpp.org/dynareport/36101.htm
26. F Calabrese, C Rosa, M Anas, P Michaelsen, K Pedersen, P Mogensen, Adaptive transmission bandwidth based packet scheduling for LTE uplink, in Proc. IEEE Vehicular Technology Conference (VTC) Fall 2008, pp. 1-4, Sept. 2008

27. C Park, L Sundström, A Wallen, A Khayrallah, Carrier aggregation for LTE-advanced: design challenges of terminals. IEEE Commun. Mag **51**(12), 76–84 (2013)
28. A Nishio, T Iwai, A Matsumoto, D Imamura, System evaluation of MU-MIMO and multi-cluster allocation in LTE-Advanced uplink, in Proc. IEEE Vehicular Technology Conference (VTC) Spring 2012, pp. 1-5, May 2012
29. B Xiong, X Zhu, W Li, G Guo, Analysis of maximum power reduction of uplink for carrier aggregation in LTE-A system, in Proc. International Conference on Connected Vehicles and Expo (ICCVE) 2012, pp. 49-54, Dec. 2012
30. 3GPP TSG RAN, R4-110955, MPR for LTE multi cluster transmission, Feb. 2011
31. 3GPP TSG RAN, R4-110265, MPR for non-contiguous allocation, Jan. 2011
32. 3GPP TS 25.213, v10.0.0, E-UTRA physical layer procedures, Oct. 2010. Available at www.3gpp.org/DynaReport/25213.htm
33. C Castellanos, D Villa, C Rosa, K Pedersen, F Calabrese, P Michaelsen, J Michel, Performance of uplink fractional power control in UTRAN LTE, in Proc. IEEE Vehicular Technology Conference (VTC) Spring 2008, pp. 2517-2521, May 2008
34. M. Boussif, C Rosa, J Wigard, R Müllner, Load adaptive power control in LTE uplink, in Proc. European Wireless 2010, April 2010
35. 3GPP TSG RAN WG1 #60bis R1-102011, Power scaling method for UL PC with carrier aggregation, April 2010
36. 3GPP TS 36.321, v10.10.0, Release 10 medium access control (MAC) protocol specification, Jan. 2014. Available at www.3gpp.org/dynareport/36321.htm
37. 3GPP TR 36.814, v9.0.0, Further advancements for E-UTRA physical layer aspects", March 2010. Available at www.3gpp.org/dynareport/36814.htm
38. ITU-R, Guidelines for evaluation of radio interface technologies for IMT-Advanced, Report ITU-R M.2135-1)", Dec. 2009. Available at www.itu.int
39. S Hämäläinen, M Reisslein, F Fitzek, A novel interface between link and system level simulations, in Proc. ACTS Mobile Telecommunications Summit, pp. 599–604, Oct. 1997

A computationally efficient method for self-planning uplink power control parameters in LTE

José Ángel Fernández-Segovia[1*], Salvador Luna-Ramírez[1], Matías Toril[1], Ana Belén Vallejo-Mora[1] and Carlos Úbeda[2]

Abstract

Uplink power control has a strong impact on the performance of mobile communication networks. In this work, an automatic parameter planning algorithm for the standardized power control scheme in the physical uplink shared channel (PUSCH) of long term evolution LTE is proposed. The method is conceived for the network design stage, when network measurements are still not available. The proposed heuristic algorithm can handle irregular scenarios at a low computational complexity. For this purpose, the parameter planning problem in a cell is formulated analytically through the combination of multiple regular scenarios built on a per-adjacency basis. Method assessment is carried out over a static system-level simulator implementing a real scenario. Results show that the proposed method can improve average user throughput and cell-edge user throughput when compared to current vendor approaches, which provide network-wide uniform parameter settings.

Keywords: Mobile network; Self-planning; LTE; Uplink power control; Real network sensitivity analysis

1 Introduction

Network planning plays a very important role in cellular communication networks. Before deployment, a thorough analysis must be done to achieve an adequate trade-off between capacity and coverage in the service area and predict future problems. A proper network planning avoids problems during the operational stage, minimizing (or, at least, delaying) subsequent capital investment [1,2].

Regardless of the radio access technology, cellular network planning is divided into core network planning and radio access network (RAN) planning. While the former is often only a dimensioning process, the latter entails the definition of RAN parameter settings that can be deployed in the roll-out stage. These parameters include not only physical parameters (e.g., site positions, antenna bearings, or power amplifier limits) but also logical parameters in radio resource management (RRM) algorithms [3-5]. In the past, due to the complexity of finding optimal settings for each specific scenario, vendors have provided operators with safe recommended values that should work reasonably well in most cases. Thus, the flexibility given by network parameters has not been fully exploited. To solve this situation, an effort has been made in the 3rd Generation Partnership Project (3GPP) long term evolution (LTE) standard to define the requirements for automating and improving network planning [6]. Likewise, automatic (or self-) planning has been identified by the industry as a key process in self-organizing networks (SON) [7-9].

Uplink power control (ULPC) is an important RRM process in mobile networks, as it has a direct impact on received signal and interference levels, as well as on user battery consumption. This makes ULPC an ideal candidate for automated parameter planning. In practice, the variability of propagation, traffic and interference conditions make it very difficult for operators to find optimal settings for these parameters before operation. For this reason, operators usually set ULPC parameters to safe values, which are deployed network wide. As a result, suboptimal network performance is obtained due to network irregularities. Even if this can be solved later during network operation, the provision of proper initial settings is greatly appreciated by operators.

*Correspondence: jfs@ic.uma.es
[1] Ingeniería de Comunicaciones, Universidad de Málaga, Campus de Teatinos S/N, 29071 Malaga, Spain
Full list of author information is available at the end of the article

In the literature, several power control schemes have been proposed. Fractional power control (FPC) has been selected for the physical uplink shared channel (PUSCH) in LTE [10,11]. A performance analysis of open-loop FPC is presented in [12,13]. Such an analysis is extended in [14-16] to closed-loop behavior. Later studies have evaluated more sophisticated power control schemes for LTE, which take into account interference and load conditions [17,18]. In [19], a parameter sensitivity analysis for the standardized FPC algorithm is carried out based on system-level simulations. As a result, a suboptimal parameter configuration is suggested for interference-limited and noise-limited macro-cellular scenarios. However, finding the best parameter configuration for every cell in the network is a more challenging task, since the underlying problem is a large-scale non-separable non-linear optimization problem. In [20], the adjustment of ULPC parameters in a single cell is formulated as a classical optimization problem with average throughput and cell throughput as alternative figures of merit. The latter analysis is extended to a multi-cell scenario in [21] by formulating ULPC as a non-cooperative game model. Then, a heuristic iterative optimization algorithm is proposed, where cells report UL power settings to the network management system and exchange power and interference information with their neighbor cells. In [22], a self-planning method is proposed for selecting the best parameter settings in FPC on a per-cell basis in an irregular LTE scenario. The proposal is based on an exhaustive search approach using Taguchi's method over a system-level simulator. A more computationally efficient planning method is presented in [23]. The core of the method is an analytical model that predicts the influence of the nominal power (a.k.a. power offset) and path-loss compensation factor on call acceptance probability for a given spatial user distribution. A suboptimal value of these parameters is computed on a per-cell basis by a randomized local greedy search algorithm. Alternatively, other studies consider tuning algorithms for the network operational stage. For instance, a self-tuning algorithm is proposed in [24] to dynamically adjust nominal power parameter based on overload indicator [25] so as to control the overall interference in the network. Similarly, in [26], a self-tuning algorithm for FPC is proposed based on fuzzy-reinforcement learning techniques. These self-tuning algorithms, conceived to be executed during network operation, can also be applied to network planning, provided that a system performance model is available. However, most self-tuning algorithms rely on iterations that require evaluating system performance for many different parameter settings. This iterative process is straight-forward for live networks due to the availability of performance measurements. However, this is not the case of network planning, where computations required to measure the quality of a new parameter plan at every iteration may jeopardize method scalability when large scenarios are considered. For this reason, most ULPC planning methods in the literature rely on simple analytical models for system performance assessment. To the authors' knowledge, few of them can handle irregular scenarios at a low computational cost and none of them considers closed-loop performance.

In this paper, a self-planning algorithm for ULPC parameters in the PUSCH of LTE is proposed. The self-planning algorithm determines the maximum uplink cell load, U_{UL}, and the nominal PUSCH power parameter in FPC, P_0. Full path-loss compensation is assumed in this work (i.e., FPC parameter α is fixed to 1), as implemented in most first vendor releases. To deal with network irregularities, the parameter tuning problem is solved on a cell-by-cell basis by aggregating the results from multiple regularized scenarios built on a per-adjacency basis. Such a regularization approach reduces the size of the solution space, thus reducing the time complexity of the method. Both open-loop and closed-loop power control performance are considered in the problem model. Method assessment is carried out over a static system-level simulator implementing a real network scenario. The main contributions of this work are as follows: a) a thorough parameter sensitivity analysis of ULPC closed-loop performance in a realistic simulated scenario and b) a self-planning method for ULPC parameters that can handle irregular scenarios at a low computational complexity and considers both open- and closed-loop ULPC performance. Similar to [23], the proposed heuristic algorithm can handle irregular scenarios while still using a semi-analytical approach. Unlike [23], the optimized parameters are the nominal power and the maximum cell load, instead of the path-loss compensation factor. More importantly, the method proposed here has better scalability and computationally efficiency due to the regularization approach. The rest of the paper is organized as follows. Section 2 outlines the LTE system model used in the planning process. Section 3 describes the design of the proposed planning algorithm. Section 4 presents simulation results obtained in a real network scenario. Finally, Section 5 presents the concluding remarks.

2 Uplink system model

LTE uplink comprises three physical channels: physical random access channel (PRACH), physical uplink shared channel (PUSCH), and physical uplink control channel (PUCCH) [27]. In particular, PUSCH is used to deliver user data and control information for active users in the uplink, which makes it an important component to be dimensioned and optimized. This section outlines how the main PUSCH features, namely the power control algorithm and the physical resource block (PRB) allocation scheme, are modeled in this work.

2.1 Power control algorithm

The aim of power control is to adjust the transmit power of the user equipment (UE) to fulfill quality of service (QoS) requirements at the base station (i.e., eNodeB, eNB). The power control scheme for PUSCH combines an open-loop and a closed-loop algorithm. The former aims to compensate for slow channel variations, while the latter adapts to changes in the inter-cell interference conditions, or measurement and power amplifier errors. In the standardized algorithm [11], the UE transmit power (in dBm) is given by

$$P_{TX} = \min \left\{ P_{tx_{\max}}, \underbrace{P_0 + \alpha \cdot PL}_{\text{basic open-loop operating point}} \underbrace{+\Delta_{TF} + f(\Delta_{TPC})}_{\text{dynamic offset}} + \underbrace{10 \cdot \log_{10} M_{PUSCH}}_{\text{bandwidth factor}} \right\}, \quad (1)$$

where $P_{tx_{max}}$ is the maximum UE transmit power, α is the channel path-loss compensation factor (set to 1 in this work), PL are the propagation losses, M_{PUSCH} is the number of PRBs assigned to the UE, and $\Delta_{TF} + f(\Delta_{TPC})$ is a dynamic term that depends on the selected modulation scheme and power-control commands sent by the eNB. Equation 1 consists of three parts: a basic open-loop operating point, a dynamic offset controlled by closed-loop operation, and a bandwidth correction factor. The open-loop term consists of a semi-static level given by the parameter P_0 (known as nominal power), defining the average received signal level target for all UEs in a cell, and a path-loss compensation term, controlled by path-loss compensation factor, α. From (1), it follows that setting P_0 controls UE transmit power, thus determining the perceived signal-to-interference-plus-noise ratio (SINR) in the serving cell and the interference level in surrounding cells.

2.2 PRB allocation scheme

The PRB allocation scheme also has a strong impact on UL performance. PRB allocation is done by a scheduler in the eNB, which assigns PRBs to users so as to maximize average cell throughput while keeping cell-edge user throughput within reasonable limits. For mathematical tractability, a simplified scheduling algorithm consisting of two stages is used in this work. It is assumed that a) full spectral sharing is used (i.e., frequency reuse is 1), b) radio resources are independently allocated between cells (i.e., PRB allocation is done on a cell basis), c) users are uniformly distributed in a cell, but unevenly distributed among cells, and d) any user has infinite data to transmit (i.e., full buffer service model). In a first stage, referred to as *naive* scheme, a preliminary PRB assignment is made based on the open-loop power control algorithm. Thus, an estimate of the maximum number of PRBs assigned for each user is obtained, from which PRB usage and average interference levels can be predicted. Then, a second scheme, referred to as *refined* scheme, considers the closed-loop power control algorithm. With the latter scheme, the final PRB assignment guaranteeing a minimum SINR is computed based on interference estimates with the PRB assignment suggested by the first naive scheme. Both schemes are described next.

2.2.1 Naive scheme - interference estimation

A theoretical PRB allocation scheme is first computed based on open-loop power control. A grid of positions in a cell is defined, so that each position represents a potential user. The number of PRBs assigned to user k, $M(k)$, is the maximum number of PRBs that can be assigned to the UE located at that point while still guaranteeing P_0 at the serving eNB, computed as

$$M(k) = \max \left\{ M_{\min}, \min \left(10^{(P_{tx\max} - \alpha \cdot PL(k) - P_0(S(k)))/10}, M_{\max} \right) \right\} \quad (2)$$

where $M_{\min}$ and $M_{\max}$ are the minimum and maximum number of PRBs that can be assigned to the UE, defined as network wide settings, $PL(k)$ are propagation losses (including antenna gains and path loss) for user k, and $P_0(S(k))$ is the value of P_0 for the cell serving user k, $S(k)$. In this work, $M_{\min}$=2 and $M_{\max}$ ranges from 6 to 100 depending on system bandwidth [28]. The value of $M(k)$ obtained by (2) is used to compute the transmit power for each user, $P_{TX}(k)$, as in (1). Then, the UL interference level received at each eNB i, $I_{UL}(i)$, is calculated as the sum of the interference from every neighbor cell j and thermal noise. The interference from each neighbor j is calculated as the average received signal level from UEs in that neighbor, weighted by the uplink cell load, $U_{UL}(j)$. $U_{UL}(j)$ is given by the PRB utilization ratio (i.e., the number of used PRBs divided by the number of available PRBs for a period of time). Thus,

$$I_{UL}(i) = N_{TH} + \sum_{\forall j \in N(i)} \left(U_{UL}(j) \cdot \frac{1}{N_u(j)} \cdot \sum_{\forall k \in A(j)} \frac{P_{TX}(k)}{PL(k,i)} \right) \quad (3)$$

where N_{TH} is the thermal noise (−119.4 dBm in this work), $N(i)$ is the set of neighbors of cell i, $A(j)$ is the set of users (i.e., points in the service area) of cell j, $N_u(j)$ is the number of users (locations) in the service area of j, and $P_{TX}(k)$ is the UE transmit power of user k.

2.2.2 *Refined scheme - SINR estimation*

Based on estimated interference levels, a second and more accurate PRB allocation defines the maximum number of PRBs while still satisfying a minimum SINR threshold, computed as

$$M(k)=\max\left\{M_{\min}, \min\left(10^{(P_{tx\max}-\alpha\cdot PL(k)-I_{UL}(S(k))-SINR_{th})/10}, M_{\max}\right)\right\} \quad (4)$$

where $I_{UL}(S(k))$ is the UL interference level in cell $S(k)$ and $SINR_{th}$ is the minimum SINR value to be satisfied (−2.8 dB in this work [28]). Equation 4 shows how the number of PRBs assigned to a user changes with radio link conditions (i.e., PL and I_{UL}). A user experiencing good radio conditions (i.e., low path-loss and low interference level) can be assigned all available PRBs, $M_{\max}$. However, as path-loss and/or interference increases, the number of PRBs allocated to the user is reduced in (4). This PRB reduction allows a higher transmission power per PRB, and thus, the minimum $SINR$ threshold is fulfilled.

2.3 Key performance indicators (KPI)

Several indicators are calculated to check user and network performance with different ULPC parameter settings. The maximum throughput per PRB achieved by a user is obtained from SINR values by the truncated Shannon bound formula [29] as

$$THperPRB(k) = \begin{cases} 0 & SINR(k) < SINR_{\min}, \\ \beta \cdot log_2(1+SINR(k)) & SINR_{\min} \le SINR(k) \le SINR_{\max}, \\ THperPRB_{\max} & SINR_{\max} < SINR(k), \end{cases} \quad (5)$$

where $THperPRB_{\max}$ is the maximum throughput per PRB achievable by the codeset, $SINR_{\max}$ and $SINR_{\min}$ are SINR values where maximum and minimum throughputs are reached, and β is a constant. Then, the maximum achievable throughput is computed by multiplying throughput per PRB by the number of assigned PRBs obtained in (4). In this work, $THperPRB_{\max}$ = 514 kbps, $SINR_{\max}$ = 14 dB, $SINR_{\min}$ = −9 dB and β=0.6.

Average user throughput, TH_{avg}, is used as a measure of cell capacity, while cell-edge user throughput, TH_{ce}, is used as a measure of cell coverage. Average user throughput is obtained as the average of individual user throughputs in the serving area of a cell, as

$$TH_{avg}(i) = \frac{\sum_{\forall k \in A(i)} M(k) \cdot THperPRB(k)}{N_u(i)} \cdot U_{UL}(i). \quad (6)$$

A user location, k, is assigned to cell i if the latter is the one providing the minimum propagation losses. Cell-edge user throughput, TH_{ce}(i), is defined as the fifth percentile of user throughput in the cell (i.e., the user throughput value exceeded by 95% of user locations in cell i).

3 Self-planning method

In this section, an algorithm for the automatic planning of the nominal power, P_0, and maximum UL cell load, U_{UL}, on a cell basis is presented. Firstly, a qualitative analysis of the influence of these parameters on network coverage and capacity is presented. Then, a preliminary parameter sensitivity analysis is performed in a regular scenario. From this quantitative analysis, the basic principles of the optimization algorithm are derived. The extension of the algorithm to deal with irregular scenarios is detailed later. With such an extension, ULPC settings can be adjusted on a per-cell basis to achieve better network performance.

3.1 General considerations

In mobile communication systems, coverage and capacity cannot be optimized independently as one affects the other. Network performance is a trade-off between cell-edge users, defining system coverage, and cell-center users, which have a larger influence on system capacity. This trade-off is the subject of the coverage-capacity optimization (CCO) problem, which has been identified as a relevant SON use case by 3GPP [30].

CCO can be performed by tuning ULPC parameters. Setting the average received signal level at the eNB, P_0, has a direct impact on the performance of both the adjusted cell and its surrounding cells. In the considered cell, a large value of P_0 enforces high transmit power for all users. Thus, users closer to the eNB should experience higher SINR values and, consequently, better UL performance. This is translated into higher peak user throughput in the cell at the expense of more user battery consumption. In contrast, cell-edge users that are power-limited cannot increase their transmit power when increasing P_0. At the same time, cell-edge users in surrounding cells experience low SINR due to increased interference level. Conversely, a low P_0 value favors cell-edge users in surrounding cells at the expense of cell-center users in the adjusted cell. Alternatively, UL cell load, U_{UL}, can be limited in a cell to avoid interference problems in surrounding cells. Control of U_{UL} can be implemented by traffic management procedures, such as call admission control or cell load sharing.

The optimal P_0 and U_{UL} settings in a cell are highly influenced by the propagation scenario and traffic distribution in both the cell under study and its neighbors. Therefore, any CCO algorithm must consider scenario irregularities and deal with neighbor cell conditions. As a consequence, CCO becomes a large scale non-separable multivariate optimization problem, where all cells must be jointly optimized. The size of the solution space is $(N_{v_{P_0}} \cdot N_{v_{U_{UL}}})^{N_c}$, where $N_{v_{P_0}}$ and $N_{v_{U_{UL}}}$ are the number of possible values for P_0 and U_{UL} parameters, and N_c is the number of cells to be planned. The large size of the solution space prevents the use of exact algorithms, which are substituted by heuristic algorithms. To reduce the search space, some approaches only evaluate a limited set of representative parameter combinations, selected by, e.g., Taguchi's method [22]. Other approaches build an initial solution by assuming that all cells have the same parameter settings, which is then refined by a randomized local search process, e.g., greedy search [23] or simulated annealing. Even the latter approach requires many iterations to obtain solutions of reasonable quality, since any parameter change in a cell affects the performance of surrounding cells, and each iteration requires evaluating the overall network performance in the whole scenario after every parameter change.

In this work, a different approach is followed to reduce the search space. The global multi-variate optimization problem is divided into N_c independent bi-variate optimization sub-problems (one per cell), each providing a sub-optimal solution for a cell i, $P_0(i)$ and $U_{UL}(i)$. To decouple sub-problems, some approximation is taken to model the irregular local scenario in the cell under study by a regular one. The idea is to approximate one real and complex scenario by many simple and regular scenarios that can be solved easily and independently. Apart from reducing the solution space, regularizing the scenario has other important advantages from the computational perspective: a) the grid resolution needed to obtain accurate propagation predictions is lower for regular scenarios than for irregular ones, and b) interference calculations for neighbor cells can be replicated due to the symmetry in a regular scenario, whereas these have to be treated individually in an irregular scenario due to the lack of symmetry. The following subsections describe first the optimization process in a regular scenario and later explain the extension to irregular scenarios.

3.2 Regular scenario

A preliminary sensitivity analysis shows the impact of tuning P_0 and U_{UL} on network performance in a regular scenario. For this purpose, a static system-level simulator is used. The simulation scenario, shown in Figure 1, consists of one central tri-sectorized site surrounded by a first tier of tri-sectorized neighbor sites. Main simulation parameters are shown in Table 1. Performance statistics are collected only for the central cell i.

Network performance is evaluated for uniform P_0 and U_{UL} settings in the scenario (i.e., all cells share the same parameter values). Figure 2 and Figure 3 show cell-edge and cell-average user throughput for cell i, $TH_{ce}(i)$ and $TH_{avg}(i)$, with different P_0 and U_{UL} settings. Increasing P_0 leads to higher $P_{TX}(k)$ for cell-center users, thus increasing average user throughput. This improvement is limited by the interference from users in other cells, also increasing their $P_{TX}(k)$ in this regular scenario. Thus, $TH_{avg}(i)$ reaches a maximum value, after which $TH_{avg}(i)$ decreases. The largest value of $TH_{avg}(i)$ is obtained for high values of P_0 (i.e., −102 dBm), where cell-center users can make the most of low propagation losses and received interference level is not much larger than the noise floor. In contrast, cell-edge performance quickly degrades as P_0 increases. This is due to the fact that cell-edge users reach their power limit, so that their transmit power does not change, but interference increases, as users in other cells increase their transmit power in the regular scenario. As a consequence, the maximum value for $TH_{ce}(i)$ is reached for low values of P_0, when $I_{UL}(i)$ is close to the noise floor

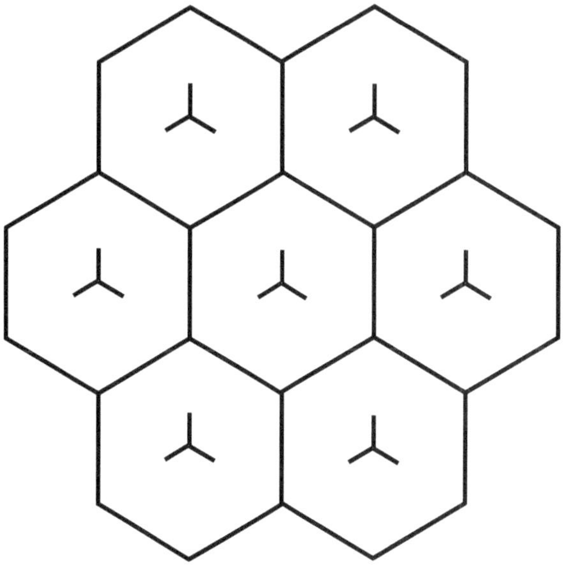

Figure 1 Regular scenario in the parameter sensitivity analysis.

Table 1 Simulation parameters

Parameters	Settings
Spectrum allocation	10 MHz (50 PRBs)
Carrier frequency	2 GHz
Cell layout	7 eNBs, 21 sectors, regular grid (200 m resolution).
Distance attenuation	$133.9 + 35.2\log_{10}(d)$, d in km
Slow fading constant	8 dB
Thermal noise density	−174 dBm/Hz
Cell radius	1.5 km (3 km inter-site distance)
eNB antenna height	30 m
eNB antenna tilt	5°
eNB antenna pattern and gain	3GPP ideal [32]
Maximum UE transmit power	23 dBm
Path-loss compensation factor, α	1 (full compensation)
Traffic model	Full buffer

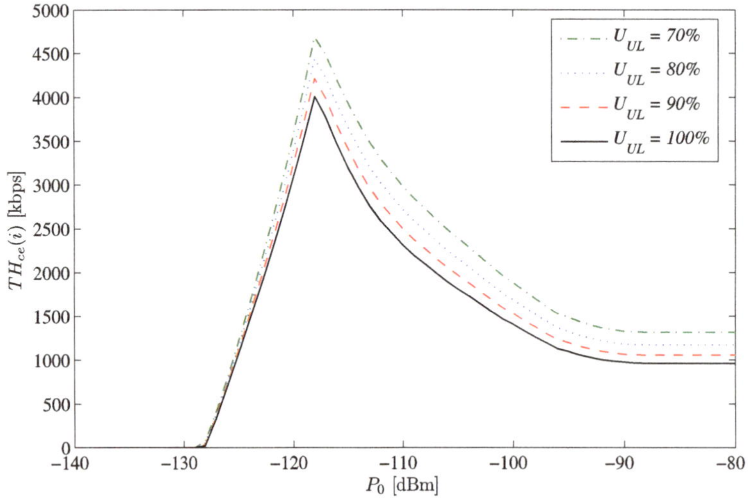

Figure 2 Cell-edge throughput performance in regular scenarios with uniform P_0 and U_{UL}.

(i.e., −119.4 dBm). Note that small P_0 deviations from the optimal value cause large deviations in $TH_{ce}(i)$, but near-optimal performance in $TH_{avg}(i)$. This is evidence that the trade-off between coverage and capacity is more critical for cell-edge users.

Based on the behavior observed in Figure 2 and Figure 3, a self-planning algorithm for P_0 and U_{UL} in a regular scenario is designed. The flow diagram of the proposed iterative algorithm is shown in Figure 4. $P_0^{(n)}$ and $U_{UL}^{(n)}$ denote parameter values in iteration n. The optimization engine is a simple gradient-based local optimization method, whose aim is to maximize the average user throughput, $TH_{avg}(i)$, while keeping cell-edge throughput over a minimum, $TH_{ce}(i) > TH_{ce,min}$. Such a formulation of the CCO problem can be found in the literature (e.g., [31]). For this purpose, P_0 and U_{UL} are initially configured to arbitrarily high values (e.g., −80 dBm and 100%, respectively). Then, the algorithm iteratively decreases parameter values. In a first stage, when minimum cell-edge throughput is not fulfilled (i.e., $TH_{ce}(i) < TH_{ce,min}$) and maximum $TH_{ce}(i)$ value has not been reached, the optimization engine tries to increase $TH_{ce}(i)$ by decreasing P_0. Once coverage constraints are fulfilled (i.e., $TH_{ce}(i) \geq TH_{ce,min}$), the optimization engine keeps decreasing P_0 with the aim of maximizing TH_{avg}, until $TH_{avg}(i)$ starts to decrease (i.e., optimal P_0 in terms of $TH_{avg}(i)$ has been surpassed) or cell-edge throughput has become less than the minimum threshold. U_{UL} is only decreased when minimum cell-edge throughput is not fulfilled. This is the case of interference-limited scenarios, consisting of very small

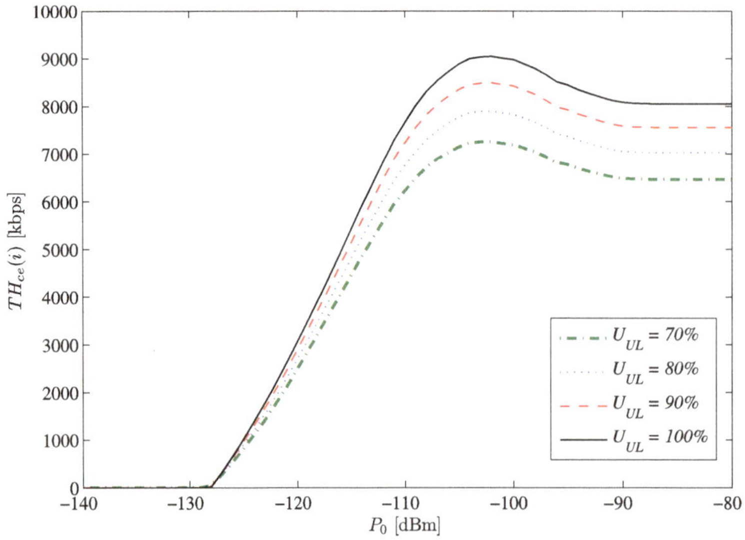

Figure 3 Average user throughput performance in regular scenarios with uniform P_0 and U_{UL}.

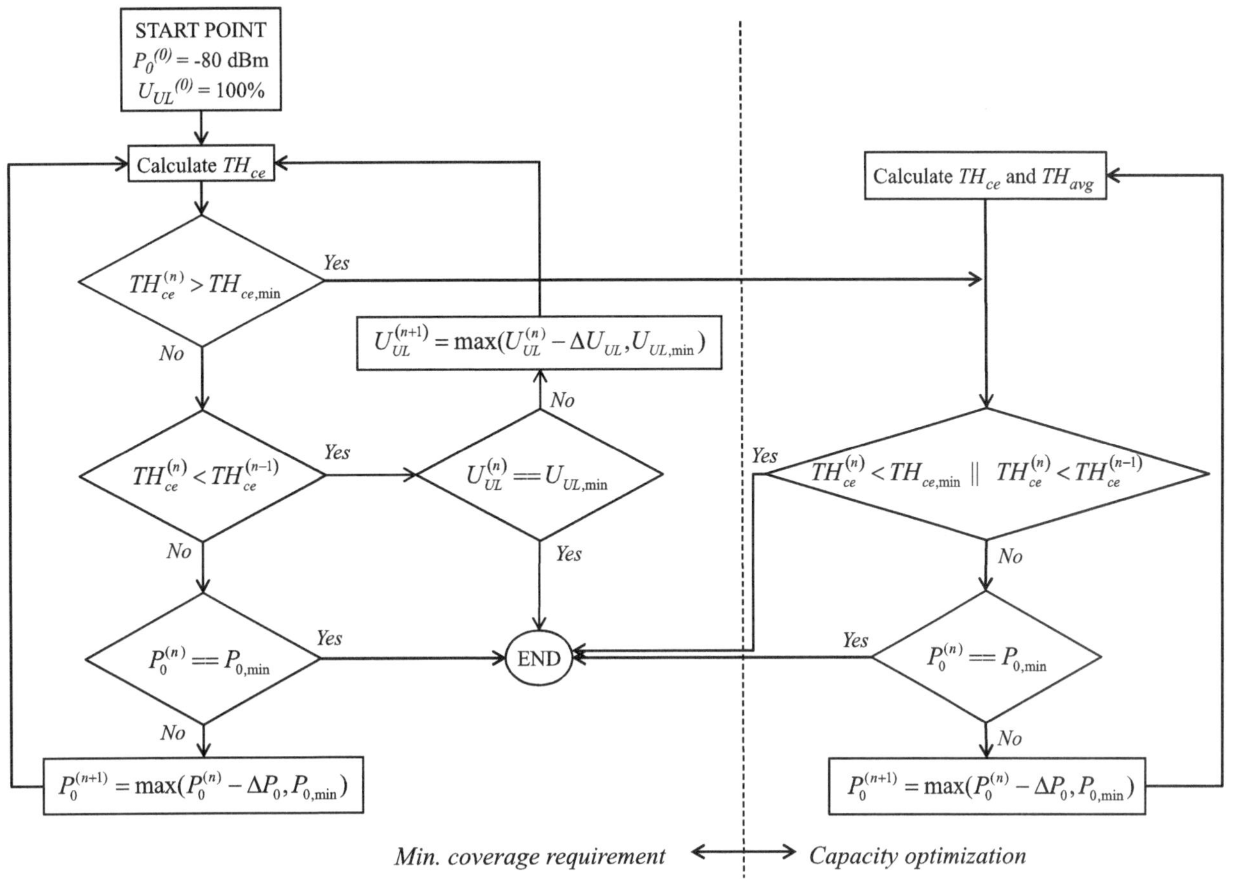

Figure 4 Flow diagram of optimization process.

cells. Reducing U_{UL} only when strictly needed is in line with operator policies, which only consider limiting the average PRB utilization in a cell under critical interference situations.

3.3 Irregular scenario

The above-described optimization algorithm is designed for a regular scenario, resulting in a solution where all cells have the same value of P_0 and U_{UL}. However, in live networks, propagation, traffic, and interference conditions greatly vary among cells. These irregularities translate into uneven parameter configuration for optimal network performance. The following paragraphs describe a method to deal with such irregularities. The method consists of four stages, namely global scenario construction, division into local regular scenarios, local problem solving, and aggregation of local solutions.

3.3.1 Global scenario construction - adjacency definition

The method starts by collecting network configuration data, consisting of site locations, antenna height, and antenna bearings (i.e., azimuth and tilt). This data is included in propagation models to define a list of relevant non co-sited neighbor cells for every cell in the scenario. Such a list is used later for regularization of local scenarios, as will be explained later.

Several options could be used to automatically define neighbors for every cell in the scenario. Relevant neighbors are usually identified by computing the contribution that every single user in each candidate neighbor cell has on the UL interference of the cell under study. In the absence of real measurements, this calculation requires computing a point-by-point path loss matrix, from which to identify cell dominance areas, which increases the computational load of the method.

For computational efficiency, a simple automatic neighbor definition algorithm is used here with planning purposes. The aim is to rank non co-sited neighbors by relevance assuming that the average path loss of all users in a neighbor cell is similar to that from neighbor eNB location. The number of neighbors must be kept within reasonable limits. An intuitive indicator for neighbor relevance in the UL, NR_{UL}, is defined as

$$NR_{UL}(i,j) = L(i,j) - A_H(i,j) - A_V(i,j)\,, \tag{7}$$

where $L(i,j)$ is the path loss between the cell under study i and cell j computed by some propagation model, and $A_h(i,j)$ and $A_v(i,j)$ are the horizontal and vertical gains for the antenna of cell i to the location of eNB j. Roughly, the closest eNBs j in terms of electrical distance tend to give the lowest $NR_{UL}(i,j)$ values and are labeled as relevant neighbors of cell i (i.e., $j \in N_r(i)$, where $N_r(i)$ is the list of relevant non co-sited neighbors of cell i). To limit the number of relevant neighbors, neighbors j having NR_{UL} 10 dB larger than that of the most relevant neighbor are discarded. For the same reason, a maximum of 12 neighbors per cell is configured.

3.3.2 *Division into local regular scenarios*

Once neighbors have been defined for each cell, the optimization process can start. Global network optimization is broken down into N_c local optimization sub-problems, which are solved independently. Every sub-problem is solved by the regularization of one or several local scenarios. Two alternatives are considered, depending on how the regular local scenario is built for a cell i:

- *Mean radius approximation* (MRA): A regular scenario, as that in Figure 1, is built for the cell under study i. Inter-site distance (ISD) in the regular scenario is taken from the global scenario by averaging ISDs between the site of cell i and the site of its neighbors in the real irregular scenario, excluding co-sited cells. Thus, MRA assumes a single 'average' regular scenario per cell. Then, optimal $P_0(i)$ and $\mathcal{U}_{UL}(i)$ values for cell i are obtained by applying the optimization method described in Figure 4 to this equivalent regular scenario.
- *Adjacency-based approximation* (AA): The optimization problem for cell i is broken down into $N_{neigh}(i)$ sub-problems, where $N_{neigh}(i)$ is the number of non co-sited neighbors of cell i (i.e., the length of $N_r(i)$). For each pair of cells (i,j), where i is the optimized cell and j is the selected neighbor, a regular scenario is built, as shown in Figure 5. ISD in the scenario is the one between cells i and j in the real network. In the figure, it is observed how the relative position (i.e., distance and antenna bearing angles) of cells in the real network is maintained. As a result of problem division, $N_{neigh}(i)$ regular scenarios are built and solved for each cell i.

Note that MRA needs a single optimization process per cell, whereas AA requires $N_{neigh}(i)$ optimization processes (one per neighbor). As a result, $N_{neigh}(i)$ pairs of sub-optimal P_0 and $\mathcal{U}_{UL}$ values are obtained for each cell i, $(P_0(i,j), \mathcal{U}_{UL}(i,j))$. Some aggregation process is then needed to obtain a single solution per cell, $(P_0(i), \mathcal{U}_{UL}(i))$.

3.3.3 *Aggregation of adjacency-level solutions in AA*

A single value of $(P_0(i), \mathcal{U}_{UL}(i))$ is computed from all sub-optimal solutions involving cell i, where cell i is either source or target in the adjacency, i.e., $(P_0(i,j), \mathcal{U}_{UL}(i,j))$ and $(P_0(l,i), \mathcal{U}_{UL}(l,i))$. Four alternative aggregation criteria are defined, namely maximum, mean, minimum, and mixed. In each criterion, the final solution per cell is computed from adjacency-level solutions as follows:

- Maximum method (MaM):

$$P_0(i) = \max \left\{ [P_0(i,j) \ P_0(l,i)] \right\}, \quad \forall j \in N_r(i), l/i \in N_r(l) \tag{8}$$

$$\mathcal{U}_{UL}(i) = \max \left\{ [\mathcal{U}_{UL}(i,j) \ \mathcal{U}_{UL}(l,i)] \right\}, \quad \forall j \in N_r(i), l/i \in N_r(l) \tag{9}$$

- Mean method (MeM):

$$P_0(i) = \text{mean} \left\{ [P_0(i,j) \ P_0(l,i)] \right\}, \quad \forall j \in N_r(i), l/i \in N_r(l) \tag{10}$$

$$\mathcal{U}_{UL}(i) = \text{mean} \left\{ [\mathcal{U}_{UL}(i,j) \ \mathcal{U}_{UL}(l,i)] \right\}, \quad \forall j \in N_r(i), l/i \in N_r(l) \tag{11}$$

- Minimum method (MiM):

$$P_0(i) = \min \left\{ [P_0(i,j) \ P_0(l,i)] \right\}, \quad \forall j \in N_r(i), l/i \in N_r(l) \tag{12}$$

$$\mathcal{U}_{UL}(i) = \min \left\{ [\mathcal{U}_{UL}(i,j) \ \mathcal{U}_{UL}(l,i)] \right\}, \quad \forall j \in N_r(i), l/i \in N_r(l) \tag{13}$$

- Mixed method (MxM). This approach uses the maximum criterion for $P_0(i)$, defined in (8), and the minimum criterion for $\mathcal{U}_{UL}$, defined in (13). The aim is to maximize SINR by selecting a large P_0 value, even if interference is thus increased. Then, the lowest $\mathcal{U}_{UL}$ is selected to mitigate such an interference increase.

Figure 6 summarizes the whole optimization process for a cell i in the AA method, including regularization on a per-adjacency basis, sub-problem solving and solution aggregation.

3.4 Computational complexity

The time complexity of the MRA self-planning algorithm is $O(N_c)$, since the number of sub-problems to be solved is the number of cells in the network. All of these sub-problems deal with a regular scenario, which can be solved by a simple gradient search method on a single variable.

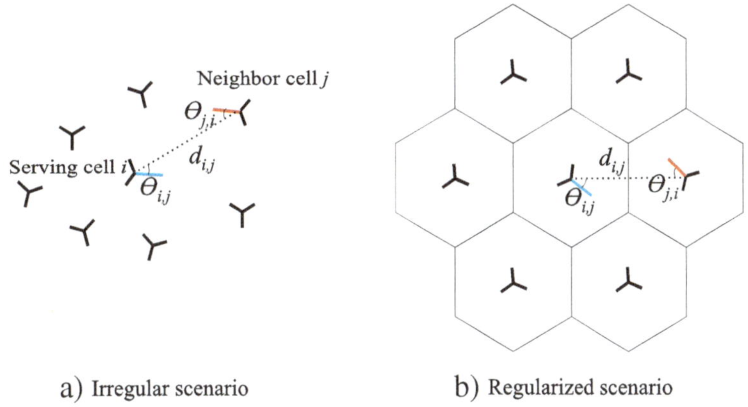

Figure 5 Scenario regularization. (a) Irregular scenario and **(b)** regularized scenario.

The time complexity of the AA self-planning algorithm is also O(N_c), since the number of sub-problems grows linear with the total number of relevant non co-sited neighbors in the scenario, which is proportional to the number of cells in the network.

4 Performance analysis

The solutions of the different variants of the proposed algorithm are tested in a system-level simulator implementing a real irregular scenario. The simulator includes all UL functionalities described in Section 2. For clarity, the analysis set-up is described first and results are presented later.

4.1 Analysis setup

The considered scenario consists of 233 sites (699 cells) in a large metropolitan area. Location, azimuth, and antenna tilts are obtained from live network configuration data. ISDs range from 0.3 to 6 km. Other simulation parameters are shown in Table 1.

Five different self-planning approaches are tested. A first method uses MRA with neighbor cell selection as described in Section 3.3.1 (referred to as MRA). The other four methods use AA for dealing scenario irregularities, considering MaM, MeM, MiM, and MxM as solution aggregation criterion, respectively. To quantify the benefit of tuning ULPC parameters on a cell-by-cell basis, uniform settings (i.e., $P_0(i) = P_0(j)$ and $\mathcal{U}_{UL}(i) = \mathcal{U}_{UL}(j), \forall i, j$) are also tested. In this case, a set of benchmarking solutions is constructed by selecting different combinations of P_0 and $\mathcal{U}_{UL}$.

To assess the value of a parameter plan, two performance indicators are used: a) as a measure of network capacity, overall average user throughput, $\overline{TH_{avg}(i)}$,

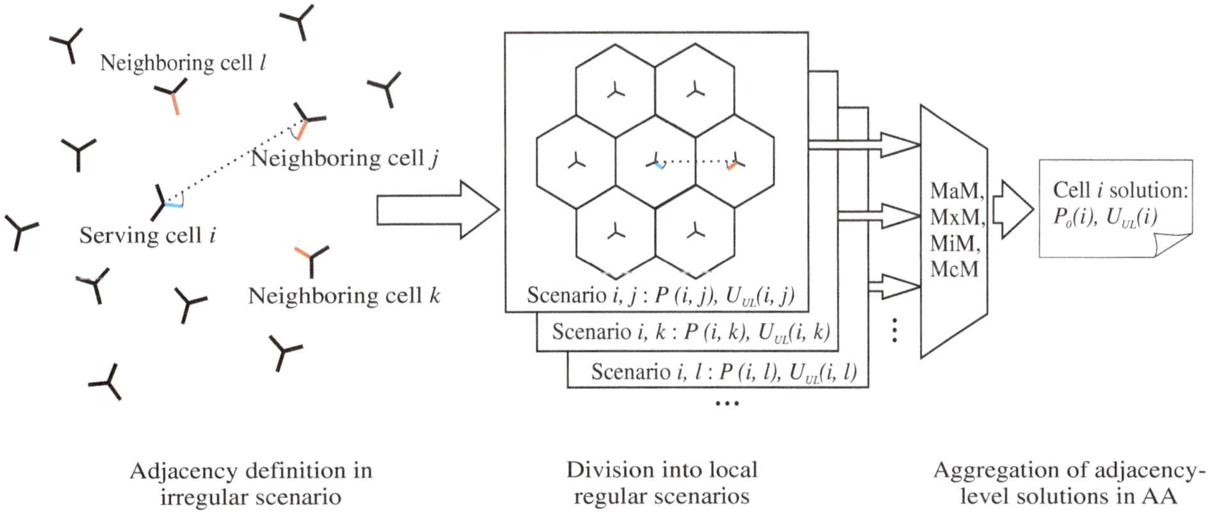

Figure 6 Optimization process for irregular scenario in the adjacency-based approximation.

computed by averaging $TH_{avg}(i)$ across cells in the scenario, and b) as a measure of coverage, the ratio of cells fulfilling $TH_{ce}(i) > TH_{ce,min}$, denoted as $R_{TH_{ce,min}}$. In this work, $TH_{ce,min} = 100$ kbps, based on typical operator demand.

4.2 Results

The analysis is first focused on network performance obtained with uniform parameter settings. Then, the analysis proceeds to cell-based solutions obtained by the proposed self-planning methods. Finally, execution time is analyzed.

4.2.1 *Uniform parameter settings*

Figure 7 shows the coverage-capacity trade-off achieved with uniform P_0 and U_{UL} settings. Each point in the figure corresponds to a network parameter plan. Different curves correspond to different U_{UL} values (i.e., 70%, 80%, 90%, and 100%) and points in the curve correspond to P_0 values ranging from −125 dBm (down and left in the figure) to −90 dBm (up and left). Note that network performance is better in the upper right area of the figure. Four regions are defined in every curve, corresponding to different network behaviors, denoted as zone 1 ($P_0 \geq -105$ dBm), zone 2 ($P_0 \in (-105, -115]$ dBm), zone 3 ($P_0 \in (-115, -119]$ dBm), and zone 4 ($P_0 < -119$ dBm).

When P_0 is extremely high (zone 1), both coverage and capacity indicators improve as P_0 decreases. In this situation, interference levels are well above the noise floor, so decreasing P_0 in a cell and its neighbors decreases interference by the same amount. As a consequence, cell-edge throughput improves, while throughput for cell-center users is maintained as both interference and desired signal level decreases. When P_0 is medium (zone 2), a trade-off between coverage and capacity exists. Cell-edge throughput improves as P_0 further decreases at the expense of degrading network average throughput. Lower values of P_0 (zone 3) cause interference levels fall below the noise level, so that there is no benefit on further decreasing interference. On the contrary, cell-edge throughput starts to degrade. Finally, a strong degradation in network performance occurs when P_0 becomes less than the noise floor, −119.4 dBm in this work (zone 4).

The influence of U_{UL} can be seen by comparing curves in Figure 7. It is observed that curves are shifted up and left when U_{UL} increases. When more UL radio resources are available, users can reach higher throughput (i.e.,

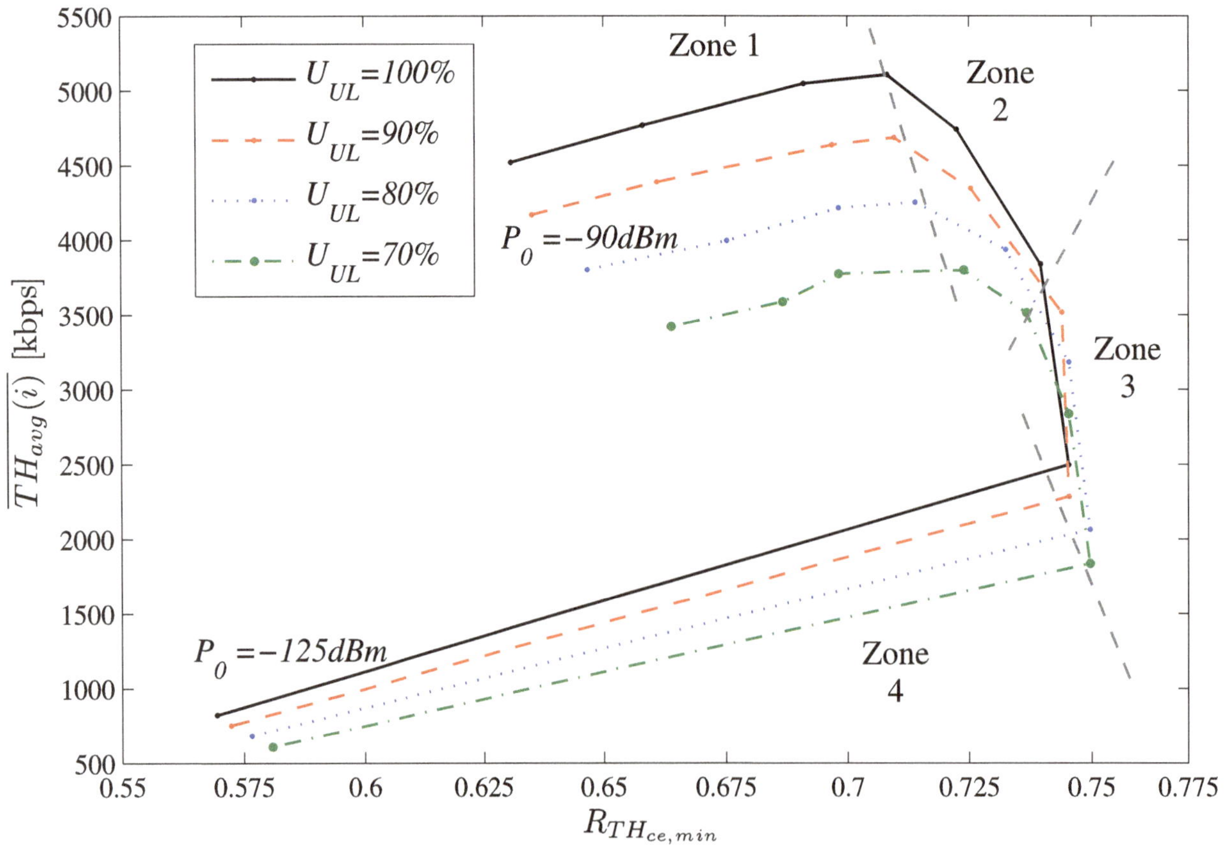

Figure 7 Overall network performance for uniform parameter settings.

$\overline{TH_{avg}(i)}$ increases), but interference levels also increase, and hence $R_{TH_{ce,min}}$ decreases.

4.2.2 Non-uniform parameter settings

Figure 8 compares the performance of the different self-planning proposals for cell-based ULPC parameter settings. For comparison purposes, results of uniform parameter plans are superimposed. Axis scales have been adjusted for optimal viewing in the figure. Note that MiM is not shown, since MiM experiences a very poor network performance (i.e., $\overline{TH_{avg}(i)}$ = 1,254 kbps and $R_{TH_{ce,min}} = 205/699 = 0.29$). This is due to the fact that MiM sets $P_0(i)$ to very low values, thus degrading SINR even for the best users. Other solutions (i.e., MaM, MeM, MxM, and MRA) perform similar to the best network performance with uniform settings (i.e., top-right in the figure). The best average throughput performance is achieved by MaM (i.e., $\overline{TH_{avg}(i)}$ = 5,113 kbps), which outperforms MeM and MRA by 11.15% and 15.27%, respectively. MaM selects the maximum value of P_0 and U_{UL} across adjacencies, thus achieving better overall throughput. As a drawback, interference problems are greater, as pointed out by the lower value of $R_{TH_{ce,min}}$ compared to other methods. The best cell-edge throughput performance is achieved by MeM (i.e., $R_{TH_{ce,min}}$ = 0.73), which outperforms MaM and MxM by 3.87% and 2.04%, respectively. Although MxM was designed as an intermediate solution between MaM and MiM, it is observed that limiting U_{UL} in MxM has a strong impact on average user throughput, which is 19.89% and 11.09% lower than those achieved by MaM and MeM, respectively. MeM outperforms all other methods, as it is the only solution achieving a coverage-capacity trade-off better than that obtained with any regular parameter setting.

4.2.3 Execution time

All methods were executed in an Intel® Xeon© computer with 3.47-GHz clock frequency and 12 GB of RAM. The time required to build the curve for uniform P_0 and U_{UL} settings in the static system-level simulator was 2,500 s. In contrast, any of the self-planning methods proposed for constructing cell-based parameter plans took only 300 s in average (i.e., 0.43 s per cell).

It is also worth mentioning that, if a multi-variate exhaustive search approach is used, the addition of a new cell to an existing (i.e., already planned) scenario needs complete re-simulation of the whole scenario to build the new curve for uniform parameter settings. In

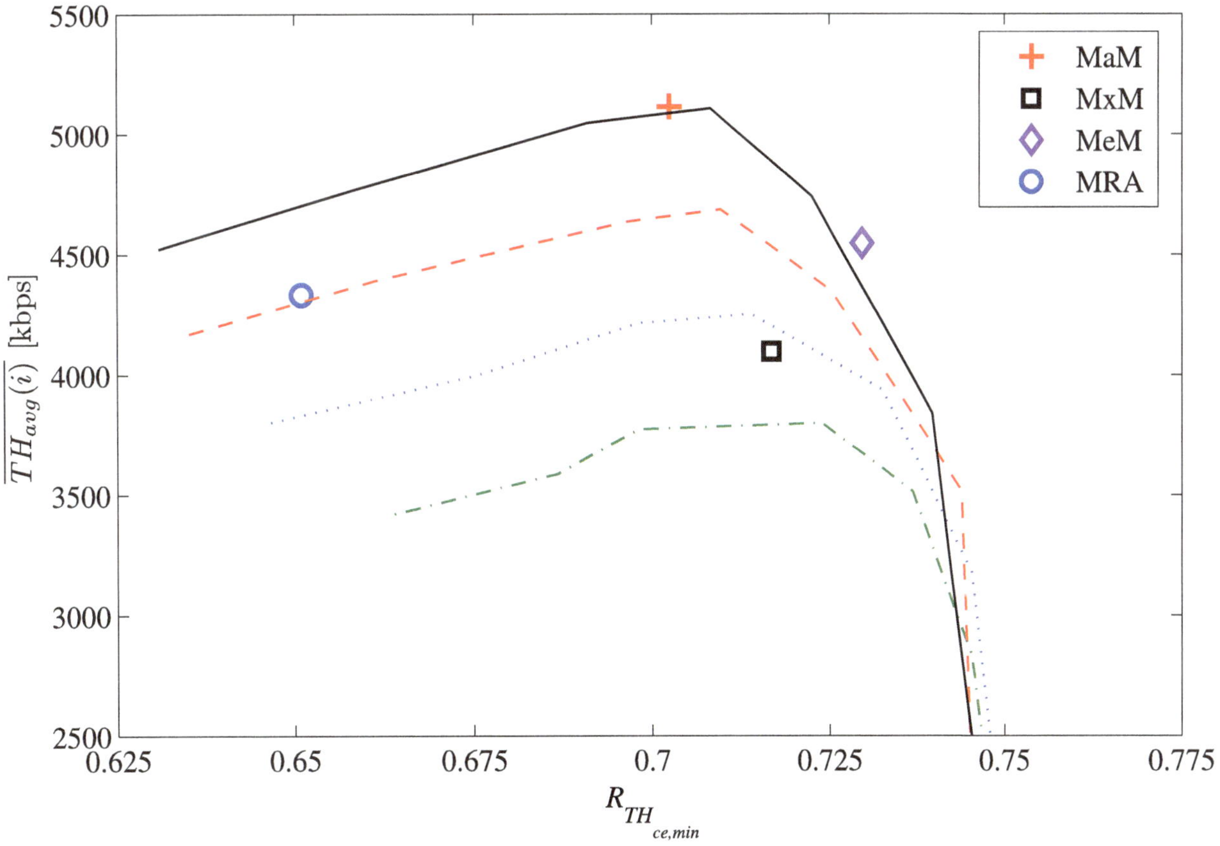

Figure 8 Overall network performance of self-planning methods.

contrast, the self-planning method proposed in this work only needs solving $1 + N_{neigh}(i)$ sub-problems (i.e., the re-computation of the new cell and its relevant neighbors) with regular scenarios.

5 Conclusions

In this work, a novel self-planning algorithm for UL power control parameters in the PUSCH of LTE has been presented. In the algorithm, the global network parameter optimization problem is divided into multiple simple optimization problems, one per adjacency, where a regular scenario is assumed. In such regular scenarios, a simple gradient descent method is used to find the optimal parameter settings that maximize average user throughput while still ensuring a minimum cell-edge user throughput. Then, the best parameter settings for each cell are calculated by aggregating adjacency-level solutions. Thus, the proposed heuristic algorithm can deal with irregular scenarios at a low computational complexity, which is critical for planning large scenarios. Performance assessment has been carried out over a static system-level simulator implementing a real scenario. Results show that averaging solutions across adjacencies (i.e., MeM aggregation method) achieves the best trade-off between coverage and capacity, outperforming uniform network configurations for nominal power and UL cell load parameters.

The proposed self-planning algorithm can be implemented as part of a centralized SON system, which communicates with the network management system (NMS). For an initial operational network stage, self-configuration can be performed by automatically retrieving ULPC parameters from the NMS. Later, every time a new eNB is added, the network self-planning algorithm can be relaunched. For this purpose, the new eNB sends its location and antenna configuration parameters to the NMS, which updates network topology in the SON server. Then, the SON server computes the new ULPC parameter configuration, which is sent to NMS so that it can be retrieved by the new eNB.

Competing interests

The authors declare that they have no competing interests.

Acknowledgements

This work has been funded by the Spanish Ministry of Economy and Competitiveness (TIN2012-36455), Optimi-Ericsson and Agencia IDEA (Consejería de Ciencia, Innovación y Empresa, Junta de Andalucía, ref. 59288) and FEDER.

Author details

[1]Ingeniería de Comunicaciones, Universidad de Málaga, Campus de Teatinos S/N, 29071 Malaga, Spain. [2]Ericsson, Via de los Poblados 13, 28033 Madrid, Spain.

References

1. A Hoikkanen, in *International Conference on Wireless and Optical Communications Networks, WOCN*. Economics of 3G Long-Term Evolution: the business case for the mobile operator (Singapore, July 2007), pp. 1–5
2. F Gordejuela-Sanchez, J Zhang, in *IEEE Global Telecommunications Conference, GLOBECOM*. LTE Access network planning and optimization: A service-oriented and technology-specific perspective (Hilton Hawaiian Village, Honolulu, Hawaii, USA, December 2009), pp. 1–5
3. J Lempiainen, M Manninen, *Radio Interface System Planning for GSM/GPRS/UMTS*. (Springer, New York, NY, USA, 2001)
4. J Laiho, A Wacker, T Novosad, *Radio Network Planning and Optimisation for UMTS*. (John Wiley & Sons, New York, NY, USA, 2002)
5. AR Mishra, *Advanced Cellular Network Planning and Optimisation*. (John Wiley & Sons, New York, NY, USA, 2006)
6. 3GPP TS 32.500, Telecommunication management; Self-Organizing Networks (SON); Concepts and Requirements. V9, ETSI, Sophia Antipolis Cedex, France, December 2009
7. NGMN Use Cases Related to Self Organising Network, Overall description. http://www.ngmn.org. Accessed 1 October 2014
8. J Ramiro, K Hamied, *Self-Organizing Networks: Self-Planning, Self-Optimization and Self-Healing for GSM, UMTS and LTE*. (John Wiley & Sons, New York, NY, USA, 2011)
9. S Hamalainen, H Sanneck, C Sartori, *LTE Self-Organising Networks (SON)*. (John Wiley & Sons, New York, NY, USA, 2011)
10. JF Whitehead, in *IEEE 43rd Vehicular Technology Conference, VTC*. Signal-level-based dynamic power control for co-channel interference management (Secaucus, New Jersey, USA, May 1993), pp. 499–502
11. 3GPP TS 36.213, Physical Layer Procedures. V8.6. ETSI, Sophia Antipolis Cedex, France, September 2009
12. W Xiao, R Ratasuk, A Ghosh, R Love, Y Sun, R Nory, in *IEEE 64th Vehicular Technology Conference, VTC*. Uplink power control, interference coordination and resource allocation for 3GPP E-UTRA (Montréal, Québec, Canada, September 2006), pp. 1–5
13. AM Rao, in *IEEE 66th Vehicular Technology Conference, VTC*. Reverse link power control for managing inter-cell interference in Orthogonal Multiple Access Systems (Baltimore, Maryland, USA, September 2007), pp. 1837–1841
14. A Simonsson, A Furuskar, in *IEEE 68th Vehicular Technology Conference, VTC*. Uplink Power Control in LTE - Overview and performance, Subtitle: Principles and benefits of utilizing rather than compensating for SINR variations (Calgary, Alberta, Canada, September 2008), pp. 1–5
15. B Muhammad, A Mohammed, in *IEEE 70th Vehicular Technology Conference, VTC*. Performance evaluation of uplink closed loop power control for LTE system (Anchorage, Alaska, USA, September 2009), pp. 1–5
16. B Muhammad, A Mohammed, in *6th International Conference on Emerging Technologies (ICET)*. Uplink closed loop power control for LTE system (Islamabad, Pakistan, October 2010), pp. 88–93
17. S Yang, Q Cui, X Huang, X Tao, in *IEEE 72nd Vehicular Technology Conference, VTC*. An effective uplink power control scheme in CoMP systems (Ottawa, Ontario, Canada, 2010), pp. 1–5
18. NJ Quintero, Advanced power control for UTRAN LTE uplink. PhD thesis, Department of Electronic Systems, Aalborg University, 2008
19. CU Castellanos, DL Villa, C Rosa, KI Pedersen, FD Calabrese, P-H Michaelsen, J Michel, in *IEEE 67th Vehicular Technology Conference, VTC*. Performance of uplink fractional power control in UTRAN LTE (Singapore, May 2008), pp. 2517–2521
20. C Suh, AT Koc, S Talwar, in *IEEE Global Telecommunications Conference, GLOBECOM*. Trade-off power control for cellular systems (Honolulu, Hawaii, USA, November 2009), pp. 1–6
21. S Xu, M Hou, K Niu, Z-Q He, W-L Wu, Coverage and capacity optimization in LTE network based on non-cooperative games. J. China Univ Posts Telecommunications. **19**(4), 14–42 (2012)
22. A Awada, B Wegmann, I Viering, A Klein, Optimizing the radio network parameters of the Long Term Evolution system using Taguchi's method. IEEE Trans. Vehicular Technol. **60**(8), 3825–3839 (2011)
23. K Majewski, M Koonert, Analytic uplink cell load approximation for planning Fractional Power Control in LTE networks. Telecommun. Syst. **52**(2), 1081–1090 (2013)
24. CU Castellanos, FD Calabrese, KI Pedersen, C Rosa, in *IEEE 68th Vehicular Technology Conference, VTC*. Uplink interference control in UTRAN LTE

based on the Overload Indicator (Calgary, Alberta, Canada, September 2008), pp. 1–5
25. 3GPP R1-074349, Overload Indicator handling for LTE, TSG RAN WG1 #50bis Meeting. V8.6, ETSI, Sophia Antipolis Cedex, France, September 2007
26. M Dirani, Z Altman, Self-Organizing Networks in next generation radio access networks: Application to Fractional Power Control. Comput. Netw. **55**(2), 431–438 (2011)
27. LTE; Evolved Universal Terrestrial Radio Access (E-UTRA); Long Term Evolution (LTE) Physical Layer; General Description. V8.3, ETSI, Sophia Antipolis Cedex, France, April 2009
28. S Sesia, I Toufik, M Baker, *LTE, the UMTS Long Term Evolution: from Theory to Practice*. (Wiley, New York, NY, USA, 2009)
29. 3GPP TR 36.942, LTE; Evolved Universal Terrestrial Radio Access (E-UTRA); Radio Frequency (RF) System Scenarios. V8.2, ETSI, Sophia Antipolis Cedex, France, July 2009
30. 3GPP TS 32.521, Technical Specification Group Services and System Aspects; Telecommunications Management; Self-Organizing Networks (SON) Policy Network Resource Model (NRM) Integration Reference Point (IRP); Requirements. V9, ETSI, Sophia Antipolis Cedex, France, March 2010
31. S Louvros, K Aggelis, A Baltagiannis, in *Proceedings of the 11th International Conference on Telecommunications (ConTEL)*. LTE cell coverage planning algorithm optimising uplink user cell throughput (Graz, Austria, June 2011), pp. 51–58
32. 3GPP TR 36.814, Technical Specification Group Radio Access Network; Evolved Universal Terrestrial Radio Access (E-UTRA); Further Advancements for E-UTRA Physical Layer Aspects. V9, ETSI, Sophia Antipolis Cedex, France, March 2010

11

The indirect self-tuning method for constructing radio environment map using omnidirectional or directional transmitter antenna

Marko Pesko[1,4*†], Tomaž Javornik[2,4†], Luka Vidmar[1], Andrej Košir[3], Mitja Štular[1] and Mihael Mohorčič[2,4†]

Abstract

In this paper, we propose a new indirect method for constructing the radio frequency layer of radio environment map (RF-REM) called self-tuning method (STM). The proposed STM takes into account the characteristics of the operating environment and performs estimation of the transmitter parameters, i.e. its location, antenna diagram, antenna azimuth, transmit power, as well the parameters of the propagation model to obtain the best match between the available measurements and the predicted signal levels. We compare STM to several most often considered existing methods using different numbers of randomly distributed measurements, which could in practice be obtained in a participatory-like manner. The performance evaluation of the methods is performed in terms of averaged root mean square error ($\overline{\text{RMSE}}$) and 95% confidence interval (CI) calculated between the constructed RF-REMs and the reference RF-REM, as well as in terms of false alarm zone ratio (FAZR) and correct detection zone ratio (CDZR_1). The analysis shows the robustness of the STM to various spatial distributions of measurements and its fast convergence and low residual root mean square error compared to the inverse distance weighted (IDW) method, the inverse distance square weighted method (IDW2), the Kriging method, and the location estimation-based method (LIvE).

Keywords: Cognitive radio; REM; RF-REM; Construction methods; Mobile network

1 Introduction

The nature of radio signal propagation imposes on wireless communication systems to share the radio frequency (RF) spectrum in time and space. Traditionally, the publicly accessible spectrum is divided into the licensed and the unlicensed frequency bands. This arrangement is becoming inefficient because more and more radio resources are needed in geographically limited areas to meet the increasing capacity demand, calling for improved spectrum utilization. In fact, several measurement campaigns revealed that many parts of the licensed frequency bands are often underutilized, while the unlicensed bands are saturated in time and space [1]. This observation led to the emergence of dynamic spectrum access (DSA) approach, which enables (through the concept of cognitive radio (CR)) more efficient spectrum usage [2], by exploiting the flexibility of software-defined radio (SDR). The basic precondition for the device that makes use of the DSA approach is spectrum awareness, which can be achieved either by the built-in spectrum sensing capability or by looking-up in a purposely deployed and maintained geolocation database of available radio resources [1].

Research and development of the spectrum sensing techniques have revealed that the low-cost spectrum sensing devices cannot yet fulfil the sensitivity criteria for a reliable detection of radio signals at signal levels required by Federal Communications Commission (FCC) [3] and Electronic Communications Committee (ECC) [4]. One proposed solution to overcome the need for high precision spectrum sensing on each device is a radio environment map (REM) comprising multi-domain information

*Correspondence: marko.pesko@telekom.si
†Equal contributors
1 Telekom Slovenije, d.d., Cigaletova 15, 1000 Ljubljana, Slovenia
4 Jozef Stefan International Postgraduate School, Jamova cesta 39, 1000 Ljubljana, Slovenia
Full list of author information is available at the end of the article

from geolocation databases [1] to allow sharing of unused spectrum primarily allocated to the television broadcast services [5]. It is also considered to support hierarchical spectrum access on licensed bands, spectrum sharing on unlicensed bands, intra-operator radio resource management, and dedicated spectrum monitoring [6]. To support such scenarios, REM must provide comprehensive up to date representation of the radio environment with its selectable and adaptable content details. This is used to depict past, current, as well as expected conditions in the area of interest and can be classified into three main categories or layers related to (i) radio elements, (ii) radio scene, and (iii) radio environment [7]. Furthermore, REM content can be divided into long-term information and short-term information. The first one is slowly changing, it may be also quasi-static or static. It is related to TV broadcast and receiver (Rx) stations, mobile operator base stations (BSs), and other similar transmitters (Tx). On the contrary, the period of updating short-term information, such as information about wireless microphones on TV frequencies and other mobile Txs, can be much shorter. The most important REM content is the interference information [7] which can be estimated from maps representing the coverage of an area with the radio signal, named RF-REM layers or RF-REMs. These are sub-layers of the radio environment layer and in the most basic form present the coverage of an area with the radio signal at a specific frequency. In the literature, RF-REMs are also referred to as RF environment maps (RFEMs), radio electric exposure level maps (REELMs), power spectral density maps (PSDs), and most often simplified as REMs, although the latter term is broadly used for the entire REM concept. Today, the signal measurements required for RF-REM construction are obtained by dedicated static sensors or by periodic measurement campaigns, which results in a static or quasi-static RF-REM. However, future radio environment when DSA is supported will be dynamic with heterogeneous wireless systems competing for the available shared spectrum. In such radio environment, the existing approaches for spectrum measurement will have to be complemented by distributed spectrum sensing. To this end, a particularly well-suited approach to the operators of wireless systems is spectrum sensing by advanced user terminals with spectrum sensing capability operating in participatory manner or dense but irregular deployments of low-cost less sensitive spectrum sensors. With such an approach, signal level measurements will be provided by heterogeneous measurement capable devices. The existing RF-REM construction methods are not designed for signal measurements obtained in participatory manner, where data is provided dynamically in irregular time intervals from random locations. They do not explicitly take into account the operating environment characteristics, and they rely on large sets of signal measurements to achieve suitable accuracy.

The main contribution of this paper is a new indirect RF-REM construction method that is well-suited also to collection of measurements points using a participatory sensing approach. Its characteristics can be summarized as follows: (i) the method takes into account the characteristics of the Tx, i.e. transmit power, antenna diagram, its azimuth and tilt, (ii) the signal measurements are applied for the propagation model calibration, and (iii) the method relies on the semi-empirical propagation model in which the empirical propagation model is complemented with the available information about the radio propagation environment including the digital elevation model and clutter information of the geographical area of interest. We perform a systematic investigation of the impact of the number and distribution of measurements in realistic directional Tx scenario and evaluate the proposed RF-REM construction method in comparison to several existing RF-REM construction methods by using averaged root mean square error ($\overline{\text{RMSE}}$) as the performance metric. Applying the $\overline{\text{RMSE}}$ metric shows that the proposed method outperforms most often cited methods for different spatial distributions of signal measurements. Furthermore, we show that the existing typically used performance metrics can yield wrong interpretation of the results when dynamic scenario is considered, since they do not take into account the different spatial distributions of measurements available for RF-REM construction in a given time moment. Nevertheless, we evaluated the considered RF-REM construction method also in terms of correct detection zone ratio (CDZR_1) and false alarm zone ratio (FAZR) for the representative number of measurements.

The rest of this paper is organized as follows. We introduce the RF-REM construction in Section 2 and describe the proposed STM in Section 3. The implementation setup and simulation scenario definition are provided in Section 4, whereas performance evaluation of the considered construction methods is presented and discussed in Section 5. Finally, Section 6 concludes the paper.

2 RF-REM construction

The construction of REM can be in a most basic form presented as a construction of RF-REM and deals with estimating the signal levels at distributed geographical locations where signal measurements are not available. The RF-REM construction methods can be classified into (i) direct methods applying interpolation, (ii) indirect methods utilizing Tx parameters and propagation modelling [8], and (iii) hybrid methods which combine both approaches.

The direct construction methods are based on local neighbourhood, geostatistical, and variational interpolation [9]. The most widely RF-REM construction applied interpolation methods for RF-REM construction found in the literature are the inverse distance weighted (IDW) method [10], the nearest neighbours (NN) method [11], the splines methods [12], modification and derivation of mentioned methods such as the modified Shepard's method (MSM) [13], the gradient plus inverse distance squared (GIDS) method [7], and the Kriging method [14]. Among these, the Kriging method is most often applied for the direct RF-REM construction, since it is the best geostatistical linear unbiased estimator that yields a zero mean residual error and minimizes the error variance.

In the indirect RF-REM construction methods, a Tx location and/or its parameters are estimated first, and then, the RF-REM is calculated using radio propagation modelling. These methods are often computationally more complex and not as commonly used as the direct methods. Nevertheless, they can be used to describe the RF coverage more realistically, especially in the long-term information REM where RF-REM construction time is not of such great importance as in the short-term information REM. The indirect methods distinguish three scenarios where (i) location and other parameters of the Tx are known; (ii) either the Tx location or its parameters are known; (iii) both, Tx location and its parameters, are unknown. In (i), the RF-REM construction is straightforward and it is reduced to the calculation of the RF signal coverage with a propagation model using known Tx parameters. This approach is also used for the planning of wireless networks. In (ii), an efficient estimation of the Tx location is performed if other parameters of Tx such as the transmit power, the antenna characteristics etc. are known. After estimating the Tx location, the approach outlined for the first case can be applied. On the other hand, if only the location is known, the Tx power, antenna characteristics and other Tx parameters are estimated by modelling the antenna diagram and suitable propagation model calibration. In (iii), localization of Tx, estimation of its parameters, and calibration of the propagation model, if considered, are performed before performing the RF-REM construction with propagation modelling. The recent research in indirect methods [1,15] follows this idea by proposing the Tx location estimation-based REM construction method (LIvE). By providing more accurate RF-REM than the direct Kriging method, LIvE indicates further research possibilities in the direction of indirect construction methods.

The hybrid RF-REM construction combines direct and indirect approaches. For example, in [11], various interpolation methods are used first, to interpolate measurements into an image, which is in the case of direct methods already the RF-REM. Then, identification of propagation and Tx features is performed by using image processing techniques. Finally, propagation modelling is used to construct final RF-REM. A different method is proposed in [16], where at first, a preliminary RF-REM is constructed by a simple numerical propagation modelling, which is in the next step corrected according to the available measurements with the direct construction approach.

The results from [1] suggest that the LIvE method, which relies on an empirical propagation model obtained from the statistical analysis of a large number of measurements and highly dependent on the adaptation to a specific geographical area of interest, outperforms most often considered direct methods IDW, IDW2, and Kriging. However, none of these methods is well suited for consideration of real operating scenarios as they do not explicitly take into account *a priori* knowledge available about the characteristics of the operating environment, nor they consider the Tx directivity. Namely, in real operating scenarios with directional Txs or with omnidirectional Txs whose radiation is heavily obstructed or completely blocked in specific azimuthal direction, appropriate modelling of Tx directivity is very important.

Both concerns can be taken into account by using deterministic and semi-empirical propagation models [17]. If properly selected with respect to the operating environment, they are usually more accurate than the empirical models and thus more often considered in practice. The deterministic models require an enormous number of operating environment information and huge computational effort, while semi-empirical models obtained as their empirical modifications have a moderate complexity.

Our hypothesis is that an indirect RF-REM construction method based on a semi-empirical propagation model, taking into account *a priori* knowledge about characteristics of the operating environment and modelling the antenna influence of the Tx, can enhance the accuracy of current RF-REM construction assuming (i) slow changing long-term information REM scenario, and (ii) that the knowledge about the operating environment is available for RF-REM construction. In the next section, we propose a new indirect self-tuning method (STM) which is following such an approach. It should be noted that STM itself is not a specific algorithm for RF-REM construction, since each step of the STM can be implemented and solved in various ways.

3 STM

The STM consists of three main steps: (i) the data collection; (ii) the Tx parameters' estimation and propagation model calibration; and (iii) the RF-REM construction.

Step 1

In the first step, signal measurements for the RF-REM construction as well as supplementary data, such as a digital model of the terrain and the corresponding clutter loss data of the geographical area of interest, are collected.

Step 2

The second step concerns approximation of the Tx location, calibration of the propagation model, and an estimation of Tx parameters. This can be in practice performed with various algorithms on different hardware platforms, so in the following, we only provide one possible implementation.

Let the pair (x, y) reflects location coordinates. Let us further assume that in the area of interest for constructing RF-REM, there are N measurements at points (pixels) which can be denoted as $R = \{R_i = (x_i, y_i), i\}$. Similarly, Tx location or an area of possible Tx locations can be represented as $T = \{T_j = (x_j, y_j), j\}$. T can be pre-known or calculated using a suitable localization algorithm (e.g. based on individual or combined localization technique such as angle of arrival (AA), timing advance (TA), time of arrival (TOA), time difference of arrival (TDOA) etc.). The output of such algorithm is usually an estimated location or an area of possible Tx locations, where the Tx can be found with a given probability. The received power P_{rx_i} [dBm] at i-th Rx location can be estimated as in (1), where P_{tx} [dBm] is the transmit power and L_{rx_i} [dB] is the corresponding path loss for the operating environment and frequency band.

$$P_{\text{rx}_i} = P_{\text{tx}} - L_{\text{rx}_i} = P_{\text{tx}} - \left(L_i + L_{\text{diff}_i} + L_{\text{clut}_i} - G_{\text{ant}_i}\right). \quad (1)$$

In this paper, the sub-GHz mobile communication systems band is considered as a reference, so we take the semi-empirical modification of the Okumura-Hata model implemented as in [18]. Thus, L_{rx_i} is a sum of the path loss of the empirical channel model L_i, diffraction loss L_{diff_i}, loss due to terrain clutter L_{clut_i}, and loss due to the directional gain G_{ant_i} of the Tx antenna. L_i follows modified Okumura-Hata model for open areas (2), where d [m] is the distance between T_j and R_i, H_{eff_i} [m] is the Tx effective antenna height [18] dependent on the Tx and Rx antenna height, f [MHz] is the carrier frequency, A_0 is the offset, and A_1, A_2, and A_3 are slope coefficients of the path loss.

$$\begin{aligned} L_i = {} & A_0 + A_1 \log_{10}(d) + A_2 \log_{10}\left(H_{\text{eff}_i}\right) \\ & + A_3 \log_{10}(d) \log_{10}\left(H_{\text{eff}_i}\right) \\ & - 3.2\left[\log_{10}(11.75 H_m)\right]^2 \\ & + 44.49 \log_{10}(f) - 4.78\left[\log_{10}(f)\right]^2. \end{aligned} \quad (2)$$

L_{diff_i} is the loss calculated as a knife edge effect on the main obstacle obstructing the radius of the first Fresnel zone, based on the elevations of the terrain. L_{clut_i} is queried directly from the REM storage or an available external database. G_{ant_i} represents generalized antenna pattern, where θ_i is the horizontal and ϕ_i is the vertical angle between Tx at T_j and i-th Rx at R_i. It assumes symmetrical or at least approximately symmetrical horizontal and vertical antenna pattern of the Tx to the bore sight direction θ_0 and ϕ_0. In Equation 3, θ_0 and ϕ_0 are the antenna boresight directions, $G_{\max}$ is the maximum antenna gain, and FBR_θ as well as FBR_ϕ are front to back ratios, whereas the antenna form is approximated with the cosine functions to the power of m and n.

$$\begin{aligned} G_{\text{ant}_i} = {} & G_{\max} - FBR_\theta + FBR_\theta \left|\cos^m\left(\frac{\theta_0 - \theta_i}{2}\right)\right| \\ & - FBR_\phi + FBR_\phi \left|\cos^n\left(\frac{\phi_0 - \phi_i}{2}\right)\right|. \end{aligned} \quad (3)$$

Now we define a vector β (4) consisting of parameters of the empirical propagation model (2) and parameters from Equation 3.

$$\vec{\beta} = \left[A_0, A_1, A_2, A_3, G_{\max}, FBR_\theta, \theta_0, m, FBR_\phi, \phi_0, n\right]^T. \quad (4)$$

As given in (5), the STM finds such j (i.e. T_j from T) and $\vec{\beta}$ that minimize the mean squared error between the measured and estimated signal strengths. In (5), N is the number of Rx locations, P_i denotes measured signal level and L_{rx_i} the semi-empirically estimated path loss at the location of the i-th Rx, considering the Tx with P_{tx} at T_j.

$$F(\vec{\beta}, j) = \frac{1}{N} \sum_{i=1}^{N} \left(P_i - \left(P_{\text{tx}} - L_{\text{rx}_i}\left(\vec{\beta}\left(T_j\right)\right)\right)\right)^2. \quad (5)$$

The non-linear optimization problem with known typical lower and upper bounds can be solved in terms of $\left(\vec{\beta}^*, j^*\right) = \min_{\vec{\beta}, j} F\left(\vec{\beta}, j\right)$ by a suitable optimization method for solving non-linear optimization problems. However, if the directivity of the Tx antenna given in (3) is neglected, i.e. if the omnidirectional Tx antenna is considered, then G_{ant_i} is constant and the optimization objective (5) becomes linear and can be solved by a suitable optimization method for linear optimization problems.

Step 3

After solving the optimization problem, the estimated values for $\vec{\beta}^*$, T_j^*, and P_{tx} are applied in the calculation of a signal level on individual RF-REM location according to (6).

$$P_{\text{rx}_i} = \widehat{P}_{\text{tx}} - \widehat{L}_{\text{rx}_i} = \widehat{P}_{\text{tx}} - \left(\widehat{L}_i + L_{\text{diff}_i} + L_{\text{clut}_i} - \widehat{G}_{\text{ant}_i}\right). \quad (6)$$

4 Implementation setup and scenario definition

In this section, we describe the reference implementation setup and operating scenario of the proposed STM as considered in subsequent performance evaluation.

4.1 Reference RF-REM

For the performance evaluation of different RF-REM construction methods, we need a reference RF-REM, i.e. a reference signal coverage distribution. To this end, we consider a representative real rectangular geographical area within Slovenia depicted in Figure 1. The size of the area is 5.15 × 6.75 km and is represented with digital elevation model with the resolution of 25 m. It consists of nine different clutter categories. In the area, we consider a single active Tx for a single channel, similar as in [1]. It is positioned at the location of a BS operating in the 900-MHz frequency band. We take into account the actual BS settings provided by the mobile operator: the transmit power of 39 dBm, Kathrein K739650 antenna pattern with 35° of azimuth, and mechanical down-tilt of 2°. The initial approximation of signal levels is depicted in Figure 2, where the reference RF-REM has been calculated down to the threshold of −120 dBm with the wireless network planning tool GRASS RaPlaT [19], which was calibrated to the drive-test measurements of the mobile operator taking into account the clutter as well as the shadowing based on the terrain elevation data. Thus, it can be assumed that the calculated signal level at each 25 × 25 m RF-REM point is very close to real situation.

4.2 Scenarios and measurements spatial distributions

In order to model distributed measurement sets of various sizes and spatial distributions, as if obtained in a participatory manner, the reference RF-REM was sampled at random locations. Since GRASS RaPlaT cannot introduce random errors at simulated RF-REM points, this has been subsequently added to the sampled signal levels at RF-REM points as random values obeying a log-normal distribution with mean 0 dB and standard deviation of 8 dB. The probability density function, its standard deviation and mean value, has been obtained by analysing measurements collected on the field by the mobile operator. We found this sufficiently approximates the real distribution of measurements at individual RF-REM points. Finally, such participatory measurement sets were used for the evaluation of the considered RF-REM construction methods.

When measurements for RF-REM construction are obtained in a participatory manner, the number and spatial distribution of available measurements in a set may vary. Consequently, for fair comparison of the performance of different RF-REM construction methods, the methods should be tested for different scenarios which can occur in practice.

Since measurements can be randomly distributed in the entire area or only in its sub-regions, we consider three simulation scenarios for modelling different situations and thus investigate the impact of spatial distribution of measurements with respect to Tx antenna main lobe on the accuracy of the RF-REM construction methods. In order to test the impact of the number and distribution of participatory measurements on the RF-REM construction methods, we sampled the reference RF-REM and generated 100 different measurement sets of the specific size by increasing the number of measurement points (and hence their density) from 10 to 1,000. In the 'reference' scenario I, measurements are randomly distributed in the entire area of interest, as depicted in Figure 3a. In the scenario II, the majority (90%) of the measurements are inside the main lobe of the antenna, as is illustrated in Figure 3b. The purpose of this scenario is to test the need for anchor measurement points in the RF-REM construction of a

Figure 1 The geographical area with nine different clutter categories.

Figure 2 The initial approximation of signal levels in the area.

directional Tx. In practice, however, the location of measurements depends on the operating environment, and thus, they can be found in certain clutter category more likely than in the others. In this respect, for the scenario III shown in Figure 3c, measurements are distributed according to clutter categories. The aim of this scenario is not to simulate real distribution of measurements over clutter but rather to investigate the impact of measurements' clustering. To this end, we define the distribution of measurements regarding the clutter categories and assume that measurements in water, i.e. rivers and lakes, and swamp areas are not available or their number can be neglected. For other clutter categories, we assume most measurements in dense urban areas (30%), followed by urban areas (25%), urban areas without buildings, i.e. mostly roads, (20%), sub-urban areas (10%), dry open land areas without special vegetation (7%), agricultural areas (5%), and forestall areas (3%).

4.3 Implementation of the methods

We implemented the IDW, IDW2, Kriging, LIvE, and STM methods in Matlab and tested them using the same measurement sets. For the STM analysis, we considered two cases, namely having a transmitter equipped with omnidirectional antenna or with directional antenna. Considering the omnidirectional antenna, we solved the linear optimization problem without optimization bounds in the well-defined least square sense by using the Matlab function *mldivide*(), similar as the LIvE method for the omnidirectional Tx case.

For the second STM case study, which includes the directivity of the Tx antenna, we considered only the horizontal part of the antenna pattern in the optimization process. This simplification significantly reduces the computational complexity and improves the convergence of optimization process; at the same time, it does not introduce significant deviation, assuming the measurements

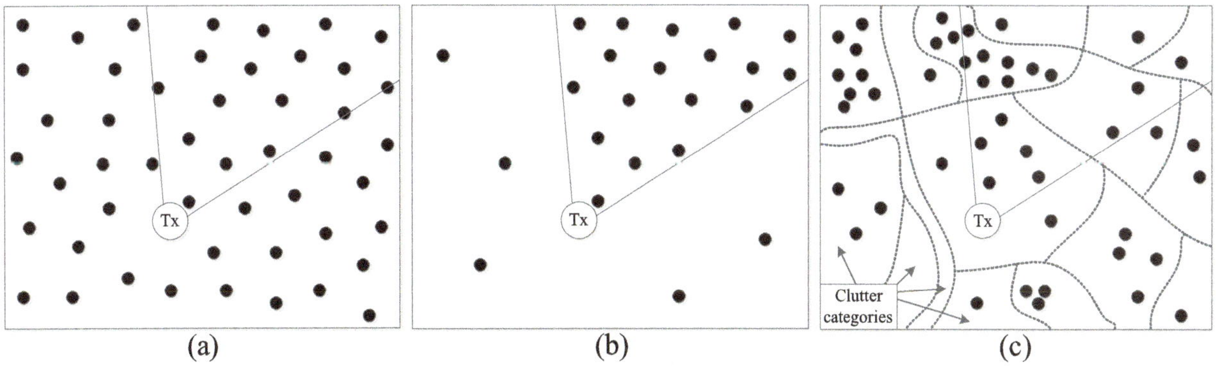

Figure 3 Distribution of measurements in considered scenarios. (a) Scenario I - randomly in space, **(b)** scenario II - 90% in the main lobe of the antenna, and **(c)** scenario III - clutter-dependent clustered distribution.

are within the main lobe of antenna vertical pattern and within the validity range of the propagation model, i.e. from 200 m up to 100 km from the Tx. The impact of introducing this simplification depends on the antenna tilt and antenna beamwidth. Figure 4 depicts vertical antenna gain vs. distance for antenna down-tilt of 2° and reveals that the impact of not considering vertical antenna pattern is smaller compared to standard deviation error between the predicted and measured signal levels, which is approximately 8 dB, for perfectly calibrated propagation models. To solve the simplified non-linear optimization problem in a non-linear least squares sense, we neglected FBR_{ϕ}, ϕ_0, and n in β (4) and used the Matlab function *lsqnonlin*() with non-equal upper and lower optimization bounds as given in Table 1. We set the parameters of *lsqnonlin*() in a way that the trust-region-reflective algorithm with lower and upper bounds was selected. This was due to the fact that, first, the nature of our objective function (a sum of weighted terms including logarithmic functions) indicated that the trust-region-reflective algorithm should be effective in finding the optimum, and second, a good initial approximation together with lower and upper bounds were available from the domain knowledge of the optimization problem. Besides, the optimization procedure turned out to provide stable results on our data in a reasonable running time. In particular, the optimization bounds for parameters A_0 to A_3 were taken from [20], and θ_0 needs to cover all possible azimuthal directions. Bounds of $G_{\max}$, FBR_{θ}, and m were extracted from antenna diagrams used by the mobile operators to cover the most typical antennas with the horizontal patterns approximately symmetrical to their bore sight direction. The starting input parameters for the function *lsqnonlin*() were set according to typical values in mobile networks.

Other considered parameters were selected or obtained as follows. The clutter loss information was read from the clutter database of a mobile operator, the diffraction loss was calculated from the digital elevation model as a knife-edge effect, and the Tx height was set to the average BS height of 20m, while the typical Rx height was considered to be 1.5 m above the ground level. Since various localization algorithms can be used to estimate the area of possible Tx locations T, we did not want to particularly emphasize any of them. As an example of estimating T in this paper, we first performed simulations of the TA localization with a typical TA step of 553.5 m by using considered measurement sets. As can be seen from Figure 5, when more than ten randomly distributed measurements are available, the area of possible Tx locations T on average comprises less than approximately 20 RF-REM points.

4.4 Performance metric

The accuracy of the constructed RF-REM reflects the performance of its construction method. This is in the literature evaluated by comparing the estimated signal levels in dBm of the constructed and reference RF-REM

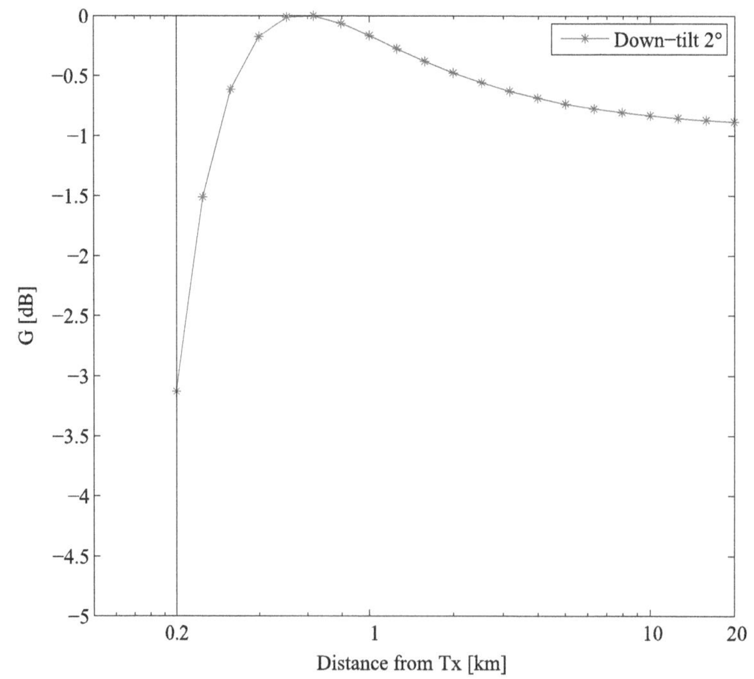

Figure 4 The impact of vertical part of antenna pattern.

Table 1 Lower and upper optimization bounds for the STM considering the directional Tx

	A_0	A_1	A_2	A_3	G_{max}[dB*i*]	FBR_θ [dB]	θ_0[°]	*m*
Min	20	40	−20	−10	2.15	0	0	0
Max	80	50	0	0	24.4	77.58	360	100

using the performance metrics such as mean squared error (MSE) [21], mean absolute error (MAE) [9], relative mean absolute error (RMAE) [22], root mean square error (RMSE), correct detection zone ratio ($CDZR_1$), and false alarm zone ratio (FAZR) [1,8,15]. These performance metrics largely depend on the number and distribution of measurements available for the construction, so they are not best suited for comparing RF-REM construction methods. For example, one RF-REM construction method can be more accurate than the other at a specific number of measurements, and if only the distribution of measurements is changed, the results can easily be just the opposite. Therefore, we start with comparison of RF-REM construction methods in terms of averaged RMSE for different measurement sets of the same size denoted as $\overline{\text{RMSE}}$ and 95% confidence interval (CI). Next, we estimate the $CDZR_1$ and FAZR curves, in order to be consistent with the works [1,15].

As described in Section 4.2, 100 different distributed measurement sets for a specific size ranging from 10 to 1,000 measurements were generated. For the comparison of the methods, we first calculate RF-REM for each set and then its RMSE. Finally, the $\overline{\text{RMSE}}$ and the 95% CI are calculated from RMSE values obtained from the measurement sets of the specific size, for all considered sizes. $CDZR_1$ and FAZR metrics are calculated only for the representative case when $N = 200$ measurements are available.

5 Performance evaluation and discussion

In this section, we provide performance evaluation of the proposed STM and compare it to other representative RF-REM construction methods pointed out in Section 2. Simulation results on the performance of the considered methods are summarized in Figure 6 for the three selected scenarios. Results for the scenario I with randomly distributed measurements in the entire area are depicted in Figure 6a,b. In terms of $\overline{\text{RMSE}}$ and the 95% CI, the results in Figure 6a show that IDW and LIvE methods have similar performance, while the IDW2 outperforms both. The performance of Kriging is comparable to the performance of IDW2, i.e. it is just slightly worse up to 1,000 measurements. STM without considering the directional antenna pattern (i.e. assuming omnidirectional transmission) in general outperforms other methods up to approximately 400 measurements, but it is outperformed by IDW2 and

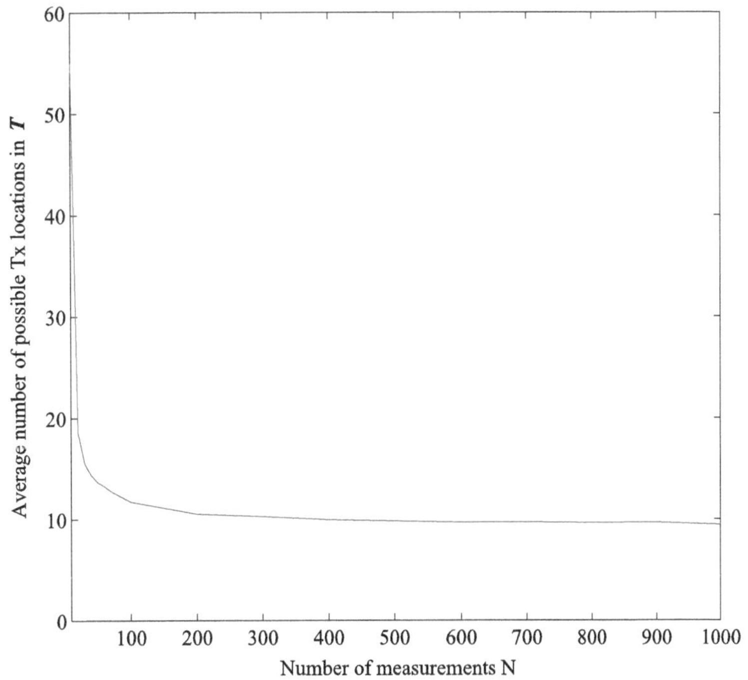

Figure 5 Average number of possible Tx locations in the area *T*. For 100 sets with increasing number of randomly distributed measurements *N*.

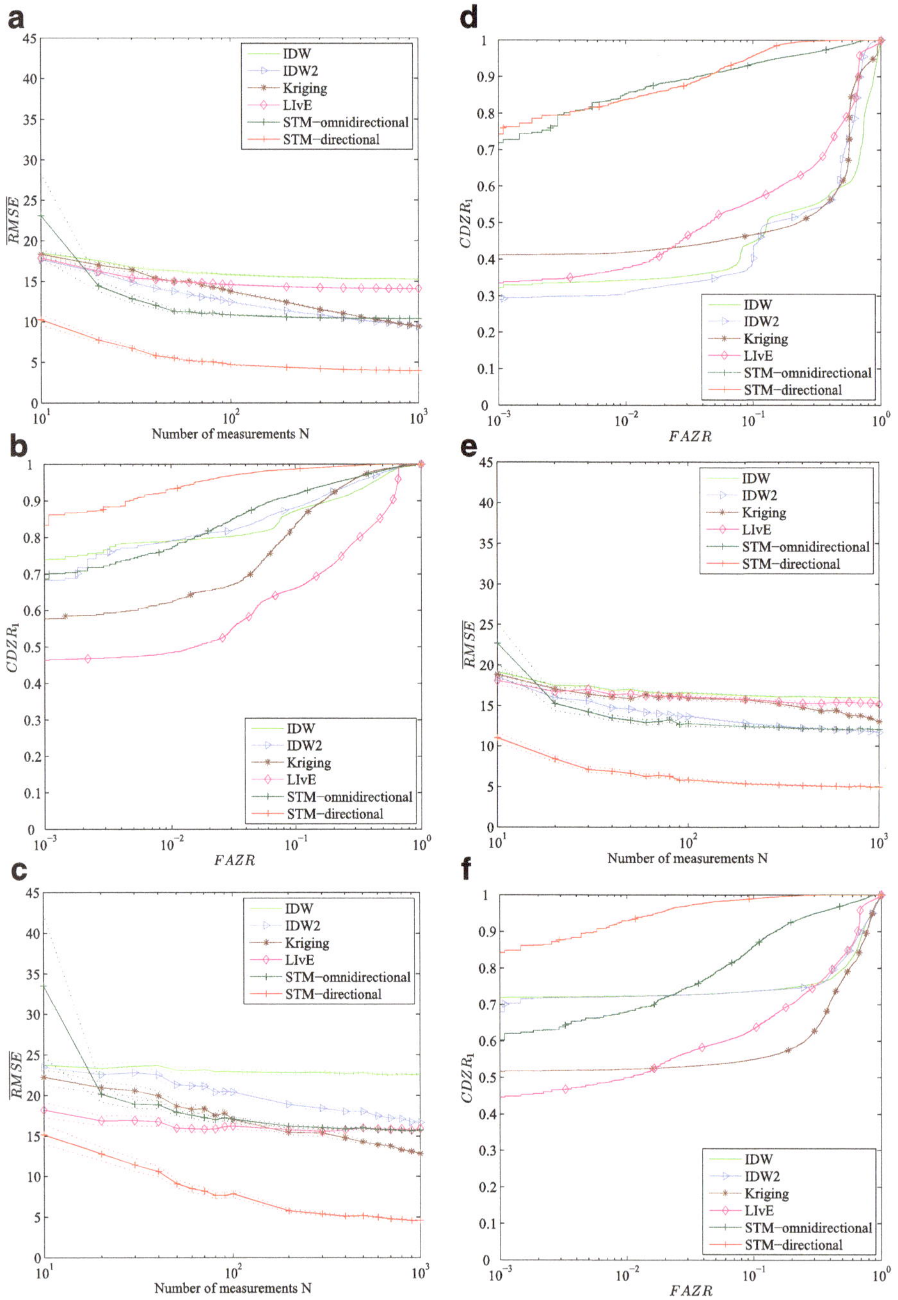

Figure 6 Results of the RF-REM construction for the sets with increasing number of randomly distributed measurements. In terms of $\overline{\text{RMSE}}$ and 95% CI for **(a)** scenario I, **(c)** scenario II, **(e)** scenario III. In terms of CDZR_1 vs. FAZR curves for **(b)** scenario I, **(d)** scenario II, **(f)** scenario III.

Kriging, when more measurement points are available. If estimating also the directional antenna pattern, STM notably outperforms all other methods for any number of measurements. The methods which do not consider antenna pattern (STM with omnidirectional antenna and LiVE) tend to reach the $\overline{\text{RMSE}}$ saturation at low number of measurements N. Similar effect is observed for the IDW direct method, which misestimates the signal decrease rate with the distance. On the other hand, STM considering antenna pattern, IDW2 and Kriging exhibit improved performance with increasing the number of measurements.

Similar conclusions can be drawn from Figure 6b, where the performance of methods is compared in terms of receiver operating characteristics, i.e. using CDZR_1 and FAZR curves. If we consider $\text{CDZR}_1 = 0.9$, meaning 90% of points in RF-REM are correctly detected as white space similar as in [1], the lowest obtained false alarm ratio is achieved by the STM considering directional antenna at FAZR = 0.0047. Notably higher FAZR values, namely 0.0654, 0.1394, 0.1641, 0.1848, and 0.5951, are achieved by STM considering omnidirectional antenna, IDW2, Kriging, IDW, and LIvE, respectively. Poor performance of direct methods is due to low number of measurements considered, i.e. $N = 200$, while bad performance of the LIvE method can attributed to the fact that it is not considering antenna directivity.

Figure 6c,d depicts the performance of investigated methods for the scenario II where the majority of the investigated measurements, i.e. 90%, are within the main lobe of the antenna, while only some of the measurements are outside of it. In this scenario, the worst performance in terms of $\overline{\text{RMSE}}$ and the 95% CI is obtained for IDW, followed by IDW2. LIvE is better or similar to STM considering omnidirectional antenna and up to approximately 100 measurements better than IDW, IDW2, and Kriging. When more measurements are available, Kriging method outperforms others, except STM with directional antenna, which regardless of the number of measurements shows the best overall performance. Similarly, as in the scenario I, the direct methods which misestimate the signal decrease rate with the distance and indirect method LIvE, which does not consider the antenna directivity, exhibit $\overline{\text{RMSE}}$ saturation with the increasing number of measurements.

The results for receiver operating characteristics in scenario II in terms of CDZR_1 and FAZR curves are given in Figure 6d. These results indicate that the STM notably outperform the other studied RF-REM construction methods at $\text{CDZR}_1 = 0.9$, which provide significantly worse but similar FAZR, with IDW performing the worst. Comparison of the results to those obtained for the scenario I shows that all methods, except the STM with omnidirectional antenna, exhibit significant performance degradation. This is a consequence of measurements clustering in the main lobe of Tx antenna only. Additionally, the results indicate that both versions of the STM have comparable performance. This is due to less accurately estimated Tx antenna in the case of the STM with directional antenna and more accurately estimated location and parameters of the Tx in the case of the STM with omnidirectional antenna.

The results in Figure 6e,f show performance comparison of the investigated methods for the scenario III where measurements are being distributed in clusters according to clutter categories in the area of interest. In terms of $\overline{\text{RMSE}}$ and the 95% CI, the IDW, LIvE, and Kriging methods are similar up to 200 measurements, when the performance of Kriging starts to improve and becomes comparable to the performance of IDW2 and STM with omnidirectional antenna at approximately 1,000 measurements. STM considering directional antenna demonstrates the smallest $\overline{\text{RMSE}}$ regardless of the number of available measurements. The CDZR_1 vs. FAZR results in Figure 6f show that also in this scenario, STM considering directional antenna guarantees the best performance in finding the white space locations. At $\text{CDZR}_1 = 0.9$, it achieves FAZR = 0.0058. It is followed by STM considering omnidirectional antenna, LIvE, IDW2, IDW, and Kriging with FAZR values equal to 0.1478, 0.6521, 0.6932, 0.7186, and 0.7720, respectively.

In the performance analysis, we showed that the RF-REM construction is highly dependent on the spatial distribution and the number of available measurements and that ignoring the Tx antenna directivity affects all methods. Consequently, the STM with estimating the antenna directivity clearly outperforms all other methods in all considered scenarios. This reconfirms the conclusions based on our preliminary results [23] that RF-REM construction can be notably enhanced by using an indirect method, in particular with semi-empirical propagation model taking into account the characteristics of Tx and operating environment. The trade-off for the performance improvement is of course the increased computational complexity compared to other methods, which can be in practice efficiently addressed by parallel processing on platforms such as multi-core central processing units (CPUs), digital signal processors (DSPs), field-programmable gate arrays (FPGAs), graphics processing units (GPUs), and massively parallel processor arrays (MPPAs) [24]. Generally speaking, the indirect methods are more suitable for low number of randomly distributed measurements, while the direct methods outperform the indirect ones at higher number of measurements. However, various spatial distributions and RF-REM resolutions may break this simple general rule. For example, clustering the majority of measurements in the main lobe of the Tx antenna, as assumed in the scenario II, notably decreases the performance of all methods compared to the scenario I with measurements distributed across the entire geographical area. This can be explained with the fact that the main lobe of the considered Tx antenna for obtaining reference RF-REM only covered approximately one quarter of the area of interest.

As shown on the case of the scenario III, clustering of measurements across the entire area of interest according to clutter categories yields very similar results to entirely random distribution of measurements, suggesting that realistic clustering as expected from mobile terminals could be potentially used for RF-REM construction of mobile networks BSs. Still, further research into this direction should be carried out by taking into account all BSs and coverage areas of the considered mobile operator and more realistic modelling of Rxs distribution over clutter categories. The usage of confidence intervals and in particular their width obtained for sets with low number of measurements indicate that the performance of the considered methods is very sensitive to the number and distribution of measurements. Moreover, evaluating the constructed RF-REM in terms of other performance metrics outlined in Section 4.4, can yield wrong interpretation of the results, as they do not take into account the different distributions of measurements which can occur in measurement sets of the specific size.

Finally, for practical implementation, it is necessary to analyse also the computational complexity of compared RF-REM construction methods. If there are N measurements available in a given time moment and M RF-REM locations for estimating signal levels, the computational time complexity of the IDW and IDW2 methods is $O(MN)$ [7]. If there are K out of N measurements within the local neighbourhood window for the ordinary Kriging interpolation, the computational time complexity of the Kriging method is $O\left(MK^2\right)$ [7]. The computational time complexity of the LIvE method is $O\left(MN^3\right)$, or $O\left(MN^{2.376}\right)$ by using Coppersmith-Winograd algorithm [8]. If the STM considering the omnidirectional antenna solves P linear optimization problems, which dominate the complexity with involving the matrix inversion, its computational-time complexity is P times of the LIvE method. P is the number of potential Tx locations in T and is dependent of the accuracy of the selected localization algorithm. On the other hand, the operation of the STM considering the Tx antenna directivity mainly depends on the selection of the non-linear optimization algorithm and its implementation. These issues were not the main focus of this paper and remain to be investigated as part of our further research, but we can note here that the price for the best performance of the STM is the highest computational-time complexity.

6 Conclusion

In this paper, we first presented the overall concept of REM and outlined the importance of its RF-REM layer, which is a strong tool for supporting DSA, radio resource management, and radio network planning. We focused on the RF-REM construction and proposed a new indirect method called STM. Its performance was compared to the IDW, IDW2, Kriging, and LIvE methods based on the simulated distributed measurements in a directional Tx scenario, which could in practice be obtained in a participatory manner. From the information provided by the mobile operator, we simulated the reference RF-REM of a single Tx to which the constructed RF-REMs were compared to obtain their performance in terms of $\overline{\text{RMSE}}$ and 95% CI, as well as with CDZR_1 vs. FAZR receiver operating characteristic curves, for the representative case of $N = 200$ measurements. Simulation results show that the performance of RF-REM construction methods clearly varies with the number of measurements and with measurement distribution across scenarios, which can occur in practice with respect to directional Tx. It has been shown that clustering of measurements in the main lobe of the Tx antenna notably decreases the performance of all construction methods, especially of the direct methods if the application goal is finding the white spaces in the entire geographical area. Moreover, although direct RF-REM construction is more often considered than the indirect construction, it has been shown that a more accurate RF-REM can be constructed already from small number of randomly distributed measurements by using an indirect construction method with a semi-empirical propagation model taking into account the characteristics of the Tx and the operating environment. Modelling of the Tx antenna proved to be of particularly great importance, although it introduces increased computational complexity, but this can be in practice efficiently addressed by using parallel processing techniques.

Abbreviations

AA: Angle of arrival; BS: Base station; CDZR_1: Correct detection zone ratio; CI: Confidence interval; CPU: Central processing unit; CR: Cognitive radio; DSA: Dynamic spectrum access; DSP: Digital signal processor; ECC: Electronic Communications Committee; FAZR: False alarm zone ratio; FCC: Federal Communications Commission; FPGA: Field-programmable gate array; GIDS: Gradient plus inverse distance squared method; GPU: Graphics processing unit; IDW: Inverse distance weighted interpolation method; IDW2: Inverse distance square weighted interpolation method; LIvE: Transmitter location estimation-based radio environment map construction method; MAE: Mean absolute error; MPPA: Massively parallel processor array; MSE: Mean squared error; MSM: Modified Shepard's interpolation method; NN: Nearest neighbour interpolation method; PSD: Power spectral density map; REELM: Radio electric exposure level map; REM: Radio environmental map; RF: Radio frequency; RFEM: Radio frequency environment map; RF-REM: Radio frequency layer of radio environment map; RMAE: Relative mean absolute error; $\overline{\text{RMSE}}$: Averaged root mean square error; RMSE: Root mean square error; Rx: Receiver; SDR: Software-defined radio; STM: Self-tuning method; TA: Timing advance; TDOA: Time difference of arrival; TOA: Time of arrival; Tx: Transmitter.

Competing interests

The authors declare that they have no competing interests.

Authors' contributions

MP, TJ, and MM carried out most of the REM and RF-REM construction studies, developed, and tested the STM, performed simulations, interpreted the results, and drafted the manuscript. LV participated in the study of localization techniques, and AK assisted with the mathematical analysis and interpretation of the results. MŠ enabled and coordinated the access to all data of the mobile operator. All authors read and approved the final manuscript.

Acknowledgements

This work has been in part funded by the European Union from Social Fund and the FP7 projects ABSOLUTE (FP7-ICT-318632) and CREW (FP7-ICT-258301).

Author details

[1]Telekom Slovenije, d.d., Cigaletova 15, 1000 Ljubljana, Slovenia. [2]Department of Communication Systems, Jozef Stefan Institute, Jamova cesta 39, 1000 Ljubljana, Slovenia. [3]Faculty of Electrical Engineering, University of Ljubljana, Tržaška 25, 1000 Ljubljana, Slovenia. [4]Jozef Stefan International Postgraduate School, Jamova cesta 39, 1000 Ljubljana, Slovenia.

References

1. HB Yilmaz, T Tugcu, Location estimation-based radio environment map construction in fading channels. Wireless Commun. Mobile Comput. **15**(3), 561–570 (2013)
2. J Mitola, G Maguire, Cognitive radio: making software radios more personal. IEEE Personal Commun. **6**(4), 13–18 (1999)
3. S Haykin, *Cognitive Dynamic Systems Perception-action Cycle, Radar and Radio*. (Cambridge University Press, Cambridge, UK, 2012)
4. ECC, Technical and operational requirements for the possible operation of cognitive radio systems in the 'white spaces' of the frequency band 470-790 mhz. Ecc report 159. Electronic Communications Committee (2010). http://www.erodocdb.dk/Docs/doc98/official/pdf/ECCREP159.PDF
5. IEEE, IEEE standard for information technology - local and metropolitan area networks - specific requirements - part 22: cognitive wireless ran medium access control (MAC) and physical layer (PHY) specifications: policies and procedures for operation in the TV bands. IEEE Std 802.22-2011, 1–680 (2011). doi:10.1109/IEEESTD.2011.5951707
6. S Subramani, J Riihijarvi, B Sayrac, L Gavrilovska, M Sooriyabandara, T Farnham, P Mahonen, in *Future Network Mobile Summit (FutureNetw)*. Towards practical REM-based radio resource management (IEEE, Warsaw, Poland, 2011), pp. 1–8
7. European FP-7 project FARAMIR reports: flexible and spectrum aware radio access through measurements and modelling in cognitive radio systems. http://www.ict-faramir.eu/index.php?id=deliverables
8. HB Yilmaz, Cooperative spectrum sensing and radio environment map construction in cognitive radio networks. PhD thesis, Bogazici University (2012)
9. D Denkovski, V Atanasovski, L Gavrilovska, J Riihijarvi, P Mahonen, in *Cognitive Radio Oriented Wireless Networks and Communications (CROWNCOM), 2012 7th International ICST Conference On*. Reliability of a radio environment map: case of spatial interpolation techniques (IEEE, Stockholm, Sweden, 2012), pp. 248–253
10. D Shepard, in *Proceedings of the 1968 23rd ACM National Conference* ACM '68. A two-dimensional interpolation function for irregularly-spaced data (ACM, New York, NY, USA, 1968), pp. 517–524
11. L Bolea, J Perez-Romero, R Agusti, in *Wireless Personal Multimedia Communications (WPMC), 2011 14th International Symposium On*. Received signal interpolation for context discovery in cognitive radio (IEEE, Brest, France, 2011), pp. 1–5
12. MA Azpurua, KD Ramos, A comparison of spatial interpolation methods for estimation of average electromagnetic field magnitude. Progress Electromagnetics Res. M (PIER M). **14**, 135–145 (2010)
13. RJ Renka, Multivariate interpolation of large sets of scattered data. ACM Trans. Math. Softw. **14**(2), 139–148 (1988)
14. EH Isaaks, RM Srivastava, *An Introduction to Applied Geostatistics*. (Oxford University Press, New York, 1989)
15. HB Yilmaz, T Tugcu, FFz Alagoz, S Bayhan, Radio environment map as enabler for practical cognitive radio networks. IEEE Commun. Mag. **51**(12), 162–169 (2013)
16. Y Ould Isselmou, H Wackernagel, W Tabbara, J Wiart, in *Antennas and Propagation, 2006. EuCAP 2006. First European Conference On*. Geostatistical interpolation for mapping radio-electric exposure levels (IEEE, Nice, France, 2006), pp. 1–6. doi:10.1109/EUCAP.2006.4584752
17. BH Fleury, PE Leuthold, Radiowave propagation in mobile communications: an overview of european research. IEEE Commun. Mag. **34**(2), 70–81 (1996)
18. Ericsson, TEMS cellplanner common features technical reference manual, EN/LZT 108 9833 R1B (June, 2008)
19. A Hrovat, I Ozimek, A Vilhar, T Celcer, I Saje, T Javornik, Radio coverage calculations of terrestrial wireless networks using an open-source GRASS system. WTOC. **9**(10), 646–657 (2010)
20. E Aarnæs, S Holm, Tuning of empirical radio propagation models effect of location accuracy. Wirel. Pers. Commun. **30**(2-4), 267–281 (2004)
21. S Grimoud, B Sayrac, S Ben Jemaa, E Moulines, in *Vehicular Technology Conference (VTC Fall), 2011 IEEE*. Best sensor selection for an iterative rem construction (San Francisco, CA, USA, 2011), pp. 1–5. doi:10.1109/VETECF.2011.6093004
22. M Angjelicinoski, V Atanasovski, L Gavrilovska, in *Telecommunications Forum (TELFOR), 2011 19th*. Comparative analysis of spatial interpolation methods for creating radio environment maps (IEEE, Belgrade, Serbia, 2011), pp. 334–337. doi:10.1109/TELFOR.2011.6143557
23. M Pesko, T Javornik, M Štular, Mohorčič, in *4th Workshop of COST Action IC0902: Cognitive Radio and Networking for Cooperative Coexistence of Heterogeneous Wireless Networks*. The comparison of methods for constructing the radio frequency layer of radio environment map using participatory measurements (Rome, Italy. 9–11 Oct 2013, 2013)
24. DB Thomas, L Howes, W Luk, in *Proceedings of the ACM/SIGDA International Symposium on Field Programmable Gate Arrays*. A comparison of CPUs, GPUs, FPGAs, and massively parallel processor arrays for random number generation (ACM, New York, NY, USA, 2009), pp. 63–72

12

Stochastic geometry modeling and analysis of cognitive heterogeneous cellular networks

Fereidoun H Panahi* and Tomoaki Ohtsuki

Abstract

In this paper, we present a cognitive radio (CR)-based statistical framework for a two-tier heterogeneous cellular network (femto-macro network) to model the outage probability at any arbitrary secondary (femto) and primary (macro) user. A system model based on stochastic geometry (utilizing the spatial Poisson point process (PPP) theory) is applied to model the random locations and network topology of both secondary and primary users. A considerable performance improvement can be generally achieved by mitigating interference in result of applying the CR idea over the above model. Novel closed-form expressions are derived for the downlink outage probability of any typical femto and macro user considering the Rayleigh fading for the desired and interfering links. We also study some important design factors which their role and importance in the determination of outage and interference cannot be ignored. We conduct simulations to validate our analytical results and evaluate the proposed schemes in terms of outage probability for different values of signal-to-interference-plus-noise ratio (SINR) target.

Keywords: Heterogeneous cellular network; Cognitive radio; Outage probability

1 Introduction

The best solution to the spectrum saturation and bandwidth availability problems in multi-tier cellular networks is to adopt technologies that make the most efficient use of existing spectrum through frequency reuse schemes [1,2]. In universal frequency reuse scheme, the existing spectrum can be aggressively and effectively reused by all of the coexisting network tiers. This will lead to higher spatial spectrum utilization and network usage capacity at the expense of an increased possibility of interference among network tiers and of a reduced quality of service (QoS). In multi-tier cellular networks, interference is increasingly becoming a major performance-limiting factor, and hence, interference modeling, coordination, and avoidance are the primary focus of interest for both the industry and academic communities. Applying the cognitive radio (CR) technology in multi-tier cellular networks to be aware of and adapt to communication environments, some of the above challenges can be tackled. In fact, CR is the key enabling technology for interference management and avoidance in multi-tier cellular networks [2,3]. On the other hand, the aggregate interference environment is more complicated to model, and evaluating the performance of communication techniques in the presence of heterogeneous interference is challenging. For interference characterization, if the base stations (BSs) of the cellular network follow a regular grid (e.g., the traditional hexagonal grid model), then the SINR characterization will be either intractable [3,4] or inaccurate due to unrealistic assumptions [5]. Moreover, as urban areas are built out, the BS infrastructure is becoming less like points on a hexagonal lattice and more random. Hence, the use of a hexagonal grid to model the BS locations is violated and is considered too idealized [6]. Furthermore, according to [3,4,6] for snapshots of a cellular network at different locations, the positions of the BSs with respect to each other follow random patterns due to the size and unpredictability of the BSs in these kind of networks. Therefore, the need for a powerful mathematical and statistical tool for modeling, analysis, and design of wireless networks with random topologies is quite obvious.

A new modeling approach called 'stochastic geometry' has been recently applied to the analysis of multi-tier cellular networks due to its ability to capture the topological randomness in the network and its aim at deriving accurate and tractable expressions for outage probability [3,6].

* Correspondence: fereidoun@ohtsuki.ics.keio.ac.jp
Graduate School of Science and Technology, Keio University, 3-14-1 Hiyoshi, Kouhoku-ku, Yokohama 223-8522, Japan

Stochastic geometry stems from applied probability and has a wide range of applications in the analysis and design of wireless networks in particular for modeling and analyzing systems with random channel access (e.g., ALOHA [7,8] and carrier sensing multiple access (CSMA) [9]), single- and multi-tier cellular networks [6], and networks with cognitive abilities [7,10]. Multi-tier cellular networks have been investigated from different perspectives such as power control [11,12], spectrum allocation [13,14], and exploiting CR techniques [15,16], and recently, many works have been done based on the similar concepts to adopt and extend the stochastic geometric approach to different network models and scenarios (see [17-20]). This paper discusses this new theoretical model to provide a better understanding of the heterogeneous cellular networks of tomorrow and their challenges (interference modeling, coordination, and avoidance) that must be tackled in order for these networks to reach their potential. We focus on a two-tier femto-macro network where low-power and small-coverage local nodes (femto nodes) are distributed in the coverage of macro nodes. We provide an insight into the role of CR in interference mitigation in two-tier heterogeneous networks. We derive closed-form expressions for the outage probability of any typical femto and macro user in the network. We also study the effect of several important design factors which play vital roles in the determination of outage and interference.

Our main contributions in this work which is an extension of [21] are therefore the following: (1) We analyze the Laplace transforms of all four types of aggregate interference between macro and CR femto networks (including the interference between macro nodes among themselves and femto nodes among themselves, the cross-interference from femto to macro network and vice versa) in perfect and imperfect spectrum sensing CR-based femto networks, considering simultaneously the Poisson point process (PPP) model, and some important design factors (such as spectrum access probability) which can play a major role in determining interference and outage. (2) This article provides an insight into the role of CR in interference mitigation in orthogonal frequency-division multiple-access (OFDMA) two-tier heterogeneous networks. (3) Closed-form expressions are derived for the outage probability of any typical femto and macro user considering the Rayleigh fading assumption for the desired and interfering links with the possibility of using the CR ability for the femto network. It should be noted that in most of the available studies in this area, none of the network tiers is equipped with the CR capability; they are mostly based on the existence of only one macro-BS (along with the macro users and the femto network); and the effect of considering multiple macro-BSs is ignored in the analysis of outage probability. Authors in [22-25] have considered the two-tier heterogeneous networks imposing the CR ability to the femto tier. Different from [22], in our work, we consider both the perfect and imperfect sensing scenarios for the CR femto-BSs, however authors in [22] ignore the effect of sensing errors on the opportunistic channel access probability and consequently the outage probability of each tier. On the other hand, in our work, the mathematical demonstration of the obtained expressions (channel access probability and outage probability expressions) is quite different from the mentioned works.

2 Downlink system model

2.1 Model description

We consider infinite spatially collocated macro-BS and femto-BS node heterogeneous networks (see Figure 1). It is assumed that the spatial distribution of the nodes is captured using two collocated and independent homogenous Poisson point processes (HPPPs) [6,7] i.e., Φ_M and Φ_F with intensities λ_M and λ_F, respectively. In other words, the locations of the macro-BS nodes constitute an HPPP Φ_M, where λ_M is the average number of the macro-BS nodes per unit area. Similar statement can be made for the HPPP formed by the femto-BS nodes Φ_F with intensity λ_F. According to Superposition Theorem [26], the overall node process over the network formed by both the macro-BS and femto-BS nodes also is an HPPP with intensity λ ($\lambda = \lambda_F + \lambda_M$). Furthermore, the macro and femto users are scattered about the plane according to some independent PPPs with different densities compared to λ_M and λ_F, respectively. However, our interference analysis is fundamentally concerned with the distribution of the transmitters (BSs).

Since femto-BSs are installed and maintained by the paying home users for better indoor performance, they are only accessible by their own mobile subscribers (femto users) (known as *closed-access* policy). On the other hand, macro-BSs can be accessed only by unauthorized users (macro users). In practice, macro network is deployed usually without awareness of the distributed femto network. To this end, wireless operators can consider giving priority to the macro users, and the femto network has to be self-optimized to mitigate its interference to the macro users. Motivated by this insight, the macro-BSs (along with the macro users) and the CR-enabled femto-BSs (along with the femto users) are analogous to primary and secondary systems in the CR model, respectively.

2.2 System structure

In OFDMA, the spectrum is orthogonally divided into time-frequency resource blocks (RBs), which increases flexibility in resource allocation, thereby allowing high spectral efficiency. As shown in Figure 2, we consider a spectrum of N RBs, out of which M ($M \leq N$) random RBs are idle or unoccupied by the macro users (primary system). With the CR capability, a femto-BS could actively

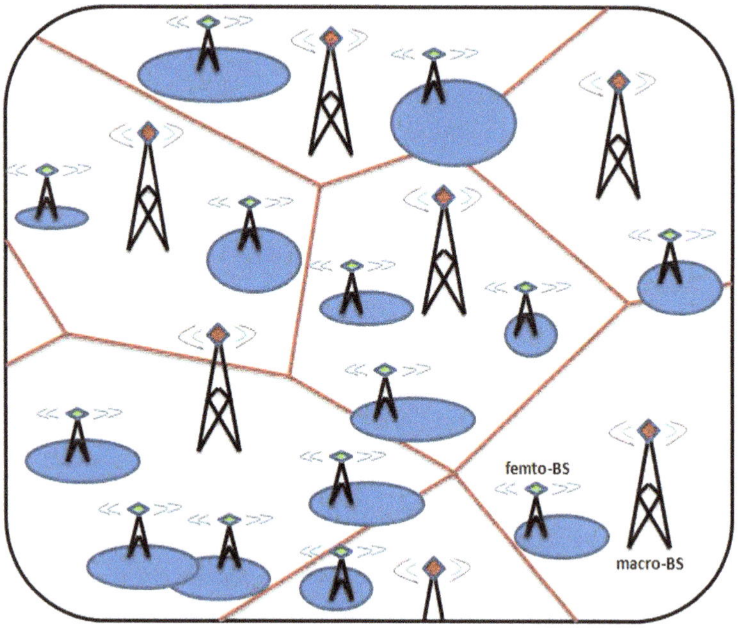

Figure 1 The heterogeneous model (femto and macro BSs).

acquire knowledge about its environment and access to the RBs without the aid of a macrocell in a decentralized fashion (clearly, no synchronization between the macro and femto network is needed any more) and automatically prevent disturbing the macro users [2].

1) As shown in Figure 2, each femto-BS's transmission strategy is divided into consecutive slots, each having a duration of T. Each slot is divided into two consecutive stages, i.e., sensing and data transmission, with durations of T_S and T_D, respectively. Each femto-BS periodically senses the spectrum to identify which RBs are occupied by the macro network.

Indeed, each femto-BS accomplishes sensing one RB in one unit slot T_{SRB} within T_S. Each femto-BS senses N_s RBs in sequence which is randomly selected from the N available RBs, and detects its idle RB set. Clearly, the time required for sensing the N_s RBs is $T_S = T_{SRB}N_s$. Note that the femto-BSs cannot perform data transmission within the sensing time T_S. We assume that all femto-BSs are perfectly synchronized and have the same time as the sensing time. Methods for implementing a perfect synchronization among the femto-BSs are outside the scope of this paper; however, a set of possible candidates exist, including GPS synchronization, the wired backhaul (IEEE 1588),

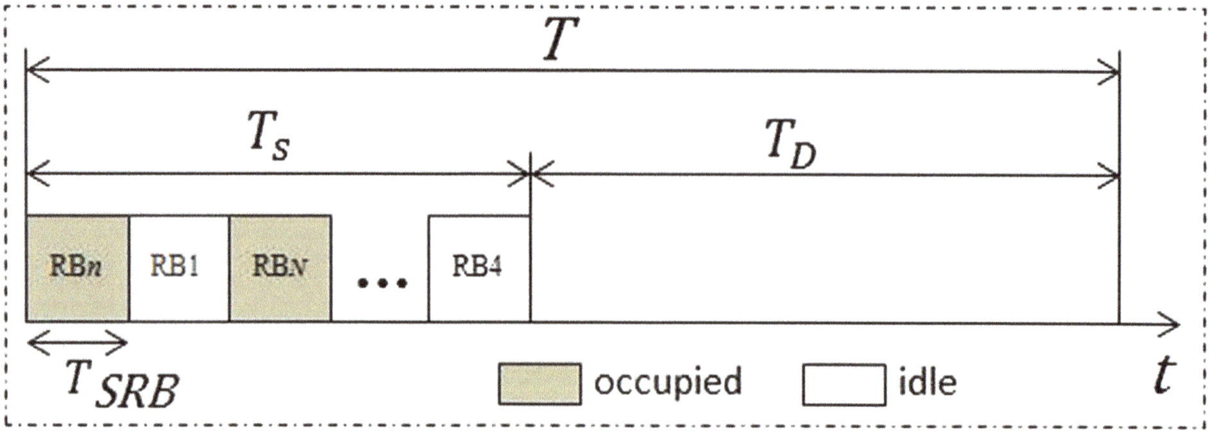

Figure 2 The CR femto-BS's transmission strategy in one time slot.

and leveraging synchronization signals broadcasted by the femto-BSs [27].

2) Each femto-BS senses the received interference power on each RB within the sensing duration:
   - If the received interference power on an RB at a typical femto-BS exceeds a certain threshold, the RB is identified as being occupied by one or more macro nodes but not by the femto network since all the femto-BSs have the same sensing time (It should be noted that if an RB is identified as being occupied at a typical femto-BS, it does not necessarily mean that it is also seen as an occupied RB at the other femto-BSs, as this status determination process depends only on the received interference power level on the RB at each individual femto-BS).
   - Otherwise, the RB is unoccupied by the macro network.
3) In the data transmission time (T_D), each femto-BS only allocates an unoccupied RB sensed in the sensing time to its user (by only utilizing these unoccupied RBs, *cross-tier* interference can be consequently avoided). Since the determination of each individual RB status as busy/idle is subject to (occasional) error, determined by the probability of (correct) detection of the presence of PUs' signals P_d and probability of false alarm P_f (probability of falsely declaring an idle RB as busy), we study the effect of both the *ideal* detection, i.e., $P_d = 1$ and $P_f = 0$, and the cases involving imperfect sensing (see [28,29]), i.e., $P_d \neq 1$ and $P_f \neq 0$ on the outage probabilities of femto and macro users.

In each realization of the point process, each macro and femto user communicates only with its nearest macro-BS and femto-BS, respectively. As shown in Figure 3, the macro users' exclusion regions with radius D are used to guarantee that the femto-BSs will, on average, not generate an aggregate interference leading to the outage of macro (primary) users. We assume that the macro users can be localized, e.g., based on pilot signals or transmitted acknowledgements. Therefore, the femto-BSs inside the macro users' exclusion regions may be able to detect the macro signals and cease their transmissions. As shown in the figure, for example, those femto-BSs located in the tagged macro user's exclusion region are not allowed to transmit data whether they pick the same RB as the tagged macro user or a different one. It should be noted that the tagged macro user is

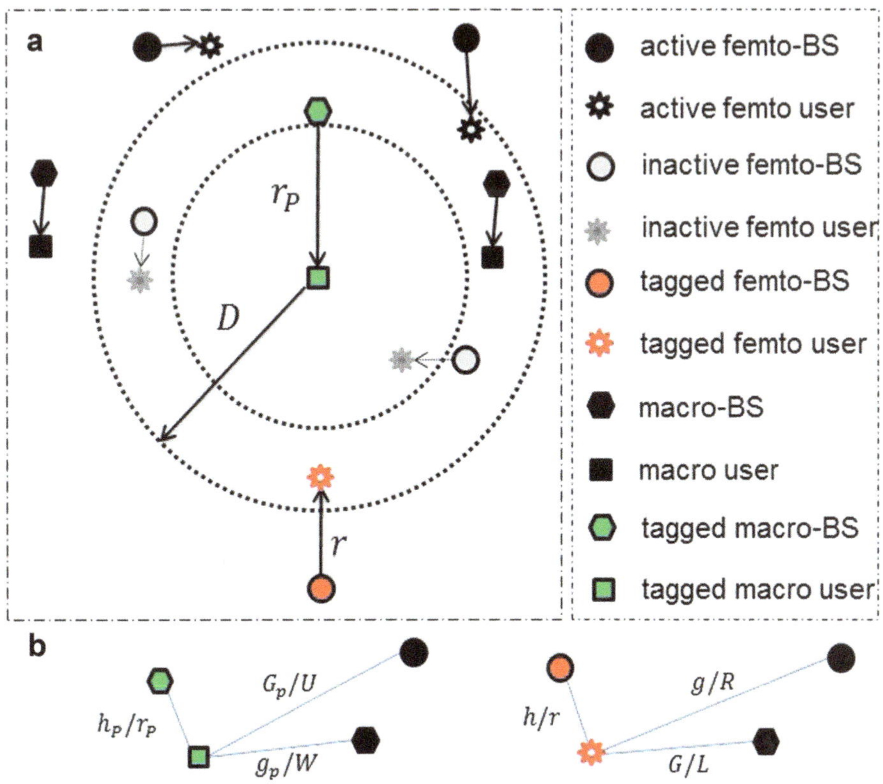

Figure 3 Bipolar network model and communication links. **(a)** The bipolar network model. **(b)** The link gains/distances.

not disturbed by the femto-BSs transmitting on different RBs (from that of the tagged macro) even if they are inside its exclusion region. However, we deactivate them to protect the tagged macro user from any harmful interference as a result of possible errors in their sensing and location detection processes (we consider this law throughout the paper even in the case of the perfect sensing scenario). For instance, if a femto-BS, located at a very close distance from the tagged macro user, wrongly (the cases involving imperfect sensing) picks the occupied RB by the tagged macro user, it will cause a severe interference to the tagged macro user if it is not able to perfectly detect the location of the tagged macro user (e.g., because of the hidden node problem in CR systems [30]) to cease its transmission.

3 Stochastic geometry-based network configuration

3.1 Femto outage probability formulation

We derive the probability of outage for a typical femto user (p_{OF}) in a downlink heterogeneous cellular network defined as the probability that a randomly chosen femto user cannot achieve a target SINR θ (or equivalently as the average fraction of femto users who do not achieve a target SINR θ, i.e., the averaged outage probability of all femto users) considering a collocated spectrum sensing CR-based femto network and macro-BSs as follows [6]:

$$p_{\mathrm{OF}} = 1 - P[\mathrm{SINR} > \theta] \tag{1}$$

In fact, the outage probability evaluates the CDF of SINR over the entire network. The experienced SINR by a typical femto user is calculated as

$$\mathrm{SINR} = \frac{P_F h r^{-\alpha}}{\sigma^2 + I_{\mathrm{FB}} + I_{\mathrm{MB}}} \tag{2}$$

where P_F is the transmission power from the nearest femto-BS (tagged femto-BS) located in the random distance r from its tagged femto user (we assume that the tagged femto user under consideration is located at the origin), and α is the path-loss exponent. I_{FB} and I_{MB} are the aggregate interference power at the origin from the other femto-BSs and macro-BSs, respectively, and σ^2 is the noise power. It should be noted that the transmission power values of all the femto-BSs in the network are kept constant, i.e., P_F.

Let $h \sim \exp(\mu)$ (with mean $1/\mu$) be a random variable accounting for the random channel gain of the link between the tagged femto user and its corresponding femto-BS, and then we have

$$\begin{aligned} P[\mathrm{SINR} > \theta] &= P\left[\frac{P_F h r^{-\alpha}}{\sigma^2 + I_{\mathrm{FB}} + I_{\mathrm{MB}}} > \theta\right] \\ &\overset{..}{=} P\left[h > (\sigma^2 + I_{\mathrm{FB}} + I_{\mathrm{MB}})\frac{\theta r^{\alpha}}{P_F}\right] \\ &\overset{...}{=} \mathrm{E}_{I_{\mathrm{FB}},I_{\mathrm{MB}}}\left[\int_{(\sigma^2+I_{\mathrm{FB}}+I_{\mathrm{MB}})\frac{\theta r^{\alpha}}{P_F}}^{\infty} [\mu \exp(-\mu x)]\mathrm{d}x\right] \\ &= \mathrm{E}_{I_{\mathrm{FB}},I_{\mathrm{MB}}}\left[\exp\left[-\mu(\sigma^2 + I_{\mathrm{FB}} + I_{\mathrm{MB}})\frac{\theta r^{\alpha}}{P_F}\right]\right] \end{aligned}$$

where $(\cdot\cdot)$ is derived by rearranging the terms in SINR; $(\cdots)$ is derived by noting that h is an exponential random variable; and $\mathrm{E}_{I_{\mathrm{FB}},I_{\mathrm{MB}}}[\cdot]$ is the expectation operator with respect to the joint distribution of the random variables I_{FB} and I_{MB}.

Note that due to the assumption of independent PPPs for the femto and macro networks, the aggregate interference received from the femto-BSs is independent of the aggregate interference received from the macro-BSs [10], therefore, we can write

$$\begin{aligned} P[\mathrm{SINR} > \theta] &= e^{-\mu\frac{\theta r^{\alpha}}{P_F}\sigma^2} . \mathrm{E}_{I_{\mathrm{FB}}}\left[e^{-\mu\frac{\theta r^{\alpha}}{P_F}I_{\mathrm{FB}}}\right] . \mathrm{E}_{I_{\mathrm{MB}}}\left[e^{-\mu\frac{\theta r^{\alpha}}{P_F}I_{\mathrm{MB}}}\right] \\ &= e^{-\mu\frac{\theta r^{\alpha}}{P_F}\sigma^2} . \mathcal{L}_{I_{\mathrm{FB}}}\left(\mu\frac{\theta r^{\alpha}}{P_F}\right) . \mathcal{L}_{I_{\mathrm{MB}}}\left(\mu\frac{\theta r^{\alpha}}{P_F}\right) \end{aligned}$$

where $\mathcal{L}_{I_{\mathrm{FB}}}(s)$ and $\mathcal{L}_{I_{\mathrm{MB}}}(s)$ are the Laplace transform of random variables I_{FB} and I_{MB} evaluated at $s\left(s = \mu\frac{\theta r^{\alpha}}{P_F}\right)$, respectively. Thus, the probability of outage averaged over the plane is derived as

$$p_{\mathrm{OF}} = 1 - \mathrm{E}_r\left[P[\mathrm{SINR} > \theta]\right]$$

$$p_{\mathrm{OF}} = 1 - \int_0^{\infty}\left[P[\mathrm{SINR} > \theta]\right] . f_r(r)\mathrm{d}r$$

where $f_r(r) = e^{-\lambda_F \pi r^2} 2\pi\lambda_F r$ (as mentioned before, r is the random distance between the tagged femto user and its corresponding femto-BS) is the probability density function (pdf) of r [6]. Then, we have

$$p_{\mathrm{OF}} = 1 - \int_0^{\infty} e^{-\pi\lambda_F r^2} . e^{-\mu\frac{\theta r^{\alpha}}{P_F}\sigma^2} . \mathcal{L}_{I_{\mathrm{FB}}}\left(\mu\frac{\theta r^{\alpha}}{P_F}\right) . \mathcal{L}_{I_{\mathrm{MB}}}\left(\mu\frac{\theta r^{\alpha}}{P_F}\right) . 2\pi r\lambda_F \, \mathrm{d}r \tag{3}$$

3.1.1 Scenario I

Ideal detection ($P_d = 1$ and $P_f = 0$):

Each secondary node (femto-BS) has perfect knowledge of each primary (macro-BS) signaling. In other words, sensing at each femto-BS is done perfectly. Therefore, an RB occupied by a macro-BS is not chosen for data transmission by the nearby femto-BSs. Under this condition, the tagged femto user, during the data transmission time, does not experience any interference from the macro-BSs since it always communicates with its corresponding femto-BS on an idle RB. In fact, we assume that the received interference power from the macro network under this scenario can be neglected if it is measured to be less than a specified threshold (if we do not neglect the received interference power under the explained condition, then the outage probability formulations will be the same as in Scenario II except for the RB selection probability expressions, p_{RB}, as we explain later. Similar arguments can be made for the outage probability of the macro tier as discussed in the next subsections).

Therefore, under this assumption, $\mathrm{E}_{I_{\text{MB}}}\left[e^{-\mu\frac{\theta r^{\alpha}}{P_F}I_{\text{MB}}}\right] = 1$ and $\mathcal{L}_{I_{\text{MB}}}\left(\mu\frac{\theta r^{\alpha}}{P_F}\right) = 1$, consequently. Then (3) is rewritten as follows

$$p_{\text{OF}} = 1 - \int_0^{\infty} e^{-\pi\lambda_F r^2}.e^{-\mu\frac{\theta r^{\alpha}}{P_F}\sigma^2}.\mathcal{L}_{I_{\text{FB}}}\left(\mu\frac{\theta r^{\alpha}}{P_F}\right).2\pi r\lambda_F dr. \tag{4}$$

The Laplace transform of the aggregate interference from all the active femto-BSs except the tagged femto-BS denoted by fbs_0 is given as follows (Note: some of the femto-BSs located in the macro users' exclusion regions with radius D are deactivated, therefore, $\lambda'_F \leq \lambda_F$ (see Figure 3))

$$\begin{aligned}\mathcal{L}_{I_{\text{FB}}}(s) &= \mathrm{E}_{I_{\text{FB}}}[\exp(-sI_{\text{FB}})]\\ &= \mathrm{E}_{\Phi_F, g_i}\left[\exp\left(-s\sum_{i\in\Phi_F\setminus\{fbs_0\}} P_F g_i R_i^{-\alpha}\right)\right]\\ &= \mathrm{E}_{\Phi_F}\left[\prod_{i\in\Phi_F\setminus\{fbs_0\}} \mathrm{E}_{g_i}\left[\exp\left(-sP_F g_i R_i^{-\alpha}\right)\right]\right]\end{aligned} \tag{5}$$

Where R_i is the distance of the ith interferer from the tagged femto receiver captured by the point process Φ_F The interference channel gains g_i are assumed to be mutually independent and have identical pdfs. Each of the active interfering femto-BSs transmits with the same power P_F. Using the definition of the Generating functional [6] for the Poisson point process (PPP), which states for some function $f(x)$ that $\mathrm{E}\left[\prod_{x\in\Phi} f(x)\right] = \exp\left(-\int_{\mathbb{R}^d}(1-f(x))\lambda dx\right)$, (5) can be rewritten as

$$\mathcal{L}_{I_{\text{FB}}}(s) = \exp\left\{-\mathrm{E}_g\left[\int_r^{\infty}\left(1-\exp(-sP_F g R^{-\alpha})\right)\lambda_I(R)dR\right]\right\} \tag{6}$$

where we flipped the order of integration and expectation. Since the closest interfering femto-BS is at least at distance r from the tagged user, the integration limits are from r to ∞. In other words, interference is encountered from all the active femto-BSs located in the area $\mathbb{R}^d\backslash b(0, r)$ (where $b(x, y)$ is ball of radius y centered at point x). However, not all the femto-BSs will contribute towards the aggregate interference, i.e., only those active femto-BSs located outside the mentioned ball which at minimum satisfy all of the following conditions qualify as potential contributors.

Before explaining these conditions ((a) and (b)), it is useful to translate the point process into polar coordinates. Therefore, according to [7,26], the intensity of the HPPP Φ_F is shown as

$$\lambda_I(R) = \lambda_F' \, dR^{d-1}b_d \tag{7}$$

where R is the distance between an arbitrary femto-BS and the tagged femto receiver. b_d is the volume of a unit sphere in $\mathbb{R}^d$ ($b_d = \frac{\sqrt{\pi^d}}{\Gamma\left(1+d/2\right)}$, $\Gamma(x) = \int_0^{\infty} t^{x-1}e^{-t}dt$ denotes the standard Gamma function).

(a) Satisfying the aforementioned condition, any active femto-BS contributes towards the interference at the tagged femto receiver, if it picks the same RB as the tagged femto-BS to communicate with its user. We show the probability of picking a same RB from a pool of all RBs as p_{RB} (the calculation of p_{RB} for this case is derived in Section 4, *Scenario I*, Case 1).

(b) We assume that the CR femto-BSs employ a slotted ALOHA MAC (medium access control) protocol to schedule their transmission. Therefore, they only transmit with probability p_{tx} in the current time slot and defer the transmission with probability $1 - p_{\text{tx}}$.

Applying both conditions (a) and (b) (which will result in reducing the number of interferers and hence outage probability at the tagged femto user), Φ_F is thinned using two independent thinning. Indeed, first, Φ_F is reconstructed by p_{RB} thinning, where each point of Φ_F is retained with probability p_{RB} (to accommodate the condition (a)), and then it is thinned again by applying the other independent thinning (to accommodate the condition (b)). Thus, the intensity of the process (the number of the interfering CR femto-BSs) becomes

$$\lambda_I(R) = \lambda_F^{'} dR^{d-1} b_d p_{\mathrm{RB}} p_{\mathrm{tx}} \tag{8}$$

Now, (6) is rewritten as

$$\exp\left\{-\mathrm{E}_g\left[\int_r^{\infty}\left(1-\exp(-sP_F gR^{-\alpha})\right)\lambda_F^{'} b_d p_{\mathrm{RB}} p_{\mathrm{tx}} dR^{d-1}\mathrm{d}R\right]\right\}.$$

By using the change of variables $R^d \to x$ and then $x^{\frac{\alpha}{d}} \to y$, and doing some simple calculations, the above expression is simplified as:

$$\mathcal{L}_{I_{\mathrm{FB}}}(s) = e^{r^d b_d p_{\mathrm{RB}} p_{\mathrm{tx}} \lambda_F^{'} - \frac{d}{\alpha} b_d p_{\mathrm{RB}} p_{\mathrm{tx}} \lambda_F^{'} (\mu\theta)^{\frac{d}{\alpha}} r^d M(\theta,\alpha)} \tag{9}$$

where

$$M(\theta,\alpha) = \mathrm{E}\left[(g)^{\frac{d}{\alpha}}\left(\Gamma(-\frac{d}{\alpha},\mu\theta g) - \Gamma(-\frac{d}{\alpha})\right)\right] \tag{10}$$

Note that $\Gamma(a,x) = \int_x^{\infty} t^{a-1}e^{-t}\mathrm{d}t$ indicates the incomplete gamma function. From (4) and (9), taking d = 2, then replacing r^2 with z, we get the final formula for the outage probability of the tagged femto user as follows

$$p_{\mathrm{OF}} = 1 - \int_0^{\infty} e^{\pi z\left(\lambda_F^{'} p_{\mathrm{RB}} p_{\mathrm{tx}} - \lambda_F\right) - \frac{\mu\theta}{P_F}\sigma^2 z^{\frac{\alpha}{2}} - \frac{2}{\alpha}\pi p_{\mathrm{RB}} p_{\mathrm{tx}} \lambda_F^{'} (\mu\theta)^{\frac{2}{\alpha}} z M(\theta,\alpha)} .\pi\lambda_F \mathrm{d}z \tag{11}$$

in which $M(\theta,\alpha) = \mathrm{E}\left[(g)^{\frac{2}{\alpha}}\left(\Gamma(-\frac{2}{\alpha},\mu\theta g) - \Gamma(-\frac{2}{\alpha})\right)\right]$. According to [13], the downlink femtocell networks are assumed to be interference-limited, i.e., the noise can be neglected as the interference dominates the whole performances of the system $\sigma^2 \to 0$.

Thus, (11) reduces to (12) as follows (α = 4)

$$p_{\mathrm{OF}} = 1 - \frac{\lambda_F}{(\lambda_F - \lambda_F^{'} p_{\mathrm{RB}} p_{\mathrm{tx}}) + \frac{\lambda_F^{'}(p_{\mathrm{RB}} p_{\mathrm{tx}})}{2}\sqrt{\mu\theta} M(\theta,4)}. \tag{12}$$

Similar with the desired link, we consider the Rayleigh fading model for the femto interfering links as well (Rayleigh fading links with equal parameter μ). Then following the derivation of $M(\theta, \alpha)$ in Appendix I, p_{OF} is re-expressed as follows

$$p_{\mathrm{OF}} = 1 - \frac{\lambda_F}{(\lambda_F - \lambda_F^{'} p_{\mathrm{RB}} p_{\mathrm{tx}}) + \sqrt{\pi}\lambda_F^{'}(p_{\mathrm{RB}} p_{\mathrm{tx}})\mu\left[\sum_{k=0}^{\infty}\frac{(\mu\theta)^k}{\Gamma(k+\frac{1}{2})(\mu+\mu\theta)^{k+1}}\Gamma(1+k)\right]}. \tag{13}$$

3.1.2 Scenario II

Imperfect detection ($P_d \neq 1$ and $P_f \neq 0$):

In this scenario, each secondary node (femto-BS) has imperfect knowledge of each primary (macro-BS) signaling. In other words, sensing at each femto-BS is done imperfectly. Therefore, an occupied RB by a macro-BS may also be wrongly considered idle by the femto-BSs, causing collision between the two networks. Two cases can take place under the imperfect sensing scenario:

Case 1. The tagged femto-BS transmits data on an idle RB (for this case, the outage probability formulations can be considered the same as in the perfect sensing scenario except for the calculation of p_{RB} (see Section 4, *Scenario II*, Case 1))

Case 2. The tagged femto-BS transmits on an occupied RB (outage probability formulation in this case is explained as follows and the calculation of p_{RB} is presented in Section 4, *Scenario II*, Case 2)

In Case 2, the tagged femto user can experience interference from both the active femto-BSs (which pick the same occupied RB as the tagged femto-BS) and macro-BSs. Indeed, the $\mathcal{L}_{I_{\mathrm{MB}}}\left(\mu\frac{\theta r^{\alpha}}{P_F}\right)$ (the Laplace transform of the aggregate interference from the macro-BSs) in (3) is not ignored. The Laplace transform of the aggregate interference power generated by the macro-BSs at the tagged femto user is given by

$$\begin{aligned} \mathcal{L}_{I_{\mathrm{MB}}}(s) &= \mathrm{E}_{I_{\mathrm{MB}}}\left[\exp(-sI_{\mathrm{MB}})\right] \\ &= \mathrm{E}_{\Phi_M,G_i}\left[\exp\left(-s\sum_{i\in\Phi_M}P_P G_i L_i^{-\alpha}\right)\right] \\ &= \mathrm{E}_{\Phi_M}\left[\prod_{i\in\Phi_M}\mathrm{E}_{G_i}\left[\exp(-sP_P G_i L_i^{-\alpha})\right]\right] \end{aligned} \tag{14}$$

where L_i is the distance of the ith interfering macro-BS from the tagged femto receiver captured by the point process Φ_M. The interference channel gains G_i are assumed to be mutually independent and have identical pdfs. Each of the active interfering macro-BSs transmits with the same power P_P.

Again, using the definition of the generating functional for the PPP, we can write

$$\mathcal{L}_{I_{\mathrm{MB}}}(s) = \exp\left\{-\mathrm{E}_G\left[\int_0^{\infty}(1-\exp(-sP_PGL^{-\alpha}))\lambda_I(L)\mathrm{d}L\right]\right\}. \tag{15}$$

The interference is encountered from all the macro-BSs located in the area $\mathbb{R}^d \backslash b(0, 0)$. It should be noted that not all the macro-BSs in $\mathbb{R}^d$ will contribute towards the aggregate interference, i.e., only those macro-BSs which are transmitting on the same RB as the tagged femto user qualify as potential interferers. The intensity of the HPPP Φ_M process can be therefore written as follows

$$\lambda_I(L) = \lambda_M^{'} dL^{d-1} b_d \tag{16}$$

where L is the distance between an arbitrary macro-BS and the tagged femto receiver and $\lambda_M^{'}$ is the intensity of those macro-BSs transmitting on the same RB as the tagged femto user at a time.

Now, (15) is rewritten as

$$\exp\left\{-\mathrm{E}_G\left[\int_0^{\infty}(1-\exp(-sP_PGL^{-\alpha}))\lambda_M^{'}L^{d-1}b_d dL^{d-1}\mathrm{d}L\right]\right\}. \tag{17}$$

By using the change of variables and doing some simple calculations, $\mathcal{L}_{I_{\mathrm{MB}}}(s)$ is obtained as follows [26]

$$\mathcal{L}_{I_{\mathrm{MB}}}(s) = e^{-b_d\lambda_M^{'}\Gamma\left(1-\frac{d}{\alpha}\right)\left(\frac{P_P\mu\theta}{P_F}\right)^{\frac{d}{\alpha}} r^d \mathrm{E}\left[(G)^{\frac{d}{\alpha}}\right]}.$$

From (3), (9), and (17), taking $d = 2$, then replacing r^2 with z, and assuming $\sigma^2 \to 0$, we get the final formula as follows

$$p_{\mathrm{OF}} = 1-\int_0^{\infty} e^{-\pi\left[\left(\lambda_F-\lambda_F^{'}p_{\mathrm{RB}}p_{\mathrm{tx}}\right)+\frac{2}{\alpha}p_{\mathrm{RB}}p_{\mathrm{tx}}\lambda_F^{'}(\mu\theta)^{\frac{2}{\alpha}}M(\theta,\alpha)-\lambda_M^{'}\left(\frac{P_P\mu\theta}{P_F}\right)^{\frac{d}{\alpha}}r^d N(\alpha)\right]z} .\pi\lambda_F \mathrm{d}z \tag{18}$$

in which $M(\theta,\alpha) = \mathrm{E}\left[(g)^{\frac{2}{\alpha}}\left(\Gamma\left(-\frac{2}{\alpha},\mu\theta g\right)-\Gamma\left(-\frac{2}{\alpha}\right)\right)\right]$ and $N(\alpha) = \Gamma\left(1-\frac{d}{\alpha}\right)\mathrm{E}\left[(G)^{\frac{d}{\alpha}}\right]$.

Finally, the closed-form expression for the outage probability of the tagged femto user, under the imperfect sensing scenario for all CR femto-BSs, is obtained as follows ($\alpha = 4$)

$$p_{\mathrm{OF}} = 1-\frac{\lambda_F}{\left(\lambda_F-\lambda_F^{'}p_{\mathrm{RB}}p_{\mathrm{tx}}\right)+\frac{\lambda_F^{'}(p_{\mathrm{RB}}p_{\mathrm{tx}})}{2}\sqrt{\mu\theta}M(\theta,4)+\lambda_M^{'}N(4)\sqrt{\frac{P_P\mu\theta}{P_F}}}. \tag{19}$$

Similar with the desired link, we consider the Rayleigh fading model for the femto interfering links as well as the macro interfering links (Rayleigh fading links with equal parameter μ for the femto interfering links and μ_p for the macro interfering links). Following the derivation of $M(\theta, \alpha)$ in Appendix I, and after the simplification of $N(\alpha)$ using the definition of expectation and the standard gamma function, p_{OF} is re-expressed as follows $\alpha = 4$

$$p_{\mathrm{OF}} = 1-\frac{\lambda_F}{\left(\lambda_F-\lambda_F^{'}p_{\mathrm{RB}}p_{\mathrm{tx}}\right)+\sqrt{\pi}\lambda_F^{'}(p_{\mathrm{RB}}p_{\mathrm{tx}})\mu\left[\sum_{k=0}^{\infty}\frac{(\mu\theta)^k}{\Gamma\left(k+\frac{1}{2}\right)(\mu+\mu\theta)^{k+1}}\Gamma(1+k)\right]+\frac{\pi\lambda_M^{'}}{2\sqrt{\mu_p}}\sqrt{\frac{P_P\mu\theta}{P_F}}}. \tag{20}$$

3.2 Macro outage probability formulation

We derive the outage probability for a typical macro user (p_{OM}) in a downlink heterogeneous cellular network defined as the probability that a randomly chosen macro user cannot achieve a target SINR γ (or equivalently as the average fraction of macro users who do not achieve a target SINR γ, i.e., the averaged outage probability of all macro users) considering a collocated spectrum sensing CR-based femto network and macro-BSs as follows:

$$p_{\mathrm{OM}} = 1-P[\mathrm{SINR} > \gamma] \tag{21}$$

The experienced SINR by a typical macro user is calculated as

$$\mathrm{SINR} = \frac{P_P h_p r_p^{-\alpha}}{\sigma^2 + I_{\mathrm{MB}} + I_{\mathrm{FB}}} \tag{22}$$

where P_P is the transmission power from the nearest macro-BS located in the random distance r_p from its tagged macro user located at the origin and α is the path-loss exponent. I_{FB} and I_{MB} are the aggregate interference power to the tagged macro user (located at the origin) from the surrounding femto-BSs and macro-BSs, respectively, and σ^2 is the noise power. It should be noted that the transmission power values of all the macro-BSs in the network are kept constant, i.e., P_P.

Let $h_p \sim \exp(\mu_p)$ (with mean $1/\mu_p$) be a random variable accounting for the random channel gain of the link between the tagged macro user and its corresponding macro-BS, then the similar approach to obtain the outage probability of the tagged femto user is used for the outage probability calculation of the tagged macro user. Therefore, from (21) and (22), we have

$$p_{\mathrm{OM}} = 1-\int_0^\infty e^{-\pi\lambda_M r_p{}^2} \cdot e^{-\mu_p \frac{\gamma r_p{}^\alpha}{P_P}\sigma^2} \cdot \mathcal{L}_{I_{\mathrm{FB}}}\left(\mu_p \frac{\gamma r_p{}^\alpha}{P_P}\right) \cdot \mathcal{L}_{I_{\mathrm{MB}}}\left(\mu_p \frac{\gamma r_p{}^\alpha}{P_P}\right) \cdot 2\pi r_p \lambda_M \mathrm{d}r_p \tag{23}$$

where $\mathcal{L}_{I_{\mathrm{FB}}}(s)$ and $\mathcal{L}_{I_{\mathrm{MB}}}(s)$ are the Laplace transform of random variables I_{FB} and I_{MB} evaluated at $s\left(s = \mu_p \frac{\gamma r_p{}^\alpha}{P_P}\right)$, respectively.

3.2.1 Scenario I

Ideal detection ($P_d = 1$ and $P_f = 0$):

Each secondary node (femto-BS) has perfect knowledge of each primary (macro-BS) signaling. In other words, sensing at each femto-BS is done perfectly. Therefore, an RB occupied by a macro-BS is not chosen for data transmission by any nearby femto-BS. Under this condition, the tagged macro user, during the data transmission time, does not experience any interference (or a negligible interference) from the surrounding femto-BSs; however, those macro-BSs operating on the same RB as the tagged macro user and located in the area $\mathbb{R}^d \backslash b(0, r_p)$ make interference to the tagged macro user. Therefore, (23) becomes

$$p_{\mathrm{OM}} = 1-\int_0^\infty e^{-\pi\lambda_M r_p{}^2} . e^{-\mu_p \frac{\gamma r_p{}^\alpha}{P_P}\sigma^2} . \mathcal{L}_{I_{\mathrm{MB}}}\left(\mu_p \frac{\gamma r_p{}^\alpha}{P_P}\right) . 2\pi r_p \lambda_M \mathrm{d}r_p \tag{24}$$

Taking a similar approach to what we had before, $\mathcal{L}_{I_{\mathrm{MB}}}(s)$ is obtained as follows

$$\mathcal{L}_{I_{\mathrm{MB}}}(s) = \exp\left\{-\mathrm{E}_{g_p}\left[\int_{r_p}^{\infty}\left(1\text{-}\exp\left(-sP_P g_p W^{-\alpha}\right)\right)\lambda_I(W)\mathrm{d}W\right]\right\} \tag{25}$$

and

$$\lambda_I(W) = \lambda'_M d W^{d-1} b_d \tag{26}$$

in which W is the distance between an arbitrary macro-BS (captured by the point process Φ_M) and the tagged macro receiver and λ'_M is the intensity of those macro-BSs transmitting on the same RB as the tagged macro user at a time. Similarly, the interference channel gains g_p (between the interfering macro-BSs and the tagged macro user) are assumed to be mutually independent and have identical pdfs. Each of the active interfering macro-BSs transmits with the same power P_P as for the tagged macro-BS. Same as before, (25) is simplified as

$$\mathcal{L}_{I_{\mathrm{MB}}}(s) = e^{r_p{}^d b_d \lambda'_M - \frac{d}{\alpha} b_d \lambda'_M (\mu_p \gamma)^{\frac{d}{\alpha}} r_p{}^d V(\gamma,\alpha)}. \tag{27}$$

From (24) and (27) and the previous assumptions, the closed-form formula is expressed as follows

$$p_{\mathrm{OM}} = 1-\frac{\lambda_M}{\lambda_M - \lambda'_M + \frac{\lambda'_M}{2}\sqrt{\mu_p\gamma}\,V(\gamma,4)} \tag{28}$$

in which $V(\gamma,\alpha) = \mathrm{E}\left[\left(g_p\right)^{\frac{2}{\alpha}}\left(\Gamma\left(-\frac{2}{\alpha},\mu_p\gamma g_p\right) - \Gamma\left(-\frac{2}{\alpha}\right)\right)\right]$.

The Rayleigh fading model is also considered for the macro interference links (Rayleigh fading links with equal parameter μ_p). The derivation of $V(\gamma, \alpha)$ is similar to the derivation of $M(\theta, \alpha)$ in Appendix I. Therefore, p_{OM} is re-expressed as follows ($\alpha = 4$)

$$p_{\mathrm{OM}} = 1-\frac{\lambda_M}{\lambda_M - \lambda'_M + \sqrt{\pi}\lambda'_M \mu_p\left[\sum_{k=0}^{\infty}\frac{(\mu_p\gamma)^k}{\Gamma\left(k+\frac{1}{2}\right)(\mu_p+\mu_p\gamma)^{k+1}}\Gamma(1+k)\right]} \tag{29}$$

3.2.2 Scenario II

Imperfect detection ($P_d \neq 1$ and $P_f \neq 0$):

In this scenario, each secondary node (femto-BS) has imperfect knowledge of each primary (macro-BS) signaling. In other words, sensing at each femto-BS is done imperfectly and subjected to (occasional) error. Therefore, an occupied RB by a macro-BS may also be wrongly considered idle by the femto-BSs. In this situation, the tagged macro user can experience interference on its RB, from both the femto and macro BSs. Indeed, the $\mathcal{L}_{I_{\mathrm{FB}}}\left(\mu_p \frac{\gamma r_p{}^\alpha}{P_P}\right)$ (the Laplace transform of the aggregate interference to the tagged macro user from the surrounding femto-BSs) in (23) is not ignored. Using the same approach as described before, $\mathcal{L}_{I_{\mathrm{FB}}}(s)$ is also given by

$$\mathcal{L}_{I_{\mathrm{FB}}}(s) = \exp\left\{-\mathrm{E}_{G_P}\left[\int_{Kr_p}^{\infty}(1-\exp(-sP_F G_P U^{-\alpha}))\lambda_I(U)dU\right]\right\}. \tag{30}$$

Since the closest interfering femto-BS is at least at distance Kr_p from the tagged macro user, the integration limits are from Kr_p to ∞. In other words, interference is encountered from all the femto-BSs located in the area $\mathbb{R}^d \backslash b(0, Kr_p)$ (see Figure 3). It should be noted that not all the femto-BSs outside this ball will contribute towards the aggregate interference, i.e., only those femto-BSs which are outside the mentioned ball and at minimum satisfy all of the following conditions are considered as potential contributors

(c) Satisfying the above condition, any arbitrary femto-BS contributes towards the interference at the tagged macro receiver, if it wrongly picks the same RB as the tagged macro-BS to communicate with its user. We show the probability of picking a same RB for data transmission from a pool of all RBs as p_{RB} (the calculation of p_{RB} for this case is seen in Section 4, *Scenario II*, Case 2).

(d) Same as the condition (b) in subsection 3.1.1.

Applying these two independent thinning, the intensity of the process (the number of the interfering CR femto-BSs at the tagged macro user) becomes

$$\lambda_I(U) = \lambda_F^{'} d U^{d-1} b_d p_{\mathrm{RB}} p_{\mathrm{tx}} \tag{31}$$

Taking the similar approach, (30) is simplified as follows

$$\mathcal{L}_{I_{\mathrm{FB}}}(s) = e^{K^d {r_p}^d b_d p_{\mathrm{RB}} p_{\mathrm{tx}} \lambda_F^{'} - \frac{d}{\alpha} b_d p_{\mathrm{RB}} p_{\mathrm{tx}} \lambda_F^{'} \left(\mu_p \frac{P_F \gamma}{P_P}\right)^{\frac{d}{\alpha}} {r_p}^d O(\gamma,\alpha)} \tag{32}$$

in which $O(\gamma,\alpha) = \mathrm{E}\left[(G_p)^{\frac{d}{\alpha}}\left(\Gamma\left(-\frac{d}{\alpha}, \frac{\mu_p \gamma P_F G_p}{K^\alpha P_P}\right) - \Gamma\left(-\frac{d}{\alpha}\right)\right)\right]$.

From (23), (27), and (32) and the previous assumptions, the closed-form expression for the outage probability of the tagged macro user is obtained as follows

$$p_{\mathrm{OM}} = 1 - \frac{\lambda_M}{\lambda_M + \lambda_M^{'}\left[\frac{\sqrt{\mu_p \gamma}\, V(\gamma,4)}{2} - 1\right] + \lambda_F^{'} p_{\mathrm{RB}} p_{\mathrm{tx}} \left[\frac{\sqrt{\frac{\mu_p \gamma}{P_P}}\, O(\gamma,4)}{2} - K^2\right]}. \tag{33}$$

Similar with the desired link, the Rayleigh fading model for the femto interfering links (with equal parameter μ) as well as the macro interfering links (with equal parameter μ_p) is considered. Considering Appendix I and after replacing $V(\gamma, 4)$ and $O(\gamma, 4)$ with their expanded versions, p_{OM} can be re-expressed as follows

$$p_{\mathrm{OM}} = 1 - \frac{\lambda_M}{\mathrm{A} + \mathrm{B}} \tag{34}$$

where

$$\mathrm{A} = \lambda_M - \lambda_M^{'} + \sqrt{\pi}\lambda_M^{'} \mu_p \left[\sum_{k=0}^{\infty} \frac{(\mu_p \gamma)^k}{\Gamma\left(k+\frac{1}{2}\right)\left(\mu_p + \mu_p \gamma\right)^{k+1}} \Gamma(1+k)\right]$$

and

$$\mathrm{B} = \lambda_F^{'} p_{\mathrm{RB}} p_{\mathrm{tx}} K^2 \left[\mu \sqrt{\frac{\pi}{P_F}} \left[\sum_{k=0}^{\infty} \frac{\left(\frac{\mu_p \gamma P_F}{K^4 P_P}\right)^k}{\Gamma\left(k+\frac{1}{2}\right)\left(\mu + \frac{\mu_p P_F}{K^4 P_P}\gamma\right)^{k+1}} \Gamma(1+k)\right] - 1\right]$$

4 Resource block selection probability (p_{RB}) calculations under perfect and imperfect sensing

In this section, we discuss how the optimal values of the RB selection probability (p_{RB}) for a secondary transmitter (femto-BS) can be determined under each femto-BS's perfect and imperfect sensing scenarios.

4.1 Scenario I

Ideal detection for all the CR femto-BSs ($P_d = 1$ and $P_f = 0$) [31,32]:

Case 1. The tagged femto-BS assigns the *i*th idle RB to its femto user.

p_{RB}: The probability that the *i*th idle RB being selected for data transmission by any of the other active CR femto-BS [32].

$$p_{\mathrm{RB}|M_s} = p_{\mathrm{idle}}(M_s) . \frac{\binom{M-1}{M_s-1}}{\binom{M}{M_s}} . \left(\frac{1}{M_s}\right) \tag{35}$$

where

$$p_{\mathrm{idle}}(M_s) = \frac{\binom{M}{M_s}\binom{N-M}{N_s-M_s}}{\binom{N}{N_s}}.$$

The first term ($p_{\mathrm{idle}}(M_s)$) indicates the probability of M_s idle RBs sensed by a femto-BS (during the sensing time, T_S), the second term is the probability that the *i*th idle RB is inside the M_s idle RBs, and finally the third term indicates that the probability of selecting the *i*th idle RB (out of the M_s idle RBs) by that femto-BS is equal to $\frac{1}{M_s}$ (since each of the idle RBs within the M_s idle RBs has an equal probability of being chosen). It should be noted that each CR femto-BS will fail to access when M_s = 0 (the maximum value of M_s is equal to min{M, N_s}). Therefore, from (35),

$$\begin{aligned} p_{\mathrm{RB}} &= p_{\mathrm{RB}|(M_s \geq 1)} = \frac{1}{M} p_{\mathrm{idle}}(M_s \geq 1) \\ &= \frac{1}{M}\left(1 - p_{\mathrm{idle}}(M_s = 0)\right). \end{aligned} \tag{36}$$

If the number of RBs sensed by a CR femto-BS (i.e., N_s) is more than or equal to $N - M + 1$, then at least we have one idle RB within the N_s detected RBs, i.e., $p_{\mathrm{idle}}(M_s = 0) = 0$. On the other hand, if N_s is smaller than or equal to $N - M$, the CR femto-BS will fail to access when the RBs within the N_s sensed RBs are all occupied by the macro-BSs. Therefore, we can write

$$p_{\text{idle}}(M_s=0)=\begin{cases}\dfrac{\binom{N-M}{N_s}}{\binom{N}{N_s}}, & \text{if } N_s\leq N-M\\ 0, & \text{if } N_s\geq N-M+1.\end{cases} \tag{37}$$

From (36) and (37), the probability that a CR femto-BS selects the ith idle RB for data transmission is obtained as follows

$$p_{\text{RB}}=\begin{cases}\dfrac{1}{M}\left(1-\dfrac{\binom{N-M}{N_s}}{\binom{N}{N_s}}\right), & \text{if } N_s\leq N-M\\ \dfrac{1}{M}, & \text{if } N_s\geq N-M+1.\end{cases} \tag{38}$$

4.2 Scenario II

Imperfect detection for all the CR femto-BSs ($P_d \neq 1$ and $P_f \neq 0$) [31,32]:

Case 1. The tagged femto-BS assigns the ith idle RB to its femto user.

p_{RB}: The probability that the ith idle RB being selected for data transmission by any of the other active CR femto-BS [32].

We show the detection result indicator of the nth RBs by D_n ($n \in \{1, 2, ..., N\}$). If $D_n = 1$, the nth RB is detected as idle RB, otherwise, $D_n = 0$. The probability of one idle RB detected with no false alarm is $1 - P_f$ and the probability for an occupied RB detected as an idle RB is $1 - P_d$. In other words,

$$\Pr(D_n=1)=\begin{cases}V_0=1-P_f, & \text{if } n\text{th RB is idle}\\ V_1=1-P_d, & \text{if } n\text{th RB is busy}\end{cases} \tag{39}$$

in which P_f is the false alarm probability and can be obtained as follows [32-34]

$$P_f(\tau)=\mathcal{Q}\left(\sqrt{2\eta+1}\mathcal{Q}^{-1}(P_d)+\sqrt{\tau f_s}\eta\right) \tag{40}$$

$\mathcal{Q}(x)=\frac{1}{\sqrt{2\pi}}\int_x^{\infty}\exp\left(\frac{-t^2}{2}\right)dt$ and P_d is the predefined detection probability. τ is the spectrum sensing time, f_s the sampling frequency, and η the received interference power on an RB to each femto-BS.

Indeed, the probability that the ith idle RB is detected with no false alarm by a CR femto-BS is $\Pr(D_i = 1) = V_0$.

To obtain the probability that the ith idle RB being selected for data transmission by a CR femto-BS (p_{RB}), first, we calculate the probability that the ith idle RB is sensed and included in the M_s idle RBs out of the N_s sensed RBs in the sensing period (T_S), and it is expressed as follows

$$\Pr(\text{the } i\text{th idle RB is sensed}|M_s)=p_{\text{idle}}(M_s).\frac{\binom{M-1}{M_s-1}}{\binom{M}{M_s}} \tag{41}$$

where

$$p_{\text{idle}}(M_s)=\frac{\binom{M}{M_s}\binom{N-M}{N_s-M_s}}{\binom{N}{N_s}}.$$

Conditioning on M_D (see Table 1) and M_s, the probability that the ith idle RB being detected as idle is obtained as follows

$$\begin{aligned}&\Pr(D_i=1|M_s,M_D)\\&=\Pr(D_i=1).\Pr\left(\sum_{n\neq i,n\in\Phi}D_n=M_D-1|M_s\right)\\&=V_0\times\left[\sum_{m_{\text{ID}}=\max\{1,M_D-(N_s-M_s)\}}^{\min\{M_D,M_s\}}\left[\binom{M_s-1}{m_{\text{ID}}-1}(V_0)^{m_{\text{ID}}-1}(1-V_0)^{M_s-m_{\text{ID}}}\right.\right.\\&\qquad\left.\left..\binom{N_s-M_s}{m_{\text{OD}}}.(V_1)^{m_{\text{OD}}}(1-V_1)^{N_s-M_s-m_{\text{OD}}}\right]\right]\end{aligned} \tag{42}$$

Table 1 Symbols used in Section 4

Symbols	Descriptions
N	Number of RBs
M	Number of idle RBs
N_s	Number of sensed RBs
M_s	Number of idle RBs within the N_s sensed RBs
M_D	Number of RBs detected as idle within the N_s sensed RBs
m_{ID}	Number of idle RBs (out of the M_D detected idle RBs) detected correctly. ($m_{\text{ID}} \in [\max\{1, M_s - (N_s - M_D)\}, \min\{M_s, M_D\}]$)
m_{OD}	Number of busy RBs (out of the M_D detected idle RBs) detected as idle. ($m_{\text{OD}} = M_D - m_{\text{ID}}$)

in which Φ is the set of the detected RBs by a femto-BS (see Table 1 for the definitions of m_{ID} and m_{OD}). Replacing m_{OD} with $M_D - m_{\mathrm{ID}}$, we have

$$\Pr(D_i = 1|M_s, M_D) = \left[\sum_{m_{\mathrm{ID}}=\max\{1,M_D-(N_s-M_s)\}}^{\min\{M_D,M_s\}} \left[\binom{M_s-1}{m_{\mathrm{ID}}-1}(V_0)^{m_{\mathrm{ID}}}(1-V_0)^{M_s-m_{\mathrm{ID}}} \cdot \binom{N_s-M_s}{M_D-m_{\mathrm{ID}}}.(V_1)^{M_D-m_{\mathrm{ID}}}(1-V_1)^{N_s-M_s-M_D+m_{\mathrm{ID}}}\right]\right] \tag{43}$$

Having the M_D detected idle RBs (including the ith idle RB), the probability of a CR femto-BS accessing the ith idle RB is equal to $\frac{1}{M_D}$. Thus, the probability that the ith idle RB is selected for data transmission by any CR femto-BS (under imperfect sensing scenario) is obtained as follows

$$p_{\mathrm{RB}|M_s,M_D} = \Pr(\text{the } i\text{th idle RB is sensed}|M_s) \times \frac{1}{M_D} \times \Pr(D_i = 1|M_s, M_D). \tag{44}$$

Finally,

$$p_{\mathrm{RB}} = \sum_{M_s=\max\{1,N_s-(N-M)\}}^{\min\{N_s,M\}} \sum_{M_D=1}^{N_s} p_{\mathrm{RB}|M_s,M_D}. \tag{45}$$

Case 2. The tagged femto-BS assigns the ith busy RB (occupied by the macro network) to its femto user.

p_{RB}: The probability that the ith busy RB being selected for data transmission by any of the other active CR femto-BS.

To obtain the probability of the ith busy RB selected for data transmission by a femto-BS (p_{RB}), first, we calculate the probability that the ith busy RB is sensed and included in the (N_s – M_s) busy RBs out of the N_s sensed RBs in the sensing period, and it is expressed as follows

$$\Pr(\text{the } i\text{th busy RB is sensed}|(N_s-M_s)) = \frac{\binom{N-M-1}{N_s-M_s-1}\binom{M}{M_s}}{\binom{N}{N_s}}. \tag{46}$$

Conditioning on M_D and (N_s – M_s), the probability that the ith busy RB being detected as idle is obtained as follows

$$\Pr(D_i = 1|(N_s-M_s), M_D) = \Pr(D_i = 1).\Pr\left(\sum_{n\neq i, n\in\Phi} D_n = M_D-1|(N_s-M_s)\right) = V_1 \times \left[\sum_{m_{\mathrm{ID}}=\max\{1,M_D-(N_s-M_s)\}}^{\min\{M_D,M_s\}} \left[\binom{M_s}{m_{\mathrm{ID}}}(V_0)^{m_{\mathrm{ID}}}(1-V_0)^{M_s-m_{\mathrm{ID}}} \cdot \binom{N_s-M_s-1}{m_{\mathrm{OD}}-1}.(V_1)^{m_{\mathrm{OD}}-1}(1-V_1)^{N_s-M_s-m_{\mathrm{OD}}}\right]\right] \tag{47}$$

in which Φ is the set of the detected RBs by a femto-BS. Again, replacing m_{OD} with $M_D - m_{\mathrm{ID}}$, we have

$$\Pr(D_i = 1|(N_s-M_s), M_D) = \left[\sum_{m_{\mathrm{ID}}=\max\{1,M_D-(N_s-M_s)\}}^{\min\{M_D,M_s\}} \left[\binom{M_s}{m_{\mathrm{ID}}}(V_0)^{m_{\mathrm{ID}}}(1-V_0)^{M_s-m_{\mathrm{ID}}} \cdot \binom{N_s-M_s-1}{M_D-m_{\mathrm{ID}}-1}.(V_1)^{M_D-m_{\mathrm{ID}}}(1-V_1)^{N_s-M_s-M_D+m_{\mathrm{ID}}}\right]\right] \tag{48}$$

Having the M_D detected idle RBs (including the ith busy RB), the probability of a CR femto-BS accessing the ith busy RB is $\frac{1}{M_D}$. Thus, the probability that the ith busy RB is selected for data transmission by any CR femto-BS (under imperfect sensing scenario) is obtained as follows

$$p_{\mathrm{RB}|(N_s-M_s),M_D} = \Pr(\text{the } i\text{th idle RB is sensed}|(N_s-M_s)) \times \frac{1}{M_D} \times \Pr(D_i = 1|(N_s-M_s, M_D). \tag{49}$$

Finally,

$$p_{\mathrm{RB}} = \sum_{M_s=\max\{1,N_s-(N-M)\}}^{\min\{N_s,M\}} \sum_{M_D=1}^{N_s} p_{\mathrm{RB}|(N_s-M_s),M_D}. \tag{50}$$

5 Simulation results and discussions

Before we present the obtained results, a brief discussion on the spatial distribution of the BSs and the macro exclusion regions is conducted as follows.

Talking about the femto outage probability, each femto user suffers from two sources of interference, i.e., macro and femto networks. For the macro network, the aggregate interference results from all macro-BSs that use the same RB as the tagged femto user (i.e., we define a homogenous PPP with intensity λ_M'). For the femto network, the aggregate interference results only from the other femto-BSs that (1) pick the same RB as the tagged femto, (2) are allowed to transmit in the current time slot, and (3) are not inside the macro users' exclusion regions. Hence, the interfering femto-BSs do not constitute a homogeneous point process anymore, and analytical characterization of interference and outage in this case is hard to characterize (the resulting point process is called Poisson hole process). In the analysis, to keep the modeling tractable, we ignored the possible correlation between the locations of the interfering femto-BSs and approximated the spatial distribution of them by a homogeneous PPP of intensity $\lambda_F^{'} p_{\mathrm{RB}} p_{\mathrm{tx}}$. Authors in [10,35] use the same approximation approach, where its accuracy is also justified by simulation in [10]. Similar arguments and approximations were considered for the macro tier outage probability.

As interferences are experienced at receivers, we centered the macro exclusion regions around the macro users. The femto-BSs inside these areas may be able to detect the macro signals and cease their transmissions. The exclusion regions are usually chosen to be centered at the location of the macro-BSs not the macro receivers based on the argument that it is easier to detect the macro-BSs than the macro receivers especially if the receivers are passive like TV receivers. However, if the macro receivers (users) can be localized, e.g., based on pilot signals or transmitted acknowledgments, our obtained results directly apply and the exclusion regions around macro users can make sense. If the macro users cannot be localized, the exclusion regions have to be formed around the macro-BSs. This scenario can be evaluated with slight changes in the proposed model. It should be noted that the location detection of the macro users is outside the scope of this paper, however, many schemes have been already proposed. Measuring the power leakage of local oscillator is a possible way to detect the presence of the macro passive users (see [10,36]). The hidden node problem in CR systems which makes it difficult to detect the macro users can be also tackled, e.g., by adding a margin to the RB access detection threshold accounting for shadow fading and receiver location uncertainty for worst-case scenarios [30].

First, the accuracy of our analytical results for the downlink analysis in the proposed model is validated by simulations, as shown in Figures 4 and 5. The simulations which are built on Matlab platform are carried out to plot the curves of outage probability versus the SINR threshold for the tagged femto and macro user, as shown in Figures 4 and 5, respectively. The considered scenario is a two-tier network (exactly following the network model described in Sections 2 and 3) over an approximately 1 × 1 km square with the locations of different classes of BSs as realizations of independent PPPs of given densities and the tagged users located at the center. To have an estimate of the outage probability at the tagged users, the simulation results are averaged over both the spatial PPP (500 different positions) and fading distribution (300 realizations) and are conducted using the parameters mentioned in the figures' captions. Analytical curves are compared with the simulations under both perfect and imperfect sensing. It is observed that the simulation results closely match our analytical model, and the curves of analytical and simulation results match fairly well, which confirms our analysis. The plots exhibit slight discrepancies between analytical

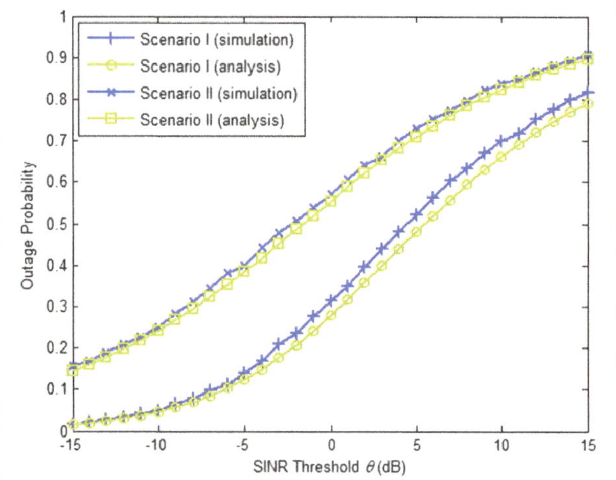

Figure 4 Outage probability of the tagged femto user. Outage probability of the tagged femto user (or any randomly chosen femto user) under perfect and imperfect sensing scenarios. The system parameters are set as $P_F = 22$ dBm, $P_P = 42$ dBm, $\alpha = 4$, $p_{\mathrm{RB}} = 0.8$, $p_{\mathrm{tx}} = 0.7$, $\mu = 0.2$, $\mu_p = 1$, $K = 1$, $\lambda_M^{'} = 12$, $\lambda_F = 800$, and $\lambda_F^{'} = 700$.

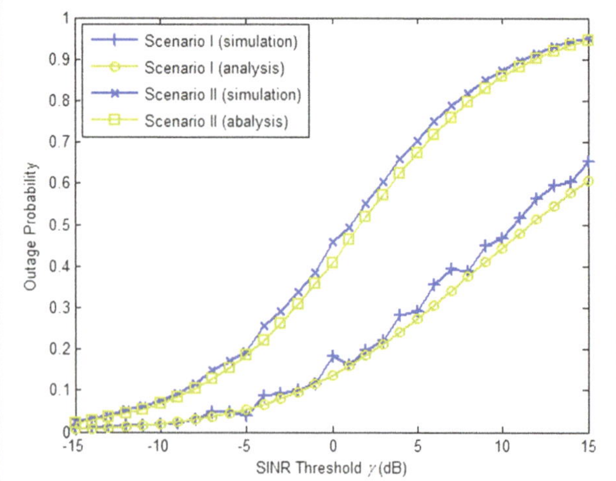

Figure 5 Outage probability of the tagged macro user. Outage probability of the tagged macro user (or any randomly chosen macro user) under perfect and imperfect sensing scenarios for the CR femto-BSs. The system parameters are set as $P_F = 22$ dBm, $P_P = 42$ dBm, $\alpha = 4$, $p_{\mathrm{RB}} = 0.8$, $p_{\mathrm{tx}} = 0.7$, $\mu = 0.2$, $\mu_p = 1$, $K = 1$, $\lambda_F^{'} = 700$, $\lambda_M^{'} = 12$, and $\lambda_M = 50$.

results and the corresponding simulation results which are mainly due to the independence assumption used in Section 3.

In Figure 6, the outage probability of the tagged femto user under perfect and imperfect spectrum sensing abilities for the CR femto-BSs is shown for different values of the target SINR θ on the horizontal axis. Our results show that the outage probability at the tagged femto receiver in the absence of a perfect spectrum sensing ability is considerably increased. In either the Scenario I, when all the femto-BSs employ perfect sensing to sense the RBs, or in *Scenario II*, Case 1, the tagged femto user does not experience any interference from the macro-BSs owing to the correct detection at the tagged femto-BS or choice of idle RBs for data transmission (RBs not occupied by the macro network). In this case, the interference seen by the tagged femto user is only the aggregate interference from the other femto-BSs which are transmitting on the same idle RB as the tagged femto. Clearly, the lowest outage probability is achieved for this case (see the red curve). Now, let us consider the imperfect sensing scenario for the CR femto-BSs. Obviously, the tagged femto user is now subject to sensing error and therefore picking an occupied RB for its data transmission period. Under this condition, it may receive interference not only from the other femto-BSs which pick the same busy RB (due to the imperfect sensing) as the tagged femto but also from those macro-BSs communicating with their own users on the same RB as the tagged femto. Therefore, the tagged femto user experiences an interference larger than before and consequently a significant increase in the outage probability. Moreover, the tagged femto user will face an outage with a higher probability whenever it picks an RB (occupied RB) already used by a larger number of macro users (a larger $\lambda_M^{'}$) (see Figure 6).

Figure 7 depicts the outage probability of the tagged macro user for different values of the target SINR γ and different situations. Considering the results obtained in Figure 6, here, we also investigate the effect of employing the two different sensing scenarios for the CR femto-BSs on the outage probability of the tagged macro user. In the case of perfect sensing, the tagged macro user does not experience any interference from the femto-BSs because only those RBs sensed to be idle (RBs not occupied by the macro network) are always chosen for data transmission by the femto network. In this case, the interference observed at the tagged macro user is only the aggregate interference received from those macro-BSs transmitting on the same RB as the tagged macro. Clearly, the lowest outage probability is obtained for this case (see the red curve). Now, the case of imperfect sensing of the CR femto-BS nodes is considered when the femto-BSs are subject to sensing error and therefore the possibility of transmitting on the RB occupied by the tagged macro. Under this condition, the tagged macro user may receive interference not only from the other macro-BSs communicating with their own users over the same RB (due to the lack of RBs) as the tagged macro but also from those femto-BSs which pick the same RB. Therefore, the tagged macro user experiences an interference larger than before and

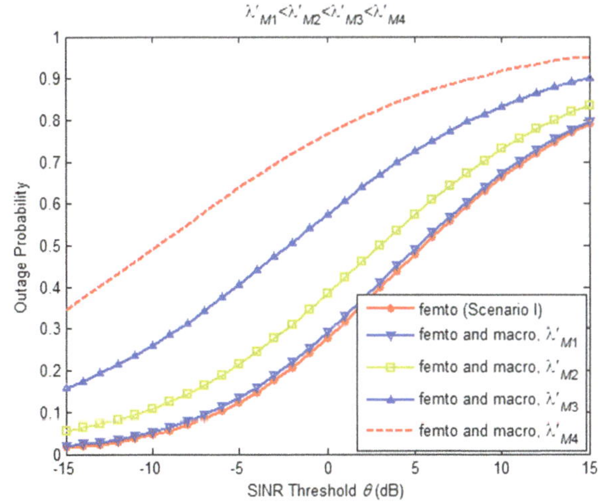

Figure 6 Outage probability of the tagged femto user with different values of $\lambda_M^{'}$. Outage probability of the tagged femto user (or any randomly chosen femto user) under perfect and imperfect sensing scenarios with different values of $\lambda_M^{'}$. The system parameters are set as $P_F = 22$ dBm, $P_P = 42$ dBm, $\alpha = 4$, $p_{RB} = 0.8$, $p_{tx} = 0.7$, $\mu = 0.2$, $\mu_p = 1$, $K = 1$, $\lambda_F = 800$, and $\lambda_F^{'} = 700$.

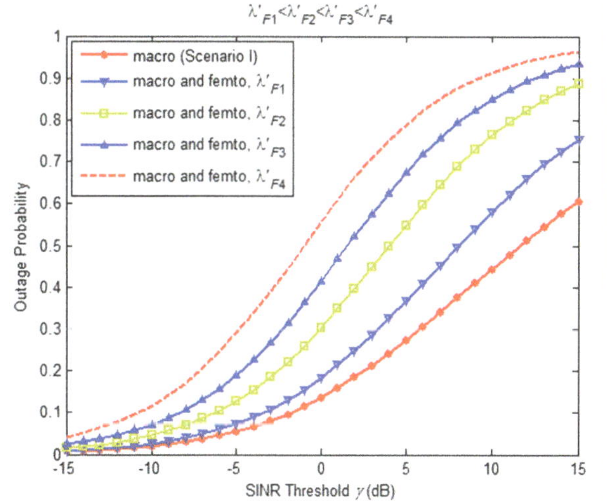

Figure 7 Outage probability of the tagged macro user with different values of $\lambda_F^{'}$. Outage probability of the tagged macro user (or any randomly chosen macro user) under perfect and imperfect sensing scenarios for the CR femto-BSs with different values of $\lambda_F^{'}$. The system parameters are set as $P_F = 22$ dBm, $P_P = 42$ dBm, $\alpha = 4$, $p_{RB} = 0.8$, $p_{tx} = 0.7$, $\mu = 0.2$, $\mu_p = 1$, $K = 1$, $\lambda_M^{'} = 12$, and $\lambda_M = 50$.

consequently a significant increase in the outage probability. Moreover, the tagged macro user will face an outage with a higher probability whenever its own RB is wrongly selected for data transmission by a larger number of femto-BSs (a larger $\lambda_F^{'}$).

Figure 8 illustrates the effect of K (in $D = Kr_p$) on the observed outage probability at the tagged macro user in the presence of both the macro and femto networks. Considering the previous explanations and Figure 3, let $\Phi_F^{'}$ include all the points (representing the femto-BSs) in Φ_F except the points inside the exclusion region D of the tagged macro user. Since $\Phi_F^{'} \subset \Phi_F$, the potential aggregate interference at the tagged macro user, caused by the active (considering slotted ALOHA) CR femto-BSs, is less than that in the case with no D. Furthermore, as the exclusion region D becomes larger (when $K = 10$, for example), the probability of outage is significantly reduced. Indeed, the bigger the value of D, the closer the outage probability curve becomes to the black curve which represents the outage probability in the case when no overall interference from the femto network is observed at the tagged macro user due to the perfect sensing ability of the CR femto-BSs. However, the reduction in the outage probability can be less when the number of macro-BSs transmitting on the same RB as the tagged macro is larger (see Figure 8b).

For a commercial network, designers must find a way to achieve a lower probability of outage for a certain SINR as the minimum quality needed for a typical femto or macro user to experience an acceptable QoS. A common way to decrease the outage probability is to reduce the number of interfering BSs encountered at the users. This can be done through applying both the (a) and (b) conditions in the perfect sensing scenario (or both the (c) and (d) in the imperfect sensing scenario). As shown in Figure 9, under these constraints, the outage probability is significantly reduced at the tagged femto user (note that the number of interfering macro-BSs at the tagged femto user is considered equal for all curves in the figure). The goal is to see the effect of both p_{RB} and p_{tx} (these two parameters are employed at each CR femto-BS) on the outage probability of the tagged femto user. As can be seen, for the case when $p_{\text{RB}} = 1$ and $p_{\text{tx}} = 1$, outage occurs with higher probability. In other words, when all the existing femto-BSs (except those who are inside the macro users' exclusion regions) pick the same RB as the tagged femto ($p_{\text{RB}} = 1$) and when they all have data to transmit in the current time slot ($p_{\text{tx}} = 1$), the tagged femto user will experience the maximum value for the outage probability derived for different SINR targets. Clearly, a significant reduction in the outage probability is occurred for the smaller values of p_{RB} and p_{tx} (see Figure 9) (the smaller the values of p_{RB} and p_{tx}, the closer the outage probability becomes to the outage probability in the case when the received interference at the tagged femto user is only the aggregate interference from the macro network). Indeed, this validates that many studies which do not consider these constraints over-estimate the interference encountered by a typical femto user.

It can be seen that even for high values of p_{tx}, the outage probability is relatively less than that in the case where there is no constraint on the femto-BS's transmission schedule ($p_{\text{tx}} = 1$). Also, it is obvious that in the presence of multiple RBs where each RB is picked with probability p_{RB}, the outage probability is further decreased. Hence,

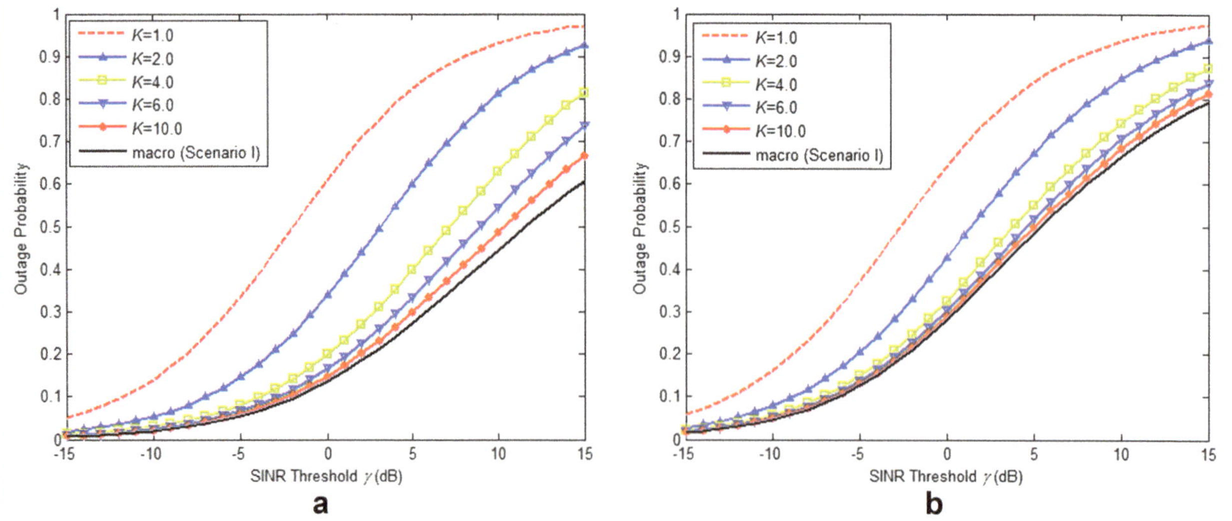

Figure 8 Effect of D on outage probability of the tagged macro user. Effect of D (the defined exclusion region around the tagged macro user) on outage probability of the tagged macro user in the presence of both the macro and femto-BSs. The system parameters are set as $P_F = 22$ dBm, $P_P = 42$ dBm, $\alpha = 4$, $p_{\text{RB}} = 0.8$, $p_{\text{tx}} = 0.7$, $\mu = 0.2$, $\mu_p = 1$, $\lambda_M = 50$, $\lambda_M^{'} = 12$ (in **(a)**) and $\lambda_M^{'} = 25$ (in **(b)**).

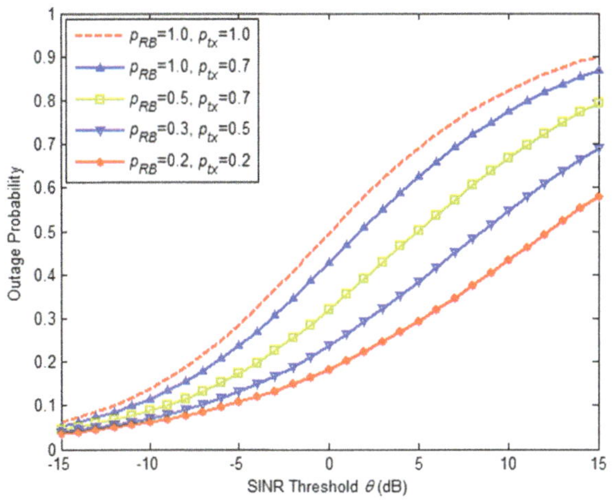

Figure 9 Effect of p_{RB} and p_{tx} on outage probability of the tagged femto user. Effect of p_{RB} and p_{tx} (the defined parameters for the CR femto-BSs) on outage probability of the tagged femto user in the presence of both the macro and femto-BSs. The system parameters are set as $P_F = 22$ dBm, $P_P = 42$ dBm, $\alpha = 4$, $\mu = 0.2$, $\mu_p = 1$, $K = 1$, $\lambda_F = 800$, $\lambda_F^{'} = 700$, and $\lambda_M^{'} = 12$.

any practical heterogeneous network designed to satisfy both of the mentioned conditions can reap the benefit of opportunistic exploitation of spectrum, while possibly causing little or no harmful interference. The same story exists when the outage probability of the tagged macro user is investigated. Figure 10 shows the effect of p_{RB} and p_{tx} (parameters which are related to the CR femto-BSs) on the outage probability of the tagged macro user. For instance, when all the active femto-BSs select the same RB as the tagged macro ($p_{RB} = 1$) and when they all have data to transmit (on this busy RB) in the current time slot ($p_{tx} = 1$), the tagged macro user will experience the maximum value for the outage probability. It should be noted that the number of interfering macro-BSs at the tagged macro is considered equal for all curves in this figure.

In Figures 11 and 12, we analyze the performance of the authorized links (femto links) in terms of throughput (achievable with a simple ARQ scheme with error-free feedback) under both the perfect and imperfect sensing scenarios. Considering the slotted ALOHA scheme, we define the following terms [37,38],

Probabilistic link throughput (τ) of a femto user:

(i) in the half-duplex (HD) communication scenario: it is defined to be the success probability of a femto user (i.e., $1 - p_{OF}$) multiplied by the probability that the corresponding femto-BS actually transmits over a specific RB (i.e., $p_{RB}p_{tx}$), and the probability that the femto receiver actually receives over that RB (i.e., $1 - p_{RB}p_{tx}$)

(ii) in the full-duplex (FD) communication scenario: it is defined to be the success probability of a femto user (i.e., $1 - p_{OF}$) multiplied by the probability that the corresponding femto-BS actually transmits over a specific RB (i.e., $p_{RB}p_{tx}$).

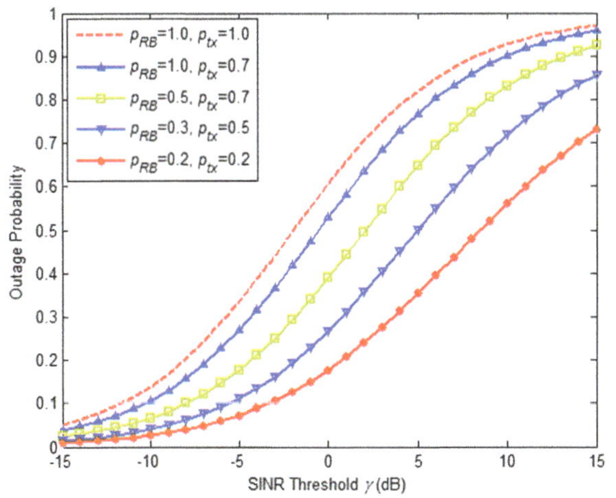

Figure 10 Effect of p_{RB} and p_{tx} on outage probability of the tagged macro user. Effect of p_{RB} and p_{tx} (the defined parameters for the CR femto-BSs) on outage probability of the tagged macro user in the presence of both the macro and femto-BSs. The system parameters are set as $P_F = 22$ dBm, $P_P = 42$ dBm, $\alpha = 4$, $\mu = 0.2$, $\mu_p = 1$, $K = 1$, $\lambda_F^{'} = 700$, $\lambda_M = 50$, and $\lambda_M^{'} = 12$.

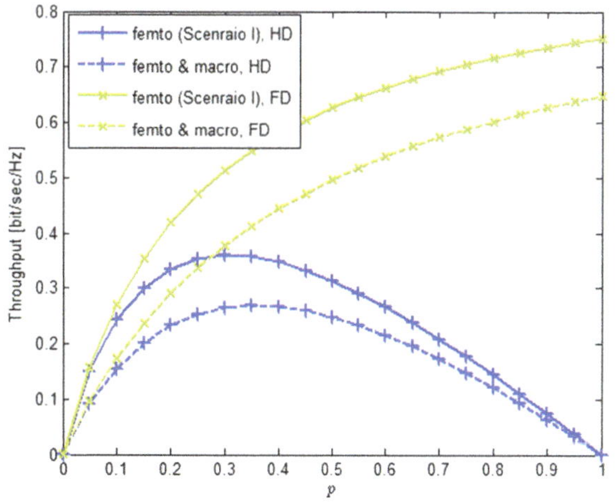

Figure 11 The femto link throughput of half- and full-duplex systems as a function of the transmission probability. The femto link throughput of half- and full-duplex systems under perfect and imperfect sensing scenarios as a function of the transmission probability over a specific RB (p). The system parameters are set as $P_F = 22$ dBm, $P_P = 42$ dBm, $\alpha = 4$, $\mu = 0.2$, $\mu_p = 1$, $K = 1$, $\lambda_M^{'} = 12$, $\lambda_F = 800$ and $\lambda_F^{'} = 700$, and $\theta = 11$ dB.

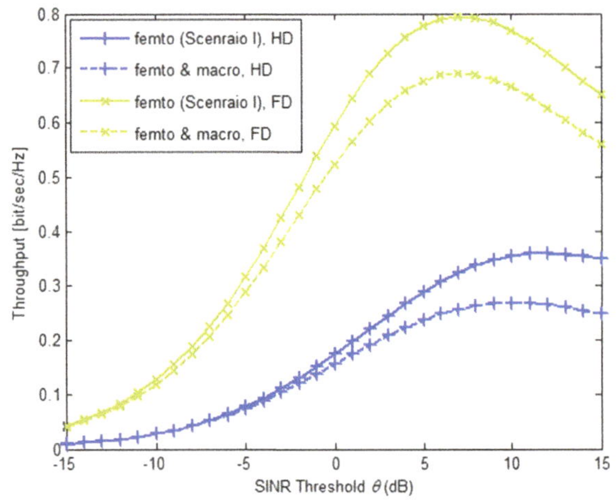

Figure 12 The femto link throughput of half and full-duplex systems as a function of the target SINR. The femto link throughput of half and full-duplex systems under perfect and imperfect sensing scenarios as a function of the target SINR. The system parameters are set as $P_F = 22$ dBm, $P_P = 42$ dBm, $\alpha = 4$, $\mu = 0.2$, $\mu_p = 1$, $K = 1$, $\lambda_M^{'} = 12$, $\lambda_F = 800$, and $\lambda_F^{'} = 700$; and for each curve, the optimal value of p obtained in Figure 11 is applied.

Femto link throughput (T):

The femto link throughput is defined as the product of the probabilistic link throughput (τ) and the rate of transmission, i.e., $T = \tau \log(1 + \theta)$. Therefore, the femto link throughput for the half and full-duplex cases is written as follows

$$\begin{aligned} T^{(\text{half})} &= p(1-p)(1-p_{\text{OF}})\log(1+\theta) \\ T^{(\text{full})} &= p(1-p_{\text{OF}})\log(1+\theta) \end{aligned}$$

in which $p = p_{\text{RB}} p_{\text{tx}}$.

In Figure 11, the performance of half- and full-duplex systems is presented for the femto users. More specifically, the link throughput of any typical femto user (e.g., the link between the tagged femto user and its corresponding femto-BS) under perfect and imperfect spectrum sensing abilities for the CR femto-BSs is shown as a function of the transmission probability over a specific RB (i.e., $p = p_{\text{RB}} p_{\text{tx}}$). It can be seen that the throughput achieved by the FD system is significantly higher, particularly when p is high. Regarding the performance of the HD system, for both the perfect and imperfect sensing cases, there is a unique optimal p which achieves the maximum throughput ($p = 0.3$ for the perfect and $p = 0.35$ for the imperfect sensing scenario). However, for high p, both throughput curves converge to zero due to over many transmissions and interferences on the RB. Obviously, for both the half- and full-duplex communications, a higher per-link throughput is achieved when the CR femto-BSs employ perfect sensing.

In Figure 12, the performance of half- and full-duplex systems is presented for femto users. More specifically, the link throughput of any typical femto user (e.g., the link between the tagged femto user and its corresponding femto-BS) under perfect and imperfect spectrum sensing abilities for the CR femto-BSs is shown as a function of the target SINR θ. It can be seen that the per-link throughput achieved by the FD system, for both the perfect and imperfect sensing scenarios, is significantly higher than the HD one. As it is seen, the link throughput curves are concave and there is an optimal point in each curve. With a high-target SINR, we can transmit the user data with high spectral efficiency; however, the outage probability of this transmission is high, too. In contrast, with a low-target SINR, we can send many packets that include little information. In other words, a high reliable transmission can be experienced at low-target SINRs, while the minimum requirements for the transmission rate cannot be met.

6 Conclusions

In this paper, utilizing the spatial Poisson point process (PPP) theory, we presented a tractable model to derive the outage probability of a typical femto and macro user in a two-tier heterogeneous network which provides insight into system design guidelines. In other words, for the case of the node locations modeled by a PPP and the desired and interfering channels are subject to Rayleigh fading, we demonstrated the use of the cognitive radio (CR)-based framework to evaluate the outage probability at any arbitrary user. Exact closed-form expressions were obtained as a result. In addition, we observed that in the downlink analysis, the outage probability is a function of the network topology and several important system design parameters such as SINR target, exclusion regions, MAC mechanisms such as ALOHA (p_{tx}), and the resource block (RB) selection constraint (p_{RB}) which is controlled by the spectrum sensing measurements.

Appendix I

DERIVATION OF $M(\theta, \alpha)$

Proof: From the expression of $M(\theta, \alpha)$ in (11), we have

$$M(\theta,\alpha) = \int_0^\infty \left[(g)^{\frac{d}{\alpha}}\left(\Gamma\left(-\frac{d}{\alpha},\mu\theta g\right)-\Gamma\left(-\frac{d}{\alpha}\right)\right)\right]\mu e^{-\mu g}\mathrm{d}g$$

We know that $\Gamma(a, x) + \gamma(a, x) = \Gamma(a)$ and $\gamma(a,x) = x^a \Gamma(a)e^{-x}\sum_{k=0}^{\infty}\frac{x^k}{\Gamma(a+k+1)}$, therefore, the above equation is simplified as follows

$$= -\Gamma\left(-\frac{d}{\alpha}\right)\mu(\mu\theta)^{-\frac{d}{\alpha}}\int_0^\infty e^{-g(\mu+\mu\theta)}\left[\sum_{k=0}^{\infty}\frac{(\mu\theta)^k g^k}{\Gamma\left(-\frac{d}{\alpha}+k+1\right)}\right]dg$$

$$= -\Gamma\left(-\frac{d}{\alpha}\right)\mu(\mu\theta)^{-\frac{d}{\alpha}}\int_0^\infty e^{-g(\mu+\mu\theta)}$$

$$\times\left[\frac{(\mu\theta)^0 g^0}{\Gamma\left(1-\frac{d}{\alpha}\right)}+\frac{(\mu\theta)^1 g^1}{\Gamma\left(2-\frac{d}{\alpha}\right)}+\frac{(\mu\theta)^2 g^2}{\Gamma\left(3-\frac{d}{\alpha}\right)}+\ldots\right]dg$$

$$= -\Gamma\left(-\frac{d}{\alpha}\right)\mu(\mu\theta)^{-\frac{d}{\alpha}}\left[\int_0^\infty\frac{(\mu\theta)^0(g)^{(1)-1}e^{-g(\mu+\mu\theta)}}{\Gamma\left(1-\frac{d}{\alpha}\right)}dg\right.$$

$$\left.+\int_0^\infty\frac{(\mu\theta)^1(g)^{(2)-1}e^{-g(\mu+\mu\theta)}}{\Gamma\left(2-\frac{d}{\alpha}\right)}dg+\ldots\right]$$

$$= -\Gamma\left(-\frac{d}{\alpha}\right)\mu(\mu\theta)^{-\frac{d}{\alpha}}\left[\frac{(\mu\theta)^0}{\Gamma\left(1-\frac{d}{\alpha}\right)(\mu+\mu\theta)}\Gamma(1)\right.$$

$$\left.+\frac{(\mu\theta)^1}{\Gamma\left(2-\frac{d}{\alpha}\right)(\mu+\mu\theta)^2}\Gamma(2)+\ldots\right]$$

Hence, we have

$$M(\theta,\alpha) = \left(-\Gamma\left(-\frac{d}{\alpha}\right)\mu(\mu\theta)^{-\frac{d}{\alpha}}\right)\times\left[\sum_{k=0}^{\infty}\frac{(\mu\theta)^k}{\Gamma(k+1-\frac{d}{\alpha})(\mu+\mu\theta)^{k+1}}\Gamma(1+k)\right]$$

Competing interests

The authors declare that they have no competing interests.

References

1. J Andrews, H Claussen, M Dohler, S Rangan, M Reed, Femtocells: past, present, and future. IEEE J Sel Areas Comm **30**(3), 497–508 (2012)
2. S Cheng, S Lien, F Hu, K Chen, On exploiting cognitive radio to mitigate interference in macro/femto heterogeneous networks. IEEE Wireless Comm Mag **18**, 40–47 (2011)
3. H ElSawy, E Hossain, M Haenggi, Stochastic geometry for modeling, analysis, and design of multi-tier and cognitive cellular wireless networks: a survey. IEEE Comm Surveys Tutorials **15**(3), 996–1019 (2013)
4. K Gilhousen, I Jacobs, R Padovani, AJ Viterbi, L Weaver, C Wheatley, On the capacity of a cellular CDMA system. IEEE Trans Veh Tech **40**(2), 303–312 (1991)
5. J Xu, J Zhang, JG Andrews, On the accuracy of the Wyner model in cellular networks. IEEE Trans Wireless Comm **10**(9), 3098–3109 (2011)
6. JG Andrews, F Baccelli, RK Ganti, A tractable approach to coverage and rate in cellular networks. IEEE Trans Comm **59**(11), 3122–3134 (2011)
7. SAR Zaidi, DC McLernon, M Ghogho, *Outage probability analysis of cognitive radio networks under self-coexistence constraint, 44th Annual Conference on Information Sciences and Systems (CISS)*, 2010, pp. 1-6-17–19
8. RK Ganti, M Haenggi, Interference and outage in clustered wireless ad hoc networks. IEEE Trans Information Theory **55**, 4067–4086 (2009)
9. A Hasan, JG Andrews, The guard zone in wireless ad hoc networks. IEEE Trans Wireless Comm **4**(3), 897–906 (2007)
10. C-H Lee, M Haenggi, Interference and outage in Poisson cognitive networks. IEEE Trans Wireless Comm **11**, 1392–1401 (2012)
11. V Chandrasekhar, JG Andrews, M Tarik, S Zukang, G Alan, Power control in two-tier femtocell networks. IEEE Trans Wireless Comm **8**(8), 4316–4328 (2009)
12. HS Jo, C Mun, J Moon, JG Yook, Interference mitigation using uplink power control for two-tier femtocell networks. IEEE Trans Wireless Comm **8**(10), 4906–4910 (2009)
13. V Chandrasekhar, JG Andrews, Spectrum allocation in tiered cellular networks. IEEE Trans Comm **57**(10), 3059–3068 (2009)
14. D Cao, S Zhou, Z Niu, Improving the energy efficiency of two-tier heterogeneous cellular networks through partial spectrum reuse. IEEE Trans Wireless Comm **12**(8), 4129–4141 (2013)
15. J Xiang, Y Zhang, T Skeie, L Xie, Downlink spectrum sharing for cognitive radio femtocell networks. IEEE Sys J **4**(4), 524–534 (2010)
16. Y Shi, YT Hou, H Zhou, SF Midkiff, Distributed cross-layer optimization for cognitive radio networks. IEEE Trans Veh Tech **59**(8), 4058–4069 (2010)
17. H Dhillon, R Ganti, F Baccelli, J Andrews, Modeling and analysis of K-Tier downlink heterogeneous cellular networks. IEEE J Selected Areas Comm **30**(3), 550–560 (2012)
18. H Jo, Y Sang, P Xia, J Andrews, *Outage Probability for Heterogeneous Cellular Networks with Biased Cell Association* (Proc, IEEE GlobeCom, 2011)
19. T Novlan, R Ganti, A Ghosh, J Andrews, Analytical evaluation of fractional frequency reuse for OFDMA cellular networks. IEEE Trans Wireless Comm **10**(12), 4294–4305 (2011)
20. T Novlan, R Ganti, A Ghosh, J Andrews, Analytical evaluation of fractional frequency reuse for heterogeneous cellular networks. IEEE Trans Comm **60**(7), 2029–2039 (2012)
21. FH Panahi, T Ohtsuki, *Stochastic Geometry Based Analytical Modeling of Cognitive Heterogeneous Cellular Networks, IEEE International Conf. on Comm. (ICC)*, 2014, pp. 5281–5286
22. H ElSawy, E Hossain, Two-tier HetNets with cognitive femtocells: downlink performance modeling and analysis in a multi-channel environment. IEEE Trans Mobile Computing **13**(3), 649–663 (2014)
23. H ElSawy, E Hossain, DI Kim, *HetNets with Cognitive Small Cells: User Offloading and Distributed Channel Allocation Techniques, IEEE Communications Magazine, Special Issue on Heterogeneous and Small Cell Networks (HetSNets), vol.* , 2013
24. H ElSawy, E Hossain, *Channel Assignment and Opportunistic Spectrum Access in Two-Tier Cellular Networks with Cognitive Small Cells, in Proc. IEEE Global Communications Conference (Globecom)*, 2013
25. H ElSawy, E Hossain, *On Cognitive Small Cells in Two-Tier Heterogeneous Networks, Workshop on Spatial Stochastic Models for Wireless Networks (SpaSWiN), in conjunction with 11th Intl. Symposium on Modeling and Optimization in Mobile, Ad Hoc, and Wireless Networks (WiOpt)*, 2013
26. M Haenggi, RK Gant, *Interference in Large Wireless Networks. Foundations and Trends in Networking, NOW Publisher*, 2008. vol. 3, no. 2
27. IEEE Instrumentation, M Society, *IEEE 1588 Standard for a Precision Clock Synchronization Protocol for Networked Measurement and Control Systems, in IEEE Std 1588*, 2008
28. Q Wu, G Ding, J Wang, Y-D Yao, Spatial-temporal opportunity detection in spectrum-heterogeneous cognitive radio networks: two dimensional sensing. IEEE Trans Wireless Comm **12**(2), 516–526 (2013)
29. Y Chen, Q Zhao, A Swami, Joint design and separation principle for opportunistic spectrum access in the presence of sensing errors. IEEE Trans Inf Theory **54**(5), 2053–2071 (2008)
30. F Paisana, N Marchetti, LA Dasilva, *Radar, TV and Cellular Bands: Which Spectrum Access Techniques for which Bands? IEEE Comm. Surveys & Tutorials, vol.16, no.3, pp.1193-1220, Third Quarter*, 2014
31. Q Chen, M Motani, W-C Wong, Y-C Liang, *Opportunistic Spectrum Access Protocol for Cognitive Radio Networks, in Proc. IEEE International Conference on Comm. (ICC)*, 2011, pp. 1–6
32. X Li, H Liu, S Roy, J Zhang, P Zhang, C Ghosh, Throughput Analysis for a Multi-User, Multi-Channel ALOHA Cognitive Radio System. IEEE Trans. on Wireless Comm. **11**(11), 3900–3909 (2012)
33. Y-C Liang, Y Zeng, E Peh, AT Hoang, Sensing-throughput tradeoff for cognitive radio networks. IEEE Trans Wireless Comm **7**(4), 1326–1337 (2008)
34. L Luo, C Ghosh, S Roy, *Joint Optimization of Spectrum Sensing for Cognitive Radio Networks, in Proc. IEEE Global Telecommunications Conference (GLOBECOM)*, 2010, pp. 1–5

35. L Wang, V Fodor, On the gain of primary exclusion region and vertical cooperation in spectrum sharing wireless networks. IEEE Trans Veh Tech **61**(8), 3746–3758 (2012)
36. T Yucek, H Arslan, *A Survey of Spectrum Sensing Algorithms for Cognitive Radio Applications, IEEE Comm. Surveys & Tutorials, vol. 11, no. 1, pp. 116–130, First Quarter*, 2009
37. M Haenggi, Outage, local throughput, and capacity of random wireless networks. IEEE Trans Wireless Comm **8**(8), 4350–4359 (2009)
38. J-M Dricot, G Ferrari, A Panahandeh, H Fr, D Ph De, *Probabilistic Coexistence and Throughput of Cognitive Dual-Polarized Networks. EURASIP Journal on Wireless Comm. and Networking*, 2010. 387625

Emulating extreme velocities of mobile LTE receivers in the downlink

José Rodríguez-Piñeiro[1*], Martin Lerch[2], José A. García-Naya[1], Sebastian Caban[2], Markus Rupp[2] and Luis Castedo[1]

Abstract

Long-Term Evolution (LTE) is expected to substitute the Global System for Mobile (GSM) Communications as the radio access technology for railway communications. Recently, considerable attention has been devoted to high-speed trains since this particular environment poses challenging problems in terms of performance simulation and measurement. In order to considerably decrease the cost and complexity of high-speed measurement campaigns, we have proposed a technique to induce effects caused by highly time-varying channels on Orthogonal Frequency-Division Multiplexing (OFDM) signals while conducting measurements at low speeds. In this work, we illustrate the performance of this technique by comparing the results of LTE measurements at different velocities in a controlled measurement environment. Additionally, we validate this technique by means of simulations, considering one of the scenarios defined as part of the Winner Phase II Channel Models, specifically designed for high-speed train scenarios.

Keywords: LTE; OFDM; Measurement; High speed

1 Introduction

Over the last few years, broadband communication between nodes moving at high speeds has attracted special attention. One of the most relevant research topics in this field is the modeling of the High-Speed Train (HST) channel. Nowadays, the most widely used communication system between trains and the elements involved in operation, control, and intercommunication within the railway infrastructure is based on the Global System for Mobile (GSM) Communications. This technology, namely, the GSM for Railways (GSM-R), is not well-suited for supporting advanced services such as automatic pilot applications or provisioning broadband services to the train staff and passengers. Besides trains, the increasing number of broadband services available for mobile devices motivated the migration from third-generation mobile networks to fourth-generation ones, mainly Long-Term Evolution (LTE). Therefore, LTE seems to be a good candidate to substitute the GSM as the fundamental technology for railway communications.

Several radio channel models have been proposed for moving radio interfaces, such as the International Telecommunication Union (ITU) channel models [1]. The first channel modeling approach for high-speed train scenarios was proposed by Siemens in 2005 [2]. More recent proposals are the Winner Phase II Model (high-speed moving networks were included in [3]) and the radio channel models approved by the ITU Radiocommunication (ITU-R) Sector for the evaluation of IMT-Advanced Technologies (high-speed train scenarios were explicitly considered in [4]). However, only a few results based on empirically obtained data which can validate the aforementioned channel models are available. One example of high-speed train channel modeling contributions based on the results obtained by means of measurement campaigns is the propagation path-loss model proposed in [5]. The same experimental results were also used for the definition of the large-scale model proposed in [6]. An efficient channel sounding method for high-speed train scenarios using cellular communications systems is proposed in [7] by considering Wideband Code Division Multiple Access (WCDMA) signals.

One of the reasons that explains the low number of measurement campaigns in high-speed environments is their complexity and cost. Furthermore, it is not possible,

*Correspondence: j.rpineiro@udc.es
[1] Department of Electronics and Systems, University of A Coruña, Facultade de Informática, Campus de Elviña, A Coruña 15071, Spain
Full list of author information is available at the end of the article

in most cases, to measure at high speeds in controlled environments in a reproducible and repeatable way. In addition, measuring in high-speed trains demands for specific hardware and software solutions (see [8]). In order to address those problems, we proposed a technique to induce the effects caused by highly time-varying channels in Orthogonal Frequency-Division Multiplexing (OFDM) signals while conducting the measurements at much lower speeds [9]. This technique consists basically in reducing the subcarrier spacing of the OFDM signal by scaling down the bandwidth of the whole OFDM signal. More specifically, we propose to interpolate the transmit OFDM signal in the time domain before its transmission. The time-interpolated signal still conveys exactly the same information as the original one but with a reduced subcarrier spacing, thus artificially increasing the sensitivity to the Inter-Carrier Interference (ICI). For example, if we time interpolate the transmit OFDM signal by a factor I, the subcarrier spacing will be reduced by the same factor I, which is similar to what would happen if the transmission was conducted at I times the original speed.

In [9], we have proposed this novel methodology for evaluating the performance of OFDM transmissions in real-world scenarios affected by large Doppler spreads while conducting the measurements at low speeds. More specifically, in [9], we considered the transmission of standard-compliant WiMAX Mobile (IEEE 802.16e) signals, both through simulations and realistic outdoor measurements. When conducting measurements with movement in realistic environments and using off-the-shelf vehicles (e.g., cars, trains), the repeatability of the results is basically lost because the following issues arise:

- It is extremely difficult to keep the vehicle speed constant during the measurements.
- Reaching high-speed values with experimental user equipment is expensive and sometimes even not possible in realistic environments.
- Impacts from measurement control can always arise (e.g., another vehicle driving nearby) leading to non-repeatable results.
- Modifications in the measurement scenarios (e.g., the path of the road is modified, a huge vehicle is parked close to the measurement path) hinder fair comparisons with future measurements.
- It is quite challenging to drive several times along exactly the same path.

In order to validate the technique proposed in [9] while, at the same time, overcoming all the abovementioned issues found when conducting non-repeatable measurements using off-the-shelf vehicles, we performed the assessment employing a setup that allows for repeatable measurements at different speeds up to 200 km/h and considered LTE as the wireless communication standard. This setup consists in an antenna that is rotated around a central pivot at a constant speed, allowing for validating our previously proposed technique under repeatable as well as controlled conditions.

We complete the validation by means of simulations considering a channel model suited for high-speed train scenarios. More specifically, we consider the channel model associated with the D2a link of the D2 scenario of the Winner Phase II Channel Models [3]. Notice that our intention is not to model the measurement environment using a channel model existing in the literature. Our intention is to prove the validity of our results in a perfectly repeatable (and controlled) environment which is widely accepted for simulating high-speed train conditions. Therefore, measurement and simulation scenarios correspond to very different wireless communication environments since the simulations consider a rural macro-cell with Line-of-Sight (LoS) between transmitter and receiver, while an outdoor-to-indoor communication without LoS is considered for the measurements.

The presented results, which were obtained under repeatable conditions, both, through measurements and simulations, show that our technique induces highly time-varying channels with excellent agreement.

The rest of the paper is organized as follows. Section 2 explains the proposed time interpolation technique to induce high-speed effects on OFDM signals while conducting measurements at much lower speeds. Section 3 describes the setup and measurement methodology for measurements as well as for simulations. Different figures of merit for the results are also introduced. Section 4 presents both, measurement and simulation results, and shows the excellent agreement obtained with the proposed technique. Finally, Section 5 is devoted to conclusions.

2 Emulating high speeds by time interpolation

Let us consider an OFDM modulation where N subcarriers are multiplexed to construct each OFDM symbol. OFDM symbols are cyclically extended by adding N_g samples, thus having a total length of $N_t = N + N_g$ samples. Accordingly, the k-th OFDM symbol can be represented as:

$$\mathbf{s}_k = \mathbf{G}_1 \mathbf{F}^H \mathbf{x}_k, \quad (1)$$

where $\mathbf{x}_k$ is a $N \times 1$ vector containing the constellation symbols corresponding to the transmitted subcarriers in the k-th symbol, $\mathbf{F}$ is the standard $N \times N$ DFT matrix, $\mathbf{G}_1$ is a $N_t \times N$ matrix which cyclically extends the OFDM symbol, and $\mathbf{s}_k$ is a $N_t \times 1$ vector with the transmitted OFDM symbol in the time domain. This signal is transmitted at a sampling period T_s, hence creating symbols with a duration $T_t = T_s N_t$ and a bandwidth $F_s = 1/T_s$. When transmitting $\mathbf{s}_k$ over a time-varying channel and assuming a sufficiently long cyclic prefix, the received signal is [10]:

$$\mathbf{r}_k = \mathbf{F}\mathbf{G}_2\left(\mathbf{H}_k^{(t)}\mathbf{G}_1\mathbf{F}^H\mathbf{x}_k + \mathbf{n}_k\right) = \mathbf{H}_k\mathbf{x}_k + \mathbf{w}_k, \quad (2)$$

where $\mathbf{H}_k^{(t)}$ is a $N_t \times N_t$ matrix with the time-varying channel impulse response experienced by $\mathbf{s}_k$, $\mathbf{G}_2$ is a $N \times N_t$ matrix to remove the cyclic prefix, $\mathbf{n}_k$ is a $N_t \times 1$ vector with uncorrelated complex-valued white Gaussian noise entries with power σ^2, $\mathbf{H}_k$ is a $N \times N$ matrix with the channel frequency response, and $\mathbf{w}_k$ is a $N \times 1$ vector containing uncorrelated complex-valued white Gaussian noise entries with variance σ^2.

If the channel is time invariant, $\mathbf{H}_k$ will be a diagonal matrix. In time-selective channels, however, non-zero entries will appear outside the main diagonal of $\mathbf{H}_k$ and ICI arises in the received signal. The amount of ICI relates to the normalized Doppler spread of the channel, which is given by $D_n = f_d T$, f_d being the maximum Doppler frequency and $T = T_s N$ the duration of the OFDM symbol excluding the cyclic prefix. As proposed in our previous work [9], parameter T can be adjusted by time interpolation by a factor I, yielding an OFDM symbol duration $T^I = IT_s N$. Therefore, given the actual velocity v of the mobile receiver, the normalized Doppler spread, impacting the time-interpolated OFDM signal can be written as:

$$D_n^I = f_d T^I = f_d I T_s N = \frac{T_s N I f_c v}{c} = \frac{T_s N f_c}{c} v^I, \quad (3)$$

with f_c the carrier frequency, c the speed of light, and $v^I = Iv$ the emulated speed as a result of an actual measurement speed v and an interpolation factor I. Consequently, enlarging the symbol length T^I by adjusting I allows for the emulation of a velocity v^I while conducting measurements at an actual speed v. Notice that this procedure is also valid for time decimation of the signal, simply taking $0 < I < 1$, leading to an emulated speed lower than the actual one. Finally, note also that I does not have to be an integer value. Fractional time interpolation and decimation factors are also possible, hence providing a great flexibility for adjusting v^I from v.

In our setup, time interpolation factors $I \geq 1$ were applied to standard-compliant downlink LTE OFDM signals before the over-the-air transmission to emulate situations with I times higher velocity than the actual receiver speed. The same principle was considered for the simulations.

3 Evaluation setup and procedure

We use the measurement setup shown in Figure 1 (measurement branch) to test the proposed technique of emulating high speeds by time interpolation of OFDM signals. The setup consists of:

1. *Signal generation (transmitter side) and signal processing (receiver side)*: at the transmitter side, standard-compliant LTE subframes are generated using the LTE Downlink Link-Level Simulator developed at the TU Wien [11]. At the receiver side, the same simulator is used to process the received signals and to estimate the following figures of merit: uncoded Bit Error Ratio (BER), coded BER, Error Vector Magnitude (EVM), and Signal-to-Noise Ratio (SNR).
2. *Time interpolation and time decimation*: the signal is time-interpolated by a factor I at the transmitter and decimated by the same factor I at the receiver side. This way, we emulate a Doppler spread similar to that obtained with a speed increase by a factor of I.
3. *Signal transmission and signal reception*: signals are transmitted over the air by using a testbed developed

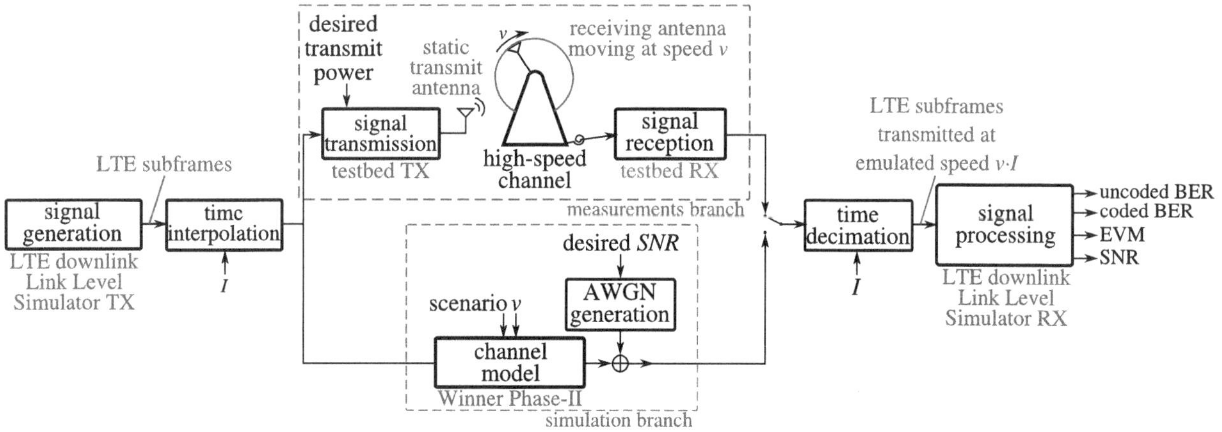

Figure 1 Block diagram of the considered setup. 'Measurement branch' is considered for the measurements, while 'simulation branch' is used for the simulations.

at the TU Wien [12,13]. The testbed transmitter is placed outdoors on a roof of a building in downtown Vienna, Austria. The receiver is placed indoors in the fifth floor of an adjacent building at a distance of 150 m (see Figure 2), hence recreating an infrastructure-to-vehicle scenario. Note that no feedback channel was used in our experiments. Consequently, no adaptive modulation and coding schemes were applied. Channel Quality Indication (CQI) values were fixed in advance. Notice also that the testbed is equipped with a highly precise time and frequency synchronization system based on Global Positioning System (GPS)-disciplined rubidium oscillators and a custom-made synchronization unit [14]. As a result, we can assume perfectly time and frequency synchronized measurements. Thus, the results are not affected by time or frequency offsets due to imperfect synchronization.

4. *Generation of high-speed conditions*: high-speed conditions are generated by transmitting from a static transmit antenna to a receive antenna that is rotated around a central pivot in a controlled and repeatable way [15,16] (see Figure 3). Different channel realizations are created by measuring at different initial positions of the receive antenna. Note that the trajectory of the antenna is well approximated by a straight line since the diameter of the trajectory is 2 m and each LTE subframe is 1 ms long (see Figure 4).

3.1 Measurement setup

In order to evaluate the impact of high-speed conditions on LTE transmissions, actual velocities ranging from 50 to 200 km/h were considered. Furthermore, interpolation factors of $I = 1$ (no interpolation), $I = 2$, and $I = 3$ were used for generating Doppler spreads equivalent to those associated to velocities ranging from 50 to 600 km/h. Note that it is possible to generate exactly the same Doppler spread value from different combinations of the actual measurement velocity and the interpolation factor (see Table 1). We considered this fact to show that our technique allows for the evaluation of wireless communication systems at high speeds while measuring at much lower speeds. In order to do that, we generated the same Doppler spread by means of different velocities and interpolation factors, and then, we compared the obtained results. Table 1 shows the combinations of actual speeds and interpolation factors which lead to equal Doppler spreads (each row of the table corresponds to a different Doppler spread factor).

3.2 Measurement procedure

Several measurements are conducted for each velocity and interpolation factor. More specifically, three different positions of the whole receiver (including the antenna rotation unit) along its rails (see Figure 3) are considered. For each of these positions, 10 measurements per velocity and interpolation factor are carried out, each starting at a given angular shift on the circumference defined by the receive antenna rotating around a central pivot. Therefore, 30 different channel realizations are measured in total for each pair of velocity and interpolation factor. Furthermore, the whole set of measurements is repeated for each SNR value, among those specified in Table 2. The transmit power employed at the testbed transmitter

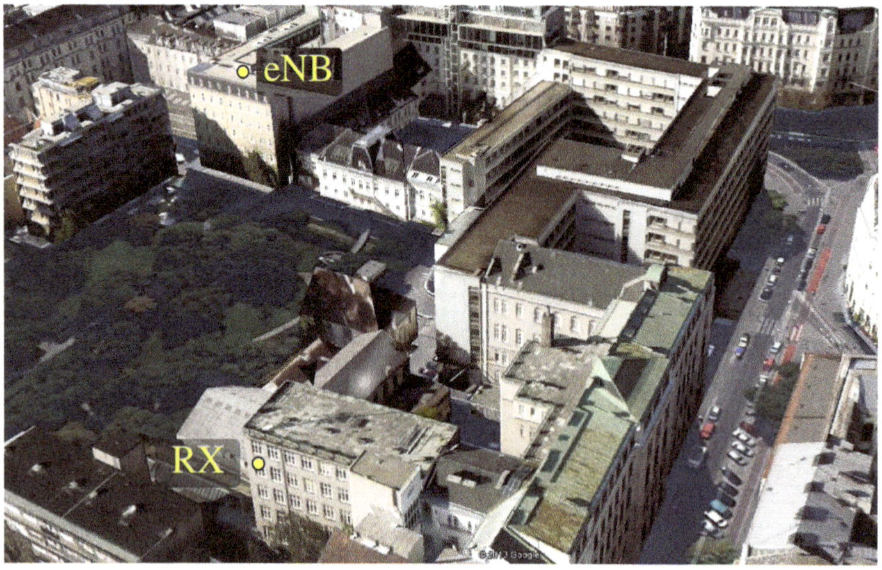

Figure 2 Location of the transmitter and the receiver. Note that the transmitter is installed outdoors on a roof of a building, while the receiver is placed indoors in the fifth floor of an adjacent building. The distance between transmitter and receiver is approximately 150 m.

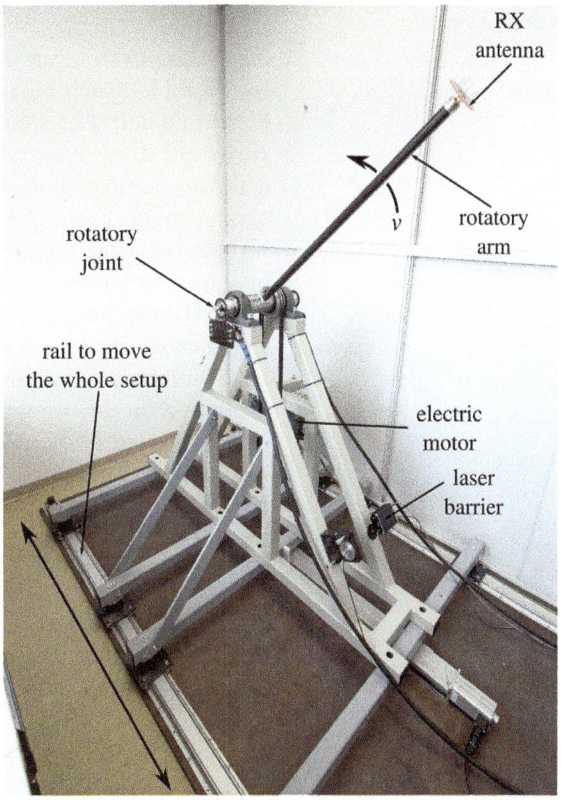

Figure 3 Setup used for generating high-speed conditions. The receive antenna is rotated around a central pivot, generating high-speed conditions in a controlled and repeatable way.

is modified in order to achieve different average SNR values.

In order to be able to compare the results gathered from different interpolation factors, we have to ensure that the receive antenna trajectory, spectrum usage, and average OFDM symbol energy remain constant for each emulated velocity $v \cdot I$:

1. *Equal receive antenna trajectory*: in order to maintain a constant Doppler spread, the rotation speed has to be divided by the interpolation factor (see Table 1). Therefore, as it is shown in Figure 4, the trajectory of the receive antenna during the transmission of one LTE subframe does not vary when changing the interpolation factor and antenna velocity as long as the emulated speed is the same.
2. *Equal spectrum usage*: when a subframe is interpolated by a factor of I, its bandwidth is decreased by the same factor, which in principle reduces the effect of the channel frequency diversity. In order to experience the same spectrum, I replicas of the interpolated signal are transmitted to ensure that the whole frequency range of the original signal is used. The results are then averaged. Figure 5 shows an example of this procedure for $I = 2, I = 3$, and for an arbitrary interpolation factor.

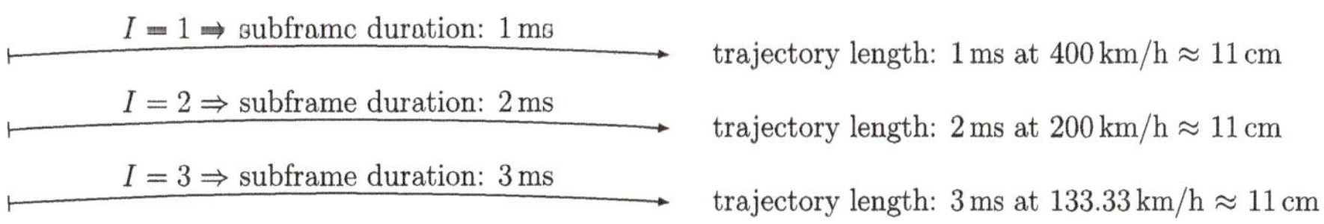

Figure 4 Trajectory followed by the antenna during the acquisition of a single LTE subframe. An emulated speed of 400 km/h was used and the three interpolation factors $I = \{1, 2, 3\}$ were considered. Evaluating the results at a speed value of 400 km/h can be done 1) by measuring at the actual velocity of 400 km/h without time interpolation ($I = 1$); 2) by measuring at half the speed with $I = 2$; and 3) by measuring at $400/3 = 133.33$ km/h with $I = 3$. In all cases, the length of the trajectory described by the antenna is the same (close to 11 cm) and can be approximated by a straight line since the antenna describes a circle with a diameter of 2 m while rotating around the central pivot.

Table 1 Emulated speed values (expressed in km/h) that can be obtained from more than an actual velocity *v*

Emulated speed, $v^I = Iv$	$I = 1$	$I = 2$	$I = 3$
100	$v = 100$	$v = 50$	–
150	$v = 150$	$v = 75$	$v = 50$
200	$v = 200$	$v = 100$	$v = 66.6$
250	–	$v = 125$	$v = 83.3$
300	–	$v = 150$	$v = 100$

Notice that these are the speed values considered in the error curves in Figures 6, 7, 8, 9, 10, and 11.

3. *Equal average transmit energy per OFDM symbol*: in order to preserve the average energy per OFDM symbol, the interpolated signals are scaled in amplitude by a factor of $\sqrt{I}$ before being transmitted.

Notice that our main objective is to validate the proposed technique considering a realistic scenario while, at the same time, keeping its complexity under reasonable terms, hence simplifying the validation process and facilitating the comprehension of the results. Therefore, we consider over-the-air standard-compliant LTE transmissions as a good example of the state of the art in wireless communications. However, due to the abovementioned simplicity reasons, we restrict the evaluation to a single-user scenario without feedback channel, so modulation and coding scheme values are fixed beforehand instead of being adapted according to the channel quality.

Table 2 Main parameters for measurements as well as for simulations

Parameters	Values		
Bandwidth, F_S (MHz)	15.36 (9 occupied)		
FFT size, N	1,024		
Used subcarriers	600 (excluding DC)		
Velocities, v (km/h)	50, 66.6, 75, 83.3, 100, 125, 150, and 200		
Carrier frequency, f_C (GHz)	2.5		
Interpolation factors, I	1, 2, and 3		
SNR (dB)	38, 31, 21, 11, 1, and ∞ (simulations only)		
CQI values	1	8	12
Modulation	4-QAM	16-QAM	64-QAM
Code rate	0.076	0.48	0.65
Efficiency	0.1523	1.9141	3.9023
Source bits	1,192	15,288	31,192
Coded bits	16,000	32,000	48,000
Peak throughput (Mbit/s)	1.192	15.288	31.192

Although three CQI values were evaluated and considered throughout the text to validate the proposed technique, results in the figures only consider CQI = 12.

Each LTE frame in the Frequency-Division Duplex (FDD) mode consists of ten subframes, each one following a different structure depending on the subframe index. For example, synchronization signals are always allocated in the first and sixth subframes, while broadcast channels are included, if necessary, in the first subframe. In order to make the comparison between channel realizations independent of the subframe structure, we always transmit the seventh subframe, considering only pilot and data symbols in subsequent evaluations. Consequently, we ensure the same payload for each channel realization, which enables us to fairly compare the results of different measurements as the subframe structure is kept invariant all the time. More specifically, the number of data symbols that can be transported in the seventh subframe according to the system configuration is 8,000, which leads to the number of coded bits, source bits, and peak throughput values specified in Table 2.

3.3 Simulation environment

We employ the setup shown in Figure 1 (simulation branch) to validate through simulations our technique of emulating high speeds by time interpolation. We select noise variance values that lead to the same SNR values estimated from the measurements. A channel model suited for high-speed train scenarios was considered, namely, the model defined by the D2a link of the D2 scenario of the Winner Phase II Channel Models [3]. This scenario models the channel between an antenna placed at the roof of a train carriage and a network of fixed eNodeBs installed in a rural environment, which usually leads to LoS propagation conditions. Speeds up to 350 km/h can be considered for the train movement.

Notice that the considered channel model does not intend to reproduce the wireless channel observed in the measurements. Our intention is to complement the evaluation of the proposed technique in a simulation environment while, at the same time, avoiding the process of characterizing the doubly selective wireless channel observed in the measurements, which is not the intention of this work. Finally, we selected the aforementioned channel model because it is specifically designed to model an outdoor-to-outdoor high-speed scenario with LoS between transmitter and receiver, while the measurement environment corresponds to an outdoor-to-indoor scenario with Non-Line-of-Sight (NLoS) between them.

3.4 Simulation procedure

The number of evaluated subframes by simulation was the same considered when measuring, i.e., 30 channel realizations for each velocity and interpolation factor pair. The values of SNR, velocity, and interpolation factors were also kept unchanged. We also guarantee an equal spectrum usage as well as equal average transmit energy per OFDM

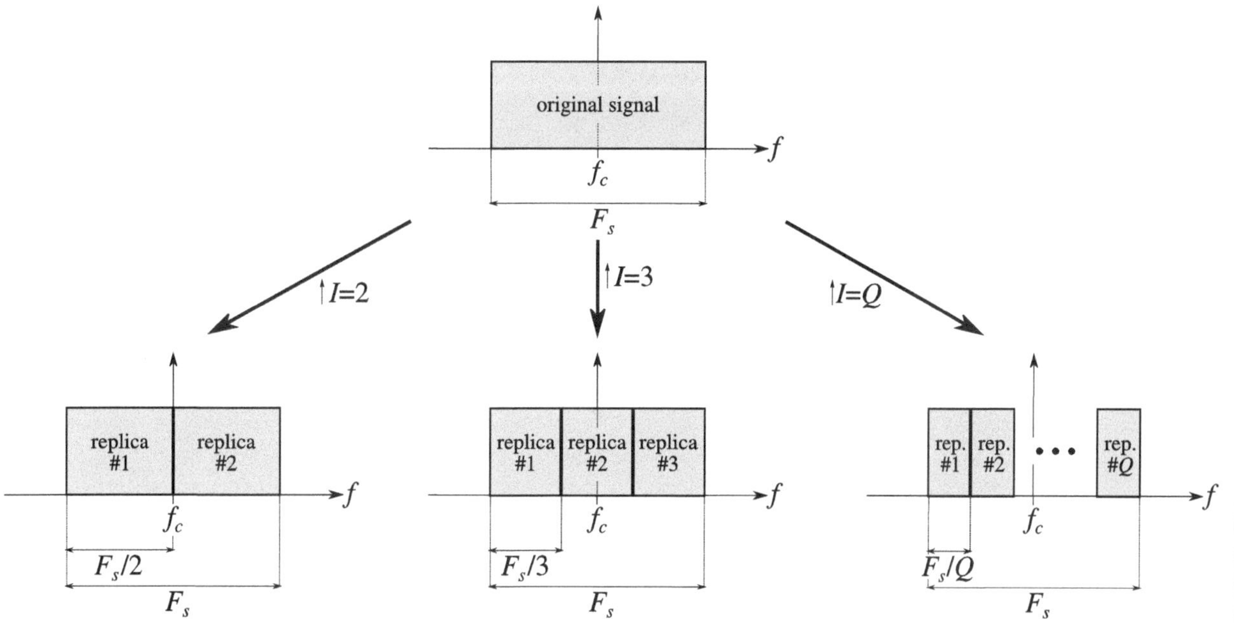

Figure 5 Ensuring equal spectrum usage for interpolation factors $I = 2$, $I = 3$, and for an arbitrary integer interpolation factor $I = Q$. I replicas of the interpolated signal are transmitted to ensure that the whole frequency range of the original signal (without interpolation) is used.

symbol. Furthermore, to fairly compare the results, the channel model was fed with identical initial conditions (e.g., delays and mean power per path) for each evaluated interpolation factor. This way, we model a situation in which the receiver moves along the same path for each interpolation factor.

Table 2 details the most relevant parameters for the experimental evaluations as well as for the simulations.

3.5 Figures of merit

We have selected three different figures of merit that evaluate the quality of the signal at different points in the receive signal processing chain: 1) EVM calculated without knowing the transmitted symbols; and 2) uncoded BER, which is the BER after the symbol decision, and coded BER, which corresponds to the BER at the output of the channel decoder.

Recall that our main objective is to inspect the level of agreement between the results obtained from those figures of merit when the wireless system is evaluated at actual speed conditions with respect to those obtained when the system is evaluated (under similar conditions) using emulated speeds by time interpolation. Given that the testbed employed for the measurements guarantees excellent time and frequency synchronization between transmitter and receiver, the results of the three figures of merit enumerated above greatly depend on the channel equalizer, which is strongly affected by ICI.

The main reason for considering both, EVM and uncoded BER, is that EVM is an unbounded and continuous metric, which is specially valuable when the SNR levels observed are high enough to saturate the BER to its minimum value of zero. On the other hand, considering uncoded BER in the evaluation is very convenient not only because it is one of the most used performance metrics in wireless communications but also because the EVM is calculated only from the received symbols. Hence, EVM results loose accuracy as the corresponding SNR value decreases. Finally, coded BER is specially relevant in our evaluation since the proposed technique causes a signal bandwidth reduction proportional to the time interpolation factor, hence reducing the frequency diversity and potentially degrading the channel decoder performance. Consequently, one could expect that, in terms of coded BER, the wireless system would perform differently when actual speeds are considered in comparison to emulated ones.

With the aim of evaluating of the level of agreement between the results obtained by means of actual speed with respect to emulated ones, we have included curves for the relative error values computed for those emulated speed values that can be obtained from more than one actual velocity (see Section 4). We considered the emulated speed values that can be obtained from at least two of the three interpolation factors considered (see Table 1).

The procedure followed to calculate each figure of merit is detailed below:

SNR. The SNR is estimated considering exclusively the data subcarriers. Thus, the guard subcarriers, Direct

Current (DC) subcarrier as well as the pilots are discarded. SNR is estimated by performing the following steps:

1. Noise samples for a given velocity v and interpolation factor I are captured directly (measurement case) or generated (simulation case). In the measurement case, noise samples in the time domain are captured with the transmitter switched off (hence, not transmitting any signal).
2. The captured noise samples in the time domain are then processed as if they were actual data samples, i.e., they are downsampled by the corresponding interpolation factor (if required), the cyclic prefix is removed, a Fast Fourier Transform (FFT) is performed, and both guard subcarriers, pilots, and also the DC subcarrier are discarded.
3. As a result, $w^{(l,k,s)}$ noise samples are obtained in the frequency domain, each one corresponding to the l-th subcarrier of the k-th OFDM symbol of the s-th subframe and for the values of l corresponding to the indexes of the data subcarriers.
4. All previous steps are repeated when the transmitter is switched on, so $r^{(l,k,s)}$ data samples are obtained in the frequency domain.
5. The average SNR for a given velocity v and interpolation factor I is then estimated by averaging out the SNR values corresponding to each channel realization. Defining:

$$\bar{r} = \frac{1}{LKS} \sum_{l=1}^{L} \sum_{k=1}^{K} \sum_{s=1}^{S} \left| r^{(l,k,s)} \right|^2, \tag{4}$$

and:

$$\bar{w} = \frac{1}{LKS} \sum_{l=1}^{L} \sum_{k=1}^{K} \sum_{s=1}^{S} \left| w^{(l,k,s)} \right|^2, \tag{5}$$

then, the SNR is calculated as follows:

$$\mathrm{SNR} = \frac{\bar{r} - \bar{w}}{\bar{w}}, \tag{6}$$

with L, K, S the total number of data subcarriers, OFDM symbols, and subframes, respectively.

EVM. The EVM is obtained based on the equalized symbols which feed the receiver decision. The following steps are considered for the EVM estimation:

1. The dynamic range of the equalized symbols is bounded. This is realistic in a practical receiver. In this sense, real and imaginary parts of the symbols are clipped to a maximum value. This avoids symbols having extremely large modulus (e.g., due to imperfect zero-forcing channel equalization), hence distorting the EVM estimation. Clipping values were selected, taking into account the mean power of the equalized symbols in perfect conditions (flat channel in the absence of noise).
2. A hard decision unit is fed with the clipped symbols. Let $\mathbf{s}_c = \left(s_c^1, s_c^2, \ldots, s_c^S\right)^T$ be the vector of S clipped symbols in a subframe and $\mathbf{s}_d = \left(s_d^1, s_d^2, \ldots, s_d^S\right)^T$ the vector of decided symbols.
3. The EVM per subframe, expressed in decibels, is obtained as:

$$\mathrm{EVM} = 10 \log_{10} \left(\frac{P_r}{P_c} \right), \tag{7}$$

with $P_c = \frac{1}{S}\mathbf{s}_c^T \cdot \mathbf{s}_c$, $P_r = \frac{1}{S}\mathbf{s}_r^T \cdot \mathbf{s}_r$, and $\mathbf{s}_r = |\mathbf{s}_c - \mathbf{s}_d|$. Notice that the aforementioned procedure for calculating the EVM does not require knowing the transmitted symbols. However, the higher the uncoded BER, the more underestimated the EVM is.

4 Results

This section presents both, measurement and simulation results. Three types of performance curves are included in the result graphs, which are:

- *Red solid lines*: they correspond to the cases with no interpolation ($I = 1$). According to the speed values considered for the measurements as well as for the simulations, the red solid curves range from 50 to 200 km/h.
- *Green dashed lines*: they correspond to the cases with $I = 2$, so the emulated velocity is twice the measured or simulated speed. Therefore, the green curves range from 100 to 400 km/h.
- *Pink dotted lines*: they correspond to the cases where $I = 3$. In this case, the emulated velocity is three times the measured or simulated speed, and the curves range from 150 to 600 km/h.

Besides the performance curves, we have also included the relative error curves for all figures although in some cases not all error curves are plotted. Three types of relative error curves are included in the graphs presenting the results (see Table 1), which are:

- *Green dashed lines*: they correspond to the relative difference between the results obtained when the interpolation factor $I = 2$ is employed and when actual speeds are used.
- *Blue dashed lines*: they correspond to the relative difference of the results between the interpolation factors $I = 3$ and $I = 2$.
- *Pink dotted lines*: they correspond to the relative difference of the results between the interpolation factors $I = 3$ and $I = 1$ (actual speeds).

The relative error values are calculated differently for EVM and for BER (coded and uncoded):

- *Relative error for EVM*: given EVM_A and EVM_B the instantaneous EVM values, corresponding to a channel realization and obtained for the interpolation factors $I = A$ and $I = B$, respectively, we define the instantaneous relative error as:

$$E_{\mathrm{EVM}}(A,B) = 100 \cdot \frac{\mathrm{EVM}_A - \mathrm{EVM}_B}{\mathrm{EVM}_B} \quad [\%]. \qquad (8)$$

- *Relative error for uncoded and coded BER*: given W_A and W_B the number of received bits estimated without errors, corresponding to a channel realization and obtained for the interpolation factors $I = A$ and $I = B$, respectively, we define the instantaneous relative error for the BER as:

$$E_{\mathrm{BER}}(A,B) = 100 \cdot \frac{W_A - W_B}{W_B} \quad [\%]. \qquad (9)$$

Mean relative error values for both error types are calculated by averaging instantaneous relative error values for all channel realizations. These mean relative error values are plotted in Figures 6, 7, 8, 9, 10, and 11 together with their corresponding 95% BCa bootstrap confidence intervals for the mean [17]. These confidence intervals are plotted as an area around each curve and in the same color as the corresponding curve. We also gauge the precision of the results by calculating the 95% confidence intervals for the mean.

Due to limitations of the inverter driving the motor used to rotate the receive antenna (see Figure 3), the actual velocity range considered for the measurements (and also for the simulations) starts at 50 km/h, while the maximum actual speed value is set to 200 km/h. Given that we consider interpolation factors $I = 1, 2, 3$, in the speed range from 50 to 300 km/h, at least two curves overlap, thus allowing for evaluating the level of agreement of the results between different interpolation factors.

4.1 Measurement results

Figure 6 shows the measured EVM for CQI = 12 (64-QAM) for SNR values ranging from 38 to 11 dB. For high SNR values, it can be seen that the higher the emulated speed is, the worse the EVM becomes, as expected. However, it can be seen that the curves for different interpolation factors (as well as their associated confidence regions) overlap, which demonstrates that the proposed technique of emulating high speeds by time interpolation performs adequately. Results for CQI = 1 (4-QAM) and CQI = 8 (16-QAM) were not included, but they have been also evaluated, and they show that the achieved performance is very similar despite the considered modulation scheme. Figure 6 also shows that the quality of the received signal, in terms of EVM, is considerably affected by the speed for high SNR values, showing that the main source contributing to the signal distortion is the ICI. As the SNR decreases, however, the performance is less affected by the

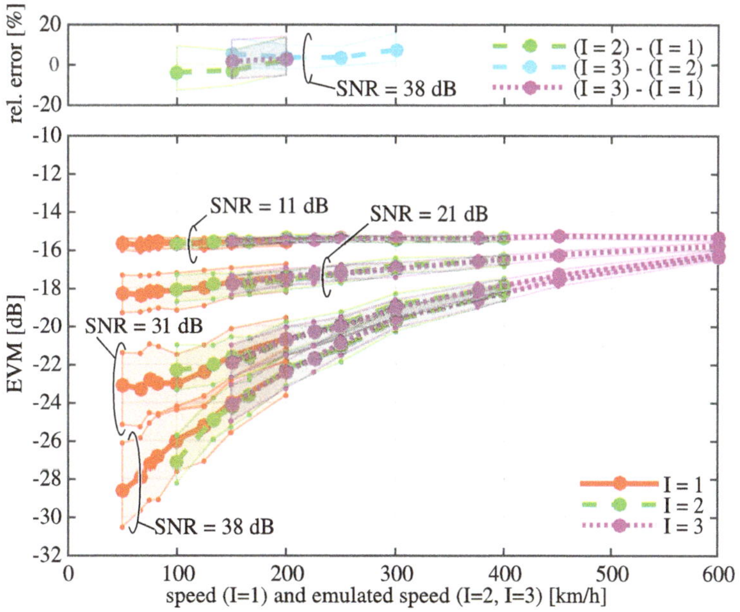

Figure 6 Measured EVM for different SNR values. EVM results obtained by measuring when CQI = 12 and SNR ranges from 11 to 38 dB for the interpolation factors $I = 1, 2, 3$. 95% confidence regions are provided. We have also evaluated the EVM for CQI = 1 (4-QAM) and CQI = 8 (16-QAM), and the results are almost indistinguishable. Corresponding relative error curves are plotted when the SNR is set to 38 dB, showing an excellent agreement.

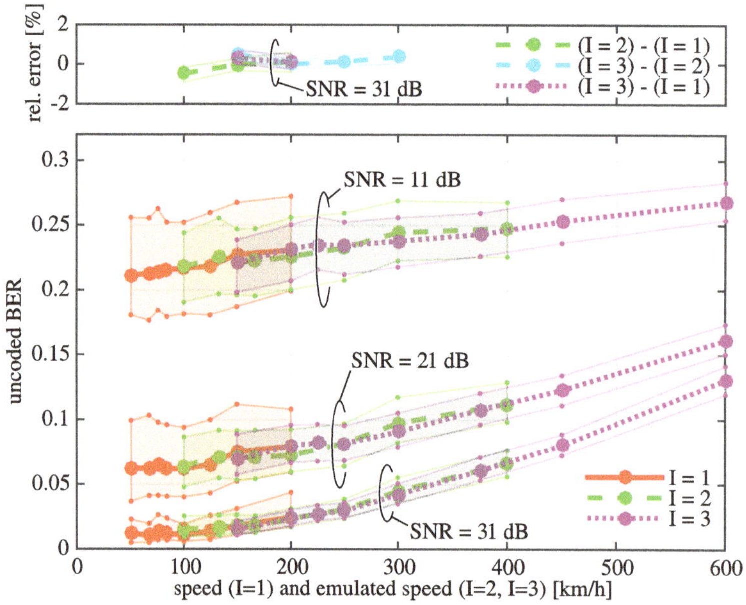

Figure 7 Measured uncoded BER for different SNR values. Uncoded BER results obtained by measuring when CQI = 12 and SNR ranges from 11 to 31 dB for the interpolation factors $I = 1, 2, 3$. 95% confidence regions are provided. Corresponding relative error curves are plotted when the SNR is set to 31 dB, showing an excellent agreement.

emulated speed because the noise is the main contributor to the signal distortion.

The relative error curves in Figure 6 also demonstrate the excellent agreement of the results independently of the source of the Doppler spread: actual speed or emulated velocity from time interpolation. Note that the confidence intervals for the relative error values are mainly influenced by the number of realizations averaged. Hence, these confidence intervals could be decreased by increasing the number of realizations measured. However, as the

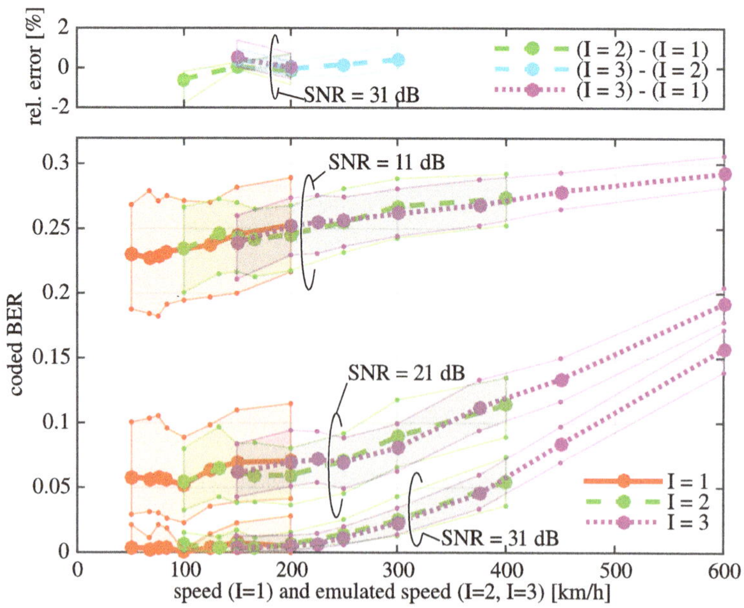

Figure 8 Measured coded BER for different SNR values. Coded BER results obtained by measuring when CQI = 12 and SNR ranges from 11 to 31 dB for the interpolation factors $I = 1, 2, 3$. 95% confidence regions are provided. Corresponding relative error curves are plotted when the SNR is set to 31 dB, showing an excellent agreement.

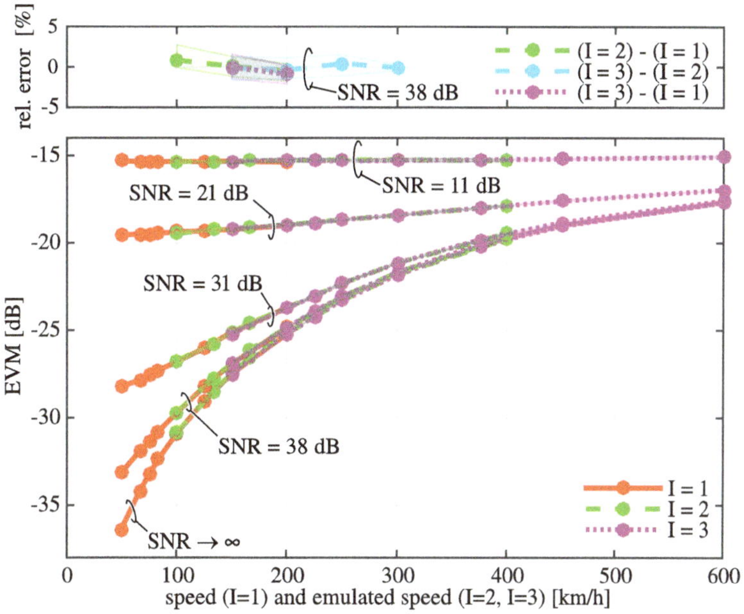

Figure 9 Simulated EVM for different SNR values. EVM results obtained by simulation when CQI = 12 and SNR ranges from 11 dB to infinity for the interpolation factors $I = 1, 2, 3$. Note that the channel model associated to the D2a link of the D2 scenario of the Winner Phase II Channel Models was applied. 95% confidence regions are provided. We have also evaluated the EVM for CQI = 1 (4-QAM) and CQI = 8 (16-QAM), and the results are almost indistinguishable. Corresponding relative error curves are plotted when the SNR is set to 38 dB, showing an excellent agreement.

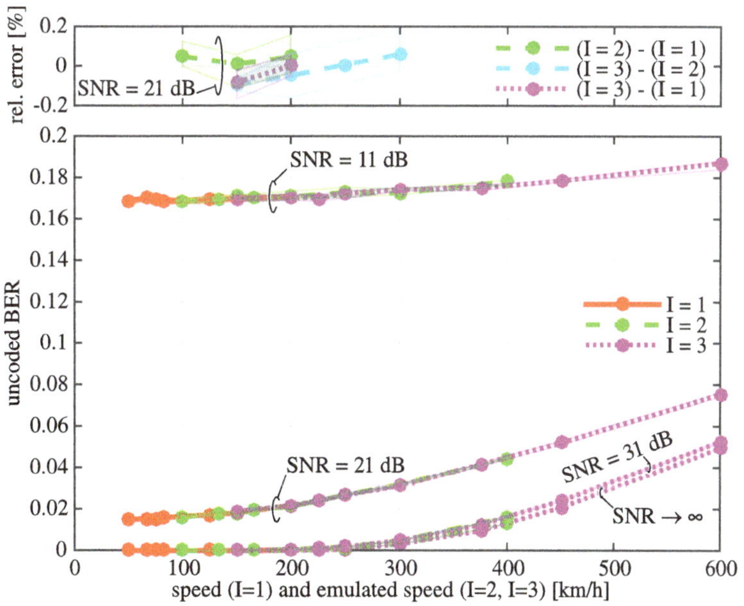

Figure 10 Simulated uncoded BER for different SNR values. Uncoded BER results obtained by simulation when CQI = 12 and SNR ranges from 11 to 31 dB for the interpolation factors $I = 1, 2, 3$. Note that the channel model associated to the D2a link of the D2 scenario of the Winner Phase II Channel Models was applied. 95% confidence regions are provided. Corresponding relative error curves are plotted when the SNR is set to 21 dB, showing an excellent agreement. Note that for SNR values greater or equal than 31 dB, the BER curves are close to zero for speeds below 250 km/h; hence, the relative error curves do not provide much information.

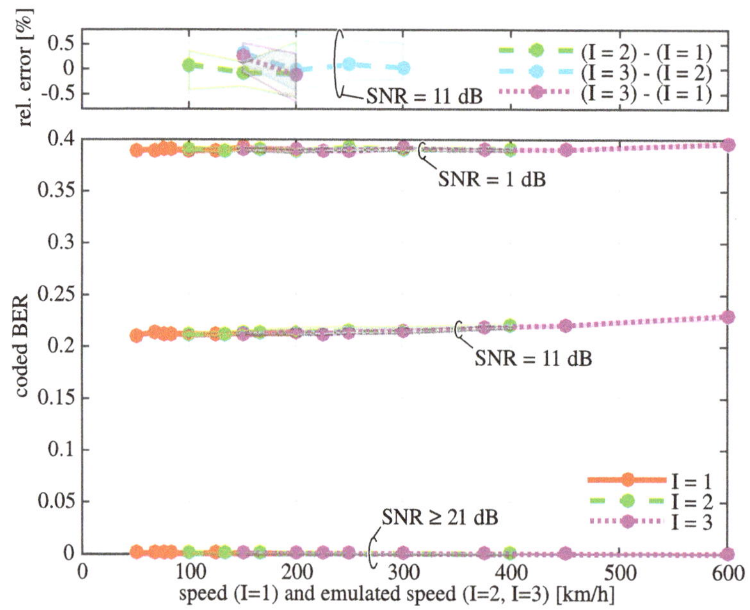

Figure 11 Simulated coded BER for different SNR values. Uncoded BER results obtained by simulation when CQI = 12 and SNR ranges from 1 dB to infinity for the interpolation factors $I = 1, 2, 3$. Note that the channel model associated to the D2a link of the D2 scenario of the Winner Phase II Channel Models was applied. 95% confidence regions are provided. Corresponding relative error curves are plotted when the SNR is set to 11 dB, showing an excellent agreement. Note that for SNR values greater or equal than 21 dB, the BER curves are close to zero for all considered speeds; hence, the relative error curves do not provide much information.

receive antenna trajectory is kept constant between different interpolation factors, the relative error is mainly influenced by the repeatability of the setup, the noise, and the interpolation itself. For this reason, we can achieve relative agreements with mean error values below 10% in the worst case (i.e., the highest SNR value of 38 dB) without increasing the number of channel realizations.

Figures 7 and 8 show the measured uncoded BER and coded BER, respectively, for different SNR values when CQI = 12. Both, uncoded BER and coded BER, are strongly dependent on the emulated speed for high SNR values, while the noise is the main source of signal distortion when the SNR decreases. It can also be observed that curves for different interpolation factors overlap both for uncoded BER and coded BER, demonstrating again the excellent results of the proposed technique. Notice also that, as shown in the corresponding relative error curves in Figures 7 and 8, the agreement of the proposed technique is below 2 %, both in terms of uncoded BER and coded BER. Therefore, such a proposed technique can be effectively used for emulating high-speed effects at any point of the receive signal processing chain.

4.2 Simulation results

The channel model of the simulations was selected because of its suitability for the simulation of high-speed conditions. Notice that such a channel model leads to frequent LoS propagation conditions [3]. However, the receiver in the measurements is located inside a building, while the eNodeB is placed on a roof of another building (see Figure 2), and hence, NLoS propagation conditions arise, leading to a much more relevant multipath effect.

Figure 9 shows the simulated EVM for CQI = 12 (64-QAM) for SNR values ranging from 38 to 11 dB. Additionally, we have also evaluated the results in the absence of noise (infinite SNR). Note that the slopes of the curves are similar to those obtained by measurements as shown in Figure 6, while the obtained EVM values are always lower for the same SNR value. This result is expected because the channel model considered for the simulations is easier to equalize due to the lower variability of the simulated channel. However, when the SNR decreases (SNR = 11 dB and SNR = 21 dB in our case), the above-mentioned effect vanishes as the noise becomes the main contribution to the signal distortion and the corresponding curves approach each other for simulations and for measurements.

The variability of the results obtained by simulations is much lower than that of the measurements because, on the one hand, the simulated channel is less variable than that observed in the measurements. On the other hand, exactly the same initial channel conditions are applied in the simulations for each interpolation factor and velocity pair, thus ensuring much less variability between the different channel realizations generated for each interpolation factor and velocity pair. In fact, the confidence region of each simulated curve completely overlaps the curve itself, which causes that the confidence regions cannot be appreciated although they are included. Additionally, curves simulated for different interpolation factors almost completely overlap, hence showing the good behavior of the proposed technique also in this case. This precise behavior is also confirmed by the corresponding relative error curves in Figure 9.

Finally, Figures 10 and 11 show the simulated uncoded BER and coded BER, respectively, for different SNR values when CQI = 12. Some level of dependency on the emulated speed for high SNR values can be still appreciated for the uncoded BER, while the coded BER is always zero for SNR values greater than or equal to 21 dB since the channel decoder is able to correct all errors. Notice that this is not the case for the measurements as the channel is much more difficult to equalize. Curves for different interpolation factors overlap both for uncoded BER and coded BER, thus showing the good behavior of the proposed interpolation technique. This is also confirmed by the corresponding error curves.

5 Conclusions

We have shown that time interpolating OFDM signals prior to their transmission followed by the corresponding decimation at the receiver is a suitable technique for inducing high-speed effects (mainly ICI) while actually measuring at much lower speeds. The key idea behind the proposed technique is that the ICI level experienced by the received signals after OFDM demodulation and before channel equalizer depends on the relative factor between subcarrier spacing and Doppler spread; thus, instead of changing the Doppler spread, one can change the subcarrier spacing by the same factor to obtain the same results. The main advantage of the proposed technique is its feature to offer experimental evaluation of OFDM-based wireless communication systems at very high velocities while conducting measurements at much lower speeds. The price to be paid is that the signal bandwidth is reduced proportionally to the time interpolation factor, and therefore, potential inaccuracies could arise due to the loss of frequency diversity. However, to combat this loss in diversity, we proposed to transmit different replicas of the interpolated signal to cover the same bandwidth as in the non-interpolated case.

We have designed a measurement methodology for validating our technique in a controlled environment and under repeatable conditions. The basic idea consists in evaluating different figures of merit for a certain range of actual speed values and to compare them with emulated speeds of the same magnitude under the same conditions. For example, when the figures of merit considered are evaluated at 150 km/h, three different cases are evaluated under the same conditions: 1) the receive antenna moves at the actual speed of 150 km/h; 2) the receive antenna moves at the actual speed of 75 km/h and a time interpolation factor $I = 2$ is used, hence the emulated speed is $75 \times 2 = 150$ km/h; and 3) the receive antenna moves at 50 km/h, and $I = 3$ is applied. Besides the measurement results, we have also evaluated the proposed technique through simulations based on a channel model specifically designed for modeling high-speed scenarios. We selected EVM, uncoded BER, and coded BER as figures of merit, all being evaluated for different SNR, CQI, and speed values. All results based on the three figures of merit have shown an excellent agreement between actual and emulated speeds for all the interpolation factors considered, hence validating the proposed technique as a good candidate to be considered not only for physical layer performance evaluations but also for higher layer ones.

Besides validating the proposed technique, we have also shed light on the effects caused by high and even extremely high speeds on OFDM signals. More specifically, we have selected LTE as a waveform example of the state of the art of current mobile wireless systems. As expected beforehand, received signals suffer a similar degradation (in terms of EVM, uncoded BER, and coded BER) when the SNR decreases or when the speed (and correspondingly the ICI level) increases. However, low SNR values can conceal the effects due to high speeds as the noise becomes the main contributor of the signal distortion. We have shown that the level of signal distortion (measured in terms of EVM) caused by high-speed conditions does not depend on the modulation scheme.

We have also shown that the proposed technique performs excellent not only in outdoor-to-indoor measurement scenarios but also in simulations for a channel model specifically designed for outdoor-to-outdoor high-speed scenarios. In both cases, the level of agreement between the results for actual speed and for emulated speeds is at the level of possible measurement accuracy, as the corresponding relative error curves have shown.

Finally, in the light of the excellent agreement of the results obtained under repeatable conditions for the three figures of merit considered (EVM, uncoded BER, and coded BER) and regardless actual or emulated speeds are considered, it can be concluded that the proposed technique is valid for inducing high-speed effects at any point in the signal processing chain at the receiver.

Competing interests

The authors declare that they have no competing interests.

Acknowledgements

This work has been funded by the Christian Doppler Laboratory for Wireless Technologies for Sustainable Mobility, KATHREIN Werke KG, and A1 Telekom Austria AG. This work was supported in part by the Xunta de Galicia, MINECO of Spain, and FEDER funds of the E.U. under Grant 2012/287, Grant IPT-2011-1034-370000, Grant TEC2013-47141-C4-1-R, Grant FPU12/04139, and Grant EST13/00272. The financial support by the Austrian Federal Ministry of

Economy, Family and Youth and the National Foundation for Research, Technology and Development is gratefully acknowledged.

Author details
[1]Department of Electronics and Systems, University of A Coruña, Facultade de Informática, Campus de Elviña, A Coruña 15071, Spain. [2]Institute of Telecommunications, TU Wien, Gusshausstrasse 25/389, A-1040 Vienna, Austria.

References

1. Radiocommunication Sector of International Telecommunication Union (ITU-R), Guidelines for evaluation of radio transmission technologies for IMT-2000. ITU-R Recommendation M.1225 (1997)
2. 3GPP TSG-RAN Working Group 4 (Radio), *High speed environment channel models (R4-050388). 3GPP TSG-RAN Working Group 4 (Radio) meeting #35*, (Athens, Greece, 2005)
3. P Kyosti, J Meinila, L Hentila, X Zhao, T Jamsa, C Schneider, M Narandzic, M Milojevic, AHJ Ylitalo, Veli-Matti, Holappa, M Alatossava, R Bultitude, Y de Jong, T Rautiainen, IST-4-027756 WINNER II D1.1.2 V1.1: WINNER II Channel Models (2007)
4. Radiocommunication Sector of International Telecommunication Union (ITU-R), Guidelines for evaluation of radio interface technologies for IMT-Advanced, Report ITU-R M.2135-1 (2009)
5. H Wei, Z Zhong, K Guan, B Ai, in *2010 IEEE 5th International ICST Conference on Communications and Networking in China*. Path loss models in viaduct and plain scenarios of the high-speed railway (IEEE, USA Beijing, China, 2011), pp. 1–5
6. K Guan, Z Zhong, B Ai, in *2011 IEEE Third International Conference on Communications and Mobile Computing (CMC)*. Assessment of LTE-R using high speed railway channel model (IEEE,USA, Quingdao, China, 2011), pp. 461–464. doi:10.1109/CMC.2011.34
7. L Liu, C Tao, T Zhou, Y Zhao, X Yin, H Chen, A highly efficient channel sounding method based on cellular communications for high-speed railway scenarios. EURASIP J. Wireless Commun. Netw. **2012**(1), 1–16 (2012)
8. J Rodríguez-Piñeiro, JA García-Naya, A Carro-Lagoa, L Castedo, in *2013 Euromicro Conference on Digital System Design*. A testbed for evaluating LTE in high-speed trains (IEEE, USA, Santander, Spain, 2013), pp. 175–182. doi:10.1109/DSD.2013.27
9. Rodríguez-Piñeiro, J, P Suárez-Casal, JA García-Naya, L Castedo, C Briso-Rodríguez, JI Alonso-Montes, in *2014 IEEE Eighth IEEE Sensor Array and Multichannel Signal Processing Workshop*. Experimental validation of ICI-aware OFDM receivers under time-varying conditions (IEEE, USA, A Coruña, Spain, 2014). doi:10.1109/SAM.2014.6882411
10. Z Wang, GB Giannakis, Wireless multicarrier communications. IEEE Signal Process. Mag. **17**(3), 29–48 (2000). doi:10.1109/79.841722
11. C Mehlführer, JC Ikuno, M Simko, S Schwarz, M Wrulich, M Rupp, The Vienna LTE simulators - enabling reproducibility in wireless communications research. EURASIP J. Adv. Signal Process. **2011**, 1–13 (2011)
12. M Lerch, S Caban, M Mayer, M Rupp, The Vienna MIMO testbed: evaluation of future mobile communication techniques. Intel Technol. J. **18**(3), 58–69 (2014)
13. S Caban, JA Garcia-Naya, M Rupp, Measuring the physical layer performance of wireless communication systems: part 33 in a series of tutorials on instrumentation and measurement. IEEE Instrum. Meas. Mag. **14**(5), 8–17 (2011). doi:10.1109/MIM.2011.6041377
14. S Caban, A Disslbacher-Fink, JA Garcia-Naya, M Rupp, in *Proc. International Instrumentation and Measurement Technology Conference (I2MTC 2011)*. Synchronization of wireless radio testbed measurements (IEEE, USA Binjiang, Hangzhou, China, 2011). doi:10.1109/IMTC.2011.5944089
15. S Caban, R Nissel, M Lerch, M Rupp, in *Proc. of 6th Extreme Conference on Communication and Computing (ExtremeCom)*. Controlled OFDM measurements at extreme velocities (Association for Computing Machinery (ACM), USA San Cristobal, Galapagos, Ecuador, 2014)
16. S Caban, J Rodas, JA Garcia-Naya, in *2011 IEEE Instrumentation and Measurement Technology Conference*. A methodology for repeatable, off-line, closed-loop wireless communication system measurements at very high velocities of up to 560 km/h (IEEE, USA Binjiang, China, 2011), pp. 1–5. doi:10.1109/IMTC.2011.5944019
17. B Efron, DV Hinkley, *An Introduction to the Bootstrap (CRC Monographs on Statistics & Applied Probability)*, 1st edn. (Chapman & Hall, USA, 1994)

14

Study of coexistence between indoor LTE femtocell and outdoor-to-indoor DVB-T2-Lite reception in a shared frequency band

Ladislav Polak*, Lukas Klozar, Ondrej Kaller, Jiri Sebesta, Martin Slanina and Tomas Kratochvil

Abstract

Nowadays, the demand for high-quality multimedia services (video, audio, image, and data) is rapidly increasing. The Digital Video Broadcasting - terrestrial (DVB-T) standard, its second-generation version (DVB-T2), and the Long-Term Evolution (LTE) standard are the most promising systems to fulfill the demand for advanced multimedia services (e.g., high-definition image and video quality), especially in Europe. However, LTE mobile services can operate in a part of the UHF band allocated to DVB-T/T2 TV services previously. The main purpose of this work is to explore the possible coexistences of DVB-T2-Lite and LTE systems in the same shared frequency band (co-channel coexistence) under outdoor-to-indoor and indoor reception conditions. Furthermore, an applicable method for evaluating coexistence scenarios between both systems is shown with a particular example. These coexistence scenarios can be noncritical and critical. In the first case, both systems can coexist without significant performance degradation. In the second one, a partial or full loss of DVB-T2-Lite and/or LTE signals can occur. We consider an indoor LTE femtocell and outdoor-to-indoor DVB-T2-Lite signal reception in a frequency band from 791 up to 821 MHz. Simulations of combined indoor and outdoor signal propagation are performed in MATLAB using 3rd Generation Partnership Project (3GPP) channel models, separately for both DVB-T2-Lite and LTE systems. Correctness of path loss simulation results is verified by measurements. Afterwards, an appropriate linear model is proposed which enables to evaluate the impact of coexistence on performance of both systems in outdoor-to-indoor and indoor-to-indoor reception scenarios. The results are related to an actual location in the building and are presented in floor plans. The floor plans include different coexistence conditions (different power imbalance and different amount of overlay of the radio channels). Service availability of both systems is verified again by measurements. The resulting maps help better understand the effect of coexistence on achievable system performance under different indoor/outdoor reception situations considering real transmission conditions.

Keywords: Channel model; Coexistence; DVB-T2-Lite; Indoor and outdoor-to-indoor propagation; LTE femtocell; Path loss; QEF; EVM; CQI; RF measurement

1 Introduction

Advanced wireless communication systems can provide users with any type of multimedia. Thanks to this, the idea to 'connect, upload, download, share and transfer anything at anytime and anywhere' is not a futuristic vision [1,2]. From the viewpoint of service providers, efficient usage of limited resources in the radio frequency (RF) spectrum is one of the biggest challenges. Hence, the increasing density of wireless networks and the increasing volume of user equipment (UE) terminals in use escalate the risk of unwanted coexistence scenarios [3,4].

In the near future, the next-generation digital terrestrial television broadcasting (DVB-T2/T2-Lite) and Long-Term Evolution (LTE) systems will be deployed to provide multimedia services for mobile and portable scenarios, mainly in Europe. DVB-T2-Lite [5-8] is a new profile which was added to the DVB-T2 system specification in April 2012. This subset within DVB-T2 is very perspective for mobile and portable TV broadcasting as it is designed to support

* Correspondence: polakl@feec.vutbr.cz
Department of Radio Electronics, SIX Research Center, Brno University of Technology, Technicka 3082/12, 616 00 Brno, Czech Republic

low-capacity applications for advanced handheld receivers [9]. It is based on the same core of technologies as the DVB-T2 standard but uses only a limited number of available modes. By avoiding the modes, which require the most computational power and memory [6], the necessary complexity of T2-Lite-only receivers is reduced. DVB-T2-Lite, compared to the first-generation DVB-T/H [10], can support TV content delivery with higher flexibility. Moreover, it can operate in VHF (from 174 up to 230 MHz) and UHF (from 470 up to 870 MHz) bands, allocated earlier for DVB-T/H. From the viewpoint of system flexibility, spectral efficiency, and available transmission scenarios, DVB-T2-Lite is the system of choice for the next-generation terrestrial mobile and portable digital TV broadcasting.

Third Generation Partnership Project (3GPP) LTE [11-13] technology brings a new concept, based on the Orthogonal Frequency Division Multiple Access (OFDMA), into mobile communications. LTE supports high data rates and flexible system configuration in order to adapt transmission parameters to the actual state of a radio link. LTE architecture involves a specific type of cells called femtocells. These short ranges, mainly indoor cells, improve coverage in desired areas, especially buildings. Femtocells are served by a special type of base station called Home eNodeB (HeNB). LTE can exploit the same UHF frequency bands which are already available for existing 2G/3G networks (e.g., bands: 800, 900, 1,800, and 2,600 MHz). Moreover, additional ranges (from 2.5 up to 2.7 GHz) and the 700-MHz band are also allocated for LTE usage. The European Union decided to harmonize the '800 MHz band' in favor of the LTE services, starting from January 2013 [4]. Consequently, DVB-T/T2 and LTE services can occupy either the same or adjacent frequency spectrum. As a result, unwanted coexistence between DVB-T/T2 and LTE services can occur [4,14].

This work deals with the study of possible co-channel coexistence between DVB-T2-Lite (outdoor-to-indoor reception) and LTE services (provided by the femtocell) under fixed indoor reception conditions.

The paper is organized as follows. An overview of related work in the field of different wireless standards' coexistence, especially DVB and LTE, is presented in Section 2. This section also includes a detailed list of aims and contributions of this work. A description of the explored coexistence scenario and the considered DVB-T2-Lite system parameters are presented in Section 3. Section 4 contains a description of the applied simulation method and the proposed measurement testbed together with its detailed setup. Results obtained from simulation and measurements are presented and discussed in Section 5. Finally, Section 6 concludes the paper.

2 Related works

Undesirable interactions between similar or different kinds of wireless communication systems, operating in adjacent or shared frequency bands, are not a new phenomenon [3,15-18]. The exploration, monitoring, measurement, and possible suppression of interferences are a hot topic. This fact is also evidenced by many published studies available. Authors of [19] studied the possible inter-band interferences between UMTS and GSM systems. In another work [20], the coexistence between advanced wireless systems and International Mobile Telecommunication-Advanced (IMT-A) services is explored. Different kinds of coexistence scenarios in LTE networks are analyzed in [21-23]. Possible methods to mitigate or suppress interferences from coexistence between two different wireless systems are outlined in [24-26].

In the last decade, researchers' attention has been devoted to the study of different coexistence scenarios between the DVB-T/T2 and LTE/LTE-A standards.

Table 1 summarizes the previously explored coexistence scenarios between such systems. From the presented works, it is clearly seen that many times the researchers use either only simulation tools or only different measuring methods to explore the coexistence scenarios. Furthermore, in most works, a scenario is considered in which macrocells are used to provide LTE service coverage, coexisting with DVB-T2-Lite services in the same or adjacent frequency band. The main aim of this research article is to explore the interaction of DVB-T2-Lite and LTE in a shared frequency band, such that femtocells (HeNB) are used to provide LTE indoor coverage. Attention is devoted to availability monitoring of DVB-T2-Lite and LTE services in different locations under fixed indoor reception conditions. For this purpose, an appropriate simulation model is proposed and verified by measurement. Based on these results, non-critical (both DVB-T2-Lite and LTE system working) and critical (partial or full loss of DVB-T2-Lite and/or LTE signals) coexistence scenarios can be identified and the general conclusions are outlined. To the best of our knowledge, no similar exploration in this form has been presented in any scientific or technical paper so far.

3 Considered coexistence scenario

The investigated coexistence scenario between the DVB-T/T2 and LTE RF signals in the fixed indoor transmission scenario is shown in Figure 1. The main system parameters of DVB-T2-Lite and LTE systems, considered in this work, can be found in Table 2. The DVB-T2-Lite TV signal is broadcast in a single frequency network (SFN) at a center frequency of 794 MHz and received by UE1 in a building. In the same building, LTE femtocells are deployed and the HeNB provides mobile connectivity in a channel belonging to Band 20 (from 791 up to 821 MHz). A user of UE2

Table 1 Comparison of explored coexistence scenarios between DVB-T/T2 and LTE systems

Reference	Coexistence scenario (TV broadcast scenario)	Type of interference	Results	Evaluation parameters
[42]	DVB-T vs. LTE (fixed)	Mutual co-channel	Simulation	SNR, SINR, QoS
[43]	LTE vs. DVB-T (fixed, portable)	Adjacent channel	Simulation	PR, CR
[44]	DVB-T vs. LTE (fixed)	Adjacent channel	Simulation	BER, PF, PR
[45]	DVB-T vs. LTE (fixed)	Intersystem	Simulation	IPL, spectral overlap
[27]	LTE-A vs. DVB-T (fixed)	Intersystem	Simulation	ADL, FO
[46]	LTE vs. DVB-T (fixed)	Intersystem (co-channel)	Measurement	Data throughput, RSRQ
[47]	DVB-T vs. LTE (fixed)	Co-channel	Simulation	I/N, C/(N+I)
[48]	DVB-T/H vs. LTE (fixed)	Co-channel	Measurement	SSIM, QEF, SIR
[30]	LTE vs. DVB-T2-Lite (mobile)	Adjacent channel	Measurement	SDR, QEF, EVM, MER
[31]	DVB-T/T2 vs. LTE (partly mobile, fixed)	Co-channel and adjacent channel	Measurement	SDR, BER, EVM, MER
This study	DVB-T2-Lite vs. LTE (fixed)	Co-channel (partial overlapping)	Simulation/Measurement	QEF, partly CQI, EVM

Abbreviations: ADL, antenna discrimination loss; PER, packet error ratio; BER, bit error ratio; PF, picture failure; CI, carrier-to-interference ratio; PR, protection ratio; CR, correction factor; QEF, quasi error-free; CQI, channel quality indicator; QoS, quality-of-service; C/(N+I), carrier-to-noise+interference ratio; RSRQ, reference signal received quality; EVM, error vector magnitude; SDR, spectral density ratio; FO, frequency offset; SIR, signal-to-interference ratio; IPL, interference power level; SINR, SIR plus noise-ratio; I/N, interference associated to new sources; SNR, signal-to-noise ratio; MER, modulation error ratio; SSIM, structural similarity.

establishes connection with HeNB at downlink frequency band from 795 (797.2 MHz) to 805 MHz (817.2 MHz). We consider that the bandwidth of the LTE signal is 10 or 20 MHz, and intersystem frequency overlapping is from 0.8 up to 3 MHz. Consequently, coexistence between HeNB (supporting 3GPP LTE Release 9) and DVB-T2-Lite system can occur. As a specific type of coexistence, a partial overlapping scenario is assumed. It means that the channel of the interferer (in this case LTE) partially overlaps with the channel of the victim (in this case DVB-T2-Lite) [27]. It is assumed that both UEs are stationary.

4 Simulation and measurement setup

In this section, the simulation method, used to explore the coexistence of digital TV and mobile RF signals under outdoor-to-indoor and indoor-to-indoor conditions, is presented. Furthermore, the proposed measurement testbed and its setup, used in this work, are introduced. The simulation and measurement campaign consists of the following:

1. Simulation (propagation loss) and measurement of LTE performance in different locations (indoor and outdoor environment);
2. Simulation (propagation loss) and measurement of DVB-T2-Lite performance in different locations (indoor and outdoor environment);
3. Simulation and measurement of simultaneous transmission (signal propagation) of both LTE and DVB-T2-Lite RF signals in order to evaluate the

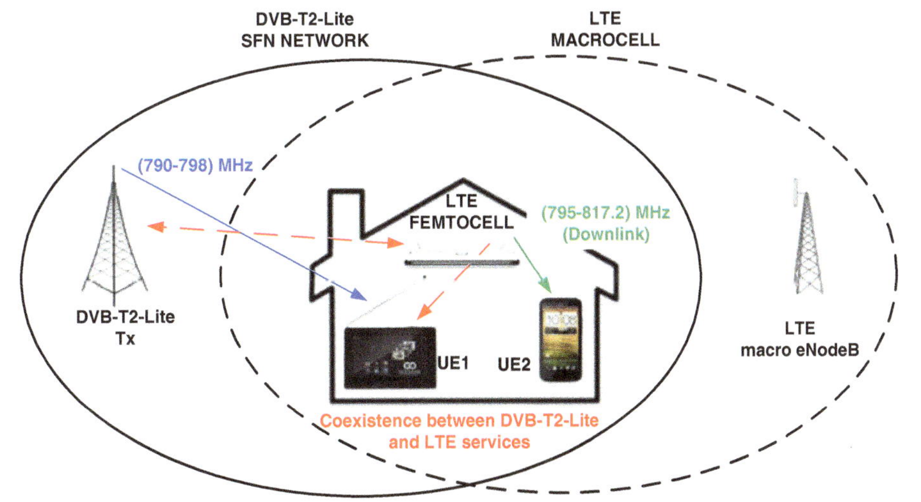

Figure 1 Unwanted coexistences between DVB-T2-Lite and LTE services at fixed indoor transmission scenario. Supposed scenario where LTE femtocell is indoors and DVB-T2-Lite signal penetrates from outdoor transmitter and affects performance.

Table 2 DVB-T2-Lite and LTE main system parameters considered in this study

Definition of parameters	DVB-T2-Lite	LTE (Release 9)
Type of FEC scheme	BCH and LDPC	Turbo
FEC code rate	2/3	1/3
Type of modulation	16QAM	QPSK 16QAM 64QAM
Constellation rotation	No	-
IFFT size	2,048 (2K)	1,024 (10 MHz) 2,048 (20 MHz)
Type of PP pattern	PP2	-
Guard interval duration	28 μs	4.7 μs
Transmission mode	SISO	SISO
Carrier frequency (MHz)	794	Downlink (791 ÷ 821)
Channel bandwidth	8 MHz	10 MHz, 20 MHz
RF power	(0.1 to 5) W	(0.01 to 0.06) W
Channel environment	Outdoor-to-indoor	Indoor (indoor-to-outdoor)
FEC decoding method	1D LLR [6]	Max Log-map
Tx antenna height (m) (above floor)	2	1
Rx antenna height (m) (above floor)	1	1
LTE user equipment	-	Huawei e389u-15 (LTE UE category 3)

BCH, Bose-Chaudhuri-Hocquenghem; LLR, log likelihood ratio; FEC, forward error correction; PP, pilot pattern; IFFT, inverse fast Fourier transform; SISO, single-input single-output; LDPC, low-density parity-check.

influence of coexistence on the performance of both systems (on physical layer (PHY) level); and

4. Identification of the noncritical (both systems can coexist) and critical (partial or full loss of DVB-T2-Lite and LTE signal) coexistence scenarios for both systems.

4.1 Simulation setup

The considered coexistence scenario was briefly outlined in the previous section. In this work, we assume that transmitters and receivers are located on the seventh floor (the top floor) in the building of Brno University of Technology (BUT), Faculty of Electrical Engineering and Communications (FEEC) in Brno. Laboratories of Digital TV Technology and Radio Communications, and Mobile Communications of the Department of Radio Electronics (DREL) are located on this floor. The floor plan of the seventh floor is shown in Figure 2. Approximate dimensions of the floor are 50 × 25 m. The HeNB is located in the Laboratory of Mobile Communication Systems (room 7107), and the DVB-T2-Lite transmitter is located outdoor on the terrace.

The whole simulation model is realized in MATLAB. Propagation of the LTE and DVB-T2-Lite RF signals are simulated separately. The simulation of separate propagation loss of LTE and DVB-T2-Lite RF signals will be used as the reference (no coexistence).

The simulation model consists of three main parts for both LTE and DVB-T systems. The first part represents the simulation of a link budget, according to the 3GPP recommendation for system level simulations [28,29] for both coexisting systems. Signal strength in the receiver can be expressed as follows:

$$P_{RX} = P_{TX} - L_{TXC} + G_{TXA} - PL + G_{RXA} - L_{RXC} \quad (1)$$

where P_{TX} is transmitter power, L_{TXC} are wiring losses, G_{TXA} is transmitting antenna gain, G_{RXA} is receiving antenna gain, L_{RXC} are wiring losses, and finally, P_{RX} is received signal level. Path losses in wireless transmission are denoted as PL (for details see Equation 3). Value of P_{TX} is known from the transmitter setup. Values of L_{TXC}, L_{RXC}, G_{TXA}, and G_{RXA} are constants depending on the used equipment (for details see Subsection 4.2). The second part represents the validation of obtained results from the simulation according to the performed measurement and their interpretation in a map. Details are in Subsections 4.1 and 4.2. The last part compares power imbalance of tested radio channels and computed achievable performance of both systems in certain locations. Details are given in Section 5.

The propagation scenario of the LTE RF signal in femtocell involves indoor-to-indoor line-of-sight (LOS) propagation for the same room where HeNB is located (room 7107) and non-LOS (NLOS) for other indoor locations. Path losses are modeled according to the 3GPP recommendation for indoor LTE femtocell as described in [28], denoted as UE to HeNB, where UE is inside the same building as HeNB. In order to model indoor-to-outdoor propagation from the HeNB to the measurement points on the terrace, the original equation was extended with outdoor wall penetration loss. On the other hand, the recommendations in [28] are generally valid for frequencies around 2 GHz, but we exploit an 800-MHz band in this study. Therefore, it is necessary to perform a correction as described in [29]. This correction defines the correction factor for 800 MHz as follows:

$$PL_{COR} = 20\log_{10}(f_c) \quad (2)$$

where f_c is carrier frequency in MHz.

The resulting path loss equation is:

$$\mathrm{PL} = 38.46 + 20\log(d) + 0.7d_{in} + \mathrm{LP}_{floor} + qL_{INwall} + nL_{OUTwall} + \mathrm{PL}_{COR} \quad (3)$$

where d is the distance between the HeNB and the UE, d_{in} is indoor propagation distance, LP_{floor} is penetration

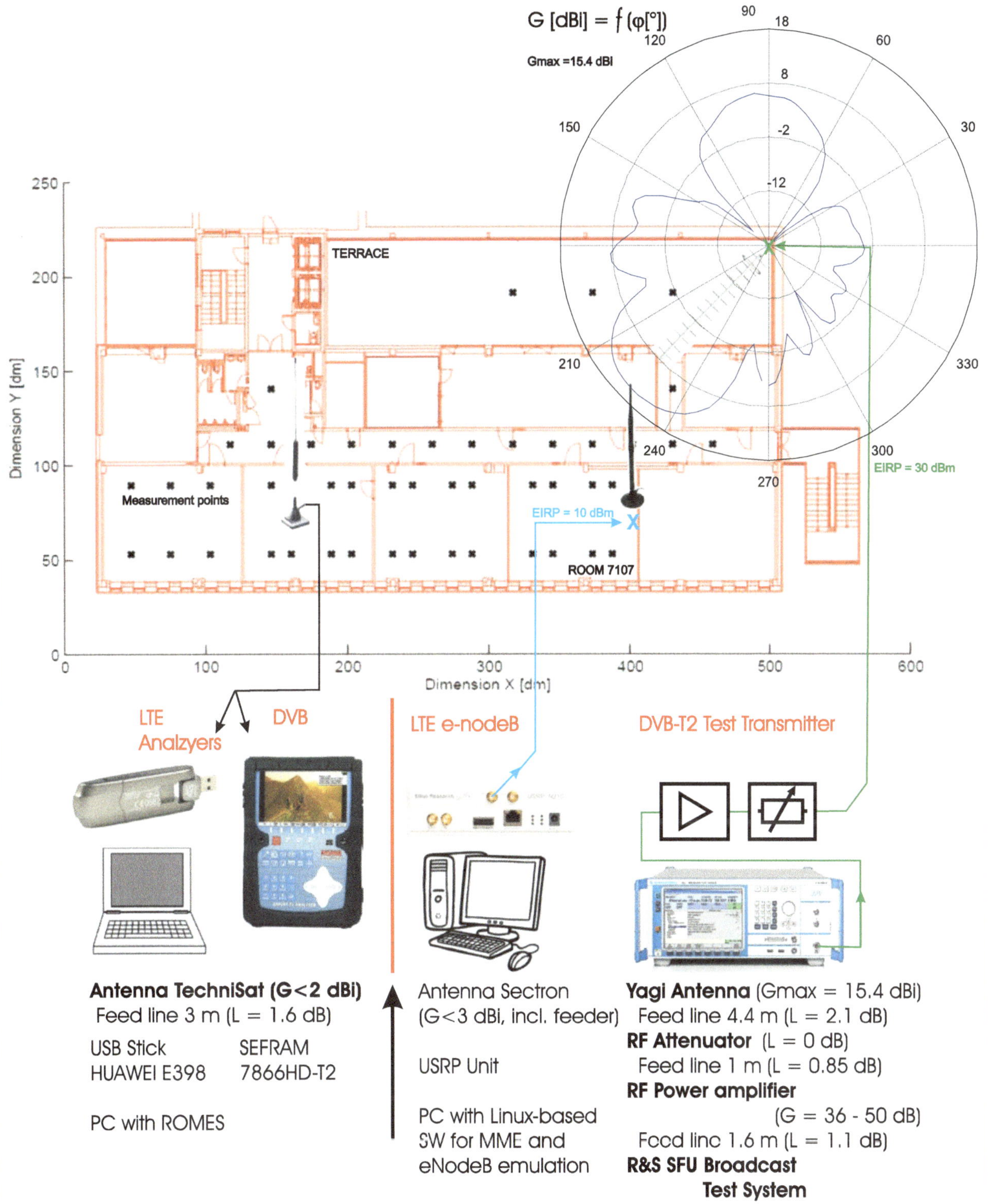

Figure 2 Floor plan and general block diagram. Floor plan of the seventh floor in the building of BUT, FEEC, DREL and general block diagram of the measurement testbed.

loss due to propagation through the floor (it is equal to zero because a single-floor propagation scenario is assumed), parameter q is the number of indoor walls separating the transmitter and receiver, L_{INwall} is the penetration loss due to walls inside the building, n is the number of outside walls, L_{OUTwall} is the penetration loss of the exterior wall, and PL_{COR} is the frequency correction factor as defined in [29] and shown in Equation 2. In our case $d_{in} = d$, $LP_{floor} = 0$ (single floor), $q > 0$ (in the case of the NLOS scenario), $L_{INwall} = 5$ dB, n is between 0 and 5 (depending on the concrete position on the floor) and $L_{OUTwall} = 10$ dB.

The propagation scenario between the DVB-T2-Lite transmitter and TV receiver is considered as outdoor-to-indoor urban femtocell propagation, where the UE is outside as described in [29]. The DVB-T2-Lite RF signal attenuation with frequency correction can be calculated similarly to Equation 3 as:

$$\begin{aligned}\mathrm{PL} &= \max\left(15.3 + 37.6\log_{10}(d), 38.46 + 20\log_{10}(d)\right) \\ &\quad + 0.7d_{\mathrm{in}} + LP_{\mathrm{floor}} + qL_{\mathrm{INwall}} + nL_{\mathrm{OUTwall}} + \mathrm{PL}_{\mathrm{COR}}\end{aligned} \tag{4}$$

where all variables have the same meaning as in Equation 3.

No fading was included in the data displayed in Figures 3 and 4, however, both fast fading and shadowing were computed according to recommendations in [29].

Figures 3 and 4 show the results of LTE and DVB-T2-Lite radio signal propagation obtained from simulation, respectively. System parameters determined according to the simulation results prove accessibility of wireless services in all tested locations.

Path loss model data provides the basis for coexistence simulations. We have provided a detailed description of LTE and DVB-T2-Lite coexistence in our previous works [30] and [31]. Based on data collected from the mentioned measurements, we made a dense description (linear model) of coexistence. There are two types of input parameters for the models: global and local. The global parameters are mainly represented by the settings of both systems' PHY, such as modulation used in DVB-T2-Lite, inverse fast Fourier transform (IFFT) size, and the Forward Error Correction (FEC) code rate of both systems. Obviously, the overlapping bandwidth is also a global parameter. Local parameters, used as the model input, are mainly local power levels of signals, background noise, and the local fading model employed. These parameters are input into the linear model, which maps them to the Quality-of-Service (QoS) parameters. More details can be found in Subsection 5.1.

4.2 Measurement setup

For evaluating the interaction of the described coexistence scenarios between DVB-T2-Lite and LTE RF signals, the same measurement testbed was used as described in our previous works ([30] and [31]). The whole measurement campaign was implemented on the seventh floor of the building of BUT, FEEC, DREL (see Figure 2). The measurement campaign and the basic principle of our measurement method are as follows.

Firstly, the parameters and performance of the 3GPP LTE network are measured in different locations on the seventh floor. At the time of LTE measurement, T2-Lite services were not broadcasted. The HeNB is located in room 7107, and its antennas are placed on top of a table (approximately 1 m above the floor). The HeNB consists of two main hardware components, namely a PC with the Fedora Linux operating system and universal software radio peripheral (USRP) N210 from Ettus, equipped with an SBX daughter card. The PC runs the commercial software package Amari LTE [32], implementing functions of LTE Mobile Management Entity (MME) and eNB (both are 3GPP LTE Release 9 compliant). A detailed configuration of the LTE network is summarized in Table 2. The receiving UE is Huawei e398-u15 (Huawei, Shenzhen, China) (LTE UE Cat. 3) [33], connected via USB port to a laptop equipped with the Rohde & Schwarz drive test software ROMES4. For receiving LTE services, the TechniSat Digiflex TT1 mobile antenna (TechniSat, Vulkaneifel, Germany) was used ($G < 2$ dBi). The length of its feed line is 3 m. The UE is connected to an external antenna placed on a wooden cart approximately 1.0 m above the floor. We set up the connection between UE and HeNB and performed simultaneous full buffer transmissions in uplink and downlink. The measurement was carried out in fixed points distributed on the seventh floor as shown in Figure 2. The receiving antenna was kept still for 2 min at each measurement point and in each location we have collected approximately 100 samples of each network parameter of interest (including RSS, Channel Quality Indicator (CQI), Error Vector Magnitude (EVM), etc.).

Secondly, we have measured the performance of the DVB-T2-Lite signal in different locations on the seventh floor. At the time of T2-Lite measurement, LTE services were not provided. By using the R&S single frequency unit (SFU) broadcast test system, an appropriate video transport stream for portable TV scenarios was generated. Then, the DVB-T2-Lite complete system configuration was set up, and the output signal was RF modulated (to the frequency of 794 MHz). For its amplification, a custom-built RF power amplifier (PA), based on hybrid module Mitsubishi RA20H8087M (Mitsubishi Electric, Tokyo, Japan) [34], was applied. This RF three-stage module is primarily destined for transmitters using FM modulation that operate in the range 806 up to 870 MHz, but it may also be applied in linear systems by setting the proper drain quiescent current with externally settable gate voltage. The PA was assembled according to the recommendations of the producers and thoroughly tested. The comprehensive measurement demonstrates that this PA can be used in a wider band, circa from 650 to 900 MHz, and can be used in the presented coexistence test. The gain of the PA strongly varies in the introduced frequency range from 36 to 50 dB, but in a

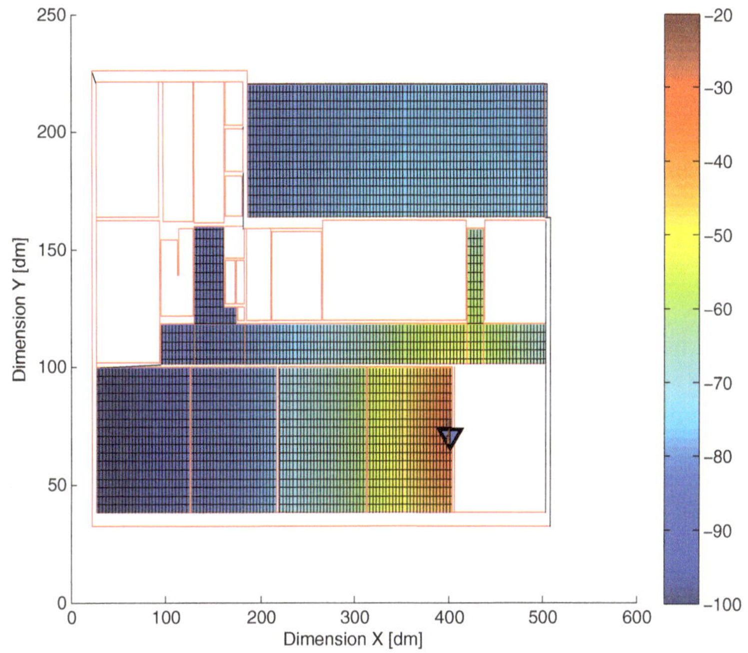

Figure 3 Simulation of LTE RF signal propagation. The HeNB is located in room 7107 (the blue triangle), and network parameters are as described in Table 2. The path loss model was adopted from [28] and extended for the 800-MHz band according to [29] (see Equations 2 and 3). All values are in dBm.

narrow band, the gain is quite stable (max. 1.5 dB in 10 MHz bandwidth). The maximum output power of this amplifier is around 30 W. However, we practically used only 1 W (5 W was used for the scenario where power imbalances were equal to 20 dB) with quiescent drain current 4 A, gate bias voltage 4.3 V, and supply voltage 13.8 V to achieve high linearity for reliable application in the mentioned setup. Accordingly, the power efficiency in this setting is only 2%. On the other hand, reaching linearity is the fundamental parameter which needs to be set for minimizing any nonlinear distortion. For the used testing DVB-T2-Lite frequency (794 MHz), the

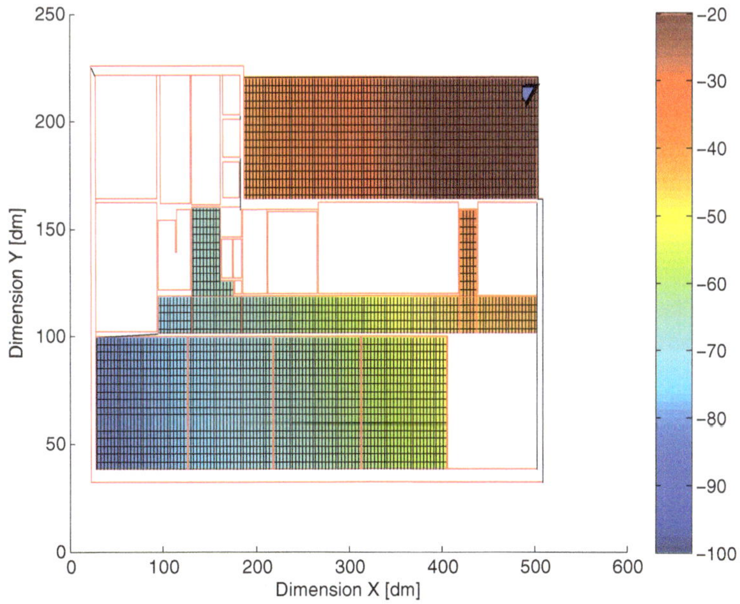

Figure 4 Simulation of DVB-T2-Lite RF signal propagation. The DVB-T2-Lite transmitter is located on the terrace (the blue triangle), and its parameters are described in Table 2. The path loss model was adopted from [28] and extended for the 800-MHz band according to [29] (see Equations 2 and 4). All values are in dBm.

measured 1 dB compression of this PA is 37.9 dBm (6.2 W), two-tone third-order intermodulation distortion (IMD3) (tone offset 1 MHz) is better than −38 dBc at the output power of 1 W, which corresponds to output intercept point (OIP3) 49 dBm. Between the PA and the antenna, there is an attenuator in the signal path. It serves as PA protection in the case of antenna switch-off or strong reflections in the antenna near the field. The JFW Industries 50BR-104 N attenuator (JFW Industries, Indianapolis, IN, USA) was used which was set to 0 dB during measurement. The mentioned nonlinear distortions caused by PA are not considered in our simulation model.

The used antenna is a multi-element Yagi antenna (G_{max} = 15.4 dBi) whose horizontal radiation pattern is shown in Figure 2. The feed line for the TV transmitter chain is a coaxial cable RG58 C/U which has a power loss of approximately 0.35 dB/m on the tested bandwidth. Attenuation of the auxiliary connection between 'N' and 'BNC' connectors is approximately 0.5 dB/m.

For the LTE system (HeNB), the Sectron AO-ALTE-MG5S antenna (Sectron Inc., Ormond Beach, FL, USA) was used. In our case, it was used as an omnidirectional antenna in vertical polarization ($G < 3$ dBi). After setting up the testbed, we moved with the Sefram 7866HD-T2 analyzer (Sefram Instruments and Systems, Saint-Étienne, France) to measure the received TV signal through all measuring points. The same antenna setup was used as is outlined above for LTE downlink. Once again, we spent 2 min at each measurement point for correctly evaluating the performance of the received DVB-T2-Lite RF signal (to avoid fast fading by averaging).

Figures 5 and 6 show measured and extrapolated values of RSS. Figure 5 shows the results of LTE radio signal propagation while Figure 6 shows the results of T2-Lite radio signal propagation obtained from measurement. System parameters determined according to the simulation results proves accessibility of wireless service in all tested locations. As we can see, results from measurement, shown in Figures 5 and 6, correspond with simulation results shown in Figures 3 and 4. This experimental result proves our simulation technique valid for coexistence applications.

Afterwards, the whole measurement campaign was repeated, but now both wireless services (DVB-T2-Lite and LTE) were provided together at the same time. The above outlined QoS parameters of both services, caused by coexistence between them, were measured separately with Rohde & Schwarz devices.

5 Experimental results

5.1 Parameters to evaluate the performance of DVB-T2-Lite and LTE

Before evaluating and discussing the obtained results, it is necessary to briefly define the most important measured parameters which were used to evaluate the performance of T2-Lite and LTE systems. To evaluate the quality of the received and decoded TV services, the Quasi Error-Free (QEF) reception conditions were monitored. QEF is a minimal limit defined in the DVB-T2-Lite standard for achieving video service availability without noticeable errors in the video. To fulfill such requirements, the bit error ratio (BER) after FEC decoding must be less than or equal to 1×10^{-7} [6].

To evaluate the performance of LTE, the RSS, CQI, and EVM parameters were monitored. The CQI contains information sent from the UE to the HeNB to indicate a suitable downlink transmission data rate. It is based on the observed signal-to-interference-plus-noise ratio (SINR) and used by the HeNB for downlink scheduling and link adaptation [28]. There are 15 different CQI values (numbered from 1 up to 15). The connection between them and the modulation scheme can be found in [35] (Table 7.2.3-1).

EVM, the second parameter, is a measure used to quantify the performance of an LTE communication link. It is the RMS value of the distance in the IQ constellation diagram between the ideal constellation point and the point received by the receiver. For each modulation, there is a defined EVM limit, for which the transmitted signal has an acceptable quality. This limit is equal to 17.5% for quadrature phase-shift keying (QPSK), 12.5% for quadrature amplitude modulation (16QAM), and 8.0% for 64QAM [11,28].

5.2 DVB-T2-Lite and LTE performance evaluation

In Subsection 4.1, it has been mentioned that the linear coexistence model maps input parameters from simulations and measurements to the area of QoS states. We have defined the following QoS states for the coexisting services. For DVB-T2-Lite, there are two states: correct reception and no reception. In the case of correct reception, the above defined condition for QEF reception is satisfied. For LTE, we have defined four QoS states which differ in user bitrate and potential radio access network (RAN) throughput. These parameters obviously increase with M in M-QAM modulation of subcarriers. The LTE system changes the modulation scheme adaptively according to the channel parameters (e.g., CQI, EVM). To be more precise, the highest useable M-state for the defined interfered radio channel sets the QoS state of LTE. Four states correspond to maximal M equaling 64 (64QAM), 16 (16QAM), and 4 (QPSK), and the state when providing LTE services is not possible.

The considered coexistence scenarios between DVB-T2-Lite and LTE services were described above. Furthermore, we also consider various system parameters. The complete list of assumed scenarios is clearly summarized in Table 3. There are three main parameters: bandwidth of the LTE RF

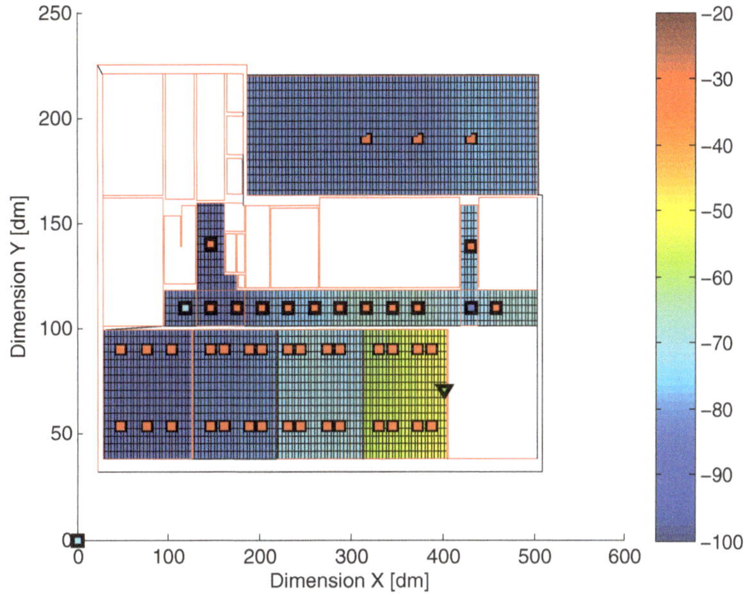

Figure 5 Measurement of LTE RF signal propagation. The HeNB is located in room 7107 (the blue triangle), and network parameters are described in Table 2. The measurement was carried out in highlighted points and extrapolated using MATLAB. All values are in dBm.

channel (marked as B_{LTE}), overlap of coexisting channel (B_{OVER}), and the power imbalance between transmitted powers (ΔP).

The last one is calculated as follows:

$$\Delta P\,[\mathrm{dB}] = \mathrm{EIRP}_{\mathrm{LTE}} - \mathrm{EIRP}_{\mathrm{TV}} \quad (5)$$

where equivalent isotropically radiated power $(\mathrm{EIRP})_{\mathrm{LTE}}$ and $\mathrm{EIRP}_{\mathrm{TV}}$ denote the channel power of LTE and T2-Lite RF signals, respectively.

Figure 7 shows the simulated results of six map representations of QoS states in DVB-T2-Lite and LTE systems. Each map (from (a) to (f)) corresponds to the

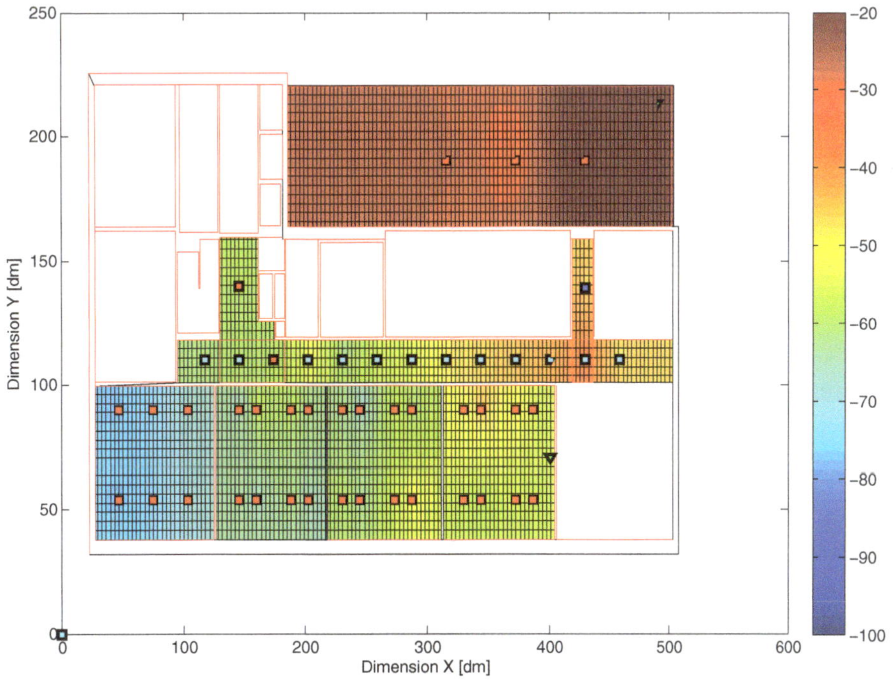

Figure 6 Measurement of DVB-T2-Lite RF signal propagation. The HeNB is located in room 7107 (the blue triangle), and network parameters are described in Table 2. The measurement was carried out in highlighted points and extrapolated using MATLAB. All values are in dBm.

Table 3 Variable parameters of DVB-T2-Lite and LTE for assumed coexistence scenarios

Map	B_{LTE} (MHz)	B_{OVER} (kHz)	Power imbalance (ΔP) (dB)
a	10	800	0
b	20	800	0
c	20	1,600	0
d	20	1,600	−10
e	20	1,600	−20
f	20	3,000	0

considered system parameters and coexistence scenarios which are presented in Table 3. In the floor plan of the university, for each point in the explored areas, the state of both coexisting systems is indicated.

Performances of T2-Lite and LTE systems can be clearly explained in the legend of Figure 7. Four colors represent the LTE maximum useable internal modulations: orange - 64QAM, yellow - 16QAM, and green - QPSK, and unavailable LTE services are indicated by a cyan color. The performance of DVB-T2-Lite services is indicated by a crosshatch in the same maps. The presence of a hatch means that the QEF limit of mobile TV reception is fulfilled. For a better explanation of the obtained results, we describe a specific example.

For example, we consider a partial overlapping coexistence scenario between T2-Lite and LTE services when B_{LTE} = 20 MHz, B_{OVER} = 1,600 kHz, and ΔP is equal to 0 dB (see line (c) in Table 3). Performances of coexisting systems for these parameters are plotted in Figure 7c. As can be seen from the legend, at the 1.6-MHz channel overlapping, in the LTE system, only sub-frames using QPSK and 16QAM modulations will be received and demodulated correctly (yellow color in the legend) on the left side of the corridor. It means that only at these modulations EVM errors do not exceed the permitted limit values [11]. In the remaining rooms, the highest 64QAM modulation (highlighted by orange color) is used in the LTE system. Consequently, CQI values can be 10 or higher. Furthermore, this field also indicates that the services of DVB-T2-Lite are highly noised and conditions for QEF reception are not fulfilled (there are no hatched parts). The situation result is the opposite on the terrace where DVB-T2-Lite services are broadcasted. At this place, the provided LTE services are not available (blue color). In this case, the LTE system could not decode the received signal and the CQI value is the lowest. Interestingly, in the small corridor, located between the terrace and the main floor corridor, partial coexistence between T2-Lite and LTE systems is possible. It means that at this place, both wireless systems can coexist. The QEF limit for DVB-T2-Lite is still fulfilled. However, in the LTE system, only sub-frames using QPSK modulation can be successfully processed. Hence, the CQI indicator values will be in the range from 1 up to 6 [35]. Similar graphical representations of considered coexistences are plotted in Figure 7a,b,c,d,e,f.

Now, let us focus on the first two charts (see Figure 7a,b). Their parameters differ just in the used LTE channel bandwidth (B_{OVER}), but the disparity in state map is high. From the point of B_{LTE} = 20 MHz LTE channel (see Figure 7b), the 800 kHz interference bandwidth is quite narrow and almost no effect can be seen on LTE inside the building. Outside, LTE works correctly with 16QAM. However, when B_{LTE} is equal to 10 MHz (see Figure 7a), then the LTE channel, affected by the same interference bandwidth (800 kHz), is occupied by almost twice the interfering RF power. In this case, the LTE system still works correctly, but only 16QAM and QPSK (indoor/outdoor) modulations can be used. Furthermore, mobile TV reception is also more affected by LTE services because LTE interference power is concentrated into a narrower channel. In real RANs, where power limits are more likely set to 1 Hz of occupied bandwidth, the impact on the reception of mobile TV services would be the same. The influence of channels overlapping and the effect of different EIRP unbalances could be investigated from the remaining charts.

Figure 8 shows six map representations of QoS states in DVB-T2-Lite and LTE systems from measurements. In general, in most measuring points, the defined states of QoS correspond with simulation results. However, there are some minor differences caused by the accumulation of two types of uncertainties. The first ones are caused by path loss channel modeling, and these are even multiplied by the second ones, caused by the proposed linear model. Most probably, the largest influences are due to inhomogeneity in walls (doors, windows and various types of material), underestimation of noise level, and impact of multipath propagation. It is obvious that the simulation and measurement results in scenario (e) have the lowest difference. This state is caused by the highest signal level (in the above mentioned scenario) which brings reduction of noise background impact and increase the influence of intersystem jamming simultaneously for all transmission paths.

6 Conclusions

The main aim of this paper is to investigate the impact of coexisting DVB-T2-Lite and LTE systems in a shared frequency band on their system performances in the outdoor-to-indoor reception scenario. To be more precise, a scenario was considered where an indoor LTE femtocell (HeNB) and outdoor-to-indoor DVB-T2-Lite services are provided in an 800-MHz frequency band (see Figures 1 and 2). We have performed separate simulations of both LTE and DVB-T2-Lite RF signal propagation in MATLAB. Further, we have carried out measurements

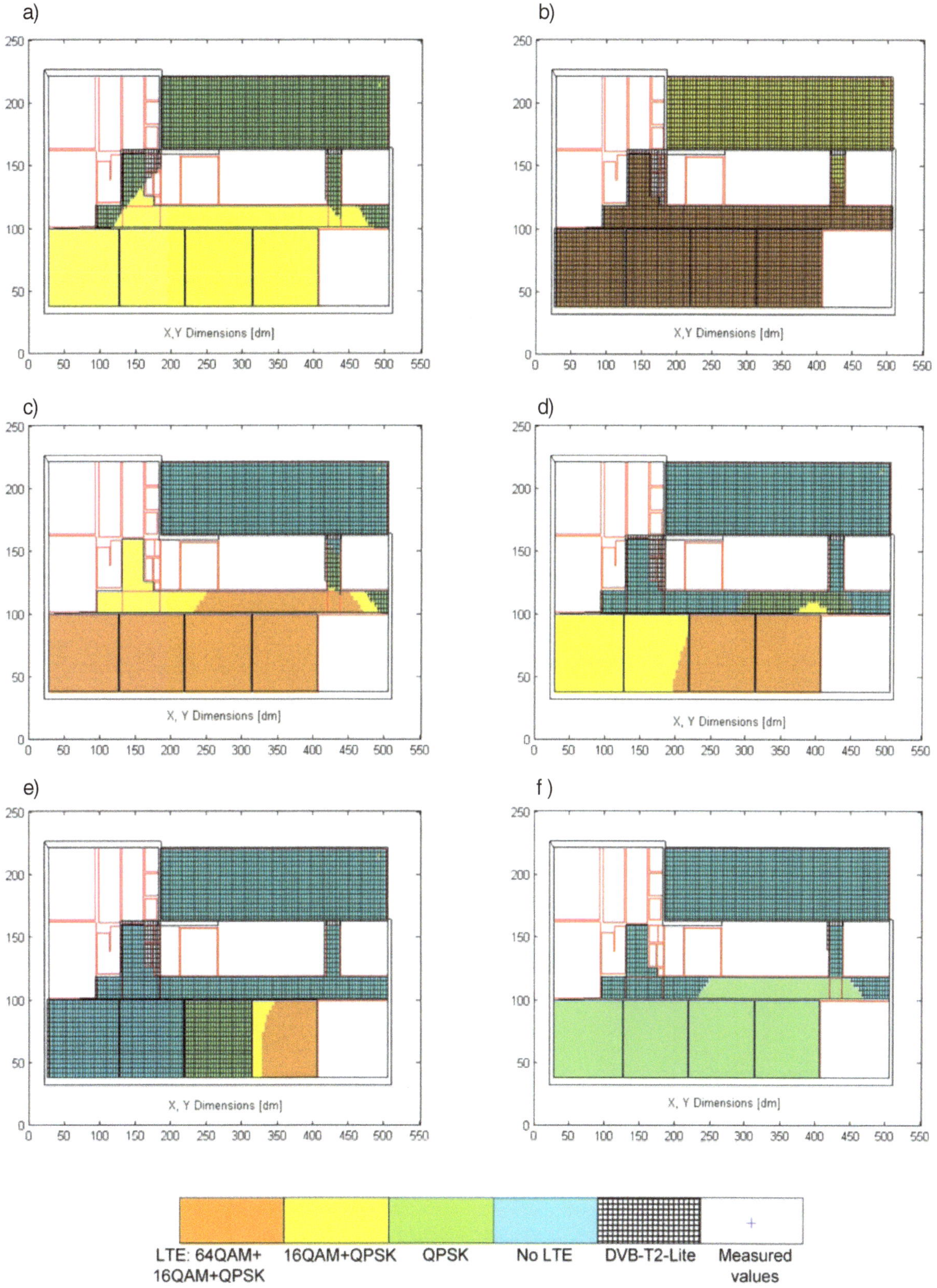

Figure 7 Simulation - the map representation of QoS states of coexisting systems **(a-f)**. Specific map parameters are summarized in Table 3.

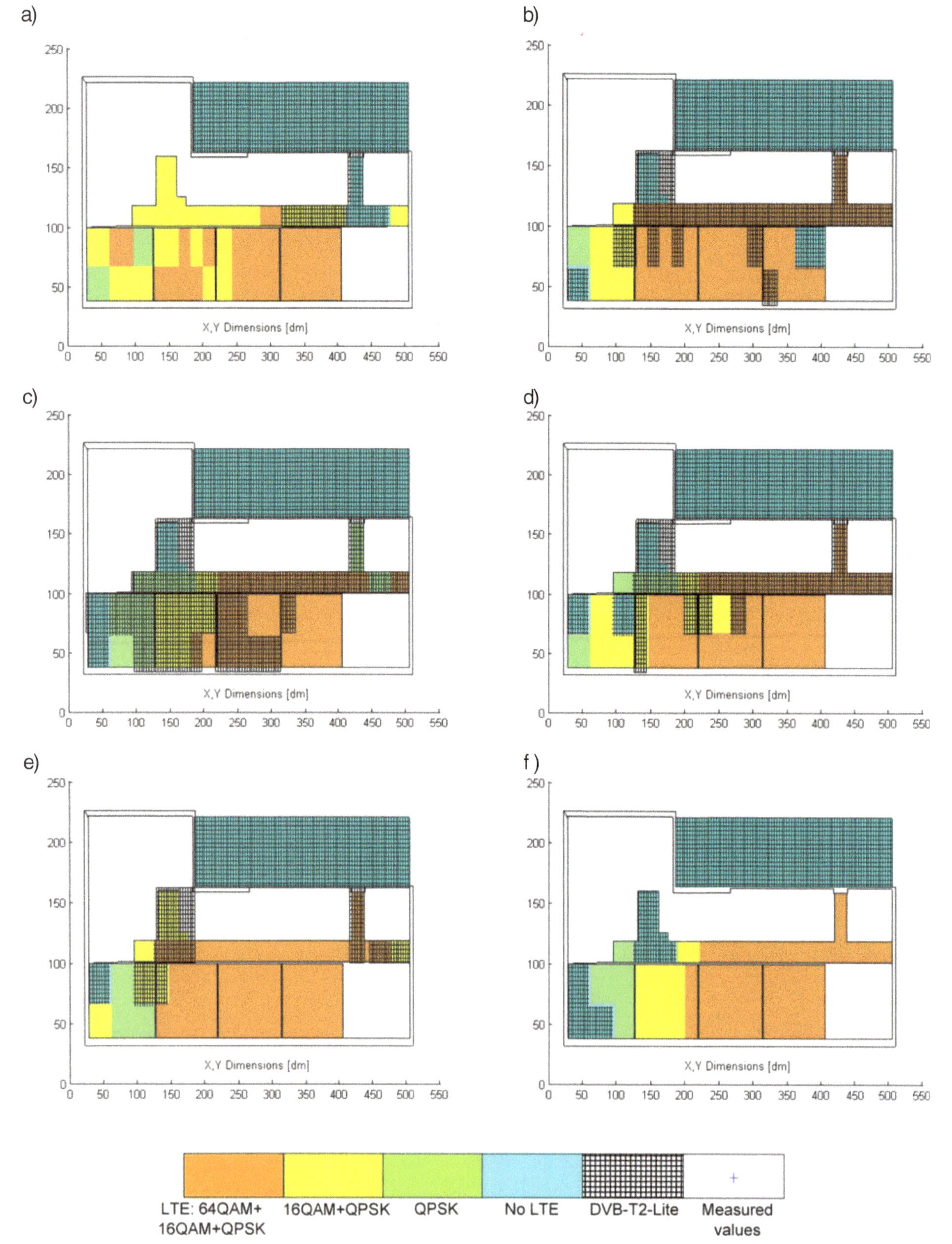

Figure 8 Measurement - the map representation of QoS states of coexisting systems **(a-f)**. Specific map parameters are summarized in Table 3.

of both wireless systems in order to evaluate the reliability of the simulation model. Results are shown in Figures 3, 4, 5, 6 and correlate well.

According to the achieved results in our previous works ([31] and [28]), we have created a linear model to map outputs of the path loss model to defined QoS states. This model considers the relation between the value of RF channels overlapping and the power imbalance of the investigated radio channels. A detailed description is outlined in Subsection 4.1.

The presented results are expressed in a set of maps (floor plans of the building) with colored areas which determine availability or non-availability of coexisting services and achievable performance. Specific values of these parameters in the considered scenarios are presented in Table 3. The effect of coexistence on valid signal reception is quantified by the change of used modulation scheme and simultaneous availability of services. A detailed description of the color maps is described in Section 5. In the proposed linear model for both systems, we assume good channel conditions (global parameters): signal-to-noise ratio (SNR) ≥35 dB for both systems and also no multipath propagation and no Doppler frequency have been set. Once again, the proposed linear model was proved by measurements (see Figures 7 and 8). In several cases, less correspondence between the simulation and measurement results is explained.

An analysis of the obtained results from the considered coexistence scenarios leads to the following general conclusions:

a) The impact of DVB-T2-Lite system performance on the LTE system performance and vice versa in their co-channel coexistence scenario in a shared frequency band highly depends on the level of their channels overlapping and on the power imbalance between RF signals.
b) The outdoor-to-indoor penetration of the T2-Lite signal is highly critical on indoor-to-indoor reception of LTE services when the power imbalance between the RF levels is high. In these cases, the T2-system acts as a co-channel interferer to indoor LTE femtocell and vice versa.
c) Digital TV fixed indoor reception is more vulnerable to interferences than fixed outdoor reception.

The main aim of our future work will be to extend our proposed linear coexistence model with more global parameters (different kinds of fading channel models and Doppler shift [36-39]) for more realistic modeling of different coexistence scenarios between DVB-T2-Lite and LTE services and vice versa. Moreover, in our future work, we will consider a larger range of system parameters (code rate, IFFT length, guard interval, and higher M-QAM modulations and bandwidth) [40,41].

Competing interests
The authors declare that they have no competing interests.

Acknowledgement
This work is supported by the Cluster for Application and Technology Research in Europe on Nanoelectronics (CATRENE) under the project named CORTIF CA116 - Coexistence of Radio Frequency Transmission in the Future, the MEYS of the Czech Republic no. LF14033, no. CZ.1.07/2.3.00/20.0007 and CZ.1.07/2.3.00/30.0005, and finally by the BUT project no. FEKT-S-14-2177. The described research was performed in laboratories supported by the SIX project; no. CZ.1.05/2.1.00/03.0072, the operational program Research and Development for Innovation. Research described in this paper was financed by Czech Ministry of Education in frame of National Sustainability Program under grant LO1401. For research, infrastructure of the SIX Center was used.

References
1. Z Raida et al., Communication subsystems for emerging wireless technologies. Radioengineering **21**(4), 1036–1049 (2012)
2. T Yos, Y Wu, N Hur, T Ikeda, P Xia, FOBTV: worldwide efforts in developing next-generation broadcasting system. IEEE Trans. on Broadcasting **60**(2), 154–159 (2014)
3. L Klozar, L Polak, O Kaller, J Prokopec, Effect of Co-Existence Interferences on QoS of HSPA/WCDMA Mobile Networks, in *Proceedings of the 23rd IEEE International Conference Radioelektronika 2013*, 2013, pp. 312–315
4. W Sami, How can Mobile and Broadcasting Networks use Adjacent Bands? in *EBU Technical Review*, 2011, pp. 1–20
5. Digital Video Broadcasting (DVB), *Framing Structure, Channel Coding, and Modulation for Digital Terrestrial Television Broadcasting System (DVB-T2)*, 2012. ETSI EN 302 755 V1.3.1
6. Digital Video Broadcasting (DVB), *Implementation Guidelines for a Second Generation Digital Terrestrial Television Broadcasting System (DVB-T2)*, 2012. 2012. ETSI TS 102 831 V1.2.1
7. ITU-Rec, *Frequency and Network Planning Aspects of DVB-T2*, 2012. ITU-R BT.2254
8. DA Samo, M Slimani, G Baruffa, L Rugini, A performance study of DVB-T2 and DVB-T2-Lite for mobile reception. Digital Signal Processing **37**(2), 35–42 (2015)
9. C Regueiro, U Gil, M Velez, I Eizmendi, P Angueira, Field trials-based planning parameters for DVB-T2 indoor reception. IEEE Trans on Broadcasting. (2015). doi:10.1109/TBC.2015.2400814)
10. Digital Video Broadcasting (DVB), *Framing Structure, Channel Coding and Modulation for Digital Terrestrial Television*, 2009. ETSI EN 300 744 v1.6.1
11. 3GPP, 3rd Generation Partnership Project, *Evolved Universal Terrestrial Radio Access (E-UTRA); User Equipment (UE) Radio Transmission and Reception (Release 8)*, 2010. Tech. Specification 3GPP TS 136.101 V8.9.0
12. E Dahlman, *3G Evolution: HSPA and LTE for Mobile Broadband*, 2nd edn. (Academic Press, Oxford, 2008)
13. D Astely, E Dahlman, A Furuskar, Y Jading, M Lindstrom, S Parkvall, LTE: the evolution of mobile broadband. IEEE Communication Magazine **47**(4), 44–51 (2009)
14. Rohde & Schwarz, Coexistence digital TV and LTE (Rohde & Schwarz, 2012), http://cdn.rohde-schwarz.com/pws/dl_downloads/dl_application/application_notes/1ma176/1MA176_3e_coexistence_digital_TV_and_LTE.pdf
15. P Stavroulakis, *Interference Analysis and Reduction for Wireless Systems* (Artech House, Norwood, 2003)
16. F Gleissner, S Hanus, Co-channel and adjacent channel interference measurement of UMTS and GSM/EDGE systems in 900 MHz radio band. Radioengineering **17**(3), 74–80 (2008)
17. J Mikulka, S Hanus, Bluetooth and IEEE 802.11b/g coexistence simulation. Radioengineering **17**(3), 66–73 (2008)
18. B Han, W Wang, Y Li, M Peng, Investigation of interference margin for the co-existence of macrocell and femtocell in orthogonal frequency division multiple access systems. IEEE Systems Journal **7**(1), 59–67 (2013)

19. A Lourerio, D Callegos, G Caldwell, Interference Analysis on UMTS-2100 Co-Existence with GSM-1900, in *Proceedings of the 74th IEEE VTC FALL 2011*, 2009, pp. 1–4
20. WA Hassan, TA Rahman, Coexistence model for compatibility between IMT-advanced and other wireless communication services. Wireless Personal Communications **79**(3), 2025–2039 (2014)
21. MH Ng, S-D Lin, J Li, S Tatesh, Coexistence studies for 3GPP LTE with other mobile systems. IEEE Communications Magazine **47**(4), 60–65 (2009)
22. D Lee, GY Li, S Tang. Inter-Cell Interference Coordination for LTE Systems, in IEEE GLOBECOM **2012**, 4828–4833 (2012)
23. T Cai, J Deng. Coexistence Study and Interference Analysis in LTE Networks, in IEEE International Conference ICCECT **2012**, 751–754 (2012)
24. L Tytgat, O Yaron, S Pollin, I Moerman, P Demeester, Avoiding collisions between IEEE 802.11 and IEEE 802.15.4 through coexistence aware clear channel assessment. EURASIP Journal on Wireless Communications and Networking **2012**, 137 (2012)
25. M Boujelben, S Benrejeb, S Tabbane, A Comparative Study of Interference Coordination Schemes for Wireless Mobile Advanced Systems, in *The 2014 International Networks, Computers and Communications Symposium (ICNC 2014)*, 2014, pp. 1–5
26. D Chen, T Jiang, Z Zhang, *Frequency partitioning methods to mitigate cross-tier interference in two-tier femtocell networks. IEEE Trans on Vehicular Technology*, 2015. doi:10.1109/TVT.2014.2335251
27. ZA Shamsan, LTE-Advanced Compatibility with Digital Broadcasting Receiver at 800 MHz, in *2013 Saudi International Electronics, Communications and Photonics Conference (SIECPC)*, 2013, pp. 1–4
28. 3GPP, 3rd generation partnership project, Technical specification group radio access network; evolved universal terrestrial radio access (E-UTRA); further advancements for E-UTRA physical layer aspects (Release 9), March 2010. Tech. Report 3GPP TR 36.814 V9.0.0, http://www.qtc.jp/3GPP/Specs/36814-900.pdf
29. 3GPP, 3rd generation partnership project, Technical specification group radio access network; evolved universal terrestrial radio access (E-UTRA); study on LTE device to device proximity services-radio aspects (Release 12), November 2013. Tech. Report 3GPP TR 36.843 V12 .0.0, http://www.3gpp.org/dynareport/36843.htm
30. L Polak, O Kaller, L Klozar, J Prokopec, Exploring and Measuring the Co-Existence Between LTE and DVB-T2-Lite Services, in *Proceedings of the 36th International Conference on Telecommunications and Signal Processing (TSP) 2013*, 2013, pp. 316–320
31. L Polak, O Kaller, L Klozar, J Sebesta, T Kratochvil, Mobile communication networks and digital television broadcasting systems in the same frequency bands - advanced co-existence scenarios. Radioengineering **23**(1), 375–386 (2014)
32. Software package Amari LTE. Amari LTE 100/Amari OTS 100 Software 4G network on PC. Online on: < http://www.amarisoft.com/?p=amarilte>
33. M Slanina, L Klozar, S Hanus, Practical Measurement of Data Throughput in LTE Network Depending on Physical Layer Parameters, in *Proceedings of the 24th International Conference Radioelektronika*, 2014, pp. 1–4
34. RA20H8087M. RF MOSFET module 806-870 MHz 20 W 12.5 V, 3 Stage Amplifier for Mobile Radio. Data Sheet. Mitsubishi Electric Corporation, July 2011, http://datasheet.eeworld.com.cn/pdf/MITSUBISHI/180155_RA20H8087M.pdf
35. 3GPP, 3rd generation partnership project, Technical specification LTE; evolved universal terrestrial radio access (E-UTRA); physical layer procedures (Release 8), October 2009. Tech. Specification 3GPP TS 136 213 V8.8.0, http://www.3gpp.org/dynareport/36213.htm
36. J Milos, S Hanus, Performance analysis of PCFICH and PDCCH LTE control channels. Radioengineering **23**(1), 445–451 (2014)
37. L Polak, T Kratochvil, Analysis and simulation of the transmission distortions of the mobile digital television DVB-SH part 2: satellite mode DVB-SH-B with TDM. Radioengineering **21**(1), 126–133 (2012)
38. L Polak, T Kratochvil, Exploring of the DVB-T/T2 Performance in Advanced Mobile TV Fading Channels, in *Proceedings of the 36th International Conference on Telecommunications and Signal Processing (TSP) 2013*, 2013, pp. 768–772
39. L Polak, O Kaller, L Klozar, J Sebesta, T Kratochvil, Exploring and measuring possible co-existences between DVB-T2-lite and LTE systems in ideal and portable fading channels. Journal of Applied Research and Technology **13**(1), 32–44 (2015)
40. H Hong, H Son, Y Chong, Protection Ratio Between DVB-T2 and LTE System, in *2014 Information and Communication Technology Convergence (ICTC)*, 2014, pp. 792–793
41. M Fuentes, C Garcia-Pardo, E Garro, D Gomez-Barquero, N Cardona, Coexistence of digital terrestrial television and next generation cellular networks in the 700 MHz band. IEEE Wireless Communications **21**(6), 63–69 (2014)
42. A Guidotti, D Guiducci, M Barbiroli, C Carciofi, P Grazioso, G Riva, Coexistence and Mutual Interference Between Mobile and Broadcasting Systems, in *Proceedings of the 73rd IEEE Vehicular Technology Conference (VTC Spring)*, 2011, pp. 1–5
43. K Sakic, S Grgic, The Influence of the LTE System on the DVB-T Reception, in *Proceedings of the 52nd International Symposium ELMAR*, 2010, pp. 235–238
44. G Baruffa, M Femminella, F Mariani, G Reali, Protection ratio and antenna separation for DVB-T/LTE coexistence issues. IEEE Communications Letters **17**(8), 1588–1591 (2013)
45. H Bawab, P Mary, J-F Hélard, YJ Nasser, O Bazzi, Global Ergodic Capacity Closed-Form Expression of Coexisting DVB-LTE-Like Systems, in *2014 IEEE VTC Spring*, 2014, pp. 1–5
46. A Tekovic, G Simac, K Sakic, LTE Downlink System Performance Measurement with Intersystem Interference Caused by DVB-T Signal, in *Proceedings of the 54th International Symposium ELMAR*, 2012, pp. 255–258
47. A De Vita, D Milanesio, B Sacco, A Scotti, Assessment of interference to the DTT service generated by LTE signals on existing head amplifiers of collective distribution systems: a real case study. IEEE Trans. on Broadcasting **60**(2), 420–429 (2014)
48. L Polak, O Kaller, L Klozar, J Prokopec, Influence of Mobile Network Interfering Products on DVB-T/H Broadcasting Services, in *Proceedings of the 5th IEEE International Conference Wireless Days 2012*, 2012, pp. 1–5

Interference characterization and mitigation benefit analysis for LTE-A macro and small cell deployments

Víctor Fernández-López[1*], Klaus I Pedersen[1,2] and Beatriz Soret[2]

Abstract

This article presents a characterization of different LTE-Advanced network deployments with regard to downlink interference and resource usage. The investigation focuses on heterogeneous networks (HetNets) with dedicated spectrum for each layer and, in particular, on cases where small cells are densely deployed. Thus, the main interference characteristics of the macro layer and the dense small cell layer are studied separately. Moreover, the potential benefit of mitigating the dominant interferer in such scenarios is quantified as an upper bound gain and its time variability is discussed and evaluated. A dynamic FTP traffic model is applied, with varying amounts of traffic in the network. The results present an uneven use of resources in all feasible load regions. The interference under the dynamic traffic model shows a strong variability, and the impact of the dominant interferer is such that 30% of the users could achieve at least a 50% throughput gain if said interferer were mitigated, with some users reaching a 300% improvement during certain time intervals. All the mentioned metrics are remarkably similar in the macro and small cell deployments, which suggests that densification does not necessarily imply stricter interference mitigation requirements. Therefore, the conclusion is that the same techniques could be applied in both scenarios to deal with the dominant interferer.

Keywords: Interference; Mitigation; Dominant interferer; LTE-Advanced; Macro cell; Small cell

1 Introduction

Interference is one of the main factors that compromise the downlink performance in LTE and LTE-Advanced (LTE-A) networks [1]. As such, it has been the focus of numerous studies since the first LTE network deployments comprising only macro cells. Research on interference management for macro-cell networks has analysed, among others, aspects such as resource partitioning in the frequency and space domains (e.g., frequency reuse, fractional frequency reuse) [2-4] to improve the signal strength at the mobile terminal or to reduce the interference. These studies were performed under static traffic models, therefore limiting the time variability of the interference. Hence, the solutions proposed in these investigations managed to bring notable benefits while using slow adaptation capabilities. More recently, coordinated multi-point transmission (CoMP) studies have tried to approach these issues in a more dynamic manner [5], by studying procedures with extensive coordination and, for the most part, under the assumption of a fast backhaul, or even fronthaul, with negligible latency [6].

Following these macro-only topologies, research turned to heterogeneous networks (HetNets) as a way to meet the increasing capacity demands in LTE-A networks. HetNets comprise a mixture of macro cells and low-power nodes known as small cells. These topologies face a challenging interference problem in cases where the macro and the small cells utilize the same carrier due to the difference in transmission power. Therefore, this inter-layer interference has been the focus of many studies [7-9]. The current trend is pointing to dedicated deployments with higher frequency bands [10] and thus shifting the focus towards intra-layer interference between the same class of nodes. Most of the research on intra-layer interference between small cells in the literature has considered femto cells (home base stations), which present a high risk for

*Correspondence: vfl@es.aau.dk
[1] Department of Electronic Systems, Aalborg University, Fredrik Bajers Vej 7, 9220 Aalborg, Denmark
Full list of author information is available at the end of the article

inter-cell interference as the nodes are commonly installed by the users, making up an unplanned network [11,12].

However, current research is contemplating increasingly denser small cell deployments in HetNets [13], where intra-layer interference becomes a considerable concern, even in a planned deployment. It has been claimed that denser scenarios exhibit unique interference characteristics and therefore will require custom-designed solutions for interference mitigation [14]. This study sets out to evaluate this hypothesis and to understand how the efforts to manage interference should be steered depending on the topology. In particular, the impact of the strongest interferer and the potential benefit from cancelling it are evaluated. This investigation continues the work begun in [15], which evaluated the behaviour of the intra-layer interference in an LTE-A dense small cell network. The analysis is extended here to a network based on a regular macro-cell deployment, and we delve deeper into the reasons for the observed interference patterns. Both the macro-only and the dense small cell scenarios are examined under a dynamic traffic model with different amounts of offered traffic. The time evolution of the interference is studied, analysing the required dynamism for interference mitigation solutions in these topologies.

The two scenarios are found to be remarkably similar with regard to these considerations, despite their very different degrees of density. This conclusion fits in with previous studies such as [8] and [16], which found that, assuming an interference-limited network with unbiased cell association and equal path loss exponents for all links, adding base stations does not modify the downlink SINR statistics. As such, similar strategies to manage the interference could be used in the two scenarios analysed in this article, potentially achieving very significant performance gains if the main interferer were ideally mitigated.

The structure of the paper is as follows: Section 2 will introduce a description of the considered network scenarios and the traffic model, together with some necessary theoretical considerations for the interference analysis; Section 3 will focus on a description of the system-level simulation settings; Sections 4 and 5 will present the analysis and discussion of the collected statistics and their significance. The article closes with a discussion on future research and the concluding remarks.

2 Setting the scene

2.1 Network model

The majority of the network models found in the literature are either variations of the Wyner model, making up an idealized, regular structure, or are based on stochastic geometry, such as Poisson point processes (PPPs) [17]. The 2D Wyner model forms a regular lattice of deterministic base station positions, in a hexagonal deployment, whereas in the case of the PPP scenarios, the network positions are random and the structure, irregular. Both methods have a series of advantages and disadvantages. The Wyner model is more tractable but highly ideal and therefore requires extensive simulations to produce realistic results. On the other hand, the PPP models account better for randomness in the network and allow us to define the notion of a typical user [18]. PPP has also been found to adequately model the user's positions, both in a macro cell when applied uniformly over the area and in hotspots when used in clustered form. The main disadvantage of stochastic models is the difficulty in modelling the correlated dependences in node positions, i.e., the fact that the location of a base station is generally dependent on the position of its neighbours [17].

A third option for network modelling comes in the form of realistic (site-specific) scenarios, generally using data from real operator deployments [19]. The main challenge of performing a study in such a scenario is that the reproducibility of the results is limited, and that it might not be easy to extract general conclusions that are applicable to other deployments. The Third Generation Partnership Project (3GPP) has adopted the use of Wyner and stochastic models in its simulation assumptions [13]. In particular, macro deployments are represented by a regular hexagonal structure, whereas small cells are deployed in clusters according to a PPP with several inter-eNodeB distance constraints.

The two LTE-Advanced network scenarios considered in this study, which are illustrated in Figure 1, follow these characteristics. The network topologies are similar to the ones described in [13]. On the left-hand side of Figure 1, the macro-cell case comprises seven sites with three sectors each. The deployment is regular, with a 500-m inter-site distance. The users are deployed uniformly over the cell area according to a PPP. All the macro cells transmit at the same frequency. The small cell scenario, depicted on the right-hand side of Figure 1, includes three non-overlapping clusters with ten small cells each, modelling areas with high traffic density. As shown in the figure, the clusters are delimited by two concentric circles. The cells are randomly placed according to a Poisson point process within the inner circle, with a 50-m radius. There is a minimum distance constraint between small cells of 20 m. The outer circle, with a 70-m radius, represents the area where the users are deployed uniformly according to a PPP (i.e., taking a clustered approach). All the small cells in the network share the same frequency band. It is assumed that the each user connects to the cell corresponding to the strongest received power.

2.2 Traffic model

Two types of traffic models are commonly used in LTE-A studies. On the one hand, closed-loop full-buffer models consider a constant number of users with unlimited

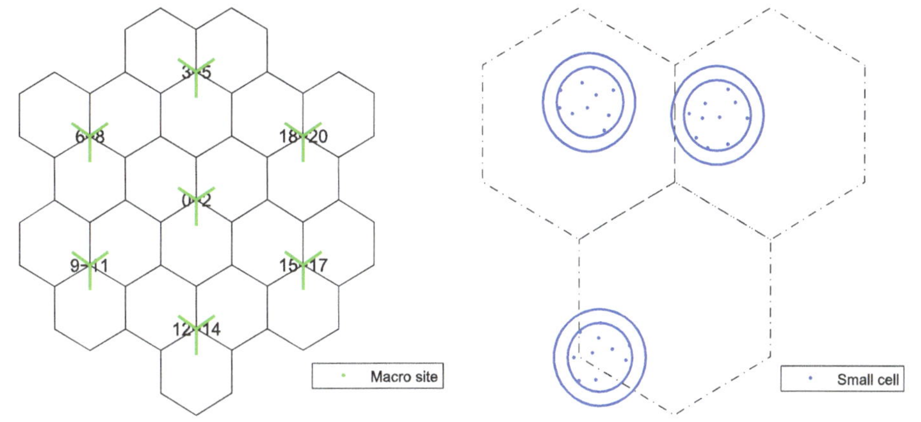

Figure 1 Network topologies. Macro-cell network (left) and dense small cell network (right).

data to transmit. In contrast, finite-buffer models (also known as FTP traffic models) include user arrival (birth) and departure (death) processes, and it is assumed that the users have a limited amount of data to transmit, and they leave the network once they have done so [20]. These models can be of the closed-loop or open-loop types, depending on whether the number of users in the network is fixed or variable, respectively. Both full- and finite-buffer models have been used in 3GPP studies [21]. The finite-buffer model with Poisson arrival has been found to adequately model the arrival of user sessions [22]. It also includes the effect of users not being simultaneously active, thereby introducing fluctuations in the interference conditions in the network. The full-buffer model is less realistic in its assumption of constantly active users and results in more stable interference patterns. The difference between the models can be significant in dense deployments where the coverage areas are reduced and the cells serve a low number of users. In such a scenario, the full-buffer model would lead to an underestimation of the interference variability, which would seemingly facilitate the scheduling decision process. In order to properly understand the challenges faced in dense deployments, this study adopts an open-loop dynamic FTP traffic model in which session arrivals are controlled by an average arrival rate, λ, following a homogeneous Poisson process, and each user demands a fixed payload of L bits. The arrival rate λ has different meanings depending on the scenario. In the macro-cell case, λ indicates the average number of users per second per cell area, whereas in the small cell case, it is defined as the average number of users per second per small cell cluster. The offered load O is defined as the product of the arrival rate and the payload, $O = L \cdot \lambda$, and will accordingly adopt different meanings depending on the scenario. Likewise, we define the carried load, C, as the average amount of supported traffic in one of the cells (macro scenario) or in one cluster (small cell scenario). The different interference and performance metrics analysed in this study will be evaluated in relation to the offered and carried loads.

The system is in equilibrium and operates in the feasible load region when the carried traffic matches the offered load, i.e., $C = O$. Congestion takes place when the system cannot support the demanded traffic ($C < O$) and the session departure rate becomes lower than the rate at which users arrive in the network. The congestion region is unstable and not of interest for the design of a practical interference management solution. Therefore, only the performance in the feasible load region will be considered in this study.

A sketch of the user performance in the feasible region with increasing traffic is presented in Figure 2. This region can be further subdivided in three sub-regions according to the occupation of the cells. Region 1 represents the low load cases, where there are plenty of available resources for the users, which in turn get served quickly and often leave the network before the next arrival. Inter-cell interference can be neglected in this case as the sessions are

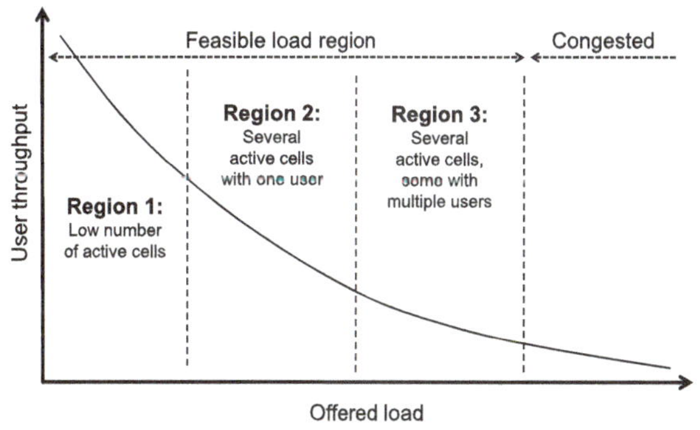

Figure 2 Characterization of system behaviour for different offered loads.

very short and the probability of having multiple active cells at the same time is low. In region 2, the offered load has increased, together with the probability of having several simultaneously active transmitters. However, the load in the cells is still fairly low, with typically one active user per cell. Finally, region 3 represents the case where the load in the network is such that the capacity limit is nearly reached, with a high number of users in some cases and a considerable number of cells transmitting at once.

2.3 Interference mitigation benefit

This section introduces the theoretical analysis that will allow us to evaluate the spectral efficiency gains from mitigating the main interferer. Starting with the most common signal quality measure, the signal-to-interference-plus-noise ratio, SINR (Γ), is defined as

$$\Gamma = \frac{P}{\sum_n I_n + N}, \quad (1)$$

where P is the power of the desired signal, $\sum_n I_n$ represents the total amount of received interference and N is the background noise. The time variation of the parameters in (1) has been kept out of the equation for the sake of notational simplicity.

The dominant interference ratio, DIR (Λ), indicates how significant the role played by the strongest interferer in the total interference profile is. The DIR is defined as

$$\Lambda = \frac{I_1}{\sum_n I_n - I_1 + N}, \quad I_1 \geq I_2 \geq \ldots \geq I_n, \quad (2)$$

where I_1 represents the strongest source of interference. We can relate the DIR to the potential performance benefit that would be obtained assuming ideal cancellation of the main interferer. Under that assumption, the SINR becomes

$$\Gamma_c = \frac{P}{\sum_n I_n - I_1 + N}. \quad (3)$$

Taking the ratio of the SINR expressions in (3) and (1), we quantify the improvement from cancelling the strongest interferer as

$$\frac{\Gamma_c}{\Gamma} = \frac{I_1}{\sum_n I_n - I_1 + N} + 1 = \Lambda + 1, \quad (4)$$

which is proportional to the DIR value. Finally, the DIR and the SINR can be related to the throughput improvement with ideal cancellation of the strongest interferer by applying Shannon's formula and calculating the ratio of the spectral efficiencies with cancellation, C_c, and without, C,

$$\frac{C_c}{C} = \frac{\log_2(1+\Gamma_c)}{\log_2(1+\Gamma)} = \frac{\log_2(1+\Gamma(\Lambda+1))}{\log_2(1+\Gamma)}. \quad (5)$$

2.4 Intra-cell packet scheduling

The packet scheduler determines how resources should be allocated among the multiple users of a cell. Because the packet scheduler performs the resource allocation in an intra-cell fashion, it can only impact the performance in region 3, where there are multiple users within the cell. This study makes use of three different scheduler algorithms, all based on the same principle of selecting a user u^* according to a metric M_u,

$$u^* = \arg\max_u \{M_u\}, \quad M_u = \frac{r_u^\alpha}{R_u^\beta}, \quad (6)$$

where u is the index of the user, r_u is the achievable throughput for user u in the current Transmission Time Interval (TTI), R_u is the past average throughput and $\alpha, \beta \in [0,1]$ are parameters which control the fairness.

The first algorithm, and one of the most commonly used, is Proportional Fair (PF) [23], obtained by applying $\alpha = 1, \beta = 1$ in (6). PF has been found to offer a good trade-off between scheduling gains and fairness, especially under full-buffer traffic models. In addition, a modified gradient search β-fair scheduler algorithm, known as Generalized PF (GPF), will be included as it was found to be more attractive for scenarios with a birth-death traffic model in [20]. Finally, we will present results for the Blind Equal Throughput (BET) scheduler, with $\alpha = 0, \beta = 1$, targeted to serving users with an equal average throughput [24].

3 Simulation methodology

The interference analysis and estimation of potential interference mitigation benefits will be based on system-level simulation results. The simulator is time based and includes all the major LTE resource management functionalities such as link adaptation, hybrid automatic repeat request (H-ARQ) and packet scheduling. In every 1-ms subframe, the SINR of each user is calculated per subcarrier according to the chosen receiver type. Subsequently, it is determined whether the transmission was successfully decoded using the effective exponential SINR model [25] for link-to-system-level mapping. H-ARQ with ideal Chase combining is applied in case of failed transmissions, and the SINRs for the different H-ARQ transmissions are linearly added. The link adaptation functionality determines the modulation and coding scheme for the first transmission based on frequency-selective feedback from the users. The simulator does not consider user mobility (for HetNet studies with mobility, the reader can refer to [26,27]). However, the user sessions are generally short and the SINR calculations include the effect of variable fast fading. Together with an open-loop traffic model, this provides a significant variability in the channel conditions. The main simulation settings for this study are summarized in this section and collected in Table 1.

Table 1 Main simulation assumptions

	Macro scenario	Small cell scenario
Network layout	7 three-sectored sites	3 clusters with 10 small cells each [13]
Bandwidth	10 MHz at 2.0 GHz	10 MHz at 3.5 GHz
Transmit power	46 dBm	30 dBm
User arrival rate, λ	0-4 users/s/cell	0 to 30 users/s/cluster
Path loss model	ITU-R UMa [21]	ITU-R UMi [21]
Antenna pattern	Directional, 70° beamwidth [21]	Omnidirectional
Receiver type	MMSE-IRC	
Traffic model	Poisson arrival, finite buffer	
Payload size, L	0.5 Mbytes	
Transmission mode	2 × 2 MIMO, single user	
OFDMA symbols/TTI	13	
β (for GPF)	0.6	

In the macro-cell scenario, the cell transmit power is 46 dBm and the antennas have a directional pattern. The carrier frequency is 2 GHz with 10-MHz bandwidth. The arrival rate will range between 0 and 4 users/cluster/s. The stochastic ITU-R urban macro-cell (UMa) radio propagation model is assumed, including different characteristics for line-of-sight (LOS) and non-LOS (NLOS) [21]. The LOS case considers shadow fading with a 4-dB standard deviation (σ) and different expressions for the path loss depending on the distance with respect to a breakpoint. The NLOS case has no breakpoint and uses $\sigma = 6$ dB for the shadow fading.

In the small cell scenario, all the small cells operate on the same carrier frequency at 3.5 GHz with 10-MHz bandwidth. The antenna pattern is omnidirectional, and the transmit power is 30 dBm. The arrival rate will be set to values between 0 and 30 users/cluster/s. The path loss model is ITU-R urban micro-cell (UMi) [21], again with different expressions for LOS and NLOS cases. The LOS expression depends on the distance to a breakpoint and applies $\sigma = 3$ dB, whereas the NLOS case assumes $\sigma = 4$ dB.

The remaining simulation parameters are common to both scenarios. The users demand a $L = 0.5$ MB payload. Closed loop 2 × 2 single-user MIMO with rank adaptation is assumed, i.e., corresponding to LTE transmission mode-4 [28,29]. Packet scheduling is performed in the time domain only, with one user per TTI [24]. This allows us to increase the number of OFDMA symbols per TTI from 11 to 13 to improve the data rate. The value of β in (6) for the GPF scheduler is fixed to 0.6 as recommended in [20]. The link to-system level modelling is according to [25]. The receiver type at the user equipment is MMSE-IRC [30].

4 Performance results

In this section, we will take a look at simulation results which illustrate the interference conditions and achievable data rates in the considered scenario, under different traffic loads and scheduling metrics. The first part of the section will focus on establishing the traffic regions as described in Section 2.2 and on a comparison of the scheduling algorithms. Next, we will take a closer look at the load behaviour of the network by examining cell occupation statistics. The magnitude and time variability of the interference in the network will be dealt with afterwards, finally offering an estimation of the potential benefits that could be obtained from interference mitigation.

4.1 Traffic load region analysis

We will begin the analysis by studying the behaviour of the traffic in terms of the different load regions in the macro-cell and small cell scenarios. The carried load and the average number of users per cell as a function of the offered traffic are presented in Figure 3 for the two scenarios and the three schedulers. The carried load plots (upper graphs) present two different segments: the feasible load region in which the carried load increases linearly and matches the offered load, and the congestion region, where the network cannot cope with the amount of demanded traffic. The feasible load region reaches approximately 7.5 Mbps offered load in the macro-only case and 75 Mbps offered load in the small cell case. The evolution of the carried load with increasing traffic is very similar in both scenarios.

The carried load can be used to find the limit of the feasible load region, but it is necessary to look at cell occupation statistics to classify the offered traffic into sub-regions as discussed in Section 2.2. The average number

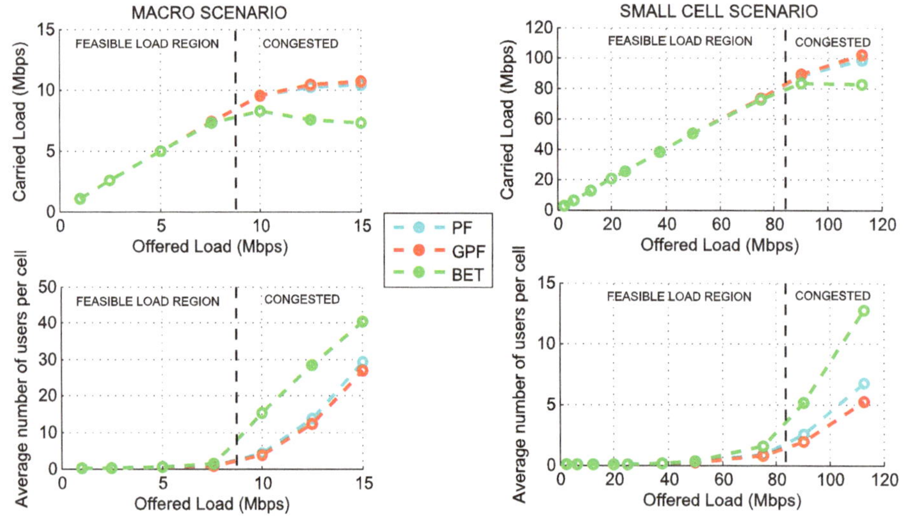

Figure 3 Scheduler performance comparison. Top: carried load vs. offered load. Bottom: average number of active users per cell over offered load.

of users per cell (bottom part of Figure 3) is kept low in the feasible load region but quickly grows beyond the congestion point. As the system is unstable in this region and the results are therefore highly dependent on parameters such as the simulation time, we will focus on the feasible region for the remainder of the article. In line with the results presented in [15] and [20] for the dense small cell case, the scheduling algorithm which provides the best results in terms of these two metrics is GPF, both for the macro-cell and the small cell scenario. The reason is that this scheduler assigns a higher priority to users under better SINR conditions than PF and BET. These users get served faster and, given the chosen open-loop traffic model, they can leave the network more quickly, reducing the generated interference. This results in an overall performance gain in the network. Therefore, we will only show the statistics obtained under the GPF scheduler in the following figures.

The behaviour observed in Figure 3, both for the carried load and the number of users per cell, can help us choose a representative offered load value for each of the three characteristic traffic regions. These values and the percentage of active cells in each region are indicated in Table 2. The table serves as a reference for the following figures in the article, where we will not refer explicitly to the offered load value but to the traffic region.

4.2 Cell occupation statistics

One result from Table 2 that immediately comes to the forefront is the low percentage of active cells in region 3, within the feasible region but close to the congestion point. We can examine the situation more closely by plotting the empirical cumulative distribution function (cdf) of the number of users in the cells in this region as shown in Figure 4. The distribution of the users in the cells in both scenarios is very uneven. As previously presented in Table 2, there is a high percentage of inactive cells, while some cells contain a fairly large number of users. The majority of the active cells, however, are only simultaneously serving one or two users. Since region 3 is within the feasible region but close to congestion, this behaviour suggests that congestion can be reached without all of the cells being active, as long as some of them are very occupied. Those cells that are very loaded cause such interference to their neighbours that the system approaches saturation while half of its resources are kept unused. This is the case not only in the dense small

Table 2 Traffic load regions according to offered load and cell occupation

	Scenario			
	Macro cell		Small cell	
Traffic region	Offered load (Mbps)	Ave. perc. of active cells (%)	Offered load (Mbps)	Ave. perc. of active cells (%)
1	2.5	8.7	25	6.9
2	5	22.8	50	20
3	7.5	43	75	41

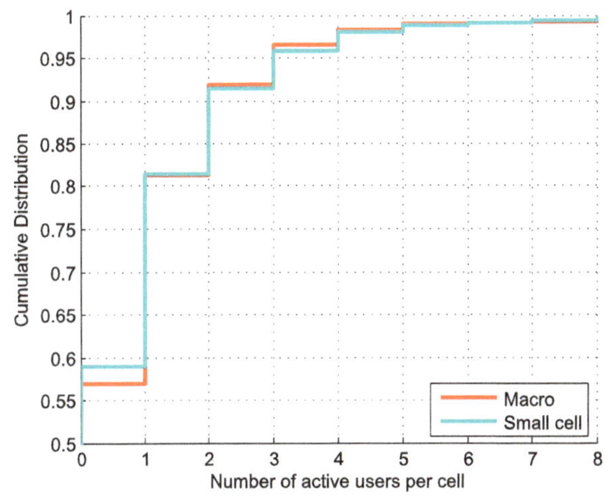

Figure 4 Cumulative distribution function of the number of active users per cell in traffic load region 3.

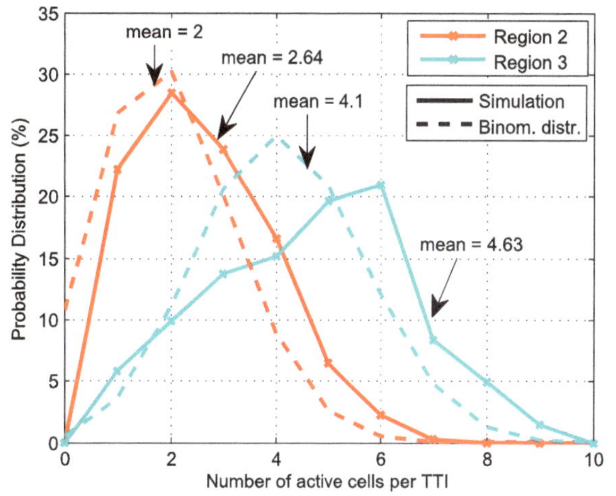

Figure 5 Probability distribution function of the number of active cells for one cluster in the dense small cell scenario.

cell cluster scenario but also in the macro-cell network, where the cell deployment is regular and the spatial user distribution is uniform. This situation is clearly undesirable and indicates the need for solutions that can reduce the congestion in the highly loaded cells. As suggested by the performance gains brought by the GPF scheduler, trying to serve the users in a faster way could have a positive impact in terms of reduced interference and cell occupation. Furthermore, inter-cell load balancing solutions could be applied to compensate for the uneven use of resources in the network [31].

Further insight on the distribution of users within the dense small cell clusters can be attained by studying the probability distribution function (pdf) of the number of active cells in each TTI. If the occupation of any given cell were statistically independent from the rest, the pdf would follow a binomial distribution [32]. The binomial distribution is the discrete probability distribution of a number of successes, X, achieved after t independent trials, each of them having a success probability p,

$$P\{X = i\} = \binom{t}{i} p^i (1-p)^{t-i}, \quad i = 0, 1, \ldots, t. \qquad (7)$$

Figure 5 presents the empirical distribution of the number of active cells in every TTI for one of the clusters in the dense small cell scenario, together with the theoretical binomial distribution. Region 1 was omitted from the figure as the number of active cells is very low. The values of p for the binomial distribution (i.e., the mean probability that a cell will be active) were obtained from the cdfs of the number of active users per cell, such as the example presented in Figure 4 (for region 3). The mismatch between the binomial and empirical distributions suggests that there is a coupling between the cells in the cluster because of mutual interference, and the occupation of the cells is not an independent process.

4.3 Signal and interference levels

The magnitude of the interference will be quantified in terms of the SINR and DIR of the users. The cdf of the users' scheduled SINR in the three traffic regions is shown in Figure 6. As expected, the values decrease with increasing traffic load as more cells start becoming active and the interference increases. Moreover, the SINR is higher for the small cell scenario than for the macro-cell case in regions 1 and 2. This is due to the larger inter-site distance in the macro-only case, making it more probable to have users located far from the serving cell and in lower SINR conditions. On the other hand, the larger number of users in region 3 increases the diversity and hence the values are very similar in both scenarios. The user throughput is directly linked to the SINR and therefore exhibits the same behaviour as the latter with regard to the traffic load regions and network scenarios, as shown in Figure 7.

A different conclusion can be drawn with respect to the DIR, which increases with the offered traffic as presented in Figure 8. At low load (region 1), having very few active cells in the network can often imply that the strongest interferer for a given user is located far in the network. When this happens, the DIR expression in (2) is dominated by the background noise component, and the DIR value becomes very low. This behaviour can also be observed for a small percentage of the cases in regions 2 and 3, but generally, as more cells start becoming active, the probability that the dominant interferer will be located closer to the user increases, and so does the DIR. An almost negligible difference between the two scenarios can be observed in region 1, but with increasing traffic, the macro-cell scenario provides the largest DIR values,

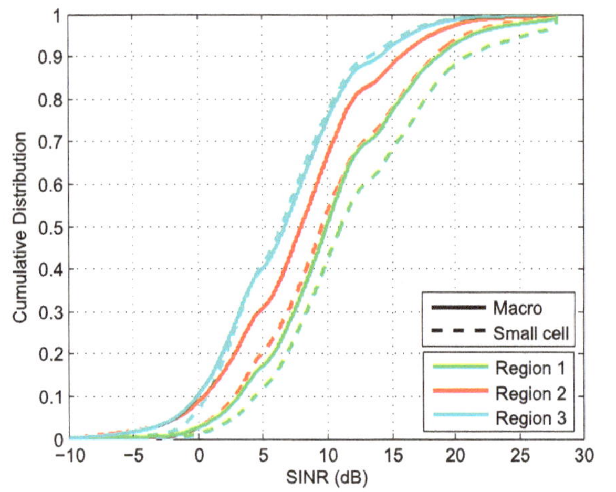

Figure 6 Cumulative distribution function of the user SINR.

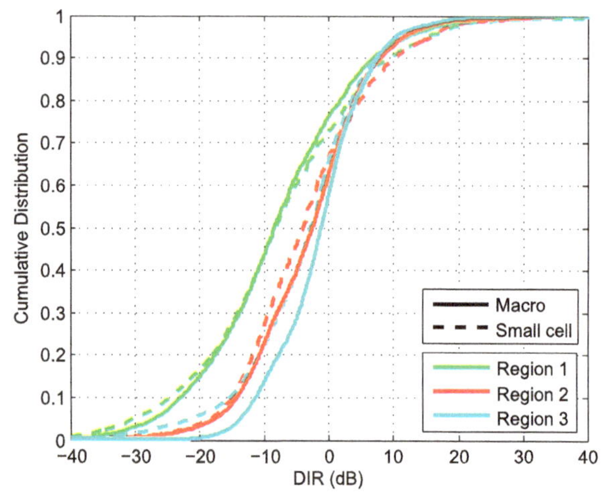

Figure 8 Cumulative distribution function of the DIR.

due to the higher transmitted power and the lower number of potentially interfering cells compared to the small cell cluster. In general, region 3 is the most interesting one in terms of applying a mechanism to mitigate the strongest interferer, as it is the one providing the largest DIR values, and the potential benefit from mitigation is related to this parameter as shown in (5).

4.4 Time variability of the interference

From the perspective of designing an interference mitigation mechanism, it is important to understand how the interference in the network evolves in time. In order to study this aspect, we present in Figure 9 an example of the time variation of the DIR for 15 of the users. The selected case is the dense small cell scenario at high load (region 3). Each horizontal bar in Figure 9 shows the values of the DIR within the lifetime of one user. The frequent colour changes indicate that this value can shift within a few TTIs, sometimes abruptly. It should be noted that our estimation of the DIR does not take into account the fast fading, which does change for every simulated TTI. Therefore, all DIR variations are due to the interference pattern changing when users enter or leave the network. The DIR changes with approximately the same frequency in both scenarios, hence the omission of macro results in Figure 9.

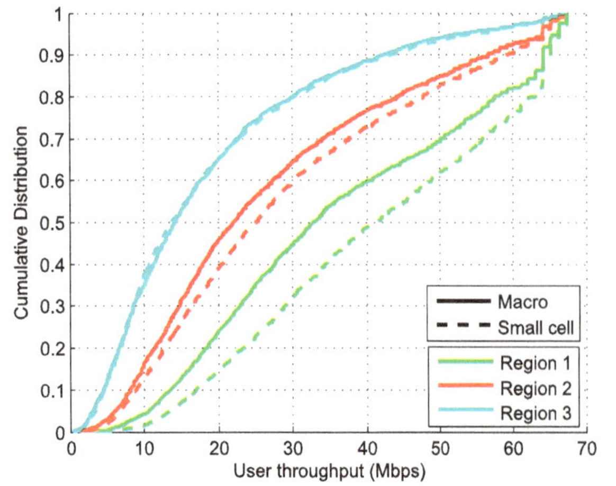

Figure 7 Cumulative distribution function of the user throughput.

To understand better the source of the DIR changes, Figure 10 shows the time variation of the strongest interferer cell index for the same set of users, in a similar fashion. Looking at Figures 9 and 10, we can see that, while there are frequent shifts in the DIR value, the strongest interferer index remains constant for a longer time. This is true not only for the few users presented in the two figures but also for the rest of the cases, as pictured in Figures 11 and 12, which show the cdf of the number of TTIs between changes in DIR value and in strongest interferer index, respectively. For example, the 50-percentile value is at 7 TTIs between DIR changes but at 100 in the case of the strongest interferer index. This indicates that the changes in the DIR are mainly due to secondary interferers becoming active or inactive.

4.5 Potential benefit from interference mitigation

The potential gains from interference mitigation in both scenarios are finally quantified in Figure 13. These gains are estimated from the empirical SINR and DIR values of the users according to (5). The improvement is more pronounced for the higher traffic loads and is overall very similar for the two scenarios. For low load (region 1), there is a high probability of having a negligible improvement, and the values are slightly higher in the small cell case. As the traffic load increases, so does the probability of

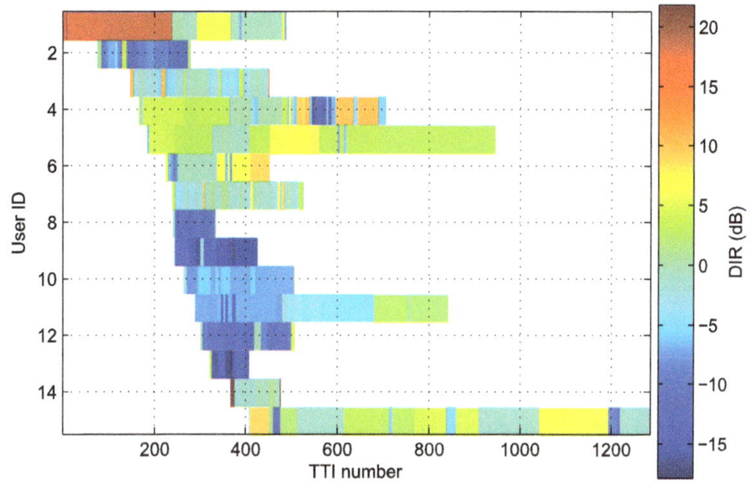

Figure 9 Time evolution of the DIR for 15 of the users. GPF scheduler and 75 Mbps offered load.

achieving more significant gains, and the macro-cell scenario starts yielding higher potential benefits. In region 3, there is around a 30% probability of having a throughput gain over 50%, and values as high as 300% can be reached for particular users and TTIs.

5 Discussion of interference mitigation options

The presented results show a comparison of the characteristics of the interference in two network scenarios. In spite of the very different nature of the macro-cell and dense small cell cases, the interference behaviour was found to be remarkably similar. Even though previous studies worked under the hypothesis that denser deployments will require the use of custom designed interference mitigation techniques [33], the findings in this article point out that the performance in such cases could be improved by applying similar solutions to those utilized in macro-cell deployments. Moreover, the gains that could be achieved seem to be comparable.

In general, interference mitigation techniques can be classified in two groups [34]: network-based coordination and user equipment-based solutions. Interference can be mitigated from the network side by limiting the resources in the cells which cause a significant portion of the interference. As explained in [35], there is an important trade-off to consider when applying resource partitioning. On the one hand, there will be a performance increase for the users that were affected by the interference. We can quantify this increase as a benefit metric. On the other hand, users served by the cells where resources have been limited will undergo a performance decrease, which can be considered as the cost metric of the solution. As long as the benefit is higher than the cost, the applied technique will bring an overall improvement in the network.

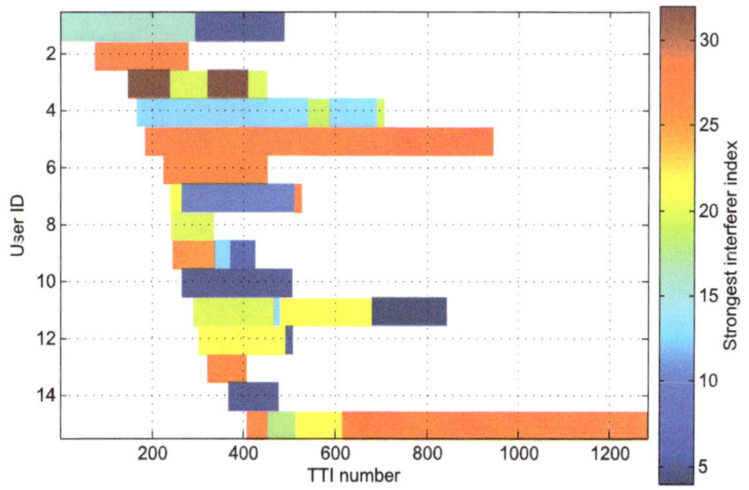

Figure 10 Time evolution of the strongest interferer index for 15 of the users. GPF scheduler and 75 Mbps offered load.

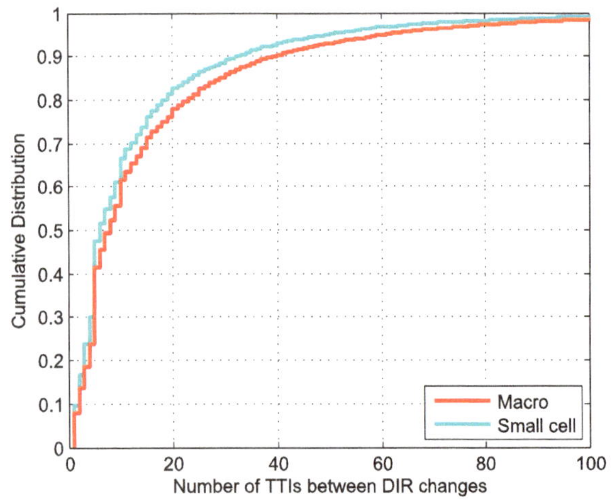

Figure 11 Cumulative distribution function of the number of TTIs between changes in DIR value.

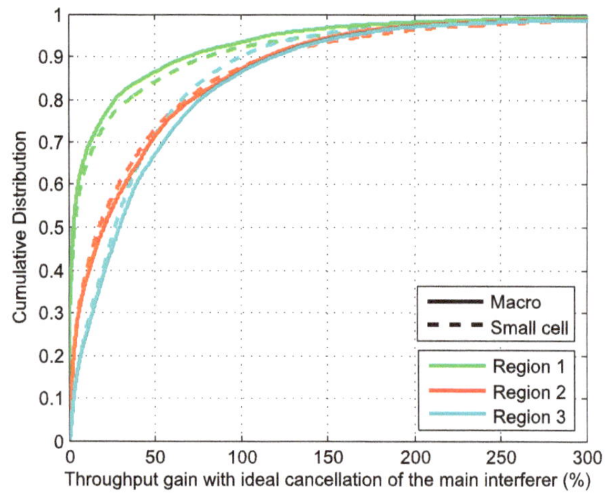

Figure 13 Potential throughput increase with ideal cancellation of the main interferer (%). Scheduling algorithm: GPF.

In addition to network-based coordination, the user equipment can play a significant role in mitigating the interference by means of advanced receivers, operating in a linear [30,36] or non-linear fashion [37]. The use of advanced receivers presents an inherent advantage in that, since the interference is mitigated at the receiver, there is no need to limit the interfering cell's resources, effectively eliminating the performance cost that network-based coordination implies. However, user equipment-based techniques are not exempt from limitations. For example, linear advanced receivers with M antennas can only suppress up to $M-r$ sources of interference [38], with r being the transmission rank, while non-linear advanced receivers can have stringent SINR requirements of the strongest interference source, to be able to reliably estimate, reconstruct and cancel it from the total received signal.

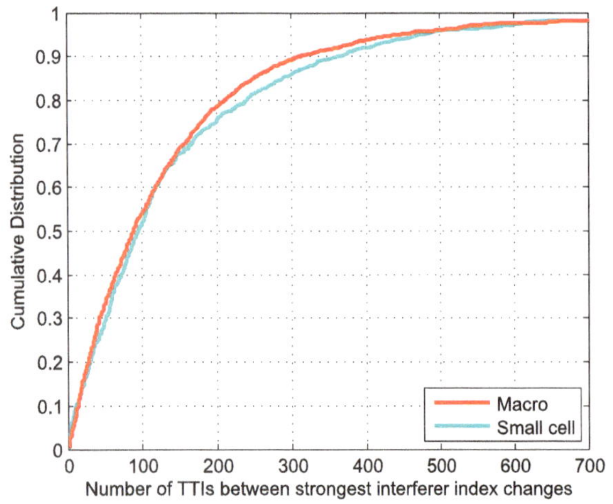

Figure 12 Cumulative distribution function of the number of TTIs between changes in strongest interferer index.

The chosen interference mitigation solution should be dynamic enough to track the changes in the interference profile as suggested by Figures 11 and 12. It is usually more important to curb the effect of the strongest interferer than of the secondary ones, and the strongest interferer index was shown to change with a median period of 100 TTIs. This period is short enough to suggest that some solutions in the literature could be re-evaluated or modified to allow for more dynamic updates. In particular, most of the studies focused on macro-only scenarios have traditionally employed rather static mechanisms (an example is frequency reuse techniques [2]). In a scenario with user mobility or with a smaller packet size, the time variability of the interference would increase, further reinforcing the need for more dynamic solutions.

6 Future work

Future research could analyse the interference and potential benefits from mitigation under different network and traffic models, to understand how the chosen simulation scenario impacts on the conclusions. Examples of network models that could be used include the ones described in Section 2.1, such as deterministic Wyner models, random models based on different point processes and site-specific scenarios based on real data. It would also be interesting to study the interference conditions under a closed-loop finite-buffer traffic model, with a fixed number of users in the network. Additionally, a model with different classes of traffic based on quality of service demands could be defined. User mobility is another aspect that might impact the results, making interference more variable and accentuating the need for sufficiently dynamic scheduling and interference mitigation solutions.

The findings presented in this paper could be applied to the design of joint inter-cell interference mitigation techniques combining some of the discussed options, including both network-based coordination and receiver-side interference suppression.

7 Conclusions

This article analysed the interference characteristics, performance and use of resources in two different LTE-A deployments: a regular macro-cell network and a network comprising dense small cell clusters. These aspects were examined under a dynamic traffic model with different amounts of offered traffic. The two deployments exhibited a strikingly similar behaviour in the different traffic load regions: both the performance figures and the time variability of the interference were comparable. The similarity became more noticeable with increasing offered loads.

The extent to which the main interferer impacts on the user performance was evaluated by means of the dominant interference ratio. This parameter was related through theoretical expressions to the potential benefit from mitigating the strongest source of interference, indicating a potential for notable performance gains in both scenarios. Furthermore, since the interference patterns in the two deployments show a strong resemblance, similar interference mitigation solutions could be applied.

Competing interests
The authors declare that they have no competing interests.

Acknowledgements
The authors would like to thank Jens Steiner from Nokia Networks for his helpful contributions to the study.

Author details
[1]Department of Electronic Systems, Aalborg University, Fredrik Bajers Vej 7, 9220 Aalborg, Denmark. [2]Nokia Networks, Alfred Nobels Vej 21C, 9220 Aalborg, Denmark.

References

1. H Holma, A Toskala, *LTE for UMTS: Evolution to LTE-Advanced*, 1st edn. (John Wiley & Sons, Ltd, Chichester, 2006)
2. A Simonsson, in *IEEE 65th Vehicular Technology Conference (VTC 2007-Spring)*. Frequency reuse and intercell interference co-ordination in E-UTRA (IEEE, New York, 2007), pp. 3091–3095
3. E Pateromichelakis, M Shariat, A ul Quddus, R Tafazolli, On the evolution of multi-cell scheduling in 3GPP LTE/LTE-A. IEEE Commun. Surv. Tutorials. **15**(2), 701–717 (2013)
4. G Boudreau, J Panicker, N Guo, R Chang, N Wang, S Vrzic, Interference coordination and cancellation for 4G networks. IEEE Commun. Mag. **47**(4), 74–81 (2009). doi:10.1109/MCOM.2009.4907410
5. J Lee, Y Kim, H Lee, BL Ng, D Mazzarese, J Liu, W Xiao, Y Zhou, Coordinated multipoint transmission and reception in LTE-advanced systems. IEEE Commun. Mag. **50**(11), 44–50 (2012). doi:10.1109/MCOM.2012.635368
6. X Wang, B Mondal, E Visotsky, A Ghosh, in *IEEE International Conference on Communications Workshops (ICC), 2014*. Coordinated scheduling and network architecture for LTE macro and small cell deployments (IEEE, New York, 2014), pp. 604–609
7. A Damnjanovic, J Montojo, Y Wei, T Ji, T Luo, M Vajapeyam, T Yoo, O Song, D Malladi, A survey on 3GPP heterogeneous networks. IEEE Wireless Commun. **18**(3), 10–21 (2011). doi:10.1109/MWC.2011.587649
8. JG Andrews, Seven ways that HetNets are a cellular paradigm shift. IEEE Commun. Mag. **51**(3), 136–144 (2013). doi:10.1109/MCOM.2013.647687
9. KI Pedersen, Y Wang, S Strzyz, F Frederiksen, Enhanced inter-cell interference coordination in co-channel multi-layer LTE-advanced networks. IEEE Wireless Commun. **20**(3), 120–127 (2013). doi:10.1109/MWC.2013.654929
10. T Nakamura, S Nagata, A Benjebbour, Y Kishiyama, H Tang, X Shen, N Yang, N Li, Trends in small cell enhancements in LTE advanced. IEEE Commun. Mag. **51**(2), 98–105 (2013). doi:10.1109/MCOM.2013.646119
11. LGU Garcia, KI Pedersen, PE Mogensen, Autonomous component carrier selection: interference management in local area environments for LTE-advanced. IEEE Commun. Mag. **47**(9), 110–116 (2009). doi:10.1109/MCOM.2009.5277463
12. Y Wang, N Marchetti, IZ Kovacs, PE Mogensen, KI Pedersen, TB Sorensen, in *IEEE 70th Vehicular Technology Conference (VTC 2009-Fall)*. An interference aware dynamic spectrum sharing algorithm for local area LTE-Advanced networks (IEEE, New York, 2009), pp. 1–5. doi:10.1109/VETECF.2009.5379036
13. 3GPP, Small cell enhancements for E-UTRA and E-UTRAN - physical layer aspects (Release 12). TR 36.872 (2013)
14. N Bhushan, J Li, D Malladi, R Gilmore, D Brenner, A Damnjanovic, R Sukhavasi, C Patel, S Geirhofer, Network densification: the dominant theme for wireless evolution into 5G. IEEE Commun. Mag. **52**(2), 82–89 (2014). doi:10.1109/MCOM.2014.673674
15. V Fernandez-Lopez, KI Pedersen, B Soret, in *IEEE 80th Vehicular Technology Conference (VTC 2014-Fall)*. Effects of interference mitigation and scheduling on dense small cell networks (IEEE, New York, 2014), pp. 1–5
16. HS Jo, YJ Sang, P Xia, JG Andrews, Heterogeneous cellular networks with flexible cell association: a comprehensive downlink SINR analysis. IEEE Trans. Wireless Commun. **11**(10), 3484–3495 (2012). doi:10.1109/TWC.2012.081612.111361
17. A Tukmanov, Z Ding, S Boussakta, A Jamalipour, On the impact of network geometric models on multicell cooperative communication systems. IEEE Wireless Commun. **20**(1), 75–81 (2013). doi:10.1109/MWC.2013.647220
18. JG Andrews, RK Ganti, M Haenggi, N Jindal, S Weber, A primer on spatial modeling and analysis in wireless networks. IEEE Commun. Mag. **48**(11), 156–163 (2010). doi:10.1109/MCOM.2010.562198
19. C Coletti, H Liang, H Nguyen, IZ Kovács, B Vejlgaard, R Irmer, N Scully, in *IEEE 76th Vehicular Technology Conference (VTC 2012-Fall)*. Heterogeneous deployment to neet traffic demand in a realistic LTE urban scenario (IEEE, New York, 2012), pp. 1–5
20. P Ameigeiras, Y Wang, J Navarro-Ortiz, P Mogensen, J Lopez-Soler, Traffic models impact on OFDMA scheduling design. EURASIP J. Wireless Commun. Netw. **2012**(1), 61 (2012). doi:10.1186/1687-1499-2012-61
21. 3GPP, Technical Specification Group Radio Access Network; Evolved Universal Terrestrial Radio Access (E-UTRA); further advancements for E-UTRA physical layer aspects (Release 9). TR 36.814 (2010)
22. V Paxson, S Floyd, Wide area traffic: the failure of Poisson modeling. IEEE/ACM Trans. Netw. **3**(3), 226–244 (1995). doi:10.1109/90.39238
23. F Kelly, Charging and rate control for elastic traffic. Eur. Trans. Telecommunications. **8**(1), 33–37 (1997)
24. G Monghal, KI Pedersen, IZ Kovacs, PE Mogensen, in *IEEE 67th Vehicular Technology Conference (VTC 2008-Spring)*. QoS oriented time and frequency domain packet schedulers for the UTRAN Long Term Evolution (IEEE, New York, 2008), pp. 2532–2536
25. K Brueninghaus, D Astely, T Salzer, S Visuri, A Alexiou, S Karger, GA Seraji, in *IEEE 16th International Symposium on Personal, Indoor and Mobile Radio Communications (PIMRC 2005)*. Link performance models for system level simulations of broadband radio access systems, vol. 4 (IEEE, New York, 2005), pp. 2306–23114
26. KI Pedersen, PF Michaelsen, C Rosa, S Barbera, Mobility enhancements for LTE-advanced multilayer networks with inter-site carrier aggregation. IEEE Commun. Mag. **51**(5), 64–71 (2013). doi:10.1109/MCOM.2013.6515048

27. D Lopez-Perez, I Guvenc, IX Chu, Mobility management challenges in 3GPP heterogeneous networks. IEEE Commun. Mag. **50**(12), 70–78 (2012). doi:10.1109/MCOM.2012.6384454
28. 3GPP, Technical Specification Group Radio Access Network; Evolved Universal Terrestrial Radio Access (E-UTRA); Physical layer procedures (Release 12). TS 36.213 (2014)
29. IZ Kovacs, M Kuusela, E Virtej, KI Pedersen, in *IEEE 67th Vehicular Technology Conference (VTC 2008-Spring)*. Performance of MIMO aware RRM in downlink OFDMA (IEEE, New York, 2008), pp. 1171–1175
30. FML Tavares, G Berardinelli, NH Mahmood, TB Sorensen, PE Mogensen, in *IEEE 78th Vehicular Technology Conference (VTC 2013-Fall)*. On the potential of interference rejection combining in B4G networks (IEEE, New York, 2013), pp. 1–5
31. JG Andrews, S Singh, Q Ye, X Lin, HS Dhillon, An overview of load balancing in HetNets: old myths and open problems. IEEE Wireless Commun. **21**(2), 18–25 (2014). doi:10.1109/MWC.2014.6812287
32. SM Ross, *Introduction to Probability and Statistics for Engineers and Scientists*. (John Wiley & Sons, 1987)
33. N Bhushan, J Li, D Malladi, R Gilmore, D Brenner, A Damnjanovic, R Sukhavasi, C Patel, S Geirhofer, Network densification: the dominant theme for wireless evolution into 5G. IEEE Commun. Mag. **52**(2), 82–89 (2014)
34. B Soret, KI Pedersen, NTK Jørgensen, V Fernández-López, Interference coordination for dense wireless networks. IEEE Commun. Mag. **53**(1), 102–109 (2015). doi:10.1109/MCOM.2015.7010522
35. R Agrawal, A Bedekar, S Kalyanasundaram, N Arulselvan, T Kolding, H Kroener, in *IEEE 79th Vehicular Technology Conference (VTC 2014-Spring)*. Centralized and decentralized coordinated scheduling with muting (IEEE, New York, 2014), pp. 1–5
36. K Pietikainen, F del Carpio, H Maattanen, M Lampinen, T Koivisto, M Enescu, in *IEEE 75th Vehicular Technology Conference (VTC 2012-Spring)*. System-level performance of interference suppression receivers in LTE system (IEEE, New York, 2012), pp. 1–5
37. 3GPP, Technical Specification Group Radio Access Network; Study on Network-Assisted Interference Cancellation and Suppression (NAICS) for LTE (Release 12). TS 36.866 (2013)
38. JH Winters, Optimum combining in digital mobile radio with cochannel interference. IEEE J. Sel. Areas Commun. **2**(4), 528–539 (1984). doi:10.1109/T-VT.1984.2400

16

Coordinated multi-point transmission for relaxation of self-backhauling bottlenecks in heterogeneous networks

Beneyam B. Haile*, Edward Mutafungwa and Jyri Hämäläinen

Abstract

The heterogeneous deployment of high-power macro cells and low-power nodes (LPNs) is now widely acknowledged as an essential requirement towards meeting the continued demand for mobile data capacity. The selection of the optimum backhaul solution for the LPNs obliges operators to consider not only the capacity of the backhaul but also other key factors so as to fully leverage the benefits provided by LPNs: the cost of the backhauling may limit the density of LPN deployments and the backhaul configuration requirements impact on the flexibility of LPN deployment. To that end, self-backhauling of LPNs via the existing macro radio access network (RAN) provides an attractive solution, particularly for deployment scenarios that are very cost-sensitive and/or require high flexibility. However, use of self-backhauling usually makes backhaul as a bottleneck due to the a) limited bandwidth allocated for legacy macro RAN, b) the need to share resources with macro user equipment (UE), and c) the high-intercell interference particularly in the macro cell edge. In this paper, we provide an overview of self-backhauled LPNs and investigate possible performance enhancements through the use of coordinated multi-point (CoMP) transmission to relax the downlink backhaul capacity bottleneck for self-backhauled LPNs. To that end, we carry out analytical studies for a practical limited-feedback CoMP technique and numerically verify the derived capacity outage expressions. Furthermore, we implement a simulation study for an exemplary heterogeneous network deployment in a realistic radio propagation environment. The results of the studies demonstrate that significant spectral efficiency and throughput gains for the LPN backhaul are achievable through the use of selected CoMP technique under realizable feedback overhead, even under feedback bit error. The achieved relaxation in the backhaul bottleneck is observed providing improved performance for the UEs served by the LPNs. Furthermore, more resources will be available for macro UEs leading to overall performance gains compared to the case without CoMP.

Keywords: Coordinated multi-point transmission; LTE-advanced; Low-power nodes; Small cells; Relays; Backhaul

1 Introduction

1.1 Background

Mobile network operators face the challenge of upgrading their networks to handle increased mobile data traffic from certain locations (hot spots or hot zones) and meet user expectations for cellular coverage in every location [1]. While the approach of scaling network capacity by increasing available spectrum and improving spectral efficiency provides considerable scaling in network capacity, even more substantial growth is possible by reusing spectrum through network densification [1-3]. The commonly considered network densification approach is through the heterogeneous deployment of low-power nodes (typically 10 W or lower) to complement traditional high-power (20 W or higher) macro sites [1-5]. The term *low-power node* (LPN) is adopted in this paper to draw a distinction between the added complementary LPNs and the legacy macro sites in a heterogeneous network (HetNet) environment.

A diverse range of LPNs exist depending on various practical considerations. These include drivers for LPN deployment (need to improve coverage, capacity, or both) and attributes of the coverage area, such as area size and

*Correspondence: beneyam.haile@aalto.fi
School of Electrical Engineering, Aalto University, P.O. Box 13000, FI-00076 Aalto, Espoo, Finland

location (indoor or outdoor). The LPNs include small cells and extensions to macro cells, such as relays, remote radio heads (RRHs), and distributed antenna systems (DAS). Small cells is itself an umbrella term for compact base stations deployed to enhance coverage and capacity in homes, enterprise environments, under-served areas, and indoor or outdoor public spaces [2]. Operator deployed small cells include metrocells, microcells, picocells, and open access femtocells. On the other hand, indoor closed-access femtocells are deployed and operated by end users in their place of residence, analogous to traditional WiFi access points.

1.2 LPN backhauling options

The reduced size and weight of LPNs allows them to be deployed rapidly, cost-effectively, and on a relatively wider range of locations compared to traditional macro sites [6]. Typical LPN deployment locations include street furniture (e.g., lamp posts or utility poles), on the side of buildings below rooftop level, on vehicle rooftops, and so on. However, while benefiting from this added deployment flexibility, the operator has still to carefully consider LPN site selection to ensure accessibility for LPN site maintenance while simultaneously securing the LPN site assets from malicious intruders. Furthermore, reliable operation of LPNs requires assured access to uninterrupted power sources and high-capacity backhaul links [5-8].

In this paper, we focus on the LPN backhauling challenge. A number of LPN backhauling solutions are possible depending on the LPN deployment scenario [6,7]. Indoor-deployed LPNs (e.g., enterprise femtos) can be backhauled using existing in-building wireline infrastructure, such as copper twisted-pair digital subscriber lines (DSL), fiber, and coaxial cables (for cable television). Outdoor-deployed LPNs in most cases do not have access to legacy cabling and the cost of targeted roll-out of cables to each LPN would be prohibitive [6]. Therefore, wireless backhauling solutions, such as microwave radio links (<7 GHz, 6 to 50 GHz), millimeter wave and E-band radio links (57 to 66 GHz, 71 to 95 GHz), and free-space optics and satellite, are usually considered for outdoor LPNs [6-8]. The differentiating attributes for the different wireless backhaul solutions include the following:

- *Deployment topology:* Configuration between LPNs and backhaul hub point. Common configurations are point-to-point (P2P), point-to-multipoint (P2MP), tree, and mesh topologies.
- *Line-of-sight (LOS) requirements:* The LOS requirements vary from demand for strict LOS to near LOS (nLOS) and non-LOS (NLOS) links.
- *Operating spectrum band:* Differs depending on spectrum licensing arrangements (licensed or unlicensed bands). Differences may also be in spectrum allocation between LPN backhaul and access links, whereby utilized spectrum bands are either overlapping (inband) or orthogonal (outband) between the access and backhaul links.
- *Capacity:* Typical capacity (bits per second) available over the backhaul link.

These aforementioned attributes strongly influence the selection of wireless backhaul solution for a particular deployment scenario. Typically, LPN operators have to consider the trade-offs among factors, such as required performance, operating costs, and feasibility or ease of deployment [6-8]. In this study, we focus on wireless backhauling solutions for LPNs that are strongly constrained by cost and demand high flexibility in terms of deployment. The highly flexible backhaul is desirable since it usually implies rapid unplanned deployment of the LPN network expansion with no intervention from the operator. Example scenarios that demand such cost-sensitive and highly flexible backhauling include the following: densely populated areas [9]; semi-permanent networks (e.g., for open air festivals) [9]; and spontaneous or rapid network deployment scenarios (e.g., public safety communications) [9-11].

1.3 Self-backhauled LPN

1.3.1 Description and benefits

Self-backhauling of LPNs via the existing macro radio access network (RAN) is an attractive backhauling solution for certain LPN deployment scenarios. In Figure 1, we provide a simple illustration to describe the difference between conventional fixed-wireless backhauling and self-backhauling for LPNs. Conventional wireless backhauling for LPNs relies on dedicated P2P or P2MP fixed-wireless technologies (e.g., millimeter wave radios) with stringent LOS requirements between the LPN and the wireless hub which may be collocated with a macro site (see Figure 1a). The LOS requirements for fixed-wireless backhauling links can be relaxed through use of multi-hop mesh or tree topologies to route the fixed-wireless links around shadowing objects (e.g., tall buildings) [6].

By contrast, in Figure 1b, self-backhauling is provided via the macro RAN for LPNs deployed within the coverage area of a particular macrocell, with the macrocell resources being shared by other LPNs and/or macro user equipment (UE) located within the same coverage area.

The LPN self-backhauling approach provides a number of benefits compared to conventional fixed-wireless backhauling. These include the following:

- Cost savings by leveraging existing macrocell site infrastructure (e.g., radio towers, standby batteries) for small cell backhauling;

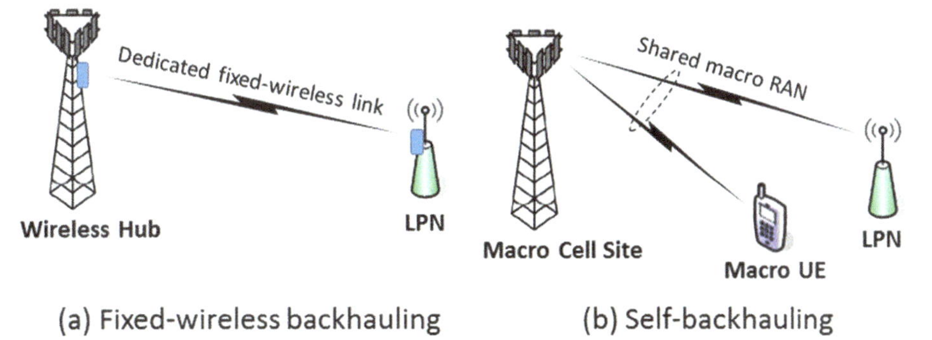

Figure 1 Backhauling of LPN using (a) fixed-wireless links and (b) self-backhauling via macro RAN.

- Further cost saving through possible reuse of spectrum bands licensed to the macro network operator and resource sharing over a number of LPNs (as opposed to having dedicated backhaul links to each LPN);
- Wireless backhaul in NLOS conditions using legacy cellular spectrum bands (typically below 3.5 GHz) and emerging lightly licensed bands (e.g., TV white space) with good RF propagation properties; and
- Possible joint radio resource management (RRM) implementation for the LPN backhaul and access links [12].

These self-backhauling benefits have motivated the development of the relaying concept introduced in Long Term Evolution (LTE)-Advanced. Relaying is a LPN solution, whereby, compact relay nodes (RNs) are wirelessly self-backhauled via a standardized LTE relay link (Un interface) towards a macro donor eNode B (DeNB) [13]. The relay access links towards the UEs are then based on the conventional LTE Uu air interface.

However, the self-backhauling approach can be extended to include all types of LPNs (beyond relays). This is based on the possible flexibility in selection of radio access technologies (RATs) and operating spectrum band strategies between the LPN backhaul (in this case, macro access) and LPN access links. The viability of this approach has been recognized in both research community and industry. For instance, Qualcomm proposes the use of so-called 'Velcro relays', whereby, macro LTE provides backhauling for evolved 3G small cells (HSPA+/EV-DO) in places with low-penetration user LTE [14]. The architectural arrangement of having a 'hub base station' as a concentration point of self-backhauled small cells has also been investigated in a recently completed project (FP7 BuNGee [15]). Notably, there have also been recent standardization activities for specification of the use of WiMAX air interface for small cell backhauling (IEEE 802.16r [16]).

1.3.2 Backhaul capacity bottleneck

The self-backhauling of Figure 1b, with use of macro resources in the backhaul, is key factor while utilizing the benefits of cost-effectiveness and flexibility in LPN deployment. However, the reuse of the macro resources usually results in a backhaul capacity bottleneck, whereby the aggregate served capacity of LPN access exceeds the achievable wireless backhaul capacity. This self-backhauling capacity bottleneck will consequently limit peak achievable throughputs for the UE served by the LPN.

Self-backhaul link capacity enhancements have been considered previously particularly for relay deployments. Proposals include strategies that optimize DeNB selection and placement of RN to provide signal-to-interference-and-noise ratio (SINR) gain for the relay backhaul link, particularly at the cell edge [17-19]. Unfortunately, these SINR gains may be difficult to achieve in practical deployments, due to the challenge of site acquisition that often limits the available set of good candidate locations. Additional capacity enhancements of self-backhauled links are being considered by increasing spectral-efficiency through higher-order modulations (256 QAM or higher) and complex multiple-input and multiple-output (MIMO) configurations [16]. However, these spectral-efficiency enhancements require proximity to macrocell center (to guarantee high SINR for the backhaul link) and rich scattering environments (for backhaul channel capacity to scale linearly with number of antennas) in order to achieve their full potential.

Other self-backhauling capacity enhancement approaches considered include the backhaul resource allocation among a group of small cells based on the relative load (number of served UEs) of each small cell [20]. The capacity gains (scheduling gains) from this approach are dependent on the differential loading among the scheduled small cells and are still constrained by the overall level of resources allocated for small cell backhaul. Additional joint optimization of resource distribution for self-backhaul and small cell access has been considered for operator-shared small cells, with additional requirements of cooperative signaling information exchange across different operators [12].

1.4 Paper contribution and structure

In this paper, we propose an approach for relaxing the self-backhauled LPN capacity bottleneck by providing backhaul link SINR gains at the cell edge through the use of coordinated multi-point (CoMP) joint transmission (JT) in a frequency-division duplexing (FDD) downlink system. CoMP JT is an LTE-Advanced enhancement proposed to improve the achievable cell edge SINR through combination of transmissions from two or more neighboring cells that simultaneously serve a selected UE [21,22]. The CoMP approach is feasible since the self-backhauled LPN appears like a regular UE over the backhaul link to the serving macrocell.

We present an analytical expression of capacity outage probability for a practical limited-feedback CoMP technique considering Rician fading channel towards the donor cell and Rayleigh fading channels towards the rest of the cells in CoMP set. The Rician fading channel towards the donor cell is based on the assumption that LPNs have LOS connection from their donor cells, which is mostly the intended scenario in practice. Previously, outage capacity analysis was carried out for CoMP systems in [23] while authors in [24] provide analysis on optimal power allocation. Distinction of our work is that analyses in both works consider Rayleigh fading channels for all base station transmissions. Furthermore, [23] presents results for CoMP methods exploiting long-term channel feedback whereas [24] provides results for CoMP methods utilizing perfect channel feedback. We provide outage capacity analysis for 3rd Generation Partnership Project (3GPP) compliant CoMP method that exploits quantized short-term phase feedback. We also present an exemplary simulation study which considers LTE-Advanced relays as the LPNs served by macrocells. In that simulation study, use of limited-feedback CoMP technique to relax relay bottlenecks is considered for a selected realistic deployment scenario. Deterministic dominant path model is used for propagation computation. We provide insight on combining the CoMP and resource scheduling for improved performance. Furthermore, we provide additional insights on possible performance degradations due to different levels of feedback errors.

Analytical results demonstrate that the CoMP method can significantly increase targeted capacity for a given outage probability; and even with low feedback overhead, the limited-feedback method performs close to the ideal maximum ratio transmission that requires full channel state information (CSI) at the transmitter. From simulation results, we observe that the CoMP method provides considerable backhaul throughput gain and narrows down the capacity gap between backhaul and access. Furthermore, we demonstrate the achieved gain either directly improving relay UE throughput or throughput of all UEs (including macro UEs) by impacting applied scheduling method. The CoMP gain results in improved throughput fairness among the whole UE population. Moreover, these benefits of CoMP at relay backhaul is sustained under feedback error although considerable reduction is observed for large ($\geq 10\%$) bit error probability.

The rest of the paper is structured as follows. In Section 2, we present the system model, the limited-feedback CoMP method, and an overview of the possible resource allocation strategies. In Section 3, we derive expression for the capacity outage probability and provide respective performance evaluation. In Section 4, we present the simulation study on CoMP applied at the backhaul of LTE-Advanced relays highlighting throughput results and observations. Finally, conclusions and suggestions for future work are given in Section 5.

2 System model, CoMP scheme, and resource allocation

2.1 System model

Figure 2 depicts a HetNet deployment scenario where macrocell resources are employed to self-backhaul LPN and to enhance coverage and/or capacity. In the non-CoMP case, the LPN backhaul is provided by the serving macrocell only, while the signals of the adjacent macrocells appear as interference. Moreover, in the CoMP case, the self-backhauling of the small cell is provided jointly by all the coordinating macrocells.

We consider flat and block-fading channel model. Thus, channel gains remain constant during each frame/block of transmitted symbols and gains from different blocks are independent. We assume Rayleigh fading channel for all links between macro base stations and the LPNs exception being the donor cell which we assume to have a LOS connection to the LPN with the channel experiencing Rician fading. Thus, channel gains $h_j = \sqrt{\xi_j} e^{j\phi_i}, j \neq 1$ are Rayleigh random variables where the channel power ξ_i follows exponential distribution with mean $\bar{\xi}_j = \mathbb{E}\{\xi_j\}$. On the other hand, donor cell channel gain $h_1 = \sqrt{\xi_1} e^{j\phi_1}$ is Rician random variable with mean channel power $\bar{\xi}_1 = \mathbb{E}\{\xi_1\}$ and Rician factor K. We note that the Rician model is most feasible for well-planned outdoor and above rooftop LPNs, but the model can also hold for other deployment scenarios with proper selection of K. We also assume that channels are non-identical (i.e., $\bar{\xi}_j \neq \bar{\xi}_i, \ i \neq j$) taking into account the distributed nature of LPN deployment and/or different main lobe directivity among antennas.

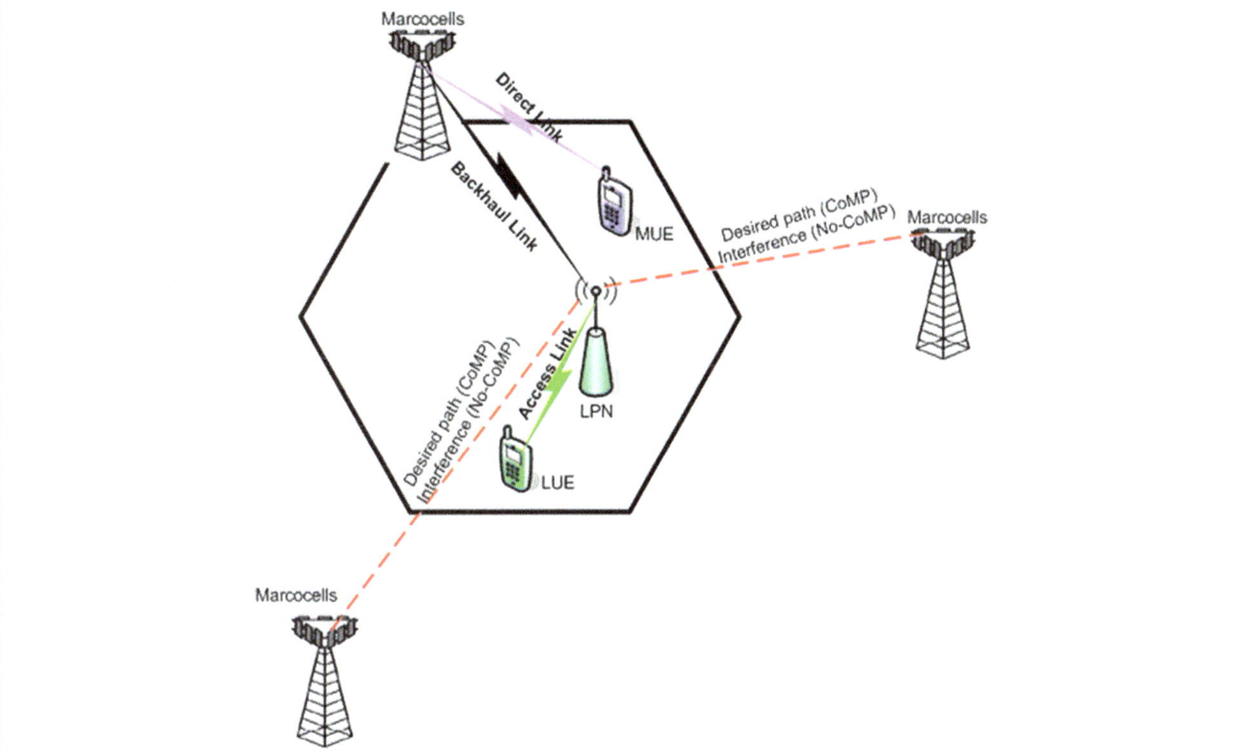

Figure 2 System illustration for considered HetNet. LUE refers to UE associated to LPN.

In the absence of coordination, the received signal in the self-backhauled LPN at a given arbitrary time instant is given by:

$$r = h_1 s_1 + \sum_{j \neq 1} h_j s_j + n, \tag{1}$$

where s_j is the transmitted information symbol from the jth macrocell such that $E[\,|s_j|^2] = P_t$, h_j is the channel from the jth macrocell, and n is zero-mean complex additive Gaussian noise with power P_N. Note that index 1 refers to the donor cell to which the LPN is associated.

When we consider coordination over the backhaul links, we apply JT from macrocells in a given CoMP set. The set comprises the donor cell and other selected macrocells based on the LPN average received signal power. We recall here that JT is one major CoMP subcategory defined in 3GPP, whereby, coordinating cells simultaneously serve a given CoMP node [22]. With JT CoMP, the received signal at the LPN becomes:

$$r = \sum_{j \in C} h_j \hat{w}_j s_1 + \sum_{j \notin C} h_j s_j + n, \tag{2}$$

where C denotes the CoMP set and $\hat{w}_j$ is selected precoding weight at the jth coordinating cell based on a given precoding method.

We consider a quantized feedback method that can be realized in FDD system due to limited feedback capacity requirement. The LPN estimates channels and selects appropriate weights based on the estimated channels and a given code book known at the coordinating cells and the LPN. The LPN then sends feedback bits identifying selected weights to the coordinating macrocells using error-free and delay-free feedback channel so that the macrocells apply selected weights in transmission. We note here that erroneous feedback will be considered later in the paper (Section 4.3) to analyze its impact on achieved results. As the same symbol s_1 is transmitted from coordinating cells, we assume that the information symbol is available among macrocells in C. Here, the assumption is that macrocells admit an ideal backhaul although they are located in different geographical locations.

2.2 Quantized co-phasing CoMP scheme

There are various precoding methods for coordination among cells under JT category with different performance and implementation requirements [22,25-28]. The JT CoMP method we consider in this work is based on short-term quantized phase information and relative phase-shifting operation referred to as quantized co-phasing (QCP) [29-31].

In QCP CoMP method, the LPN considers the channel from donor macrocell h_1 as a reference channel and set the corresponding weight $\hat{w}_1 = 1/\sqrt{b}$, where b is a parameter that determines transmit power of macrocells in C. It is set to 1 when we assume full transmission power from each cell whereas it is set to the number of macrocells in C when power is normalized. Then, for each of remaining cells in C, the LPN selects weights $\hat{w}_j$ based on:

$$\hat{w}_j = \arg_{w_n \in \mathbf{W}} \max |h_1 + h_j w_n|, \tag{3}$$

where the code book $\mathbf{W}_j$ is given as:

$$\mathbf{W}_j = \left\{ w_n = \frac{1}{\sqrt{b}} e^{j\varphi_n} : \varphi_n = \pi n / 2^{N_j - 1}, n = 0, \ldots, 2^{N_j} - 1 \right\}. \tag{4}$$

The code book $\mathbf{W}_j$ is known by the LPN and respective coordinating macrocells, and its size depends on value of N_j which is number of bits used to report selected codeword to the macrocell.

This QCP method is suboptimal while it can be implemented in FDD systems with realizable feedback capacity requirement. If $N = 2$ and only two cells coordinate, then the method resembles closed-loop transmit diversity that is applied in HSDPA and LTE [32,33].

2.3 Resource allocation

The co-deployment of macrocells providing umbrella coverage over LPN hot spots is the commonly considered two-tier heterogeneous network deployment approach [1-5]. The implementation of self-backhauling in these two-tier networks presents a number of strategic options for RAN resource allocation (time resources, frequency resources, etc.) across three layers, namely macro access, LPN backhaul, and LPN access (see Figure 3). The orthogonal resource allocation (see Figure 3a) eliminates cross-layer co-channel interference but is inefficient due to lack of resource reuse. On the other hand, full sharing (see Figure 3f) achieves the high resource utilization at the cost of excessive cross-layer interference.

The alternative for co-channel deployment of macro access with resources shared between either LPN access or backhaul (see Figure 3(b),(c)) provides a trade-off between interference reduction and resource reuse efficiency. However, the reuse of the macro resources in both cases results in a backhaul capacity bottleneck, whereby the aggregate served capacity of LPN access exceeds the achievable LPN backhaul capacity. This will consequently limit peak achievable throughputs for the UE served by the LPN. The backhaul capacity bottleneck

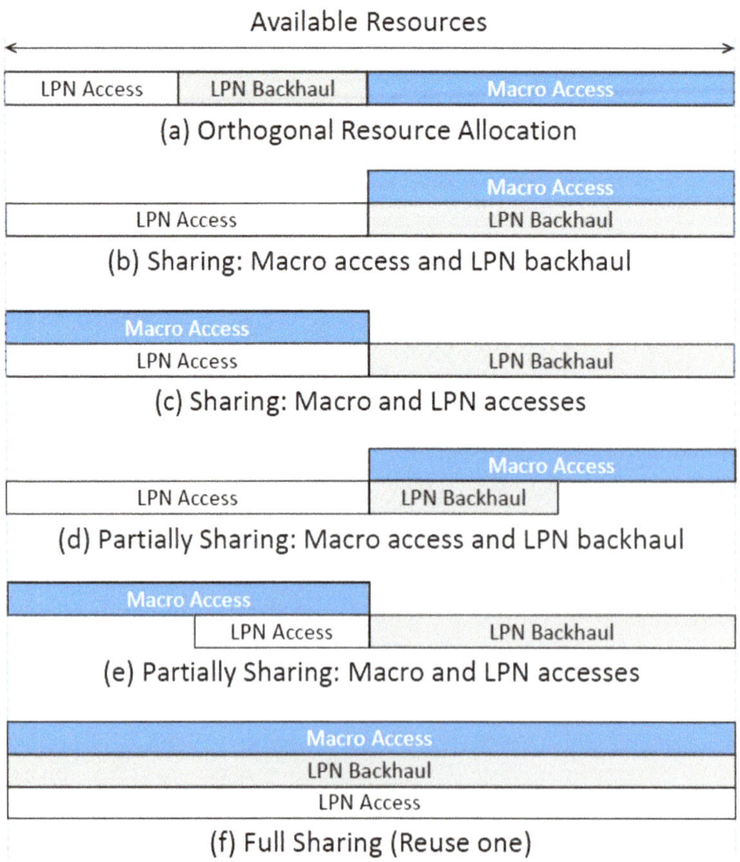

Figure 3 Possible resource allocation strategies in heterogeneous networks with self-backhauled LPNs.

is attributed to radio resource contention between LPN backhaul and existing macro UE (for resource allocation scheme of Figure 3(b)) or limits on RAN resource that can be exclusively dedicated for LPN backhauling (Figure 3(c)). Further backhaul capacity limitations occur due to inter-cell interference between adjacent macrocells which reduces SINR, particularly for self-backhauled LPNs deployed in the macrocell edge. Partial sharing of macro access resources (see Figure 3(d),(e)) may provide opportunities for reduced inter-cell interference but with the consequence of decreased resource utilization efficiency.

In this work, we assume that macro access share time resource with LPN access as shown in Figure 3(c) while frequency resource is fully reused in all layers. This provides best interference scenario for the backhaul link that it is not interfered by both the macro and the LPN accesses. This assumption is consistent with the resource-sharing strategy adopted in 3GPP for LTE-Advanced type I relays. However, we note here that even with these interference mitigation measures and relatively large resource allocation, the backhaul link could create a bottleneck as discussed in further detail later in Section 4.

3 Outage capacity analysis

In this section, we present analysis for the CoMP method performance gain in terms of outage capacity. For a required capacity $C_r = \log_2(1 + Z_r)$ in the backhaul, the outage probability is defined as:

$$P_o = P(Z < Z_r), \tag{5}$$

where Z is instantaneous SINR at the LPN and Z_r is the required SINR to meet the targeted capacity.

3.1 Outage probability analysis

In the absence of CoMP, all macrocell transmissions except transmission from donor cell are interfering the LPN backhaul link; hence the instantaneous SINR denoted by Z_{nc} attains the form:

$$Z_{nc} = \frac{\gamma_1}{\sum_{j\neq 1} \gamma_j + P_n}, \tag{6}$$

where received powers are $\gamma_j = P_t\xi_j$. We note here that γ_1 is power of Rician channel with mean value $\bar{\gamma}_1 = P_t\bar{\xi}_1$ and γ_j, $j \neq 1$ follows exponential distribution with mean $\bar{\gamma}_j = P_t\bar{\xi}_j$. If we assume network is interference limited, then Equation 6 becomes Rician/Rayleigh scenario, whereby the closed-form expression for outage probability is presented in [34]. Accordingly, we find that for non-CoMP case, an outage probability expression of the form:

$$P_o^{nc} = 1 - \sum_{j\neq 1} D_j \left[1 - \frac{\bar{\gamma}_j(K+1)Z_r}{\bar{\gamma}_1 + \bar{\gamma}_j(K+1)Z_r} * \exp\left(\frac{-K\bar{\gamma}_1}{\bar{\gamma}_1 + \bar{\gamma}_j(K+1)Z_r}\right)\right], \tag{7}$$

where $D_j = \prod_{i\neq j} \bar{\gamma}_j / \left(\bar{\gamma}_j - \bar{\gamma}_i\right)$.

When we apply QCP CoMP, some previously interfering links are now exploited to enhance the backhaul link; and the SINR denoted by Z_c takes the form:

$$Z_c = \frac{|\sum_{j\in C} \sqrt{\gamma_j} e^{j\phi_j} \hat{w}_j|^2}{\sum_{j\notin C} \gamma_j + P_n}. \tag{8}$$

Without loss of generality, we focus on the case where the number of macrocells in the CoMP set is limited to 3 although analysis for more than three macrocells is doable but long and impractical. In well-planned mobile network, strong signals are usually received only from two or three base stations while signals from other base stations are weaker. Let coordination takes place among donor cell and two other selected cells (indexed by 2 and 3) to serve the LPN. Then, utilizing weights $\hat{w}_j$ determined from Equation 3, we write the tight lower-bound outage probability P_o^c in the form:

$$P_o^c = P\left(\hat{Z}_c < Z_r\right), \tag{9}$$

where:

$$\hat{Z}_c = \frac{|\sqrt{\gamma_1} + \sqrt{\gamma_2} e^{j\theta_2} + \sqrt{\gamma_3} e^{j\theta_3}|^2}{b\sum_{j\notin C} \gamma_j}. \tag{10}$$

We note that $\theta_j = \phi_j - \phi_1 + \varphi_n$ is uniformly distributed on the range $\left(-\pi/2^{N_j}, \pi/2^{N_j}\right)$. As the channels are independent, we can write the outage probability in the form:

$$P_o^c = 1 - \int_0^\infty F_{\eta_2}(z/Z_r) f_{\eta_1}(z)dz, \tag{11}$$

where $F_{\eta_2}(\cdot)$ is the cumulative distribution function (CDF) of $\eta_2 = b\sum_{j\notin C} \gamma_j$ and $f_{\eta_1}(\cdot)$ is the probability distribution function (PDF) of $\eta_1 = |\sqrt{\gamma_1} + \sqrt{\gamma_2} e^{j\theta_2} + \sqrt{\gamma_3} e^{j\theta_3}|^2$. The CDF of η_2 is well-known that [35]:

$$F_{\eta_2}(z) = \sum_{j\notin C} D_j \left(1 - e^{\frac{-z}{b\bar{\gamma}_j}}\right), \tag{12}$$

where $D_j = \prod_{i\neq j, i\notin C} \bar{\gamma}_j / \left(\bar{\gamma}_j - \bar{\gamma}_i\right)$. We then substitute Equation 12 into Equation 11 and obtain:

$$P_o^c = 1 - \sum_{j\notin C} D_j \left(1 - \int_0^\infty e^{-\frac{z}{b\bar{\gamma}_j Z_r}} f_{\eta_1}(z)dz\right). \tag{13}$$

To solve the integral in Equation 13, we approximate PDF of η_1 using error corrected chi-squared distribution with 6 degrees of freedom (see Appendix A in [36]). The approximation is based on the fact that η_1 follows chi-squared distribution when $K = 0$, and full CSI is available at the coordinating macrocells. Thus:

$$f_{\eta_1}(z) = \frac{m^3 e^{-mz}}{2} \sum_{i=1}^{3} a_i z^{i+1}, \tag{14}$$

where $m = 3/\mathrm{E}[\eta_1]$, $a_3 = 2m^2\left(m^2\mathrm{E}\left[\eta_1^2\right]/48 - 0.25\right)$, $a_2 = -8a_3/m$, and $a_1 = -\mathrm{E}\left[\eta_1\right]a_2/2 + 1$. Let expand η_1 as:

$$\begin{aligned}\eta_1 = \gamma_1 + \gamma_2 + \gamma_3 + 2\sqrt{\gamma_1}\sqrt{\gamma_2}\cos\theta_2 + 2\sqrt{\gamma_1}\sqrt{\gamma_3}\cos\theta_3 \\ + 2\sqrt{\gamma_2}\sqrt{\gamma_3}\cos\theta_2\cos\theta_3 + 2\sqrt{\gamma_2}\sqrt{\gamma_3}\sin\theta_2\sin\theta_3,\end{aligned} \tag{15}$$

and from Equation 15 and its square, we compute first and second moments of η_1 based on expected values of γ_n^{δ} $(n = 1, 2, 3,\ \delta = 1/2, 1, 3/2)$, $\cos^{\iota}\theta_i$ and $\sin^{\iota}\theta_i$ $(i, \iota = 2, 3)$. Accordingly, we find:

$$\begin{aligned}\mathrm{E}\left[\eta_1\right] =& \bar{\gamma}_1 + \bar{\gamma}_2 + \bar{\gamma}_3 + \frac{\pi}{2}\sqrt{\frac{\bar{\gamma}_1}{K+1}}M(-1/2, 1; -K) * \\ & \left[\sqrt{\bar{\gamma}_2}C_2 + \sqrt{\bar{\gamma}_3}C_3\right] + \frac{\pi\sqrt{\bar{\gamma}_2\bar{\gamma}_3}C_2C_3}{2},\end{aligned} \tag{16}$$

where $M(a, b, z)$ is the confluent hypergeometric function defined in ([37], [13.2.1]), and $C_j = \mathrm{E}[\cos\theta_j] = \mathrm{sinc}(\pi/N_j)$. For simplicity, we do not present the attained expression for $\mathrm{E}[\eta_1^2]$ as it has a longer form. Finally, we substitute Equation 14 into Equation 13 and solve the integral to find an outage probability expression of the form:

$$\begin{aligned}P_o^c =& 1 - \sum_{j\notin C} D_j \\ & \times \left(1 - \frac{m^3}{2}\sum_{i=1}^{3} a_i (i+1)!\left(\frac{b\bar{\gamma}_j Z_r}{1 + mb\bar{\gamma}_j Z_r}\right)^{i+2}\right).\end{aligned} \tag{17}$$

3.2 Validation of outage capacity analysis

Figure 4 illustrates outage probability as a function of targeted capacity. Solid curves refer to analytical results and markers refer to numerical results. For benchmarking purposes, we also present numerical results for the ideal maximum ratio transmission that requires full CSI at coordinating cells, depicted with dashed and dash-dotted curves in Figure 4. All results are obtained for LPN1 shown in Figure 5 that depicts an irregular but realistic HetNet deployment scenario. The scenario will be explained further in Section 4, but for now, we note here that average channel powers computed based on a dominant path loss model. Furthermore, the Rician factor K is set to 3 dB; and the number of feedback bits N_1 and N_2 are both set to 3.

We can see from Figure 4 that the outage probability analysis and approximation presented in Subsection 3.1 are well validated. The QCP CoMP technique significantly improves the outage probability performance of the LPN relative to the non-CoMP case. For instance, 4.37- and

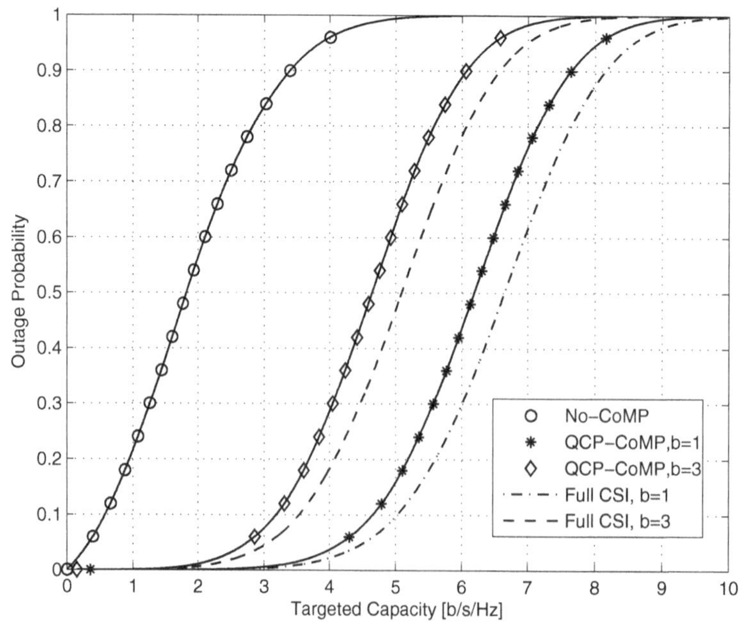

Figure 4 Outage probability as a function of required capacity. Solid curves refer to analytical results, markers refer to numerical results, and dashed and dash-dotted curves refer to full CSI case.

Figure 5 Deployment scenario for the CoMP-enhanced-relay-based HetNet.

2.83-b/s/Hz capacity increments are seen at 50th percentile when $b = 1$ and $b = 3$, respectively. Furthermore, it can be seen that QCP CoMP with a few number of feedback bits (that is, $N_2 = N_3 = 3$) achieves performance close to MRT that assumes full CSI.

4 Simulation case study

Although Section 3 presents analytical expressions to quantify the spectral efficiency gain that can be achieved by applying QCP CoMP at the backhaul of a given LPN, it does not illustrate the trade-off between the CoMP capacity gain and resource scheduling. In this section, we present a stimulation case study that presents the overall throughput gains of CoMP when applied at the backhaul of LTE-Advanced relays deployed in a realistic exemplary deployment scenario.

4.1 Simulation scenario and parameters

The simulation case study focuses on the performance investigation of CoMP-enhanced self-backhauling in a densely populated area. The densely populated area scenario is particularly interesting for low-income underserved urban areas in most emerging markets which will eventually constitute the majority in emerging countries due to rapid urbanization [38]. In these informal settlements, the fixed line-penetration has remained virtually flat over the last few decades [39]. Therefore, mobile network operators in those regions are rapidly upgrading or rolling-out mobile broadband networks, as these provide the only economically feasible means for universal broadband connectivity [39]. In addition to RAT upgrades, operators are obliged to consider network densification with outdoor LPN deployment to meet capacity demands in these fast-expanding urban settlements. However, the low-revenue potential and limited supporting infrastructure means that the cost of operating the LPN (including cost of backhauling the LPNs) has to be minimized and therefore innovative end user-deployed shared-access models should be considered [40].

For the simulation study, we assume that three tri-sector macro eNBs and three cell-edge RNs are deployed in a selected simulation area as illustrated in Figure 5. The label SxCy in Figure 5 is used to refer to cell (or sector) y of tri-sectored macro site x. We note that the LPN in Figure 5 hereafter refers to an LTE-Advanced RN. Furthermore, we deliberately increase the number of macrocell coordination scenarios by placing the 3 RNs at the edges of three adjacent cells, such that each RN is served by a unique cell of a different site from the other two. To be specific using Figure 5, the deployment is such that S1C1 is donor cell for LPN1 (RN1); S2C1 is donor cell for LPN2 (RN2); and S3C1 is donor cell for LPN3 (RN3). The radio coverage estimations behind the deployment are based on realistic three-dimensional (3D) building vectors and topographical data for a densely settled area (Hanna Nassif in Tanzania) [30]. The radio link loss is evaluated using the dominant path model implemented in the WinProp ray tracing tool [41]. Static system-level simulations are performed to investigate network performance for CoMP and non-CoMP scenarios on the relay link, with UEs being dropped at random locations. In this case, we refer to UEs attached to RNs and macrocells as RUEs and MUEs, respectively. The downlink CoMP simulation parameters and assumptions follow commonly used 3GPP guidelines [22], and key values are listed in Table 1.

4.2 Scheduling and throughput formula

In typical FDD LTE-Advanced type 1 inband relay scenarios, the MUE, RN, and RUE are served on a common frequency band [13]. Time-domain half-duplexing is used to isolate backhaul and access link transmissions, with the RN continuously switching between the relay and access links to avoid self-interference, as illustrated in Figure 6. Note that parameter τ_r denotes time ratio allocated to the backhaul link.

Optimal resource utilization on average is achieved at each RN when τ_r is semi-statically selected by DeNBs such

Table 1 Basic simulation parameters and assumptions

Parameters	Values/assumptions		
Air interface	LTE FDD		
Carrier frequency/bandwidth	2,110/10 MHz		
Simulation	Radio propagation modeling (WinProp) Static sytem level simulation (Matlab) 5-m resolution 1.5-m UE and 5-m RN prediction height		
CoMP	Three cooperating cells Downlink JP CoMP QCP with $b = 1$ and $N = 3$ feedback bits		
Throughput	$B_{\text{eff}} = 0.62, Z_{\text{eff}} = 1.8, Z_{\text{min}} = -10$ dB, $S_{\text{max}} = 4.4$ b/s/Hz		
Macro (DeNB/eNB) parameters			
Macro sites	Site 1	Site 2	Site 3
Transmit power per antenna	46 dBm		
Antenna height	18 m	20 m	25 m
Antenna patterns	Kathrein 741984		
Sector azimuths	20, 140, 260	0, 120, 240	10, 130, 250
Intersite distance	Approximately 500 m		
RN parameters			
Transmit power/antenna gain/noise figure	30 dBm/5 dBi/5 dB		
Deployment height	5 m		
RN locations	250 m from DeNB, 140 m inter-RN distance		
Antenna configuration	1 × 2		
UE (RUE and MUE) parameters			
Noise figure	9 dB		
Antenna configuration	1 × 2		
Mean UE number	15 UE per macrocell		
UE mobility	Mobility off 50% UEs dropped randomly in whole area 50% UEs cluster-dropped in 40-m radius of RNs		
Buildings and fading characteristics			
Buildings	Building of variable dimensions, height 3 to 6 m Building penetration loss: 10 dB		
Fading	Shadow fading: dominant path model (WinProp) Fast fading: Rician fading (RN-DeNB, $K = 3$ dB) and Rayleigh fading (other links)		

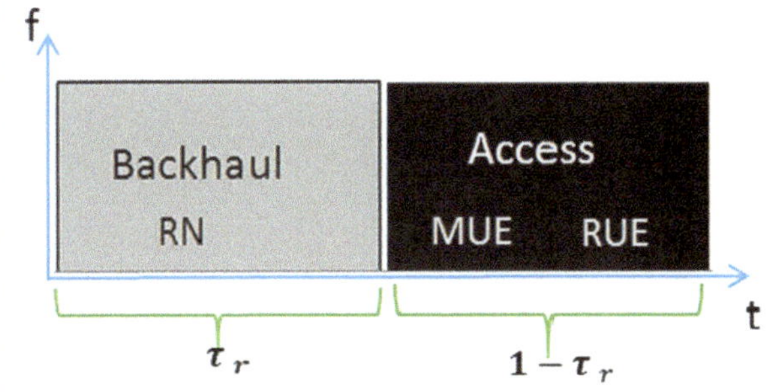

Figure 6 Resource allocation among relay, macro, and relay users.

that $\tau_r \bar{R}_r = (1 - \tau_r)\bar{R}_a$, where $\bar{R}_r$ is the average rate at the backhaul and $\bar{R}_a$ is the aggregate rate at the access, both assuming transmission in whole time frame. Accordingly, optimal τ_r and corresponding end-to-end rate can be given as:

$$\hat{\tau}_r = \frac{\bar{R}_a}{(\bar{R}_r + \bar{R}_a)}, \quad \bar{R}_{e2e} = \frac{\bar{R}_r \bar{R}_a}{(\bar{R}_r + \bar{R}_a)} \tag{18}$$

We note here that selection of τ_r semi-statically is attractive as it requires less capacity to feedback long-term feedback information than dynamic selection that requires short-term feedback information although the latter potentially delivers better resource utilization with more overhead. In practice, downlink backhaul radio resources are achieved by allocating RNs up to six multicast-broadcast single frequency network (MBSFN) subframes (frame numbers 0, 1, 2, 3, 6, 7, 8) out of the ten subframes (each of 1 ms duration) specified in the 3GPP Release 8 LTE frame structure [42]. Thus, τ_r can take granular values such that $\tau_r \in \{0.1, 0.2, 0.3, 0.4, 0.5, 0.6\}$.

We have three ranges to which values of $\hat{\tau}_r$, computed according to Equation 18, can lie: $\hat{\tau}_r > 0.6$, $0.1 < \hat{\tau}_r < 0.6$, and $\hat{\tau}_r < 0.1$. If $\hat{\tau}_r > 0.6$, then backhaul is a bottleneck. If $0.1 < \hat{\tau}_r < 0.6$, then we can balance the backhaul and access. Moreover, $\hat{\tau}_r < 0.1$ means access is a bottleneck. In conventional relay deployment scenarios including this case study, the former is the most likely case. Therefore, we select values of τ_r statically in our simulation. Note here that the higher values of selected τ_r, the better the backhaul capacity but the worst MUE (allocated $1 - \tau_r$ fraction of time resources) throughput. We will revise the discussion on τ_r selection and performance trade-off in Subsection 4.3 based on the simulation results.

In the relay access, the RN performs scheduling decision in each subframe or transmission time interval (TTI), whereby, RUEs are allocated a fraction of the total physical resource blocks (PRBs) available in the subframe. It should be noted that no RUEs are scheduled during the backhaul subframes when the RN will have reverted to relay link operation in the half-duplex configuration. The actual scheduler implementation (round robin (RR), proportional fair, etc.) for RN access resources is vendor specific as in the case of eNBs [42]. We initially apply here RR scheduling to divide the resources equally among RUEs so that we can observe throughput gap between backhaul and access. In the case that the backhaul presents a capacity bottleneck ($R_r < R_a$), the initially allocated resources using RR scheduling are redistributed among the RUEs using max-min fair scheduling which priorities RUEs with lower throughputs [43]. Otherwise, if backhaul is not a bottleneck ($R_r \geq R_a$), the resource allocation for the RUEs from RR scheduling is retained.

During coordination among cells in CoMP set for joint relay link transmissions in each backhaul subframe, orthogonal frequency resource allocation is applied among the set of served RNs. We apply here basic RR resource partitioning strategy where relay link resources (subcarriers) are partitioned equally among relay link transmissions. Note that advanced CSI-based scheduling methods, such as proportional fair, can be applied to achieve better performance.

Throughput (TP) in both the backhaul and access links is computed through mapping SINRs results using a modified Shannon formula [44]:

$$\mathrm{TP} = \tau N_{\mathrm{PRB}} B_{\mathrm{PRB}} * \begin{cases} \min\{B_{\mathrm{eff}} \log_2\left(1 + \frac{Z}{Z_{\mathrm{eff}}}\right), S_{max}\}, & Z \geq Z_{\min} \\ 0, & Z < Z_{\min}, \end{cases} \tag{19}$$

where τ reflects the amount of time resource allocated for the transmission, N_{PRB} is the number of PRBs, and B_{PRB} is the bandwidth per PRB. Note also that $S_{\max}$ is the maximum spectral efficiency, $Z_{\min}$ is the minimum required SINR, B_{eff} adjusts the system bandwidth efficiency to fit with LTE, and Z_{eff} adjusts for the SINR implementation efficiency. We recall that $\tau = \tau_r$ for backhaul links and $\tau = 1 - \tau_r$ for access links. Parameters B_{eff}, Z_{eff}, $S_{\max}$, and $Z_{\min}$ are obtained from link-level simulations and curve fitting. We consider a 1×2 MIMO configuration for the simulation and values of the parameters computed for the configuration are listed in Table 1 [45].

4.3 Simulation results and performance evaluation

The CDF of the backhaul and access throughput considering all relays is illustrated in Figure 7. The value of τ_r is set to the maximum 0.6 for both CoMP and non-CoMP cases.

We can clearly see from Figure 7 that QCP-CoMP significantly improves the backhaul throughput (end-to-end throughput) and diminishes the gap between the backhaul and access. It reduces the 219.12%, 112.43%, and 52.62% relative throughput gap for no-CoMP case at 10-ile, 50%-ile and 90%-ile to 5.67%, 8.00%, and 22.81% relative throughput gap, respectively.

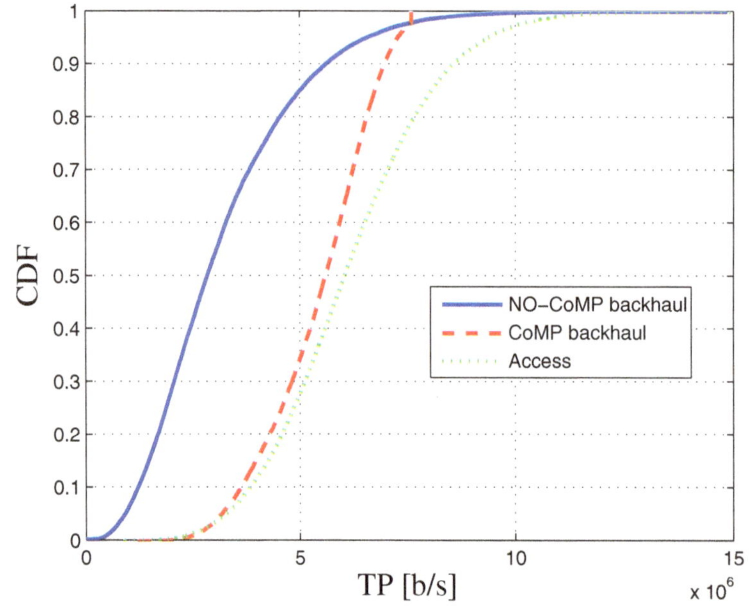

Figure 7 Cumulative distribution function of backhaul and access throughput. Solid curve refers to results for non-CoMP backhaul whereas dashed and dotted curves refer to results for CoMP backhaul and access, respectively.

We illustrate contribution of CoMP-enhanced backhaul on the RUE throughput using a bar plot depicted in Figure 8. The plot shows CoMP gains over non-CoMP scenario when $\tau_r = 0.6$ and $\tau_r = 0.3$. We see that the applied CoMP scheme for both $\tau_r = 0.6$ and $\tau_r = 0.3$ significantly improves all the 10%-ile, 50%-ile, and 90%-ile RUE throughputs. Unlike $\tau_r = 0.6$ case, the scheme helps more the 10%-ile and 50%-ile when $\tau_r = 0.3$. Furthermore, the plot in Figure 8 shows throughput gains of CoMP when $\tau_r = 0.3$ over non-CoMP case when $\tau_r = 0.6$. We see that there are still gains at 10%-ile and 50%-ile although loss is experienced at 90%-ile. Note that the reduction in τ_r allows for more time resources to be allocated for MUEs thus enhancing MUE throughput performance.

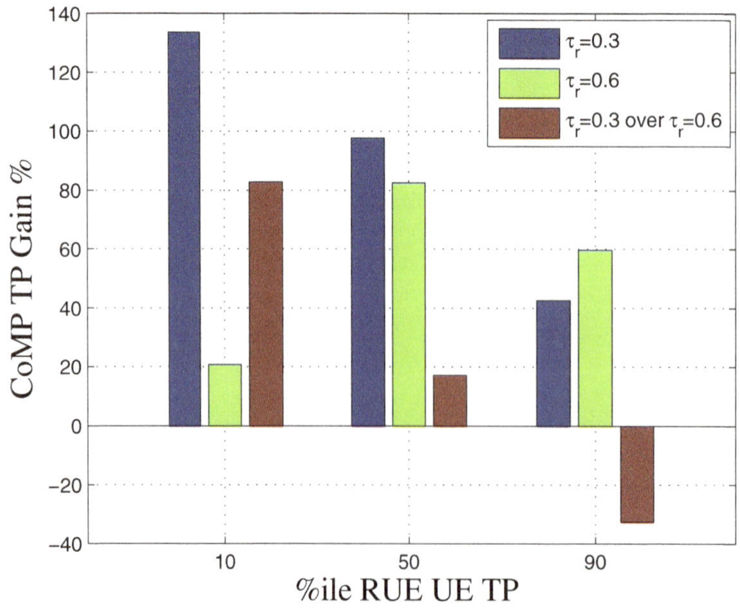

Figure 8 CoMP gain in terms of RUE 10%-ile, 50%-ile, and 90%-ile throughput when $\tau_r = 0.6$ and $\tau_r = 0.3$.

Performance impact of the CoMP technique in all UEs (both RUEs and MUEs) is illustrated in Figure 9. We see that CoMP delivers 10%-ile, 50%-ile, and 90%-ile percentage gains of 85.96%, 56.98%, and 0.15% when $\tau_r = 0.3$, respectively. The percentage gain values are 8.48%, 43.83%, and 33.59% when $\tau_r = 0.6$. Note that CoMP improvements are more for the lower percentile throughput when $\tau_r = 0.3$ unlike the case when $\tau_r = 0.6$ where improvements are more for the higher percentile throughputs. If we compare CoMP with $\tau_r = 0.3$ and non-CoMP with $\tau_r = 0.6$, we see that CoMP with $\tau_r = 0.3$ provides 81.89%, 26.38%, and 24.07% gains at 10%-ile, 50%-ile, and 90%-ile, respectively, relative to the case of non-CoMP with $\tau_r = 0.6$.

In order to show the impact of CoMP on the consistency of UE throughput experience, we also present in Figure 10 the CDF of fairness index defined as [46]:

$$\mathrm{FI} = \frac{\left[\sum_{i=1}^{N_{\mathrm{ue}}} \mathrm{TP}_i\right]^2}{N_{\mathrm{ue}} \sum_{i=1}^{N_{\mathrm{ue}}} \mathrm{TP}_i^2}, \tag{20}$$

where N_{ue} is total number of UEs. The figure clearly shows that CoMP provides more consistent user experiences for both $\tau_r = 0.3$ and $\tau_r = 0.6$.

4.4 Performance evaluation under feedback error

Figure 11 depicts the CDF of the backhaul throughput considering errors in CoMP phase feedback bits when $\tau_r = 0.6$. The throughput CDF of the non-COMP case is presented as a benchmark considering a similar value $\tau_r = 0.6$. Solid curve refers to results for non-CoMP backhaul while dashed, dotted, and dash-dotted curves refer to results for CoMP backhaul under 0%, 10%, and 20% bit error probability, respectively. We assume 3-bit Gray coding and uniformly distributed errors in the simulation. We observe from Figure 11 that large feedback bit-error probability can considerably diminish the CoMP benefit of narrowing the capacity gap between backhaul and access. For instance, the 92.19% gap reduction obtained at 10%-ile in the absence of feedback error is reduced to 72.42% and 51.64% for 10% and 20% feedback bit-error probability, respectively. As a result, the feedback error negatively impacts the CoMP throughput gain of RUEs as illustrated in Figure 12. The gain reduction at 10%-ile RUE throughput due to feedback bit error is considerably higher when $\tau_r = 0.3$. On the other hand, the reduction due to feedback error when $\tau_r = 0.3$ and $\tau_r = 0.6$ is almost the same at 50%-ile.

5 Conclusions

In this paper, we discussed some of the key operator considerations in the selection of LPN backhauling solutions. The use of self-backhauling for LPNs by exploiting the existing macro RAN was highlighted as a promising method both in terms of cost-effectiveness of backhaul implementation and flexibility of LPN deployment. The backhaul capacity bottleneck which limits the LPN capacity was underlined as the primary limitation for the self-backhauling approach. To that end, we investigated the use of CoMP transmission for relaxing the downlink backhaul capacity bottleneck for self-backhauled LPNs. We approached this work from two

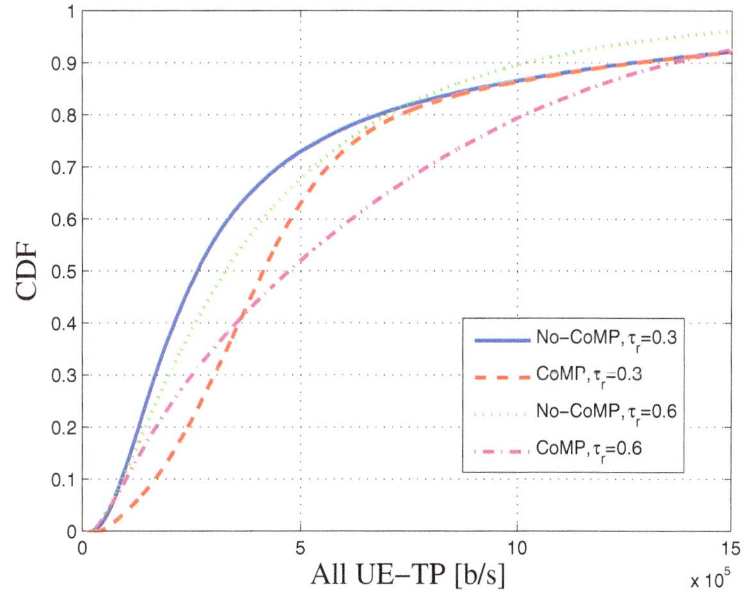

Figure 9 Cumulative distribution function of all UE throughput. Solid and dashed curves refer to results for non-CoMP and CoMP cases, respectively, when $\tau_r = 0.3$ whereas dotted and dash-dotted curves refer to non-CoMP and CoMP cases, respectively, when $\tau_r = 0.6$.

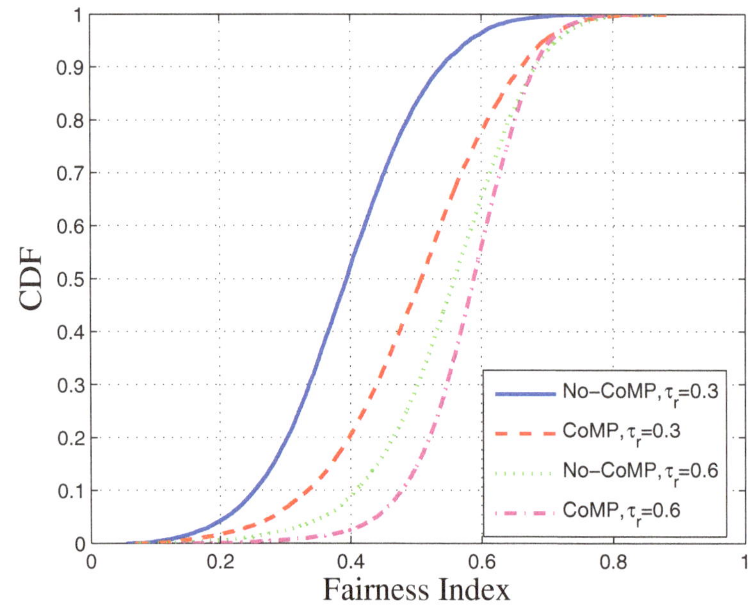

Figure 10 Cumulative distribution function of fairness index. Solid and dashed curves refer to results for non-CoMP and CoMP cases, respectively, when $\tau_r = 0.3$ whereas dotted and dash-dotted curves refer to results for non-CoMP and CoMP cases, respectively, when $\tau_r = 0.6$.

ways. First, an analytical study was conducted to find analytical expression for capacity outage probability for a practical limited-feedback CoMP technique. In this context, CoMP was considered for relaxing backhaul capacity bottlenecks in general self-backhauled LPNs. Secondly, we carried out an extensive simulation campaign for an exemplary HetNet deployment in a realistic dense settlement scenario, whereby the aforementioned CoMP technique was used for relaxing downlink backhaul capacity bottlenecks for LTE-Advanced type 1 relays.

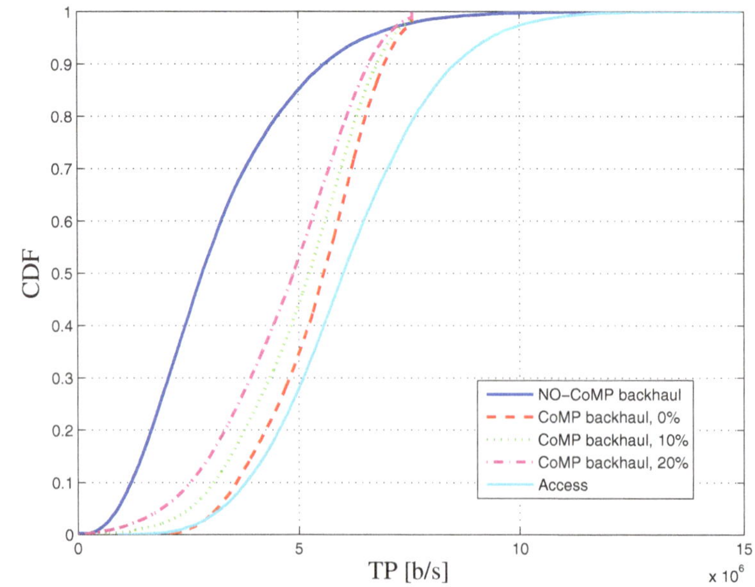

Figure 11 Cumulative distribution function of backhaul and access throughput under feedback bit error. Solid curves refers to results for non-CoMP backhaul and access, whereas dashed, dotted, and dash-dotted curves refer to results for CoMP backhaul assuming 0%, 10%, and 20% bit error probability, respectively, when $\tau_r = 0.6$.

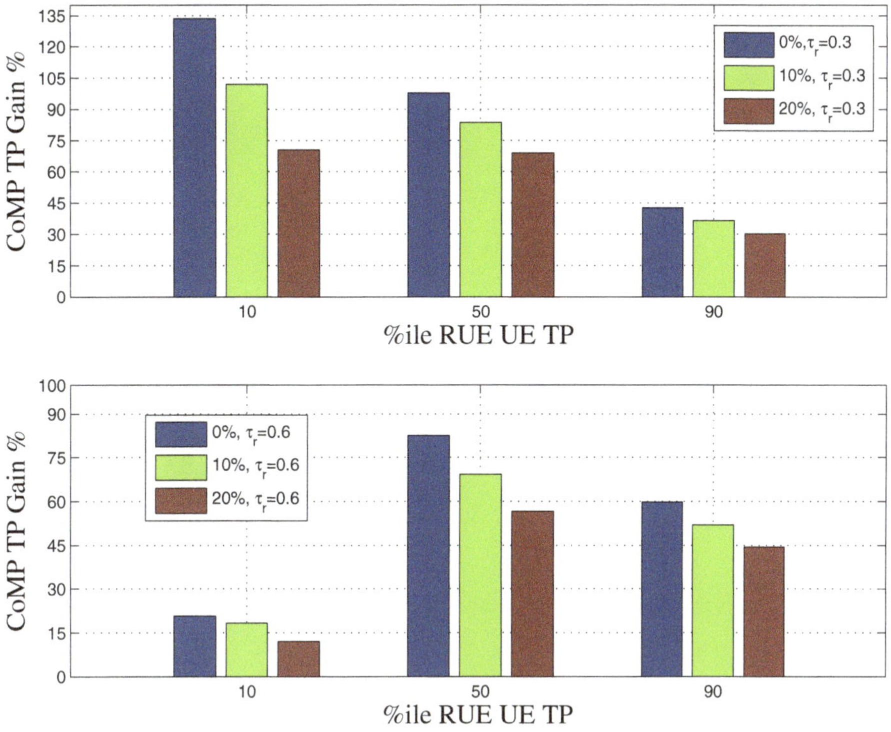

Figure 12 CoMP gain in terms of RUE 10%-ile, 50%-ile, and 90%-ile throughput. For 0%, 10% and 20% bit error probability cases when $\tau_r = 0.6$ and $\tau_r = 0.3$.

Analytical results show that we can considerably increase expected capacity for a certain outage probability by the CoMP method. For instance, a capacity increase by 4.37 b/s/Hz was noted at the 50%-ile when each cell in the CoMP set transmits with full power, and the LPN applies a feedback word of 3 bits for each cell in the CoMP set (except for the donor cell). Furthermore, the CoMP method performs close to the ideal maximum ratio transmission even with the 3-bit feedback word.

Throughput results obtained from the simulation campaign assert that the CoMP scheme significantly improves the capacity gap between backhaul and access. For instance, the gap is reduced by 86.00% at the 50%-ile when we allocate 60% of the LTE subframes for the backhaul. The enhancement is accompanied by an approximately 80% RUE throughput gain at 50%-ile. When 30% of the subframes are allocated for the backhaul, we note a relatively higher RUE throughput gain at the 10%-ile and 50%-ile (compared to the case with 60% of subframes allocated for the backhaul). Besides, we can use the CoMP gain at the backhaul to increase the percentage of subframes for MUE (decrease subframes for backhaul) so that throughput is improved not only for RUEs but also for MUEs. We observe that CoMP with 30% backhaul subframes provides 26.38% throughput gain for all UEs at 50%-ile relative to the case of non-CoMP with 60% backhaul subframes. Furthermore, the fairness index shows that the use of CoMP for LPN backhaul provides more even throughput distribution among UEs in both 60% and 30% subframe configurations for the backhaul. Although these CoMP performance gains are sustained under feedback error, it was also noted that considerable reduction in the gains are observed for large bit-error probability ($\geq$ 10%).

Possible future work includes computation of analytical expressions for multi-antenna scenarios and performance measures (such as average capacity) that were not considered herein, and comparative analysis of various CoMP techniques with different requirements (amount of overhead, shared information, implementation requirements, and so on). Also of interest would be the consideration of various clustering methods (static, dynamic, or hybrid) to define the CoMP set. Furthermore, we could consider semi-static resource allocation (backhaul subframe selection) to balance backhaul and access capacity. The overall research question that needs to be addressed is how to use CoMP efficiently and adaptively for the right LPN and at the right moment.

Competing interests

The authors declare that they have no competing interests.

Acknowledgements

This work was performed within the ICT&E project funded by the Aalto University School of Electrical Engineering programme on energy efficiency and EWINE-S project, funded in part by the Finnish Funding Agency for Technology and Innovation, European Communications Engineering and Efore Oyj. Hanna Nassif GIS data was kindly provided by Prof. R. Sliuzas of the ITC-Faculty of Geo-Information Science and Earth Observation, University of Twente.

References

1. J Zander, Mahonen P, Riding the data tsunami in the cloud: myths and challenges in future wireless access. IEEE Commun. Mag. **51**(3), 145–151 (2013)
2. Small Cell Forum. Small cells - what is the big idea? TR Doc. 030.01.01, SCF Release 1 (February 2012). http://smallcellforum.org/smallcellforum/Files/File/SCF-Small_Cells_White_Paper.pdf. Accessed 03 March 2015
3. I Hwang, B Song, SS Soliman, A holistic view on hyper-dense heterogeneous and small cell networks. IEEE Commun. Mag. **51**(6), 20–27 (2013)
4. JG Andrews, H Claussen, M Dohler, S Rangan, MC Reed, Femtocells: Past, present, and future. IEEE J. Sel. Areas Commun. **30**(3), 497–508 (2012)
5. Andrews JG, Seven ways that HetNets are a cellular paradigm shift. IEEE Commun. Mag. **51**(3), 136–144 (2013)
6. J Robson, Small cell deployment strategies and best practice backhaul. white paper, Cambridge broadband networks, (August 2012). http://cbnl.com/resources/small-cell-strategies. Accessed 03 Macrch 2015
7. NGMN, Small cell backhaul requirements. white paper, NGMN Alliance, (June 2012). http://www.ngmn.org/uploads/media/NGMN_Whitepaper_Small_Cell_Backhaul_Requirements.pdf. Accessed 03 March 2015
8. NSN, Outdoor 3G/LTE small cells deployment strategy: 'the race to the pole'. white paper (2013). http://networks.nokia.com/sites/default/files/document/nokia_outdoor_3glte_small_cells_deployment_strategy_whitepaper.pdf. Accessed 03 March 2015
9. FP7 METIS, Scenarios, requirements and KPIs for 5G mobile and wireless system. Project deliverable d1.1 (April 2014). https://www.metis2020.com/wp-content/uploads/deliverables/METIS_D1.1_v1.pdf. Accessed 03 March 2015
10. Flyyn K, Delivering public safety communications with LTE, 3gpp news (July 2013). http://www.3gpp.org/news-events/3gpp-news/1455-Public-Safety
11. T Sakano, ZM Fadlullah, T Ngo, H Nishiyama, M Nakazawa, F Adachi, N Kato, A Takahara, T Kumagai, H Kasahara, Kurihara S, Disaster-resilient networking: a new vision based on movable and deployable resource units. IEEE Netw. **27**(4), 40–46 (2013)
12. P Blasco, M Bennis, M Dohler, in *Paper presented at the 2013 IEEE international conference on communications*. Backhaul-aware self-organizing operator-shared small cell networks (IEEE, Budapest, 9–13 June 2013)
13. C Hoymann, W Chen, J Montojo, A Golitschek, C Koutsimanis, X Shen, Relaying operation in 3GPP LTE: challenges and solutions. IEEE Commun. Mag. **50**(2), 156–162 (2012)
14. Qualcomm, Rising to meet the 1000x mobile data challenge. white paper (2012). http://www.qualcomm.com/media/documents/
15. Z Roth, M Goldhamer, N Chayat, A Burr, M Dohler, N Bartzoudis, C Walker, Y Leibe, C Oestges, M Brzozowy, I Bucaille, in *Paper presented at the 2010 future network and mobile summit*. Vision and architecture supporting wireless GBit/sec/km^2 capacity density deployments (IEEE, Florence, 16-18 June 2010)
16. IEEE, Architecture and requirements for small cell backhaul (2013). Technical report, IEEE 802.16r
17. O Bulakci, S Redana, B Raaf, J Hamalainen, in *Paper presented at the 2010 71st vehicular technology conference (VTC 2010-Spring)*. Performance enhancement in LTE-Advanced relay networks via relay site planning (IEEE, Taipei, 16–19 May 2010)
18. O Bulakci, A Bou Saleh, J Hamalainen, S Redana, Performance analysis of relay site planning over composite fading/shadowing channels with cochannel interference. IEEE Trans. Veh. Technol. **62**(4), 1692–1706 (2013). doi:10.1109/TVT.2012.2233506
19. A Saleh, O Bulakci, J Haĺmaĺlaĺinen, S Redana, B Raaf, Analysis of the impact of site planning on the performance of relay deployments. IEEE Trans. Veh. Technol. **61**(7), 3139–3150 (2012). doi:10.1109/TVT.2012.2202253
20. I Marić, B Boštjančič, A Goldsmith, in *Paper presented at the 2011 information theory and applications workshop (ITA)*. Resource allocation for constrained backhaul in picocell networks (IEEE, 6–11 February 2011)
21. M Sawahashi, Y Kishiyama, A Morimoto, D Nishikawa, M Tanno, Coordinated multipoint transmission/reception techniques for LTE-Advanced. IEEE Wireless Commun. Mag. **17**(3), 26–34 (2010). doi:10.1109/MWC.2010.5490976
22. 3GPP: Coordinated multi-point operation for LTE physical layer aspects. TR 36.819, Ver. 11.0.0, 3GPP Technical Report (2011). http://www.3gpp.org/dynareport/36819.htm. Accessed 03 March 2015
23. V Garcia, N Lebedev, J-M Gorce, Capacity outage probability for multi-cell processing under Rayleigh fading. Commun. Lett. IEEE. **15**(8), 801–803 (2011)
24. B Luo, Q Cui, X Tao, D Alexis, in *Paper presented at the 2013 IEEE international conference on communications*. On the optimal power allocation for coordinated wireless backhaul in OFDM-based relay systems (IEEE, Budapest, 9–13 June 2013)
25. DJ Love, RW Heath, VKN Lau, D Gesbert, BD Rao, M Andrews, An overview of limited feedback in wireless communication systems. IEEE J. Sel. Areas Commun. **26**(8), 1341–1365 (2008)
26. S Sun, Q Gao, Y Peng, Y Wang, L Song, Interference management through CoMP in 3GPP LTE-advanced networks. IEEE Trans. Wireless Commun. **20**(1), 59–66 (2013)
27. J Zhao, TQS Quek, Z Lei, Coordinated multipoint transmission with limited backhaul data transfer. IEEE Trans. Wireless Commun. **12**(6), 2762–2775 (2013)
28. R Irmer, H Droste, P Marsch, M Grieger, G Fettweis, S Brueck, H-P Mayer, L Thiele, V Jungnickel, Coordinated multipoint: concepts, performance, and field trial results. IEEE Commun Mag. **29**, 102–111 (2011). doi:10.1109/MCOM.2011.5706317
29. J Hämäläinen, Wichman R, in *Proc. Asilomar Conf. on Signals, Systems and Computers*, vol. 1. Closed-loop transmit diversity for FDD WCDMA systems, (2000), pp. 111–115. doi:10.1109/ACSSC.2000.910927
30. BB Haile, E Mutafungwa, J Hamalainen, in *Paper presented at the 2013 AFRICON*. Coordinated multipoint transmission for LTE-Advanced networks in dense informal settlements (IEEE Pointe-Aux-Piments, 9–12 September 2013)
31. A Narula, MJ Lopez, MD Trott, GW Wornell, Efficient use of side information in multiple-antenna data transmission over fading channels. IEEE J. Sel. Areas Commun. **16**(8), 1423–1436 (1998). doi:10.1109/49.730451
32. 3GPP, Physical layer procedures (FDD). TS 25.214, Ver. 10.4.0, 3GPP Technical Specification (2011). http://www.3gpp.org/DynaReport/25214.htm. Accessed 03 March 2015
33. 3GPP, Physical channels and modulation (2011). TS 36.211, Ver. 10.3.0, 3GPP Technical Specification
34. Y-D Yao, AUH Sheikh, Outage probability analysis for microcell mobile radio systems with cochannel interferers in Rician/Rayleigh fading environment. Electron. Lett. **26**(13), 864–866 (1990)
35. SM Ross, *Introduction to Probability Models*. (Academic Press, San Diego, 2000)
36. AA Dowhuszko, G Corral-Briones, J Hämäläinen, Wichman R, On throughput-fairness tradeoff in virtual MIMO systems with limited feedback. EURASIP J. Wireless Commun. Netw. **2009**, 1–17 (2009). doi:10.1155/2009/102064
37. M Abramowitz, IA Stegun, *Handbook of Mathematical Functions with Formulas, Graphs, and Mathematical Tables*, Ninth dover printing, tenth gpo printing. (Dover, New York, 1964)
38. UN-HABITAT, Slum dwellers to double by 2030. report (April 2007). http://www.un.org/apps/news/story.asp?newsid=8427&cr=&cr1=. Accessed 03 March 2015
39. GSMA, Sub-Saharan Africa Mobile Observatory 2012. report November 2012. http://www.gsma.com/publicpolicy/wp-content/uploads/2012/03/SSA_FullReport_v6.1_clean.pdf. Accessed 03 March 2015
40. P Amin, NS Kibret, E Mutafungwa, BB Haile, J Hamalainen, JK Nurminen, in *Paper presented at the 25th IEEE international symposium on personal, indoor and mobile radio communications (PIMRC)*. Performance study for off-grid self-backhauled small cells in dense informal settlements (IEEE, Washington DC, 2–5 September 2014)

41. WinProp Software Package. http://www.awe-communications.com/
42. H Holma, A Toskala, *LTE for UMTS - OFDMA and SC-FDMA Based Radio Access*. (Wiley, Chichester, 2009)
43. A Coluccia, D A'Alconzo, F Ricciato, On the optimality of max-min fairness in resource allocation. Ann. Telecommun. **67**(1–2), 15–26 (2012)
44. P Mogensen, W Na, IZ Kovacs, F Frederiksen, A Pokhariyal, KI Pedersen, T Kolding, K Hugl, M Kuusela, in *Paper presented at the IEEE 2007, 65th vehicular technology conference*. LTE capacity compared to the Shannon bound (IEEE, Dublin, 22–25 April 2007)
45. E Lahetkangas, K Pajukoski, E Tiirola, J Hamalainen, Z Zheng, in *Poznan*. On the performance of LTE-Advanced MIMO: How to set and reach beyond 4G targets (Poland, 18–20 April 2012)
46. R Jain, D Chiu, W Hawe, A quantitative measure of fairness and discrimintation for resource allocation in shared computer systems. TR 301, Digital Equipment Corporation Technical Report September 1984. http://www.cs.wustl.edu/~jain/papers/ftp/fairness.pdf. Accessed 03 March 2015

Permissions

All chapters in this book were first published in EURASIP-JWCN, by Springer; hereby published with permission under the Creative Commons Attribution License or equivalent. Every chapter published in this book has been scrutinized by our experts. Their significance has been extensively debated. The topics covered herein carry significant findings which will fuel the growth of the discipline. They may even be implemented as practical applications or may be referred to as a beginning point for another development.

The contributors of this book come from diverse backgrounds, making this book a truly international effort. This book will bring forth new frontiers with its revolutionizing research information and detailed analysis of the nascent developments around the world.

We would like to thank all the contributing authors for lending their expertise to make the book truly unique. They have played a crucial role in the development of this book. Without their invaluable contributions this book wouldn't have been possible. They have made vital efforts to compile up to date information on the varied aspects of this subject to make this book a valuable addition to the collection of many professionals and students.

This book was conceptualized with the vision of imparting up-to-date information and advanced data in this field. To ensure the same, a matchless editorial board was set up. Every individual on the board went through rigorous rounds of assessment to prove their worth. After which they invested a large part of their time researching and compiling the most relevant data for our readers.

The editorial board has been involved in producing this book since its inception. They have spent rigorous hours researching and exploring the diverse topics which have resulted in the successful publishing of this book. They have passed on their knowledge of decades through this book. To expedite this challenging task, the publisher supported the team at every step. A small team of assistant editors was also appointed to further simplify the editing procedure and attain best results for the readers.

Apart from the editorial board, the designing team has also invested a significant amount of their time in understanding the subject and creating the most relevant covers. They scrutinized every image to scout for the most suitable representation of the subject and create an appropriate cover for the book.

The publishing team has been an ardent support to the editorial, designing and production team. Their endless efforts to recruit the best for this project, has resulted in the accomplishment of this book. They are a veteran in the field of academics and their pool of knowledge is as vast as their experience in printing. Their expertise and guidance has proved useful at every step. Their uncompromising quality standards have made this book an exceptional effort. Their encouragement from time to time has been an inspiration for everyone.

The publisher and the editorial board hope that this book will prove to be a valuable piece of knowledge for researchers, students, practitioners and scholars across the globe.

List of Contributors

Fereidoun H Panahi
Graduate School of Science and Technology, Keio University, 3-14-1 Hiyoshi, Kouhoku-ku, 223-8522 Yokohama, Japan

Tomoaki Ohtsuki
Graduate School of Science and Technology, Keio University, 3-14-1 Hiyoshi, Kouhoku-ku, 223-8522 Yokohama, Japan

Dongwan Kim
The School of Electrical Engineering, Korea University, Anam-dong, Seongbuk-Gu, Seoul 136-701, South Korea
Samsung Electronics Co., Ltd., Suwon-si, Gyeonggi-do 443-742, South Korea

Sunshin An
The School of Electrical Engineering, Korea University, Anam-dong, Seongbuk-Gu, Seoul 136-701, South Korea

Zhanji Wu
School of Telecommunication Engineering, Beijing University of Posts and Telecommunications, Xitucheng Road 10, Beijing, China

Xiang Gao
School of Telecommunication Engineering, Beijing University of Posts and Telecommunications, Xitucheng Road 10, Beijing, China

Jie Wu
School of Information Science and Engineering, Southeast University, No. 2 Sipailou, Nanjing 210096, China

Shi Jin
School of Information Science and Engineering, Southeast University, No. 2 Sipailou, Nanjing 210096, China

Lei Jiang
NEC Laboratories China, 11F, Bldg A, Innovation Plaza, Tsinghua Science Park, No. 1 Zhongguancun East Road, Haidian District, Beijing 100084, China

Gang Wang
NEC Laboratories China, 11F, Bldg A, Innovation Plaza, Tsinghua Science Park, No. 1 Zhongguancun East Road, Haidian District, Beijing 100084, China

Reben Kurda
Laboratoire de Recherche en Informatique (LRI), University of Paris-Sud 11 - CNRS, 15, rue Georges Clémenceau, 91405 Orsay Cedex, France

Lila Boukhatem
Laboratoire de Recherche en Informatique (LRI), University of Paris-Sud 11 - CNRS, 15, rue Georges Clémenceau, 91405 Orsay Cedex, France

Megumi Kaneko
Graduate School of Informatics, Kyoto University, Yoshida-Honmachi, Sakyo-ku, Kyoto 606–8501, Japan

Jiajia Song
Department of Electrical and Computer Engineering, Yokohama National University, 79-5 Tokiwadai, Hodogaya, Yokohama 240-8501, Japan

Hideki Ochiai
Department of Electrical and Computer Engineering, Yokohama National University, 79-5 Tokiwadai, Hodogaya, Yokohama 240-8501, Japan

Cen Ling
College of Electronics and Information Engineering, Tongji University, 4800 Caoan Road, Shanghai, 201804, China

Xuefeng Yin
College of Electronics and Information Engineering, Tongji University, 4800 Caoan Road, Shanghai, 201804, China

Silvia Ruiz Boqué
Department of Signal Theory and Communications, University of Polytechnic in Catalunya, Building C4 C. Esteve Terradas, 7 Castelldefels, 08860, Spain

Mario García-Lozano
Department of Signal Theory and Communications, University of Polytechnic in Catalunya, Building C4 C. Esteve Terradas, 7 Castelldefels, 08860, Spain

Jose F Monserrat
Universitat Politecnica de Valencia, Camino de Vera, s/n, Valencia 46022, Spain

Genevieve Mange
Alcatel-Lucent Bell Labs, Stuttgart, Germany

Volker Braun
Alcatel-Lucent Bell Labs, Stuttgart, Germany

Hugo Tullberg
Ericsson, Stockholm, Sweden

Gerd Zimmermann
Deutsche Telekom, Darmstadt, Germany

Ömer Bulakci
Huawei ERC, Munich, Germany

Hua Wang
Aalborg University, Fredrik Bajers Vej 7, DK-9220 Aalborg Ø, Denmark

Claudio Rosa
Nokia Networks, Alfred Nobel Vej 21, DK-9220 Aalborg, Ø, Denmark

Klaus I Pedersen
Aalborg University, Fredrik Bajers Vej 7, DK-9220 Aalborg Ø, Denmark
Nokia Networks, Alfred Nobel Vej 21, DK-9220 Aalborg, Ø, Denmark

José Ángel Fernández-Segovia
Ingeniería de Comunicaciones, Universidad de Málaga, Campus de Teatinos S/N, 29071 Malaga, Spain

Salvador Luna-Ramírez
Ingeniería de Comunicaciones, Universidad de Málaga, Campus de Teatinos S/N, 29071 Malaga, Spain

Matías Toril
Ingeniería de Comunicaciones, Universidad de Málaga, Campus de Teatinos S/N, 29071 Malaga, Spain

Ana Belén Vallejo-Mora
Ingeniería de Comunicaciones, Universidad de Málaga, Campus de Teatinos S/N, 29071 Malaga, Spain

Carlos Úbeda
Ericsson, Via de los Poblados 13, 28033 Madrid, Spain

Marko Pesko
Telekom Slovenije, d.d., Cigaletova 15, 1000 Ljubljana, Slovenia
Jozef Stefan International Postgraduate School, Jamova cesta 39, 1000 Ljubljana, Slovenia

Tomaž Javornik
Department of Communication Systems, Jozef Stefan Institute, Jamova cesta 39, 1000 Ljubljana, Slovenia
Jozef Stefan International Postgraduate School, Jamova cesta 39, 1000 Ljubljana, Slovenia

Luka Vidmar
Telekom Slovenije, d.d., Cigaletova 15, 1000 Ljubljana, Slovenia

Andrej Košir
Faculty of Electrical Engineering, University of Ljubljana, Tržaška 25, 1000 Ljubljana, Slovenia

Mitja Štular
Telekom Slovenije, d.d., Cigaletova 15, 1000 Ljubljana, Slovenia

Mihael Mohorčič
Department of Communication Systems, Jozef Stefan Institute, Jamova cesta 39, 1000 Ljubljana, Slovenia
Jozef Stefan International Postgraduate School, Jamova cesta 39, 1000 Ljubljana, Slovenia

Fereidoun H Panahi
Graduate School of Science and Technology, Keio University, 3-14-1 Hiyoshi, Kouhoku-ku, Yokohama 223-8522, Japan

Tomoaki Ohtsuki
Graduate School of Science and Technology, Keio University, 3-14-1 Hiyoshi, Kouhoku-ku, Yokohama 223-8522, Japan

José Rodríguez-Piñeiro
Department of Electronics and Systems, University of A Coruña, Facultade de Informática, Campus de Elviña, A Coruña 15071, Spain

Martin Lerch
Institute of Telecommunications, TU Wien, Gusshausstrasse 25/389, A-1040 Vienna, Austria

José A. García-Naya
Department of Electronics and Systems, University of A Coruña, Facultade de Informática, Campus de Elviña, A Coruña 15071, Spain

Sebastian Caban
Institute of Telecommunications, TU Wien, Gusshausstrasse 25/389, A-1040 Vienna, Austria

Markus Rupp
Institute of Telecommunications, TU Wien, Gusshausstrasse 25/389, A-1040 Vienna, Austria

Luis Castedo
Department of Electronics and Systems, University of A Coruña, Facultade de Informática, Campus de Elviña, A Coruña 15071, Spain

Ladislav Polak
Department of Radio Electronics, SIX Research Center, Brno University of Technology, Technicka 3082/12, 616 00 Brno, Czech Republic

Lukas Klozar
Department of Radio Electronics, SIX Research Center, Brno University of Technology, Technicka 3082/12, 616 00 Brno, Czech Republic

Ondrej Kaller
Department of Radio Electronics, SIX Research Center, Brno University of Technology, Technicka 3082/12, 616 00 Brno, Czech Republic

Jiri Sebesta
Department of Radio Electronics, SIX Research Center, Brno University of Technology, Technicka 3082/12, 616 00 Brno, Czech Republic

Martin Slanina
Department of Radio Electronics, SIX Research Center, Brno University of Technology, Technicka 3082/12, 616 00 Brno, Czech Republic

Tomas Kratochvil
Department of Radio Electronics, SIX Research Center, Brno University of Technology, Technicka 3082/12, 616 00 Brno, Czech Republic

Víctor Fernández-López
Department of Electronic Systems, Aalborg University, Fredrik Bajers Vej 7, 9220 Aalborg, Denmark
Nokia Networks, Alfred Nobels Vej 21C, 9220 Aalborg, Denmark

Klaus I Pedersen
Department of Electronic Systems, Aalborg University, Fredrik Bajers Vej 7, 9220 Aalborg, Denmark
Nokia Networks, Alfred Nobels Vej 21C, 9220 Aalborg, Denmark

Beatriz Soret
Nokia Networks, Alfred Nobels Vej 21C, 9220 Aalborg, Denmark

Beneyam B. Haile
School of Electrical Engineering, Aalto University, P.O. Box 13000, FI-00076 Aalto, Espoo, Finland

Edward Mutafungwa
School of Electrical Engineering, Aalto University, P.O. Box 13000, FI-00076 Aalto, Espoo, Finland

Jyri Hämäläinen
School of Electrical Engineering, Aalto University, P.O. Box 13000, FI-00076 Aalto, Espoo, Finland